Hydrosphere Under the Driving of Human Activity and Climate Change: Status, Evolution and Strategies

Hydrosphere Under the Driving of Human Activity and Climate Change: Status, Evolution and Strategies

Topic Editors

Yong Xiao
Jianping Wang
Jinlong Zhou

MDPI

Basel • Beijing • Wuhan • Barcelona • Belgrade • Novi Sad • Cluj • Manchester

Topic Editors
Yong Xiao
Faculty of Geosciences
and Engineering
Southwest Jiaotong
University
Chengdu
China

Jianping Wang
Qinghai Institute of
Salt Lakes
Chinese Academy
of Sciences
Xining
China

Jinlong Zhou
College of Hydraulic and
Civil Engineering
Xinjiang Agricultural
University
Urumqi
China

Editorial Office
MDPI AG
Grosspeteranlage 5
4052 Basel, Switzerland

This is a reprint of the Topic, published open access by the journals *Land* (ISSN 2073-445X), *Sustainability* (ISSN 2071-1050), *Water* (ISSN 2073-4441) and *Hydrology* (ISSN 2306-5338), freely accessible at: https://www.mdpi.com/topics/S7E8GSPDLR.

For citation purposes, cite each article independently as indicated on the article page online and as indicated below:

Lastname, A.A.; Lastname, B.B. Article Title. *Journal Name* **Year**, *Volume Number*, Page Range.

ISBN 978-3-7258-4519-4 (Hbk)
ISBN 978-3-7258-4520-0 (PDF)
https://doi.org/10.3390/books978-3-7258-4520-0

Contents

About the Editors

Yong Xiao

Yong Xiao is an Associate Professor of Hydrogeology at Southwest Jiaotong University, China, and was a Visiting Scholar (2017) in the Department of Earth and Environmental Sciences at the University of Waterloo, Canada. He received his bachelor's (2009–2013) and doctoral (2013–2018) degrees in Hydrogeology from China University of Geosciences (Beijing). His research focuses on groundwater circulation, surface water–groundwater interactions, geothermal water, groundwater environments, and ecohydrology. He has presided over more than 20 research projects of various types, including those funded by the National Natural Science Foundation of China and the Natural Science Foundation of Sichuan Province (China). He employs a wide range of tools including field investigations, numerical modeling, and hydrogeochemical and isotopic tracers to understand the complex behavior and processes of groundwater systems, both natural and human-driven. His findings are documented in approximately 80 peer-reviewed publications in international journals, including the *Journal of Hydrology, Hydrology and Earth System Sciences, Catena, Agricultural Water Management, Journal of Cleaner Production, Science of the Total Environment, Journal of Hydrology: Regional Studies, Process Safety and Environmental Protection*, etc., achieving an H-index of 25. He was awarded the Second Prize for Science and Technology Progress of the Ministry of Natural Resources of China. He serves as an Editorial Board Member/Associate Editor for *Scientific Reports, Sustainability, Hydrology, China Geology*, etc., and as a Guest Editor for Special Issues in *Remote Sensing, Frontiers in Environmental Science, Land, Geofluids, Marine and Freshwater Research*, etc.

Jianping Wang

Jianping Wang graduated with a Ph.D. in Natural Resource Management from the University of Birmingham, UK, in 2009. She currently is working at the Qinghai Institute of Salt Lakes, Chinese Academy of Sciences, as a Researcher and Doctoral Supervisor. Her primary research focuses on 1) hydro-saline dynamics and mineral dissolution mechanisms in salt lake regions and 2) resource-environment carrying capacity and efficient water resource utilization in salt lake areas. In recent years, she has published over 90 peer-reviewed papers, authored 2 monographs, obtained 10 authorized patents and software copyrights, and spearheaded the establishment of China's largest saline lake resource-environment scientific database. Her notable honors include 1) being selected for the Chinese Academy of Sciences "BR Program"; 2) being recognized as the Leading Talent of Qinghai Province; 3) receiving the State Council Special Government Allowance; 4) being awarded the Qinghai Provincial Scientific Achievement Award for Innovation-Driven Development; 5) and being the winner of the Qinghai Provincial Natural Science Award (Second Class).

Jinlong Zhou

Jinlong Zhou is a second-level professor of Hydrogeology and a member of the academic committee of Xinjiang Agricultural University. He obtained a bachelor's degree in Hydrogeology from China University of Geosciences (Wuhan) (1980–1984) and a doctorate degree in geological engineering from University of Chinese Academy of Sciences (2007–2010). His research mainly focuses on groundwater balance, groundwater circulation, surface water–groundwater interaction, soil and water environmental geochemistry, groundwater and ecological environments, and groundwater quality evolution in inland arid areas. He has presided over more than 60 scientific research projects, including the National Natural Science Foundation of China, the National Science and Technology Support Program, the National Key R&D Program "Comprehensive Treatment of Atmospheric and Soil and Groundwater Pollution", the National Science and Technology Basic Resources Survey Special "Third Xinjiang Comprehensive Scientific Investigation" Project, the Key R&D Special Project of Xinjiang Autonomous Region, and the Major Science and Technology Special Project of Xinjiang Autonomous Region. As the primary researcher, he has participated in more than 50 scientific research projects. In his work, he uses a wide range of methods to understand the complex behaviors and processes of groundwater systems in inland arid regions. His more than 40 academic papers have been published in roughly 20 international journals, including the *Journal of Environmental Management, Environmental Pollution, Applied Geochemistry, Journal of Contaminant Hydrology, Journal of Hydrology, International Journal of Remote Sensing, Journal of Earth Science, Exposure and Health, Environmental Earth Science*, etc. He is a member of the editorial committee of the *Journal of Groundwater Science and Engineering* and a Guest Editor of *Atmosphere, Land, Remote Sensing*, etc.

sustainability

Editorial

Hydrosphere Under the Driving of Human Activity and Climate Change: Status, Evolution, and Strategies

Yong Xiao [1,2,*], Jianping Wang [3] and Jinlong Zhou [4]

1 MOE Key Laboratory of Groundwater Circulation and Environmental Evolution, China University of Geosciences (Beijing), Beijing 100083, China
2 Faculty of Geosciences and Engineering, Southwest Jiaotong University, Chengdu 611756, China
3 Qinghai Institute of Salt Lakes, Chinese Academy of Sciences, Xi'ning 810008, China
4 College of Hydraulic and Civil Engineering, Xinjiang Agriculture University, Urumqi 830052, China
* Correspondence: xiaoyong@swjtu.edu.cn

1. The Critical Role of the Hydrosphere in the Earth's System

The hydrosphere is a foundational component of the Earth's systems, sustaining biological processes, regulating environmental stability, and enabling socioeconomic development [1,2]. As the primary medium for biogeochemical cycles, it governs global energy distribution, nutrient transport, and habitat provision for both aquatic and terrestrial ecosystems [3]. While 71% of the Earth's surface is covered by water, accessible freshwater resources constitute less than 1% of total reserves [4], highlighting its irreplaceable value in supporting agriculture, industry, and domestic needs. Historically, human civilizations have thrived along river valleys and coastal zones, relying on water for irrigation, transportation, and cultural practices [5]. The development of groundwater abstraction techniques has facilitated the establishment of human settlements and socio-economic activities in regions historically constrained by the availability of surface water resources. Beyond its water supply functions, the hydrosphere modulates climate patterns through ocean–atmosphere interactions and acts as a critical buffer against temperature extremes [6]. Its intrinsic linkages to the atmosphere, lithosphere, and biosphere underscore its centrality in maintaining planetary equilibrium [7], making its preservation imperative for ecological integrity and human survival.

2. Emerging Threats to Hydrospheric Stability

Anthropogenic activities and climatic disruptions have accelerated the systemic degradation of water resources, marked by both quantitative depletion and qualitative deterioration [8–10]. Overexploitation of groundwater resources, driven by agricultural intensification and urbanization, has led to aquifer depletion rates that exceed natural recharge capacities in many regions worldwide such as China, Chile, India, Mexico, and Iran [11–15]. Concurrently, industrial effluents, agricultural runoff, and plastic pollution have compromised water quality, posing threats to aquatic biodiversity and human health [16–20]. Climate change exacerbates these challenges by altering precipitation regimes, accelerating glacial melt, and amplifying extreme hydrological events, including floods and droughts [21]. Projections indicate a potential 40% global water deficit by 2030 under current consumption trends [22], with transboundary water conflicts intensifying geopolitical tensions. Additionally, rising sea levels [23] and ocean acidification [24] jeopardize coastal ecosystems, further compromising food security and disaster resilience. These cascading impacts underscore the urgency of addressing hydrospheric vulnerabilities within integrated sustainability frameworks.

3. Knowledge Gaps in Hydrospheric Research

Despite significant progress in hydrological sciences, critical knowledge gaps persist in understanding how the hydrosphere responds to combined stressors of rapid and drastic human-induced climatic changes. Previous studies have inadequately represented feedback mechanisms between anthropogenic water use, climate variability, and ecosystem dynamics, particularly across various scales from inland to coastal regions. Limited long-term monitoring of data hinders the assessment of water resource depletion trends and contaminant dispersion pathways, especially in developing countries with fragmented hydrological networks. Additionally, socio-hydrological interactions, such as the interplay between water governance, economic policies, and community adaptation, remain under-researched, hampering the design of context-specific mitigation strategies. The integration of emerging technologies with traditional methodologies is still in its infancy, limiting the accuracy of predictions for extreme hydrosphere events. Future research must prioritize interdisciplinary approaches, improve the spatiotemporal resolution of datasets, and incorporate indigenous knowledge systems to address these limitations and inform resilient water management paradigms.

4. Papers in This Topic

A total of seven journals, including *Atmosphere, Land, Remote Sensing, Sustainability, Water, Earth,* and *Hydrology,* participated in this Topic. Numerous manuscripts were submitted to these journals, and ultimately, 17 manuscripts passed the rigorous peer-review process and were accepted for publication by four journals including *Land, Sustainability, Water,* and *Hydrology* (as shown in the list of contributions).

The studies cover a range of environments from continental interiors to transitional coastal zones, systematically examining hydrological compartments across both surface water systems and subsurface groundwater reservoirs. Multiple contributions focused on characterizing groundwater systems in arid regions, particularly on how hydrogeochemical processes fundamentally govern long-term resource sustainability under persistent water scarcity. Current investigations systematically examine the complex interplay between geochemical evolution pathways and hydrological constraints that shape water accessibility patterns in moisture-deficient environments [25–28]. Meanwhile, some studies assessed hydrochemical dynamics within terrestrial-marine ecotones, particularly in coastal transition zones experiencing saline intrusion and synergistic anthropogenic pressures [29–31]. Human-induced organic contaminants in groundwater systems were investigated using approaches such as hydrochemistry, isotopes, microbial techniques, and numerical modeling to assess their distribution, migration, natural attenuation, and threats [32,33]. Geochemically distinct groundwater resources, notably mineral-enriched aquifers, were systematically investigated for their spatial heterogeneity and hydrogeological formation mechanisms, with particular emphasis on their roles in potable supply and therapeutic applications [34]. Additionally, the geochemical signatures, evolution, and formation of deeply buried continental brines have been explored, providing a valuable hydrogeological "window" into these liquid resources enriched in rare metal minerals [35].

The interconnectedness of water systems and ecosystems under anthropogenic perturbations was explored, with a particular focus on how human-induced surface and subsurface hydraulic modifications mediate hydro-ecological feedback mechanisms. River runoff dynamics in semiarid basins have been examined, revealing that anthropogenic activities are the predominant factors leading to runoff reduction [36]. Additionally, the hydrological connectivity between coastal terrestrial ecosystems (e.g., wetlands) and adjacent marine aquifers has been extensively investigated, with particular emphasis on their bidirectional interactions with human activities in coastal zones [37,38]. These studies reveal significant anthropogenic modifications to hydrological exchange processes that fundamentally alter both ecological integrity and geological stability within these transitional environments. Contemporary research increasingly documents how human-induced perturbations disrupt natural biogeochemical cycles while exacerbating geological vulnerabilities in coastal interface systems. Furthermore, the hydrological impacts of land-use change in semiarid inland basins have been explored from the Production–Living–Ecological Space perspective [39]. Results indicate that the spatiotemporal configuration of territorial functions (production, living, and ecological) fundamentally regulates watershed hydrodynamics. Specifically, the reduction in ecological zones is quantitatively correlated with decreased aquifer recharge efficiency and compromised moisture retention capacities in semiarid terrestrial systems. Comprehensive analyses of soil moisture retention attributes reveal that textural composition exerts dominant control over unsaturated hydraulic property relationships, with distinct soil textural patterns governing pore-water retention behavior in variably saturated subsurface systems [40]. Anthropogenic infrastructure development, particularly traffic tunnel construction, has also been systematically investigated and evaluated for its potential to trigger perturbations in groundwater systems and natural hydrogeological landscapes [41].

5. Final Remarks

This collection significantly advances our understanding of hydrological systems across continental to coastal environments, with a focus on hydrogeochemical processes, human impacts, and resource sustainability. Key findings reveal how groundwater evolution in arid regions is governed by geochemical–hydrological interactions under water scarcity, while coastal zones face compounded threats from saline intrusion, contaminants, and anthropogenic pressures. Multidisciplinary approaches—integrating isotopes, microbial analyses, and modeling—prove vital for tracking contaminant dynamics and natural attenuation.

The studies underscore the profound anthropogenic role in reshaping hydrological-ecological connectivity. Semiarid regions exhibit reduced runoff due to human activities, coastal ecosystems experience altered hydraulic feedback, and land-use shifts disrupt aquifer recharge and soil moisture retention. Infrastructure development, particularly tunnels, further highlights risks of groundwater system perturbations.

Future priorities include refining predictive models for multi-stress scenarios, unlocking insights from deep brines and mineralized aquifers, and mitigating anthropogenic disruptions through evidence-based safeguards. As environmental pressures intensify, fostering interdisciplinary collaboration and science-policy synergy remains critical to securing equitable, sustainable water futures. Collectively, these works chart a path for innovative research and actionable solutions to protect the Earth's vital water resources.

Author Contributions: Conceptualization, Y.X., J.W. and J.Z.; investigation, Y.X.; writing—original draft preparation, Y.X.; writing—review and editing, Y.X., J.W. and J.Z.; project administration, Y.X.; funding acquisition, Y.X. All authors have read and agreed to the published version of the manuscript.

Funding: This research was funded by MOE Key Laboratory of Groundwater Circulation and Environmental Evolution, China University of Geosciences (Beijing), Grant No. 2023-004; National Natural Science Foundation of China, Grant No. 42477059; Key Lab of Environmental Geology of Qinghai Province (2023-KJ-15; 2024-KJ-06); Applied Basic Research Project of Science and Technology Program of Qinghai Province, Grant No. 2024-ZJ-771; Innovative Practice Bases of Geological Engineering and Surveying Engineering of Southwest Jiaotong University, Grant No. YJG-2022-JD04.

Acknowledgments: As guest editors, we extend our sincere gratitude to the global hydrological community for their contributions to this Topic Collection. The competitive nature of this scholarly endeavor is reflected in the relatively low acceptance rate, driven by substantial submission volumes and stringent quality benchmarks. We commend the Contributors whose robust submissions successfully navigated multi-round peer review, while also acknowledging unpublished works that have advanced our understanding of the hydrosphere. Our international reviewers provided essential critiques that enhanced analytical precision and theoretical coherence, supported by the editorial team's efficient manuscript management. This collective effort exemplifies the collaborative scholarship necessary to address hydrological-environmental complexities. We believe that such synergistic interactions among researchers, evaluators, and institutions are foundational to hydrological security research. We hope these scholarly exchanges will catalyze innovative water governance strategies, emphasizing our professional responsibility to protect the hydrosphere from intensifying climatic and anthropogenic pressures. Through disciplined interdisciplinary cooperation, this initiative reaffirms academia's crucial role in achieving sustainable water resource stewardship. We further acknowledge the editorial support staff and publishing personnel for their diligent efforts during manuscript assessment and publication phases, maintaining rigorous quality standards and facilitating timely dissemination through editorial workflows.

Conflicts of Interest: The authors declare no conflicts of interest.

List of Contributions:

1. Wang, F.; Yang, H.; Zhang, Y.; Wang, S.; Liu, K.; Qi, Z.; Chai, X.; Wang, L.; Wang, W.; Banadkooki, F.B.; et al. Solute Geochemistry and Water Quality Assessment of Groundwater in an Arid Endorheic Watershed on Tibetan Plateau. *Sustainability* **2022**, *14*, 15593.
2. Guo, B.; Yang, P.; Zhou, Y.; Ai, H.; Li, X.; Kang, R.; Lv, Y. Numerical Simulation of Carbon Tetrachloride Pollution-Traceability in Groundwater System of an Industrial City. *Sustainability* **2022**, *14*, 16113.
3. Deng, S.; Li, C.; Jiang, X.; Zhao, T.; Huang, H. Research on Surface Water Quality Assessment and Its Driving Factors: A Case Study in Taizhou City, China. *Water* **2023**, *15*, 26.
4. Wang, L.; Nie, Z.; Yuan, Q.; Liu, M.; Cao, L.; Zhu, P.; Lu, H.; Feng, B. Spatiotemporal Oasis Land Use/Cover Changes and Impacts on Groundwater Resources in the Central Plain of the Shiyang River Basin. *Water* **2023**, *15*, 457.
5. Xu, M.; Hu, C.; Zhu, L.; Song, G.; Peng, W.; Yang, S.; Song, J. Spatial Distribution Characteristics and Genetic Mechanism of the Metasilicate-Rich Groundwater in Ji'nan Rock Mass Area, Shandong Province, China. *Water* **2023**, *15*, 713.
6. Yang, C.; Wu, J.; Li, P.; Wang, Y.; Yang, N. Evaluation of Soil-Water Characteristic Curves for Different Textural Soils Using Fractal Analysis. *Water* **2023**, *15*, 772.
7. An, G.; Kang, H.; Fu, R.; Xu, D.; Li, J. Investigation on the Hydrogeochemical Characteristics and Controlling Mechanisms of Groundwater in the Coastal Aquifer. *Water* **2023**, *15*, 1710.
8. Xu, R.; Gu, C.; Qiu, D.; Wu, C.; Mu, X.; Gao, P. Analysis of Runoff Changes in the Wei River Basin, China: Confronting Climate Change and Human Activities. *Water* **2023**, *15*, 2081.
9. Zhu, Y.; Yang, H.; Xiao, Y.; Hao, Q.; Li, Y.; Liu, J.; Wang, L.; Zhang, Y.; Hu, W.; Wang, J. Identification of Hydrochemical Characteristics, Spatial Evolution, and Driving Forces of River Water in Jinjiang Watershed, China. *Water* **2024**, *16*, 45.
10. Yang, S.; Zhao, Z.; Wang, S.; Xiao, S.; Xiao, Y.; Wang, J.; Wang, J.; Yuan, Y.; Ba, R.; Wang, N.; et al. Hydrogeochemical Insights into the Sustainable Prospects of Groundwater Resources in an Alpine Irrigation Area on Tibetan Plateau. *Sustainability* **2024**, *16*, 9229.
11. Wang, X.; Gong, L.; Liu, Y.; Wang, Y.; Wang, Q.; Song, M.; Xiao, P.; Shi, Z. Investigating the Hydrological Relationship between the North Taihang Tunnel and Tianshengqiao Nine Falls. *Water* **2024**, *16*, 1549.
12. Zhu, Y.; Liu, Y.; Xiao, Y.; Liu, J.; Zhao, Z.; Li, Y.; Hao, Q.; Liu, C.; Li, J. Construction of Ecological Security Patterns Incorporating Multiple Types of Ecological Service Functions for Developed Coastal Regions: A Case Study in Jinjiang Watershed, China. *Land* **2024**, *13*, 1227.

13. Zheng, Z.-l.; Xie, B.; Wu, C.-m.; Zhou, L.; Zhang, K.; Zhang, B.-c.; Yang, P.-h. Geochemical Characteristics and Genesis of Brine Chemical Composition in Cambrian Carbonate-Dominated Succession in the Northeastern Region of Chongqing, Southwestern China. *Water* **2024**, *16*, 2859.
14. Gana, B.; Rodes, J.M.A.; Díaz, P.; Balboa, A.; Frías, S.; Ávila, A.; Rivera, C.; Sáez, C.A.; Lavergne, C. Geoenvironmental Effects of the Hydric Relationship Between the Del Sauce Wetland and the Laguna Verde Detritic Coastal Aquifer, Central Chile. *Hydrology* **2024**, *11*, 174.
15. Yang, H.; Wei, J.; Shi, K. Hydrochemical and Isotopic Characteristics and the Spatiotemporal Differences of Surface Water and Groundwater in the Qaidam Basin, China. *Water* **2024**, *16*, 169.
16. Zhang, J.; Laghari, A.A.; Guo, Q.; Liang, J.; Kumar, A.; Liu, Z.; Shen, Y.; Wei, Y. Evolution of Land Use and Its Hydrological Effects in the Fenhe River Basin Under the Production–Living–Ecological Space Perspective. *Sustainability* **2024**, *16*, 11170.
17. Shi, J.; Zhang, Y.; Lai, Y.; Yang, R.; Cai, M.; Fan, S.; Gu, X. Study on Natural Attenuation of Groundwater Organic Pollutants by Integrating Microbial Community Dynamics and Isotope Analysis. *Water* **2025**, *17*, 555.

References

1. Peters, N.E.; Böhlke, J.K.; Brooks, P.D.; Burt, T.P.; Gooseff, M.N.; Hamilton, D.P.; Mulholland, P.J.; Roulet, N.T.; Turner, J.V. 2.11—Hydrology and Biogeochemistry Linkages. In *Treatise on Water Science*; Wilderer, P., Ed.; Elsevier: Oxford, UK, 2011; pp. 271–304. [CrossRef]
2. Xiao, Y.; Zhang, Y.; Yang, H.; Wang, L.; Han, J.; Hao, Q.; Wang, J.; Zhao, Z.; Hu, W.; Wang, S.; et al. Interaction regimes of surface water and groundwater in a hyper-arid endorheic watershed on Tibetan Plateau: Insights from multi-proxy data. *J. Hydrol.* **2024**, *644*, 132020. [CrossRef]
3. Kuang, X.; Liu, J.; Scanlon, B.R.; Jiao, J.J.; Jasechko, S.; Lancia, M.; Biskaborn, B.K.; Wada, Y.; Li, H.; Zeng, Z.; et al. The changing nature of groundwater in the global water cycle. *Science* **2024**, *383*, eadf0630. [CrossRef] [PubMed]
4. Fitts, C.R. 1—Groundwater: The Big Picture. In *Groundwater Science*, 2nd ed.; Fitts, C.R., Ed.; Academic Press: Boston, MA, USA, 2013; pp. 1–22. [CrossRef]
5. Wang, Y.; Ni, J.; Wan, J.; Xu, J.; Zheng, C.; Borthwick, A.G.L. Global river economic belts can become more sustainable by considering economic and ecological processes. *Commun. Earth Environ.* **2024**, *5*, 18. [CrossRef]
6. Zardi, D. Atmosphere and ocean interactions. *Rend. Lincei. Sci. Fis. E Nat.* **2024**, *35*, 311–325. [CrossRef]
7. Trofimov, V.T.; Korolev, V.A.; Kharkina, M.A. Ecological Functions as Fundamental Integral Characteristics of Peculiarities of Abiotic Environments of Ecosystem: Lithosphere, Pedosphere, Atmosphere, and Hydrosphere. *Mosc. Univ. Geol. Bull.* **2024**, *79*, 467–476. [CrossRef]
8. Li, Q.; Li, P.; Elumalai, V. Identification and apportionment of groundwater pollution sources in the Guanzhong region based on PMF model. *Hum. Ecol. Risk Assess. Int. J.* **2025**, *31*, 30–52. [CrossRef]
9. Yu, X.; Luo, L.; Hu, P.; Tu, X.; Chen, X.; Wei, J. Impacts of sea-level rise on groundwater inundation and river floods under changing climate. *J. Hydrol.* **2022**, *614*, 128554. [CrossRef]
10. Hao, Q.; Shao, J.; Cui, Y.; Zhang, Q.; Huang, L. Optimization of groundwater artificial recharge systems using a genetic algorithm: A case study in Beijing, China. *Hydrogeol. J.* **2018**, *26*, 1749–1761. [CrossRef]
11. Ou, J.; Ding, B.; Feng, P.; Chen, Y.; Yu, L.; Liu, D.L.; Srinivasan, R.; Zhang, X. How to stop groundwater drawdown in North China Plain? Combining agricultural management strategies and climate change. *J. Hydrol.* **2025**, *647*, 132352. [CrossRef]
12. Taucare, M.; Viguier, B.; Figueroa, R.; Daniele, L. The alarming state of Central Chile's groundwater resources: A paradigmatic case of a lasting overexploitation. *Sci. Total Environ.* **2024**, *906*, 167723. [CrossRef]
13. Kumar, H.; Syed, T.H.; Amelung, F.; Agrawal, R.; Venkatesh, A.S. Space-time evolution of land subsidence in the National Capital Region of India using ALOS-1 and Sentinel-1 SAR data: Evidence for groundwater overexploitation. *J. Hydrol.* **2022**, *605*, 127329. [CrossRef]
14. Bockstiegel, M.; Richard-Cerda, J.C.; Muñoz-Vega, E.; Haghighi, M.H.; Motagh, M.; Lalehzari, R.; Schulz, S. Simulation of present and future land subsidence in the Rafsanjan plain, Iran, due to groundwater overexploitation using numerical modeling and InSAR data analysis. *Hydrogeol. J.* **2024**, *32*, 289–305. [CrossRef]
15. Salcedo-Sánchez, E.R.; Esteller, M.V.; Garrido Hoyos, S.E.; Martínez-Morales, M. Groundwater optimization model for sustainable management of the Valley of Puebla aquifer, Mexico. *Environ. Earth Sci.* **2013**, *70*, 337–351. [CrossRef]
16. Yang, H.; Xiao, Y.; Yang, S.; Zhao, Z.; Wang, S.; Xiao, S.; Wang, J.; Zhang, Y.; Wang, J.; Yuan, Y.; et al. Geochemical fingerprints, evolution, and driving forces of groundwater in an alpine basin on Tibetan Plateau: Insights from unsupervised machine learning and objective weight allocation approaches. *J. Hydrol. Reg. Stud.* **2024**, *56*, 102054. [CrossRef]

17. Qu, S.; Wang, C.; Liang, X.; Luo, A.; Shi, Z.; Wang, G.; Yu, R. Regional characteristics of groundwater sulfate source and evolution in the multi-layer aquifer system of the northern Shaanxi coal mine base, northwestern China: Evidence from geochemical and isotopic fingerprints. *J. Hazard. Mater.* **2024**, *480*, 135866. [CrossRef]
18. Xiao, Y.; Liu, K.; Hao, Q.; Xiao, D.; Zhu, Y.; Yin, S.; Zhang, Y. Hydrogeochemical insights into the signatures, genesis and sustainable perspective of nitrate enriched groundwater in the piedmont of Hutuo watershed, China. *CATENA* **2022**, *212*, 106020. [CrossRef]
19. Hu, W.; Xiao, Y.; Wang, L.; Zhang, Y.; Feng, M.; Shi, W.; He, C.; Wen, Y.; Yang, H.; Han, J.; et al. Spatial variability, source identification, and partitioning of groundwater constituents in a typical lakeside plain on Yungui Plateau. *Process Saf. Environ. Prot.* **2024**, *191*, 2402–2415. [CrossRef]
20. Yu, Y.; Yan, B.; Tuo, Y.; Zhan, X.; Xie, X. Chemical evolution characteristics and influencing factors of groundwater in the saline and fresh water funnel area in Hengshui City, North China. *Sci. Rep.* **2025**, *15*, 3276. [CrossRef]
21. Li, C.; Liu, J.; Du, F.; Zwiers, F.W.; Feng, G. Increasing certainty in projected local extreme precipitation change. *Nat. Commun.* **2025**, *16*, 850. [CrossRef]
22. Boccaletti, G. Framing water problems with global statistics is both powerful and misleading. *Nat. Water* **2023**, *1*, 660–661. [CrossRef]
23. Majumder, S.; Fatma, R. Alteration of shoreline in a macrotidal setting induced by rising sea level: A case study of Purba Medinipur coastal stretch. *Mar. Pollut. Bull.* **2025**, *213*, 117657. [CrossRef] [PubMed]
24. Caillon, C.; Fleury, E.; Di Poi, C.; Gazeau, F.; Pernet, F. Food availability, but not tidal emersion, influences the combined effects of ocean acidification and warming on oyster physiological performance. *Aquaculture* **2025**, *604*, 742459. [CrossRef]
25. Wang, F.; Yang, H.; Zhang, Y.; Wang, S.; Liu, K.; Qi, Z.; Chai, X.; Wang, L.; Wang, W.; Banadkooki, F.B.; et al. Solute Geochemistry and Water Quality Assessment of Groundwater in an Arid Endorheic Watershed on Tibetan Plateau. *Sustainability* **2022**, *14*, 15593. [CrossRef]
26. Wang, L.; Nie, Z.; Yuan, Q.; Liu, M.; Cao, L.; Zhu, P.; Lu, H.; Feng, B. Spatiotemporal Oasis Land Use/Cover Changes and Impacts on Groundwater Resources in the Central Plain of the Shiyang River Basin. *Water* **2023**, *15*, 457. [CrossRef]
27. Yang, H.; Wei, J.; Shi, K. Hydrochemical and Isotopic Characteristics and the Spatiotemporal Differences of Surface Water and Groundwater in the Qaidam Basin, China. *Water* **2024**, *16*, 169. [CrossRef]
28. Yang, S.; Zhao, Z.; Wang, S.; Xiao, S.; Xiao, Y.; Wang, J.; Wang, J.; Yuan, Y.; Ba, R.; Wang, N.; et al. Hydrogeochemical Insights into the Sustainable Prospects of Groundwater Resources in an Alpine Irrigation Area on Tibetan Plateau. *Sustainability* **2024**, *16*, 9229. [CrossRef]
29. An, G.; Kang, H.; Fu, R.; Xu, D.; Li, J. Investigation on the Hydrogeochemical Characteristics and Controlling Mechanisms of Groundwater in the Coastal Aquifer. *Water* **2023**, *15*, 1710. [CrossRef]
30. Deng, S.; Li, C.; Jiang, X.; Zhao, T.; Huang, H. Research on Surface Water Quality Assessment and Its Driving Factors: A Case Study in Taizhou City, China. *Water* **2023**, *15*, 26. [CrossRef]
31. Zhu, Y.; Yang, H.; Xiao, Y.; Hao, Q.; Li, Y.; Liu, J.; Wang, L.; Zhang, Y.; Hu, W.; Wang, J. Identification of Hydrochemical Characteristics, Spatial Evolution, and Driving Forces of River Water in Jinjiang Watershed, China. *Water* **2024**, *16*, 45. [CrossRef]
32. Guo, B.; Yang, P.; Zhou, Y.; Ai, H.; Li, X.; Kang, R.; Lv, Y. Numerical Simulation of Carbon Tetrachloride Pollution-Traceability in Groundwater System of an Industrial City. *Sustainability* **2022**, *14*, 16113. [CrossRef]
33. Shi, J.; Zhang, Y.; Lai, Y.; Yang, R.; Cai, M.; Fan, S.; Gu, X. Study on Natural Attenuation of Groundwater Organic Pollutants by Integrating Microbial Community Dynamics and Isotope Analysis. *Water* **2025**, *17*, 555. [CrossRef]
34. Xu, M.; Hu, C.; Zhu, L.; Song, G.; Peng, W.; Yang, S.; Song, J. Spatial Distribution Characteristics and Genetic Mechanism of the Metasilicate-Rich Groundwater in Ji'nan Rock Mass Area, Shandong Province, China. *Water* **2023**, *15*, 713. [CrossRef]
35. Zheng, Z.-l.; Xie, B.; Wu, C.-m.; Zhou, L.; Zhang, K.; Zhang, B.-c.; Yang, P.-h. Geochemical Characteristics and Genesis of Brine Chemical Composition in Cambrian Carbonate-Dominated Succession in the Northeastern Region of Chongqing, Southwestern China. *Water* **2024**, *16*, 2859. [CrossRef]
36. Xu, R.; Gu, C.; Qiu, D.; Wu, C.; Mu, X.; Gao, P. Analysis of Runoff Changes in the Wei River Basin, China: Confronting Climate Change and Human Activities. *Water* **2023**, *15*, 2081. [CrossRef]
37. Gana, B.; Rodes, J.M.A.; Díaz, P.; Balboa, A.; Frías, S.; Ávila, A.; Rivera, C.; Sáez, C.A.; Lavergne, C. Geoenvironmental Effects of the Hydric Relationship Between the Del Sauce Wetland and the Laguna Verde Detritic Coastal Aquifer, Central Chile. *Hydrology* **2024**, *11*, 174. [CrossRef]
38. Zhu, Y.; Liu, Y.; Xiao, Y.; Liu, J.; Zhao, Z.; Li, Y.; Hao, Q.; Liu, C.; Li, J. Construction of Ecological Security Patterns Incorporating Multiple Types of Ecological Service Functions for Developed Coastal Regions: A Case Study in Jinjiang Watershed, China. *Land* **2024**, *13*, 1227. [CrossRef]
39. Zhang, J.; Laghari, A.A.; Guo, Q.; Liang, J.; Kumar, A.; Liu, Z.; Shen, Y.; Wei, Y. Evolution of Land Use and Its Hydrological Effects in the Fenhe River Basin Under the Production–Living–Ecological Space Perspective. *Sustainability* **2024**, *16*, 11170. [CrossRef]

40. Yang, C.; Wu, J.; Li, P.; Wang, Y.; Yang, N. Evaluation of Soil-Water Characteristic Curves for Different Textural Soils Using Fractal Analysis. *Water* **2023**, *15*, 772. [CrossRef]
41. Wang, X.; Gong, L.; Liu, Y.; Wang, Y.; Wang, Q.; Song, M.; Xiao, P.; Shi, Z. Investigating the Hydrological Relationship between the North Taihang Tunnel and Tianshengqiao Nine Falls. *Water* **2024**, *16*, 1549. [CrossRef]

sustainability

Article

Solute Geochemistry and Water Quality Assessment of Groundwater in an Arid Endorheic Watershed on Tibetan Plateau

Fenglin Wang [1,2,3], Hongjie Yang [4], Yuqing Zhang [4], Shengbin Wang [1,2,3,*], Kui Liu [4], Zexue Qi [1,2,3,*], Xiaoran Chai [1,2,3], Liwei Wang [4], Wanping Wang [1,2,3], Fatemeh Barzegari Banadkooki [5], Venkatramanan Senapthi [6] and Yong Xiao [4]

1 Bureau of Qinghai Environmental Geological Prospecting, Xi'ning 810007, China
2 Key Laboratory of Geo-Environment of Qinghai Province, Xi'ning 810007, China
3 Qinghai 906 Engineering Survey and Design Institute Co., Ltd., Xi'ning 810007, China
4 Faculty of Geosciences and Environmental Engineering, Southwest Jiaotong University, Chengdu 611756, China
5 Faculty of Engineering, Payam Noor University, Tehran P.O. Box 76169-14111, Iran
6 Department of Disaster Management, Alagappa University, Karaikudi 630002, India
* Correspondence: wsb13897651096@126.com (S.W.); qzx13639722055@126.com (Z.Q.)

Abstract: Understanding groundwater geochemistry is crucial for water supply in arid regions. The present research was conducted in the arid Mo river watershed on the Tibetan plateau to gain insights into the geochemical characteristics, governing processes and quality of groundwater in arid endorheic watersheds. A total of 28 groundwater samples were collected from the phreatic and confined aquifers for hydrochemical analysis. The results showed that the groundwater was slightly alkaline in all aquifers of the watershed. The phreatic groundwater samples (PGs) and confined groundwater samples (CGs) had the TDS value in the ranges of 609.19–56,715.34 mg/L and 811.86–2509.51 mg/L, respectively. PGs were salter than CGs, especially in the lower reaches. Both the PGs and CGs were dominated by the Cl-Na type, followed by the mixed Cl-Mg·Ca type. The toxic elements of NO_2^- (0.00–0.20 mg/L for PGs and 0.00–0.60 mg/L for CGs), NH_4^+ (0.00–0.02 mg/L for PGs and 0.00–0.02 mg/L for CGs) and F^- (0.00–4.00 mg/L for PGs and 1.00–1.60 mg/L for CGs) exceeded the permissible limits of the Chinese guidelines at some sporadic sites. Water–rock interactions, including silicates weathering, mineral dissolution (halite and sulfates) and ion exchange, were the main contributions to the groundwater chemistry of all aquifers. The geochemistry of PGs in the lower reach was also greatly influenced by evaporation. Agricultural sulfate fertilizer input was responsible for the nitrogen pollutants and salinity of PGs. All CGs and 73.91% of PGs were within the Entropy-weighted water quality index (EWQI) of below 100 and were suitable for direct drinking purposes. Precisely 8.70 and 17.39% of PGs were within the EWQI value in the range of 100–150 (medium quality and suitable for domestic usage) and beyond 200 (extremely poor quality and not suitable for domestic usage), respectively. The electrical conductivity, sodium adsorption ratio, sodium percentage and permeability index indicated that groundwater in most parts of the watershed was suitable for irrigation, and only a small portion might cause salinity, sodium or permeability hazards. Groundwater with poor quality was mainly distributed in the lower reaches. CGs and PGs in the middle-upper reaches could be considered as the primary water resources for water supply. Agricultural pollution should be paid more attention to safeguard the quality of groundwater.

Keywords: groundwater; hydrochemistry; geochemical process; water quality assessment; arid region

1. Introduction

Although our earth is called the "blue planet", as more than 70% of its surface is covered by water, most of the water is salty and not suitable for various usages of human society and continental ecosystems [1,2]. Freshwater is considered one of the most important natural resources supporting the sustainable development of a human community and has been treated as the strategic natural resource for national development by many counties over the world [3,4]. Our world is experiencing unprecedented, drastic climate change and facing serious and complex environmental challenges [5–9]. Among all of the environmental challenges, water shortage is the most widespread and profound across the world [10]. This challenge is much more severe in arid and semiarid regions where there is a significantly poorer endowment of water resources [7,11,12]. Groundwater is an invaluable water resource for various usages such as municipal, industrial and agricultural purposes for many people in the world, particularly in water-scarce regions [13–15]. Groundwater has been a crucial water resource supporting the development of human communities and ecosystems worldwide [16–18]. Thus, a deep understanding of groundwater endowment and its suitability is essential for the scientific and rational utilization of groundwater resources and for maintaining the sustainable development of groundwater [19].

The availability of groundwater is dependent on two major aspects, including quantity and quality [20]. Quantity is the first factor being considered in groundwater resource exploration and utilization. Vast works and research have been conducted in terms of groundwater quantity and the yield of aquifers worldwide [21,22]. A lot of hydrogeological achievements have been obtained during this process and supported the exploitation of many potential groundwater reservoirs across the world [3,23–25]. Aside from the water quantity, the chemical quality of groundwater also significantly determines the availability of groundwater resources [26]. It is widely known that water is fundamental to life, and water quality determines the health of flora and fauna [27,28]. Access to clean groundwater resources is related to the preservation of the health and sustainability of earth systems, including human and animal communities and ecosystems [8]. Therefore, revealing the hydrogeochemical mechanisms and quality of groundwater resources is significantly important for guaranteeing the availability of quality groundwater.

It is widely known that water in nature is not pure and contains many substances. The chemical substances of groundwater originate from numerous sources [18,29]. Initially, the chemicals in groundwater are brought by the recharge of water into the aquifers from external sources [30]. The chemical substances in recharge waters make up the fundamental framework of groundwater hydrochemistry and quality. After water enters the aquifers, various hydrogeochemical processes occur in groundwater which further changes its composition [31,32]. Water–rock interactions are the predominant substances contributing to groundwater quality [14,33]. This contribution varies spatially due to the various lithologies and hydrogeological conditions of the aquifers [10]. The residence time of groundwater in the aquifers is an important factor in determining the degree of water–rock interactions [34,35]. Additionally, the hydrodynamic of groundwater is also reported to affect the hydrogeochemical processes and finally regulate the composition and quality of groundwater [36]. The climatic factors, particularly the strong evaporation in arid regions, can greatly influence groundwater chemistry and are reportedly responsible for the salinity of phreatic groundwater in many parts of the world [17,37].

Aside from the natural processes, anthropogenic disturbances to groundwater chemistry have become more and more significant due to the rapid growth of world populations and intensive human activities in the past decades [38–40]. The most important and direct influence induced by human communities on groundwater quality is the direct input of pollutants into aquifers, which pollutes the clean water beneath the ground surface [7,41]. The hydrogeological environment has become more and more complex as the overlay of natural and anthropogenic mechanisms [42,43]. In order to ensure the safety of the water supply, the mechanisms of groundwater chemistry should be distinguished, and then corresponding measures in water environment protection and groundwater exploitation should be conducted [14,44].

Arid and semiarid regions are important parts of the earth and account for more than 40% of the global land [45]. These regions contain a large number of world populations, cities and agricultural lands [12,31,35,46]. The conflict of water resources is much more significant in these regions due to the scarcity of water resources [10,39]. The special climatic features in these regions, such as rare precipitation and strong evaporation, create poor water endowment and complex hydrogeochemical conditions. Endorheic watersheds are more special in arid and semiarid regions since the chemical salts in water cannot leave the watersheds but accumulate in the water of rivers, underground reservoirs and tail lakes [34,47]. The strong evaporation and accumulation of chemical salts would further aggravate the water crisis in these arid and semiarid watersheds [48,49]. The available freshwater resource is limited in small endorheic watersheds due to the short distance of groundwater evolution from freshwater in the headwater region to salty water in the tail lake region [29]. Thus, focusing on groundwater chemistry and its availability is significantly important for ensuring the long-term sustainable utilization of water resources for small endorheic watersheds [50]. Although a great number of research works have been conducted regarding groundwater availability with both quantity and quality aspects, research concerned with small endorheic watersheds is rare [51–53].

The present research takes a typical arid endorheic watershed, namely the Mo river watershed on the Tibetan plateau, as an example to gain insights into the hydrogeochemical characteristics, governing processes and quality of groundwater in small arid endorheic watersheds. The results and findings could provide useful references and scientific support for the utilization and management of groundwater resources in hyper-arid endorheic watersheds worldwide.

2. Materials and Methods

2.1. Description of the Study Area

The Chaka basin is one of the most significantly arid basins located in the northern Tibetan plateau (Figure 1). The present watershed, i.e., the Mo river watershed, is the largest watershed in this basin. The study area extends between 36°37′51.92″ N and 36°57′18.67″ N latitudinally and between 98°45′16.16″ E and 99°15′03.36″ E longitudinally, covering an area of 1593.98 km^2. The Mo river is the largest river running through the watershed. It originates from the northwest mountainous area of the Chaka basin and ends at the center of the basin. Chaka salt lake, enjoying a reputation as the "sky mirror", is the tail lake of the watershed [29]. All surface and groundwater in the watershed finally discharge into the Chaka salt lake. As a result, the Mo river watershed is an endorheic watershed. This watershed features a semi-desert arid continental climate. The annual average temperature is 2.9 °C. The precipitation is limited and less than 200 mm per year, as most of the precipitation occurs during the wet season from May to September. The evaporation is significantly strong and reaches about ~2500 mm per year. Due to the scarcity of local precipitation, river water in the watershed is also not developed. As a result, human societies and ecosystems in the watershed are heavily dependent on groundwater resources.

The Chaka basin is a closed fault basin that was fundamentally formed during the Cenozoic era. It is bounded by the Nanshan mountains in the north and the Ela mountains in the south and presents in the shape of a long strip in the NW direction. The lithology of the basin ranges from Paleozoic bedrocks to Quaternary deposits. The Mo river watershed is a typical part of the Chaka basin. The pre-Quaternary strata are mainly distributed in the upper reaches of the mountainous area and dominated by the Paleozoic and Triassic metamorphic rocks. The strata in the basin are predominantly Quaternary deposits with various lithologies. The lithologies of the Quaternary deposits in the basin vary from gravel in the piedmont to fine sand and clay in the lower reaches (Figure 2). Groundwater in the basin mainly occurs in the Quaternary aquifers. Due to the spatial variation of Quaternary deposits lithology, the Quaternary aquifers in the basin vary from a single thick phreatic aquifer in the piedmont to a semi-confined aquifer in the middle reaches, and finally, the multiple layers of phreatic-confined aquifers in the basin center. The groundwater of the basin predominantly originated from the leakage of river water at the piedmont. The lateral flow from the bedrock at the mountain pass also contributes some groundwater to the Quaternary aquifers. The contribution of local precipitation to the Quaternary aquifers in the basin is limited. Groundwater naturally discharges through springs and flows laterally into the Chaka salt lake. With the rapid development of the watershed, human pumping has become one of the dominant ways groundwater discharges into the watershed. Groundwater exploration is mainly concentrated in the Mo river and Chaka farms, and Chaka town (Figure 1c).

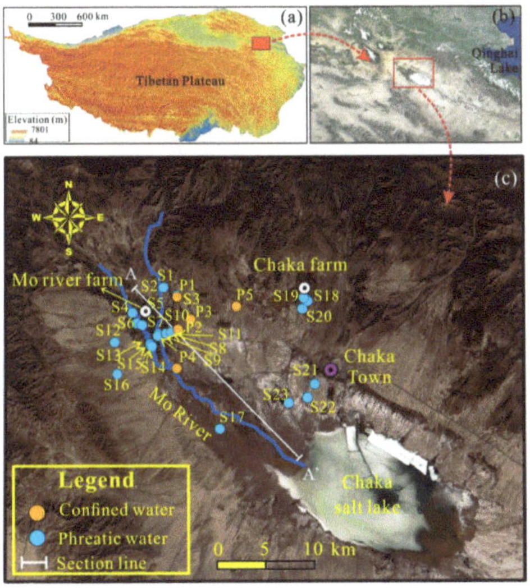

Figure 1. Location map of (**a**) the Tibetan Plateau, (**b**) the Chaka basinand (**c**) the Mo River watershed and sampling sites.

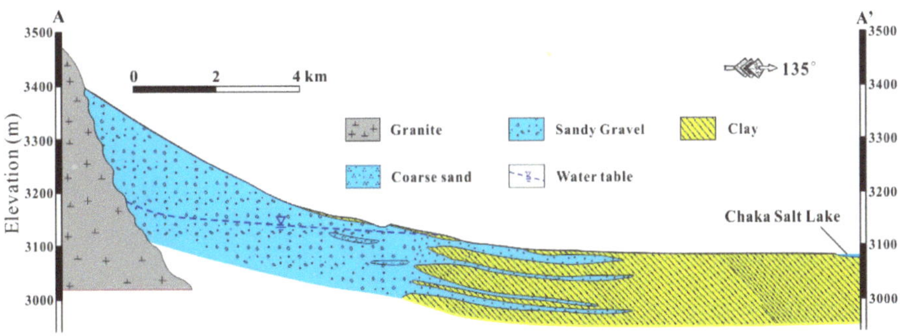

Figure 2. Hydrogeological cross-section of the study area along the A–A′.

2.2. Sampling and Analytical Techniques

A total of 28 groundwater samples were collected from the Mo river watershed during the months of June and July of 2022. These groundwater samples were sampled from 23 phreatic wells and five confined boreholes. The depths of phreatic boreholes are in the range of 40–160 m and are mostly greater than 100 m in the middle-upper area of the watershed and within 60 m in the lower reach of the watershed. All confined boreholes have greater depth and range from 120–195 m. In order to eliminate the influence of stagnant waters in the representative groundwater samples, the wells and boreholes were pumped for more than three volumes. Water temperature, pH and electrical conductivity

(EC) were monitored in situ, and the sampling strategy was conducted after these in-situ monitoring parameters were stable. The sample containers consisted of 2.5 L high-density polyethylene bottles and were thoroughly washed at least three times using the targeted groundwater before sampling. The groundwater samples were stored at 4 °C and sent to the laboratory for analysis within two days.

The in-situ parameters (pH, EC) of groundwater were determined in the field with the aid of a portable multiparameter device (Multi350i/SET, Munich, Germany). The major cations of Ca^{2+}, Mg^{2+}, Na^{+} and K^{+} and toxic elements (Cu, Pb, As, Al, Fe, Mn, Zn, Cr, etc.) were measured using the inductively coupled plasma-mass spectrometry (Agilent 7500ce ICP-MS, Tokyo, Japan). Ions, including NH_4^{+}, Cl^{-}, SO_4^{2-}, NO_3^{-}, NO_2^{-} and F^{-}, were analyzed by ion chromatography (Shimadzu LC-10ADvp, Kyoto, Japan). The HCO_3^{-} was determined by acid-base titration. The gravimetric analysis was conducted to obtain the total dissolved solids (TDS) of groundwater. The accuracy of groundwater analysis was checked using the ionic charge balance error percentage (%ICBE). The results showed that all groundwater samples had an %ICBE of ± 5%, implying the hydrogeochemical data of groundwater were reliable and could be used for research.

3. Results and Discussion

3.1. General Chemistry of Groundwater

The physicochemical parameters of phreatic and confined groundwater sampled from the study area are statistically presented in Table 1. The measured pH values varied from 7.42 to 8.66 in the phreatic aquifers and between 7.78 and 8.52 in the confined aquifers, suggesting that groundwater in the watershed was of a slightly alkaline nature regardless of the depth. The phreatic groundwater samples had a wider range of EC than confined groundwater samples. The EC values of phreatic groundwater ranged from 1033.50 to 44,205.00 µS/cm, and the confined groundwater ranged from 1309.00 to 3608.10 µS/cm. The TDS values also showed a similar state at different depths. The TDS values were in the range of 609.19–56,715.34 mg/L for phreatic groundwater and 811.86–2509.51 mg/L for confined groundwater. All of the above suggests that the phreatic groundwater in the watershed is saltier in nature than the confined groundwater.

Table 1. Summary statistics for the physicochemical parameters of groundwater in the study area.

	Index	Unit	Min	Max	Mean	SD *	Guideline	% of the Sample Exceeding the Guideline
Phreatic groundwater	pH	/	7.42	8.66	7.98	0.31	6.5–8.5 **	8.70%
	EC	µs/cm	1033.50	44,205.00	5932.70	11,271.57	/	
	TDS	mg/L	609.19	56,715.34	6246.78	14,213.90	1000 **	47.83%
	K^{+}	mg/L	1.50	68.81	13.51	14.44	/	
	Na^{+}	mg/L	85.00	18,610.00	2012.05	4972.36	200 **	43.48%
	Ca^{2+}	mg/L	50.84	1102.20	160.48	221.62	75 ***	69.57%
	Mg^{2+}	mg/L	18.95	1358.37	135.53	277.33	50 ***	47.83%
	Cl^{-}	mg/L	172.29	31,550.50	3177.79	8061.68	250 **	56.52%
	SO_4^{2-}	mg/L	33.62	3866.42	595.96	960.04	250 **	43.48%
	HCO_3^{-}	mg/L	137.91	634.61	278.44	96.42	/	
	NO_3^{-}	mg/L	0.92	37.30	17.12	11.62	50.0 ***	0%
	NO_2^{-}	mg/L	0.00	0.08	0.01	0.02	0.02 **	8.70%
	NH_4^{+}	mg/L	0.00	0.20	0.07	0.09	0.2 **	17.93%
	F^{-}	mg/L	0.00	4.00	0.48	1.03	1.0 **	13.04%
Confined groundwater	pH	/	7.78	8.52	8.05	0.32	6.5–8.5 **	20%
	EC	µs/cm	1309.00	3608.10	2011.00	934.77	/	
	TDS	mg/L	811.86	2509.51	1325.94	690.32	1000 **	60%
	K^{+}	mg/L	4.50	17.02	8.26	5.25	/	
	Na^{+}	mg/L	145.25	635.29	295.11	197.46	200 **	60%
	Ca^{2+}	mg/L	76.55	160.32	100.52	35.37	75 ***	100%
	Mg^{2+}	mg/L	41.07	80.68	56.77	16.21	50 ***	60%
	Cl^{-}	mg/L	232.55	1153.54	485.95	381.41	250 **	80%
	SO_4^{2-}	mg/L	180.59	379.44	248.80	81.83	250 **	40%
	HCO_3^{-}	mg/L	146.45	335.61	244.08	71.03	/	
	NO_3^{-}	mg/L	4.00	16.00	9.20	4.60	50.0 ***	0%
	NO_2^{-}	mg/L	0.00	0.06	0.02	0.03	0.02 **	40%
	NH_4^{+}	mg/L	0.00	0.20	0.08	0.08	0.2 **	20%
	F^{-}	mg/L	1.00	1.60	1.27	0.31	1.0 **	60%

* Standard deviation; ** Chinese guideline [54]; *** WHO guideline [55].

Among the major cations, Na^+ is the predominant ion in both phreatic and confined aquifers, with concentrations ranging from 85.00 to 18,610.00 mg/L (averaging 2012.05 mg/L) and from 145.25 to 635.29 mg/L (averaging 197.46 mg/L), respectively. Ca^{2+} ranks second regardless of the depth and is in a range of 50.84–1102.20 mg/L (averaging 160.48 mg/L) for phreatic groundwater and 76.55–160.32 mg/L (averaging 100.52 mg/L) for confined groundwater. Mg^{2+} ranks third in both phreatic and confined aquifers, followed by K^+. Phreatic groundwater has an Mg^{2+} concentration in the range of 18.95–1358.37 mg/L (averaging 135.53 mg/L) and a K^+ concentration in the range of 1.50–68.81 mg/L (averaging 13.51 mg/L). The confined groundwater samples show an Mg^{2+} concentration varying between 41.07 and 80.68 mg/L (averaging 100.52 mg/L) and a K^+ concentration varying between 4.50 and 17.02 mg/L (averaging 8.26 mg/L). For all major cations, phreatic groundwater shows greater variation in ion concentrations than confined groundwater, according to the standard deviation values of each parameter.

Among the major anions, Cl^- was the dominant ion in both phreatic and confined aquifers. Its concentration varied from 172.29 to 31,550.50 mg/L (averaging 3177.79 mg/L) for the phreatic groundwater samples and ranged between 232.55 and 1153.54 mg/L (averaging 485.95 mg/L) for the confined groundwater samples. $SO_4{}^{2-}$ ranked second with a concentration ranging from 33.62 to 3866.42 mg/L (averaging 595.96 mg/L) for the phreatic groundwater samples and varied between 180.59 and 379.44 mg/L (averaging 248.80 mg/L) for the confined groundwater samples. $HCO_3{}^-$ was the lowest ion in all aquifers. It had a concentration in the range of 137.91–634.61 mg/L (averaging 278.44 mg/L) for the phreatic groundwater and 146.45–335.41 mg/L (averaging 244.08 mg/L) for the confined groundwater. The major anions also showed greater variation in the phreatic aquifers than that in the confined aquifers.

Nitrogen, fluoride and other toxic element concentrations (Cu, Pb, As, Al, Fe, Mn, Zn, Cr, etc.) were determined in phreatic and confined aquifers. Overall, most toxic elements (Cu, Pb, As, Al, Fe, Mn, Zn, Cr, etc.), except nitrogen and fluoride, had relatively low concentrations and were even below the detection limits; thus, they were not included in the following analysis. For nitrogen, the concentration of all nitrogen pollutants in phreatic groundwater was higher than that in confined groundwater. The $NO_3{}^-$ concentration in phreatic groundwater varied from 0.92 to 37.30 mg/L (averaging 17.12 mg/L), and confined groundwater ranged between 4.00 and 16.00 mg/L (averaging 9.20 mg/L). All collected phreatic and confined groundwater samples tested for $NO_3{}^-$ concentrations were below the permissible limits of 50 mg/L, as determined by the WHO guidelines [55]. The $NO_2{}^-$ and $NH_4{}^+$ in phreatic aquifers were in the range of 0.00–0.20 mg/L (averaging 0.01 mg/L) and 0.00–0.02 mg/L (averaging 0.07 mg/L), respectively. Additionally, $NO_2{}^-$ and $NH_4{}^+$ in confined aquifers were in the range of 0.00–0.60 mg/L (averaging 0.02 mg/L) and 0.00–0.02 mg/L (averaging 0.08 mg/L), respectively. Both phreatic and confined aquifers were found with $NO_2{}^-$ and $NH_4{}^+$ pollutants exceeding the permissible limits of 0.02 and 0.2 mg/L determined by the Chinese guideline [54] at some sampling locations. The F^- in phreatic and confined groundwater was in the range of 0.00–4.00 mg/L (averaging 0.48 mg/L) and 1.00–1.60 mg/L (averaging 1.27 mg/L), respectively.

3.2. Hydrogeochemical Facies

The overall hydrochemical characteristics of groundwater can be illustrated by the hydrogeochemical facies [42], which could be graphically represented by the Piper trilinear diagram. The Piper trilinear diagram involves two triangles below and one diamond above. The two lower triangles are used to separate the cations and anions of water and represent their dominant ions. The diamond is used to map the combination of cations and anions and reveals the hydrogeochemical facies of water [35,56,57]. Six hydrogeochemical facies, including HCO_3-Ca, mixed HCO_3-Na·Ca, HCO_3-Na, mixed Cl-Mg·Ca, Cl-Ca and Cl-Na, are categorized by the Piper trilinear diagram.

As presented in Figure 3, all collected groundwater samples except two phreatic groundwater samples are observed plotting in the [Na^+] type dominance in the lower left of the Piper trilinear diagram, suggesting that Na^+ is the predominant cation of groundwater in most sites of the watershed regardless of the sampling depth. The other two phreatic groundwater samples are situated in the [Na^+] type dominance of the lower left trilinear, indicating Ca^{2+} is the dominant cation in these two groundwater samples. For the anions, all confined groundwater samples and almost all phreatic groundwater samples are situated in the [Cl^-] type dominance in the lower right of the Piper trilinear diagram, indicating Cl^- is the dominant anion for both phreatic groundwater and confined groundwater. Only one phreatic groundwater is observed plotting in the [$SO_4{}^{2-}$] type dominance, and $SO_4{}^{2-}$ is the predominant anion. The diamond of the Piper trilinear diagram shows that all groundwater in the watershed plot from the mixed Cl-Mg·Ca type dominance to the Cl-Na type dominance. It can be seen that most of the collected groundwater samples, regardless of the depth, belonged to the Cl-Na type, and only a small part of phreatic (26.09%) and confined (20%) groundwater samples were mixed Cl-Mg·Ca type water. Overall, groundwater in phreatic aquifers is salty to some degree in terms of the hydrogeochemical facies, and confined groundwater is presented as fresher in nature compared to phreatic groundwater.

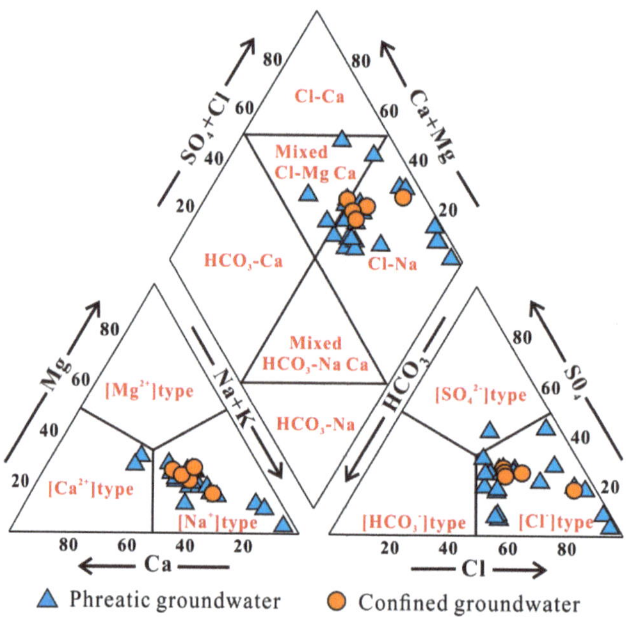

Figure 3. Piper trilinear diagram of groundwater samples in the study area.

3.3. *Hydrogeochemical Mechanisms Governing Groundwater Chemistry*

Generally, the solutes in groundwater are mainly originated from natural sources. Principally, the hydrogeochemical composition of groundwater is governed by three natural factors, i.e., precipitation, water–rock interactions and evaporation. These three main natural mechanisms governing groundwater geochemistry can be virtually illustrated by the Gibbs diagrams, which are constructed by the TDS versus $Na^+/(Na^++Ca^{2+})$ and $Cl^-/(Cl^-+HCO_3{}^-)$ [58].

As shown in Figure 4, all groundwater samples, regardless of the sampling depth, displayed an evaporative trend in both of the two sub-Gibbs diagrams, implying that recharge water had experienced significant evaporation effects during the recharging process. This is in accordance with hyper-arid climate features in the study area. Specifically, all sampled confined groundwater samples are plotted in the rock dominance of both these two sub-Gibbs diagrams, suggesting that water–rock interactions are the predominant mechanism regulating the hydrogeochemical composition of groundwater in the confined aquifers of the Mo river watershed after the water enters the aquifers. For phreatic groundwater, most of the samples are situated in the rock dominance. This implies that the hydrochemistry of phreatic groundwater in the watershed is also dominantly governed by the water–rock interaction. Some phreatic groundwater samples (S17, S21, S22 and S23) are observed plotting in the evaporation dominance, indicating evaporation is the main process regulating the hydrogeochemical composition of phreatic groundwater in some sites (S17, S21, S22 and S23). In fact, these evaporation-governed samples (S17, S21, S22 and S23) are located in the lower reaches of the Mo river watershed. All of these four groundwater samples are from shallow, buried depths and are easily affected by the strong evaporation in the arid climatic condition of the watershed. As a result of the strong evaporation, the collected phreatic groundwater samples in the lower reaches of the Mo river watershed are with relatively high salinity and a TDS value beyond thousands of milligrams per liter.

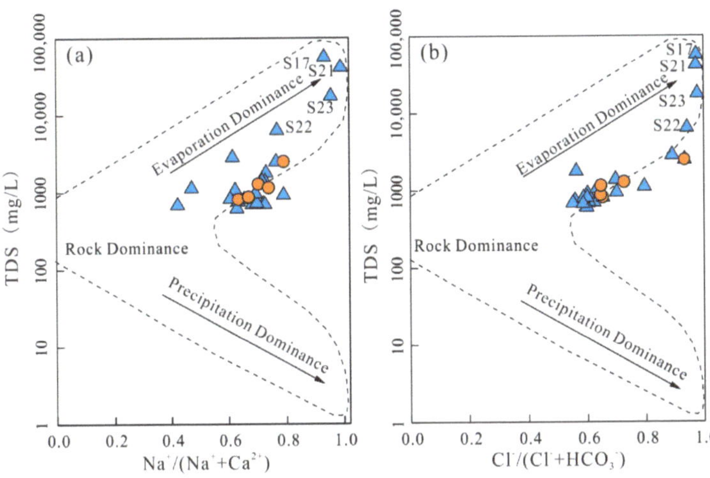

Figure 4. Gibbs diagram illustrating the natural principle mechanisms governing groundwater chemistry. (**a**) TDS versus $Na^+/(Na^++Ca^{2+})$; (**b**) TDS versus $Cl^-/(Cl^-+HCO_3^-)$.

As mentioned above, water–rock interactions are the predominant processes regulating the hydrogeochemical composition of both phreatic and confined groundwater samples in the watershed. There are various rocks and minerals existing in nature that are potentially involved in the water–rock interaction processes. In fact, not all of the rocks and minerals in nature would be involved in the interactions. The involved rocks/minerals in a specific aquifer are determined by the existing rocks/minerals and the hydrogeochemical conditions of aquifers along the groundwater flow path. To further illustrate the rocks and minerals involved in the water–rock interactions, the end-member diagrams that are constructed by the Ca^{2+}/Na^+ versus Mg^{2+}/Na^+ and HCO_3^-/Na^+ [59] were introduced in the present research. These end-member diagrams separate into three dominances, i.e., the evaporite, silicate and carbonate dominances, and can be used to virtually identify the main rock types (Figure 5) involved in the water–rock interactions. As shown in Figure 5, all confined groundwater samples and most of the phreatic groundwater samples are located in and around the silicate dominance, indicating that silicate weathering (Equation (1)) is the predominant source of solutes for groundwater regardless of the depth in the Mo river watershed. A small portion of the phreatic groundwater samples was observably situated in and around the evaporite dominance, suggesting that evaporite dissolution is the dominant source for the major ions of phreatic groundwater at some locations. No phreatic or confined groundwater was situated in or around the carbonate dominance, implying carbonate dissolution is not the main process contributing to groundwater chemistry in the watershed.

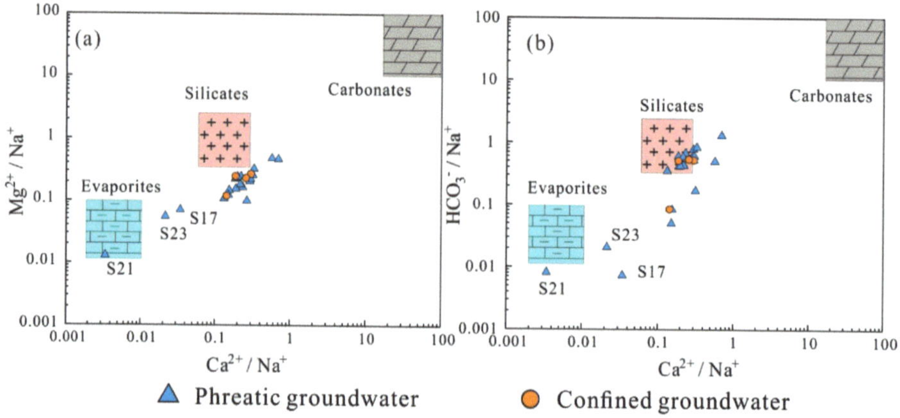

Figure 5. End-member diagrams representing the main rock type involved in the water–rock interactions. (**a**) Mg^{2+}/Na^+ versus Ca^{2+}/Na^+; (**b**) HCO_3^-/Na^+ versus Ca^{2+}/Na^+.

$$2NaAlSi_3O_8 + 2CO_2 + 11H_2O \rightarrow Al_2Si_2O_5(OH)_4 + 4H_4SiO_4 + 2Na^+ + 2HCO_3^- \quad (1)$$

In order to further determine the contribution of specific evaporites and silicates, the correlation matrix and scatter plots are employed in the present research. It can be seen from Figure 6a and Table 2 that a significant positive correlation exists between Cl^- and Na^+ for both phreatic and confined groundwater, with a correlation coefficient of 0.71 and 0.99, respectively. This suggests a high possibility of halite dissolution contributing to the major ions (Cl^- and Na^+) in the hydrogeochemical components of groundwater in phreatic and confined aquifers of the watershed. The mineral saturation status was simulated to confirm the potential contribution of halite dissolution further. As shown in Figure 7, all collected phreatic and confined groundwater samples had a halite saturation index below zero, confirming the contribution of halite dissolution to major ion contents in phreatic and confined aquifers. The SO_4^{2-} and Ca^{2+} ions were observed to have a significant positive correlation for both phreatic groundwater samples (with a correlation coefficient of 0.70) and confined groundwater samples (with a correlation coefficient of 0.97) (Figure 6b and Table 2). This implies that the dissolution of gypsum and anhydrite is the potential contribution of SO_4^{2-} and Ca^{2+} to phreatic and confined groundwater, as evidenced by the saturation index of gypsum (less than zero) and anhydrite (less than zero) (Figure 6). Significant positive correlations were also found between SO_4^{2-} and Na^+, Mg^{2+} and K^+ for phreatic groundwater, suggesting that the dissolution of other sulfates, including Na_2SO_4, $MgSO_4$ and K_2SO_4, could contribute to major ions in the phreatic groundwater if these minerals exist in the aquifers. For confined groundwater, significant positive correlations are also found between SO_4^{2-}, Na^+ and Mg^{2+}, implying the potential contribution of sulfate (Na_2SO_4 and $MgSO_4$) dissolution. It should be noted that all groundwater samples, regardless of the sampling depth, were observed and plotted along the 1:1 line of Cl^-/Na^+ and SO_4^{2-}/Ca^{2+} (Figure 6). This evidenced the aforementioned conclusion that water had experienced evaporation during its recharge process. As discussed in the end-member diagram, only several phreatic groundwater samples (S17, S21 and S23) are located in the evaporite dominance. Thus, although the dissolution of evaporites (halite and sulfates) could contribute solutes to both phreatic and confined groundwater in the whole watershed, they are dominated only in the lower reaches adjacent to the Chaka salt lake.

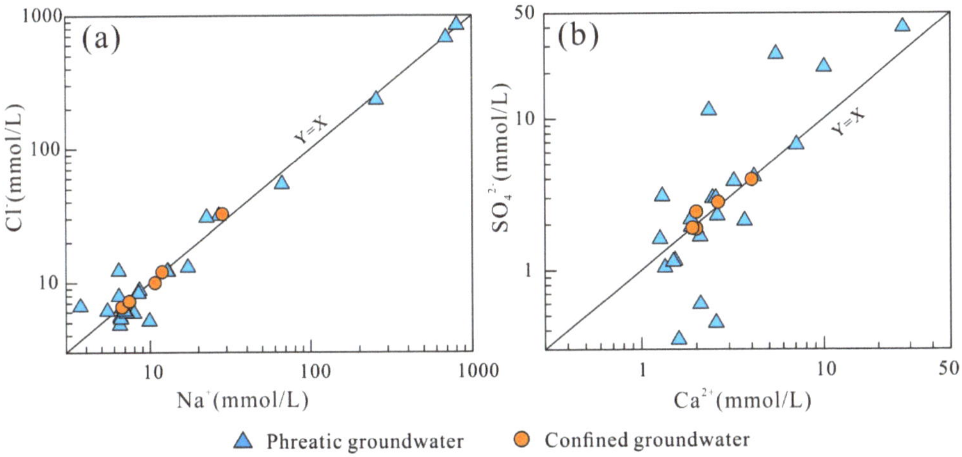

Figure 6. Scatter plots of (**a**) Cl^- versus Na^+ and (**b**) SO_4^{2-} versus Ca^{2+} of groundwater in the Mo river watershed.

Table 2. Correlation matrix of groundwater physicochemical parameters in the study area.

		pH	**TDS**	**K^+**	**Na^+**	**Ca^{2+}**	**Mg^{2+}**	**Cl^-**	**SO_4^{2-}**	**HCO_3^-**	**NO_2^-**	**NO_3^-**	**NH_4^+**	**F^-**
Phreatic groundwater	pH	1.00												
	TDS	−0.07	1.00											
	K^+	0.19	**0.67**	1.00										
	Na^+	−0.17	**0.87**	0.48	1.00									
	Ca^{2+}	−0.17	**0.77**	**0.78**	**0.56**	1.00								
	Mg^{2+}	−0.08	**0.90**	**0.85**	**0.69**	**0.85**	1.00							
	Cl^-	−0.08	**0.87**	**0.86**	**0.71**	**0.86**	**0.96**	1.00						
	SO_4^{2-}	−0.17	**0.95**	**0.55**	**0.91**	**0.70**	**0.78**	**0.74**	1.00					
	HCO_3^-	0.42	0.21	0.17	0.27	0.20	0.24	0.28	0.12	1.00				
	NO_2^-	0.11	0.30	−0.19	0.20	0.25	0.20	0.18	0.20	0.34	1.00			
	NO_3^-	0.00	0.24	0.03	0.29	0.28	0.16	0.26	0.24	0.26	0.13	1.00		
	NH_4^+	−0.23	−0.27	−0.42	0.03	−0.41	−0.34	−0.33	−0.12	−0.26	−0.11	−0.29	1.00	
	F^-	−0.12	0.08	−0.37	0.19	−0.08	−0.18	−0.20	0.20	−0.08	0.26	0.31	0.58	1.00
Confined groundwater	pH	1.00												
	TDS	0.15	1.00											
	K^+	0.66	0.84	1.00										
	Na^+	0.15	**0.99**	0.84	1.00									
	Ca^{2+}	0.22	**0.98**	0.85	**0.97**	1.00								
	Mg^{2+}	0.04	0.92	0.71	0.92	0.84	1.00							
	Cl^-	0.21	**0.99**	0.87	**0.99**	**0.98**	0.90	1.00						
	SO_4^{2-}	0.05	**0.98**	0.76	**0.98**	**0.97**	0.93	**0.97**	1.00					
	HCO_3^-	−0.68	−0.60	−0.84	−0.60	−0.69	−0.29	−0.66	−0.49	1.00				
	NO_2^-	0.86	−0.15	0.37	−0.15	−0.03	−0.39	−0.07	−0.26	−0.67	1.00			
	NO_3^-	−0.59	−0.08	−0.46	−0.10	−0.03	0.02	−0.15	0.10	0.54	−0.62	1.00		
	NH_4^+	−0.86	−0.28	−0.72	−0.29	−0.28	−0.14	−0.35	−0.12	0.76	−0.77	0.90	1.00	
	F^-	0.84	−0.11	0.65	−0.07	−0.68	0.54	−0.25	−0.24	0.74	−0.50	−0.50	−0.50	1.00

The bolded values represent a significant relation.

The mineral saturation status simulation results also presented that the carbonates, including aragonite, calcite and dolomite, were within a saturation index greater than zero in both phreatic and confined aquifers (Figure 7), confirming the conclusion drawn from end-member diagrams that the dissolution of carbonates is not the significant contribution of groundwater solutes in phreatic and confined aquifers of the watershed (Figure 5). The results of the hydrogeochemical simulation showed that all collected confined groundwater samples and almost all phreatic groundwater samples (except one) had a fluoride mineral saturation index below zero (Figure 7), suggesting that the dissolution of fluoride minerals is one of the dominant processes regulating the hydrogeochemical composition of both phreatic and confined groundwater. This is the reason that some collected phreatic and confined groundwater samples are with relatively high fluoride content and even exceed the permissible limit of 1 mg/L (Table 1).

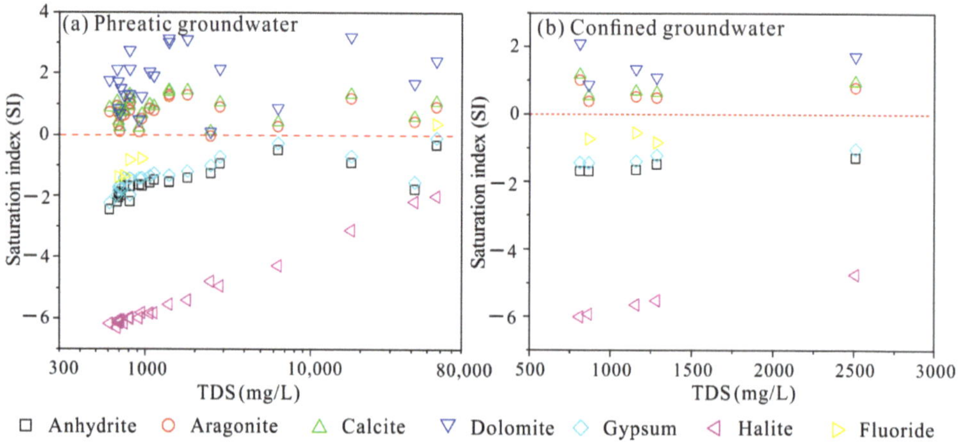

Figure 7. Plots of TDS versus minerals' saturation index for (**a**) phreatic groundwater and (**b**) confined groundwater.

Aside from the silicate weathering and evaporite dissolution, the ion exchanges are usually important processes contributing to groundwater solutes, especially in sedimentary aquifers. The ion processes can be revealed by the chloro-alkaline indices. The chloro-alkaline indices of CAI-1 and CAI-2 can be obtained by Equations (2) and (3). The combination of these two chloro-alkaline indices can be used to reveal the ion exchange processes occurring in aquifers. If these two indices are below zero, then the cation-exchange reaction (Equation (4)) is the dominant reaction occurring in the aquifer. However, if both CAI-1 and CAI-2 are positive, a reverse cation-exchange reaction (Equation (5)) is implied. As shown in Figure 8, most of the phreatic groundwater samples and confined groundwater samples are situated in the left lower dominance of the diagram and with both CAI-1 and CAI-2 below zero, implying that the cation-exchange reaction (Equation (4)) is one of the dominant processes regulating the hydrogeochemical composition of groundwater in phreatic and confined aquifers at most of the sites in the Mo river watershed. In addition, several collected phreatic groundwater samples and one confined groundwater sample had positive CAI-1 and CAI-2 values, suggesting the occurrence of reverse cation-exchange reactions at these sites.

$$CAI-1=\frac{Cl^{-}-(Na^{+}+K^{+})}{Cl^{-}} \tag{2}$$

$$CAI-2=\frac{Cl^{-}-(Na^{+}+K^{+})}{SO_4^{2-}+CO_3^{2-}+HCO_3^{-}+NO_3^{-}} \tag{3}$$

$$\mathrm{Ca^{2+}}\left(\mathrm{or\ Mg^{2+}}\right)+\mathrm{2NaX\ (solids)}\rightarrow \mathrm{2Na^{+}}+\mathrm{CaX_2\ (or\ MgX_2)\ (solids)} \tag{4}$$

$$\mathrm{2Na^{+}}+\mathrm{CaX_2\ (or\ MgX_2)\ (solids)}\rightarrow \mathrm{Ca^{2+}}\left(\mathrm{or\ Mg^{2+}}\right)+\mathrm{2NaX\ (solids)} \tag{5}$$

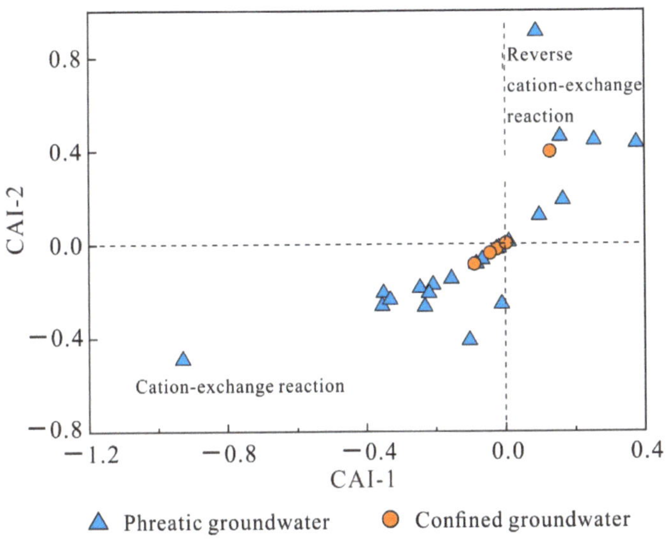

Figure 8. Plots of chloro-alkaline indices CAI-1 versus CAI-2 for phreatic and confined groundwater.

Apart from natural sources, groundwater hydrochemical solutes can be originated from anthropogenic sources, such as domestic effluents, industrial sewages and agricultural fertilizers. NO_3^- is one of the most common indicators of groundwater solutes in human societies. Generally, the geological limit of NO_3^- in nature is within 10 mg/L. Any water with a NO_3^- concentration beyond this limit is regarded as anthropogenic pollution. As presented in Table 1, both phreatic and confined groundwaters were found with a NO_3^- concentration exceeding this geological limit at some sites, indicating that the anthropogenic inputs contribute some chemicals to both phreatic and confined aquifers. Additionally, the maximum NO_3^- concentration in phreatic aquifers (37.3 mg/L) was higher than that in confined aquifers (16.00 mg/L). This vertical distribution trend of NO_3^- concentrations also confirmed the external inputs of chemicals into phreatic and confined aquifers.

To further illustrate the specific anthropogenic factors influencing groundwater chemistry, the relationship between the molar ratios of NO_3^-/Na^+ and SO_4^{2-}/Ca^{2+} was determined. As shown in Figure 9, no phreatic and confined groundwater is plotted in the lower right dominance region of the diagram, indicating that the groundwater in the watershed is out of the significant influence of domestic sewage and agricultural nitrogen fertilizers. Some phreatic groundwater samples are situated in the upper left dominance of the diagram, implying the potential influence of industrial wastes or agricultural sulfate fertilizers. The land use and cover showed there are no industrial areas but agricultural lands and human settlements in the watershed. Thus, the agricultural application of sulfate fertilizers rather than nitrite fertilizers in the middle reaches of the watershed is an important source of chemicals in groundwater. This was evidenced by the correlation matrix, in which TDS showed no significant relation with NO_3^- but SO_4^{2-} (Table 2).

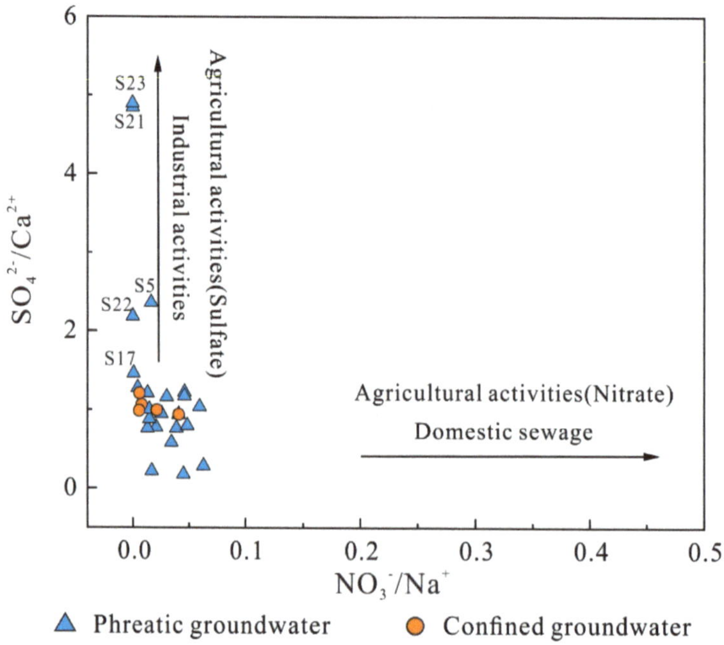

Figure 9. Plots of molar ratio NO_3^-/Na^+ versus molar ratio SO_4^{2-}/Ca^{2+} for phreatic and confined groundwater in the study area.

3.4. *Groundwater Quality Assessment*

3.4.1. Drinking Purposes

Groundwater is the only water resource for domestic usage in the study area due to the shortage of precipitation and surface water. Understanding the quality and suitability of groundwater for domestic purposes is essential for the residents' health and sustainable development of the study area. The quality of groundwater for domestic usage can be comprehensively revealed by the Entropy-weighted water quality index (EWQI) approach. The EWQI is an improved water quality assessment method based on the water quality index (WQI). It avoids the subjective issues in the assignment of hydrogeochemical parameters' weights involved in the assessment. Entropy weight is introduced to replace the empirical weight in the water quality assessment. The Entropy weight can avoid the loss of valuable information on water hydrochemistry during the groundwater quality appraisal. As a result, much more accurate results can be obtained by the EWQI than the traditional WQI. The procedure of the EWQI assessment can be conducted according to the following Equations (6)–(12) [60,61]. All of the hydrogeochemical parameters listed in Table 1 were considered in the EWQI assessment for the present research. The permissible quality of drinking water recommended by the WHO [55] and Chinese guidelines [54] is listed in Table 1.

$$Y = \begin{bmatrix} y_{11} & y_{12} & \cdots & y_{1n} \\ y_{21} & y_{22} & \cdots & y_{2n} \\ \vdots & \vdots & \ddots & \vdots \\ y_{m1} & y_{m2} & \cdots & y_{mn} \end{bmatrix} \tag{6}$$

$$y_{ij} = \frac{x_{ij} - (x_{ij})_{\min}}{(x_{ij})_{\max} - (x_{ij})_{\min}} \tag{7}$$

$$w_j = \frac{1 - e_j}{\sum\limits_{i=1}^{n} (1 - e_j)} \quad (8)$$

$$e_j = -\frac{1}{\ln m} \sum_{i=1}^{m} (P_{ij} \times \ln P_{ij}) \quad (9)$$

$$P_{ij} = \frac{y_{ij}}{\sum\limits_{i}^{m} y_{ij}} \quad (10)$$

$$q_j = \frac{C_j}{S_j} \times 100 \quad (11)$$

$$EWQI = \sum_{j=1}^{m} (w_j \times q_j) \quad (12)$$

where x_{ij} represents the value of the sample i's jth hydrogeochemical parameter. $(x_{ij})_{\min}$ and $(x_{ij})_{\max}$ are the minimum and maximum values of the jth hydrogeochemical parameter, respectively.

The EWQI assessment results of the collected phreatic and confined groundwater samples are presented in Figure 10. It can be seen that the EWQI values of groundwater, especially in confined aquifers, had a relatively wide range and varied from less than 30 to more than 1000. This indicates that groundwater in the Mo river watershed has a large variation in groundwater quality. Specifically, the collected confined groundwater samples had relatively low EWQI values. All of the EWQI values of the confined groundwater samples were below 100. According to the water quality classification based on EWQI values, water quality can be classified into five categories, i.e., excellent quality ($EWQI \leq 50$), good quality ($50 < EWQI \leq 100$), medium quality ($100 < EWQI \leq 150$), poor quality ($150 < EWQI \leq 200$) and extremely poor quality (> 200). Thus, the collected confined groundwater samples are within qualities ranging from excellent to good and suitable for drinking purposes. For phreatic groundwater, about 73.91% of samples were within an EWQI value below 100, suggesting excellent to good water quality. Two phreatic groundwater samples (accounting for 8.70%) were observed situating in the medium quality category with an EWQI in the range of 100–150, indicating suitable domestic usage but not for direct drinking. Four phreatic groundwater samples (accounting for 17.39%) had an EWQI value exceeding 200 and belonged to the extremely poor groundwater category.

The distributions of groundwater quality in various aquifers are presented in Figure 11. As shown in Figure 11a, most of the collected phreatic groundwater samples in the watershed are of excellent or good quality. It can be seen that the phreatic groundwater of relatively good quality is dominantly distributed in the middle-upper reaches of the watershed. The two medium-quality groundwater samples are located on the southwestern margin of the basin and the Chaka farm. All of the extremely poor-quality phreatic groundwater samples are situated in the lower parts of the watershed and adjacent to the Chaka salt lake. The phreatic groundwater samples that presented a worse quality trend are in the upper to lower reaches of the watershed. All of the medium and extremely poor -quality phreatic groundwater samples are featured with high salinity. The relatively poor quality of phreatic groundwater in the watershed is ascribed to the high salinity rather than the high concentration of potentially toxic elements. Although the middle-lower reaches have worse groundwater quality than the upper reach, all collected confined groundwater samples are of excellent or good quality based on the EWQI values (all less than 100) (Figure 11b). Overall, the confined groundwater at all sampling sites and the phreatic groundwater in the middle-upper reaches of the watershed are of relatively good water quality and are suitable for various domestic usages. Contrastingly, phreatic groundwater in the lower reaches adjacent to the salt lake is not suitable for domestic purposes due to the high salinity.

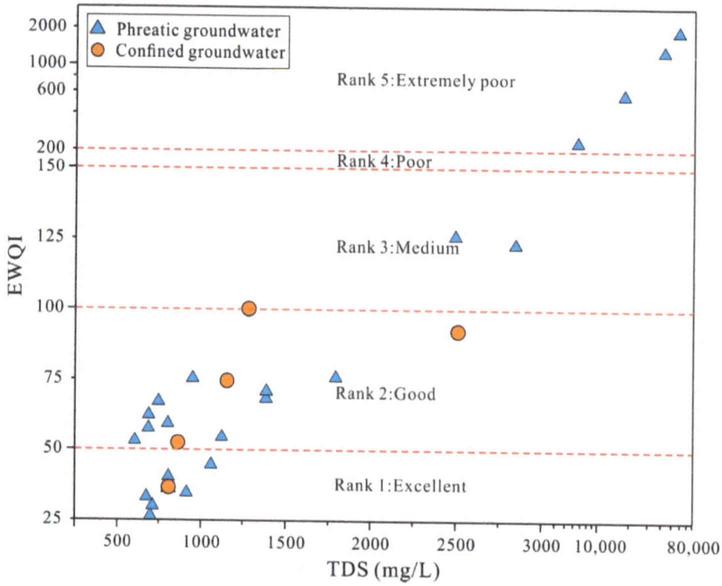

Figure 10. Plots of TDS versus EWQI values of phreatic and confined groundwater.

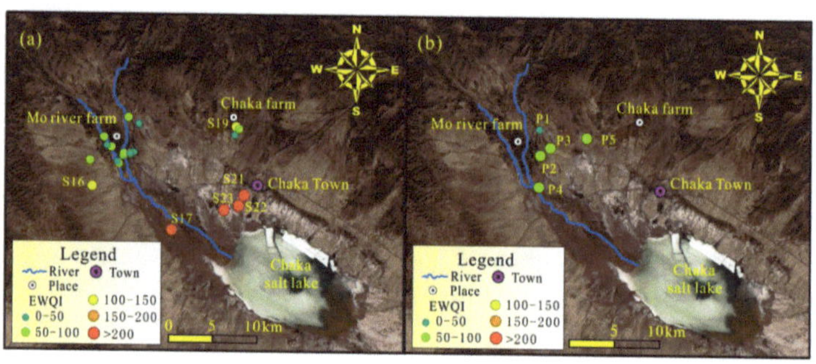

Figure 11. Distribution of groundwater quality in the (**a**) phreatic and (**b**) confined aquifers based on the EWQI assessment.

3.4.2. Irrigation Purposes

As shown in Figure 1a, there are two continuously distributed farmlands in the study area, i.e., the Mo river and the Chaka farms. Due to the scarcity of local precipitation, all farmlands in the basin need to be irrigated by human beings. All of the agricultural irrigation waters are solely dependent on the exploitation of groundwater resources rather than surface water due to the shortage of surface water. Thus, the irrigation quality of groundwater is critical to the sustainable development of agriculture in the basin.

Irrigation water with high content of salinity would pose negative effects on the growth of agricultural plants by weakening their capacities to uptake nutrients and water [46]. The rich ions, such as Na^+ and Ca^{2+}, in irrigation waters would lead to a change in soil structure and result in compact and dry soil, reducing the production of agricultural plants [29,62]. The EC, sodium adsorption ratio (SAR), sodium percentage (%Na) and permeability index (PI), together with the Wilcox and USSL charts, were employed in the present research to illustrate the irrigation quality of phreatic and confined groundwater in the watershed. The SAR, %Na and PI can be obtained by the following equations [12,28,35,46].

$$SAR = \frac{Na^+}{\sqrt{\frac{Ca^{2+}+Mg^{2+}}{2}}} \quad (13)$$

$$\%Na = \frac{Na^+}{Ca^{2+} + Mg^{2+} + Na^+ + K^+} \quad (14)$$

$$PI = \frac{Na^+ + \sqrt{HCO_3^-}}{Ca^{2+} + Mg^{2+} + Na^+} \times 100\% \quad (15)$$

The EC is an integrated parameter representing the salt content in water. The EC values of the collected phreatic and confined groundwater samples were in the range of 1033.50–44,205.00 μS/cm (averaging 5932.70 μS/cm) and 1309.00–3608.10 μS/cm (averaging 2011.00 μS/cm), respectively (Table 1). Generally, water quality can be classified into three categories, i.e., no restriction (EC < 700 μS/cm), slight-to-moderate restriction (700–3000 μS/cm), and severe degree of restriction to irrigation (>3000 μS/cm) [63]. According to this criterion, no sampled phreatic and confined groundwater falls into the none restriction category. Approximately 73.91% of the collected phreatic groundwater samples belonged to the slight-to-moderate restriction classification (700–3000 μS/cm), and 23.09% of the phreatic samples were considered as the tolerant class (>3000 μS/cm). For confined groundwater, only one sample had an EC slightly exceeding the limit of 3000 μS/cm and fell into the tolerant class. All other confined groundwater samples (80%) belonged to the slight-to-moderate restriction classification (700–3000 μS/cm). Overall, most of the phreatic and confined groundwater samples are relatively good quality for irrigation in terms of salinity.

The sodium adsorption ratio (SAR) is a widely used parameter to represent the potential hazard of alkali/sodium to agricultural plants. The values of SAR for the collected phreatic and confined groundwater samples varied from 1.78 to 206.01 and between 3.44 and 10.45, respectively. According to the irrigation criterion recommended by USSL, water with a SAR value < 10, 10–18, 18–26 or $\geq$ 26 is considered excellent, good, fair and unsuitable for irrigation, respectively [64]. For phreatic groundwater, approximately 82.61% of samples had a SAR value of less than 10, indicating excellent quality for irrigation. One sample (4.35%) was found with a SAR value in the range of 10–18, suggesting good irrigation quality. About 13.04% of the phreatic groundwater samples had a SAR value greater than 26, implying an unsuitable quality for irrigation.

The Wilcox chart constructed by the EC and SAR is an integrated and useful approach to illustrate the water quality for irrigation usage. It divides the potential salinity hazard of irrigation water into four categories, including low salinity hazard (C1), medium salinity hazard (C2), high salinity hazard (C3) and very high salinity hazard (C4). The sodium hazard of irrigation water is also classified into four categories, i.e., low sodium hazard (S1), medium sodium hazard (S2), high sodium hazard (S3) and very high sodium salinity hazard (S4). As shown in Figure 12, most of the collected phreatic and confined groundwater samples are situated in the C3S1 and C3S2 dominances, indicating low to medium sodium and high salinity hazards. Three phreatic groundwater samples (13.04%) are plotted in the C4S2 category, suggesting medium sodium hazard and very high salinity hazard. One confined groundwater, but no phreatic groundwater, is plotted in the C4S3 dominance, indicating both high sodium and very high salinity hazard risks. In addition, four phreatic groundwater samples fall in the C4S4 category, implying the existing risks of very high sodium and very high salinity hazards.

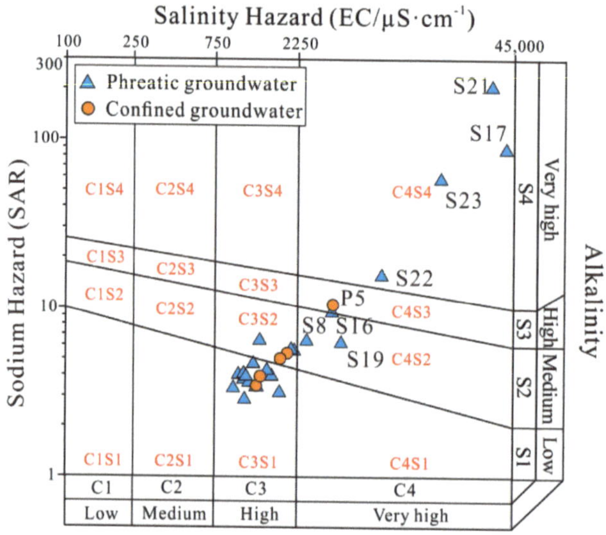

Figure 12. Wilcox chart presenting the water quality for irrigation purposes.

Sodium percentage (%Na) is another useful and robust parameter representing the sodium hazard of irrigation water to soil structure [46,65]. The phreatic groundwater samples had a relatively large variation of %Na ranging from 32.75 to 96.84% with a mean of 57.18%. The collected confined groundwater samples had %Na values varying from 46.88 to 65.80%, with an average of 54.57%, and presented with a relatively small range of %Na. The USSL chart constructed by the EC and sodium percent was introduced to illustrate the suitability of groundwater for irrigation in terms of the %Na. As shown in Figure 13, most of the collected phreatic and confined groundwater samples are situated in the good-to-permissible category and permissible-to-suitable category, indicating it is safe for irrigation in terms of the potential sodium hazard. Three phreatic groundwater samples were plotted in the doubtful-to-unsuitable category. Six phreatic groundwater samples (S16, S19, S21, S22 and S23) and one confined groundwater (P5) fell into the unsuitable category. Groundwater samples situated in these two categories have a potential sodium hazard if used for irrigation and are, therefore, not suitable for irrigation. Groundwater samples with relatively poor quality for irrigation in terms of sodium percentage are also mainly distributed in the lower reach of the watershed, especially the phreatic groundwater adjacent to the salt lake. Groundwater, regardless of the depth in the middle-upper reaches of the watershed, had relatively good quality for irrigation and would not potentially pose a sodium hazard to the soil.

Permeability index (PI) is an effective and widely used parameter assessing the possibility of irrigation water destroying the permeability of the soil. The PI values of the collected phreatic groundwater samples varied from 41.27 to 97.18 with a mean of 67.45, and the confined groundwater samples ranged from 59.94 to 69.40 with an average of 64.73. The Doneen chart constructed by the PI and the total concentration of ions was employed in the present research. Three classifications, i.e., Class-I (good), Class-II (doubtful) and Class-III (unsuitable) of irrigation water quality, can be identified from this chart. As presented in Figure 14, most of the phreatic groundwater samples (86.96%) and all confined groundwater samples were plotted in Class-I, indicating they are good for agricultural irrigation. Two phreatic groundwater samples (S17 and S23) fell into Class-II, and one phreatic groundwater sample (S21) belonged to Class-III, suggesting doubtful and unsuitable for irrigation, respectively. All of the poor irrigation-quality groundwater samples are distributed adjacent to the salt lake. Groundwater in the middle-upper reaches of the watershed, regardless of the depth, had good irrigation quality and would not pose permeability damages to soil structure.

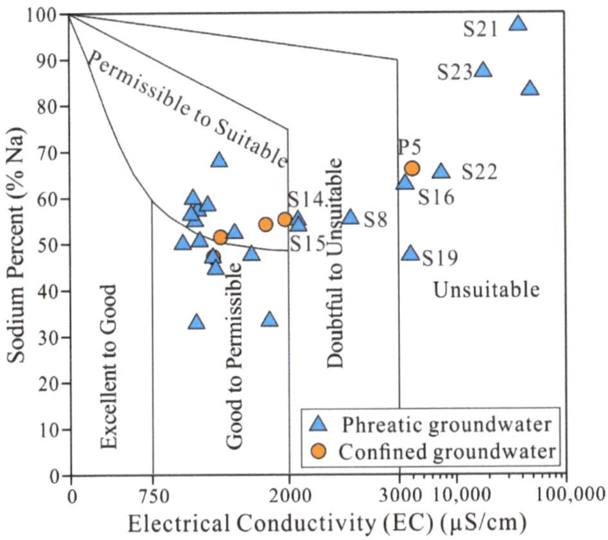

Figure 13. USSL chart presenting the water quality for irrigation usage.

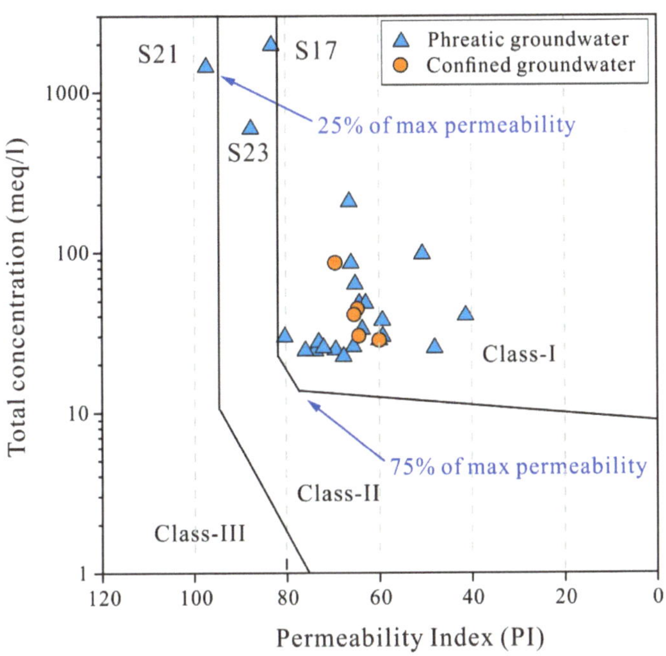

Figure 14. Doneen chart illustrating the irrigation quality of groundwater.

4. Conclusions

Groundwater is essential for the development of the socio-economy and eco-environment in many parts of the world, especially in arid and semiarid regions. Groundwater geochemistry is the key factor determining its suitability for various purposes and greatly influences the health of water consumers. The present research takes the Mo river watershed on the Tibetan plateau as an example to gain insights into the geochemistry, governing mechanisms and quality of groundwater in arid endorheic watersheds. The main findings are as follows:

Groundwater, regardless of the depth in the Mo river watershed, is slightly alkaline. Phreatic groundwater (TDS in the range of 609.19–56,715.34 mg/L) is presented as much saltier than confined groundwater (TDS in the range of 811.86–2509.51 mg/L) in the watershed. Na^+ and Cl^- are the dominant ions for groundwater in both phreatic and confined aquifers. The hydrogeochemical facies of both phreatic and confined groundwater are of Cl-Na type, followed by the mixed Cl-Mg·Ca type. Toxic elements of $NO_2{}^-$, $NH_4{}^+$ and F^- are beyond the permissible limits of the Chinese guidelines.

The chemical compositions of groundwater, regardless of the depth, are controlled by water–rock interactions in nature. The main rocks involved in the water–rock interactions are silicates and evaporites in both phreatic and confined aquifers. Most groundwater sample compositions are mainly contributed by silicate weathering. Only a small portion of phreatic groundwater samples was dominantly contributed by the dissolution of evaporites (such as halite, gypsum, anhydrite and sulfates). Fluoride mineral dissolution was the reason that groundwater in some locations had relatively high fluoride content. The cation-exchange reaction occurred in most sites for both phreatic and confined groundwater. The reverse cation-exchange reaction also occurred at a few locations. Aside from the natural hydrogeochemical processes, anthropogenic inputs of agricultural sulfate fertilizers have significantly influenced the hydrochemistry of groundwater in the farmlands in the middle reaches of the watershed.

Most groundwater samples, regardless of the depth, had relatively good quality. All confined groundwater samples and most of the phreatic groundwater samples had an EWQI value within 100, and could be used for direct drinking purposes. About 8.70% of phreatic groundwater samples had an EWQI value in the range of 100–150 and were suitable for domestic usage. In addition, 17.39% of phreatic groundwater samples had an EWQI value beyond 200 and were not suitable for various domestic purposes. The EC, SAR, %Na and PI suggested that most of the groundwater samples, regardless of the depth, were of relatively good irrigation quality. Only a small number of phreatic groundwater samples may cause salinity, sodium or permeability hazards to the soil. The relatively poor-quality groundwater samples were mainly located in the lower parts of the watershed. Thus, phreatic groundwater in the middle-upper reaches and confined groundwater of the whole watershed are the primary water resources for domestic and irrigational usage. The present research can provide references for understanding the characteristics, genesis and quality distribution of groundwater in arid and semiarid regions and benefit the rational utilization of groundwater resources.

Author Contributions: Conceptualization: F.W., Y.X. and S.W.; Methodology: H.Y., K.L. and Y.Z.; Formal analysis: F.W., X.C., W.W., Z.Q., K.L., F.B.B. and V.S.; Investigation: S.W., X.C., Z.Q., K.L., H.Y. and L.W.; Data curation: F.W., S.W., L.W., Z.Q., K.L. and H.Y.; Software: F.W., H.Y., Y.Z., K.L. and Z.Q.; Writing-original draft: F.W.; Writing-review and editing: Y.X., F.B.B. and V.S.; Supervision: F.W., X.C., Y.X., W.W. and S.W. All authors have read and agreed to the published version of the manuscript.

Funding: This research was funded by the Key Lab of Geo-environment of Qinghai Province (grant number 2021KJ011, 2021KJ00102), the Natural Science Foundation of China (grant number 42007183), the Natural Science Foundation of Sichuan Province (grant number 2022NSFSC1084), the Fundamental Research Funds for the Central Universities (grant number 2682021ZTPY063) and the Student Research Training Program of Southwest Jiaotong University (grant number 2022174).

Data Availability Statement: The raw data supporting the conclusions of this article will be made available by the authors without undue reservation upon request.

Acknowledgments: The authors appreciate the editors and reviewers for their critical comments and suggestions, which greatly helped us improve the present manuscript.

Conflicts of Interest: The authors declare that the research was conducted in the absence of any commercial or financial relationships that could be construed as potential conflicts of interest.

References

1. Avci, H.; Dokuz, U.E.; Avci, A.S. Hydrochemistry and groundwater quality in a semiarid calcareous area: An evaluation of major ion chemistry using a stoichiometric approach. *Environ. Monit. Assess.* **2018**, *190*, 641. [CrossRef]
2. Zhang, Q.; Xu, P.; Qian, H. Groundwater Quality Assessment Using Improved Water Quality Index (WQI) and Human Health Risk (HHR) Evaluation in a Semi-arid Region of Northwest China. *Expo. Health* **2020**, *12*, 487–500. [CrossRef]
3. Banadkooki, F.B.; Xiao, Y.; Malekinezhad, H.; Hosseini, M.M. Optimal allocation of regional water resources in an arid basin: Insights from Integrated Water Resources Management. *J. Water Supply Res. Technol.-Aqua* **2022**, *71*, 910–925. [CrossRef]
4. Yin, S.; Gu, X.; Xiao, Y.; Wu, W.; Pan, X.; Shao, J.; Zhang, Q. Geostatistics-based spatial variation characteristics of groundwater levels in a wastewater irrigation area, northern China. *Water Supply* **2017**, *17*, 1479–1489. [CrossRef]
5. Xiao, Y.; Liu, K.; Zhang, Y.; Yang, H.; Wang, S.; Qi, Z.; Hao, Q.; Wang, L.; Luo, Y.; Yin, S. Numerical investigation of groundwater flow system and its evolution under the climate change in the arid Golmud river watershed on Tibetan Plateau. *Front. Earth Sci.* **2022**, *10*, 943075. [CrossRef]
6. Díaz-Alcaide, S.; Martínez-Santos, P. Review: Advances in groundwater potential mapping. *Hydrogeol. J.* **2019**, *27*, 2307–2324. [CrossRef]
7. Hao, Q.; Xiao, Y.; Chen, K.; Zhu, Y.; Li, J. Comprehensive Understanding of Groundwater Geochemistry and Suitability for Sustainable Drinking Purposes in Confined Aquifers of the Wuyi Region, Central North China Plain. *Water* **2020**, *12*, 3052. [CrossRef]
8. Barbieri, M.; Barberio, M.D.; Banzato, F.; Billi, A.; Boschetti, T.; Franchini, S.; Gori, F.; Petitta, M. Climate change and its effect on groundwater quality. *Environ. Geochem. Health* **2021**. [CrossRef]
9. Zhao, W.; Lin, Y.-z.; Zhou, P.-p.; Wang, G.-c.; Dang, X.-y.; Gu, X.-f. Characteristics of groundwater in Northeast Qinghai-Tibet Plateau and its response to climate change and human activities: A case study of Delingha, Qaidam Basin. *China Geol.* **2021**, *4*, 377–388. [CrossRef]
10. Gu, X.; Xiao, Y.; Yin, S.; Hao, Q.; Liu, H.; Hao, Z.; Meng, G.; Pei, Q.; Yan, H. Hydrogeochemical Characterization and Quality Assessment of Groundwater in a Long-Term Reclaimed Water Irrigation Area, North China Plain. *Water* **2018**, *10*, 1209. [CrossRef]
11. Li, P. Meeting the environmental challenges. *Hum. Ecol. Risk Assess.* **2020**, *26*, 2303–2315. [CrossRef]
12. Adimalla, N. Groundwater Quality for Drinking and Irrigation Purposes and Potential Health Risks Assessment: A Case Study from Semi-Arid Region of South India. *Expo. Health* **2019**, *11*, 109–123. [CrossRef]
13. Cui, W.; Hao, Q.; Xiao, Y.; Zhu, Y.; Li, J.; Zhang, Y. Combining river replenishment and restrictions on groundwater pumping to achieve groundwater balance in the Juma River Plain, North China Plain. *Front. Earth Sci.* **2022**, *10*, 902034. [CrossRef]
14. Li, P.; He, X.; Li, Y.; Xiang, G. Occurrence and Health Implication of Fluoride in Groundwater of Loess Aquifer in the Chinese Loess Plateau: A Case Study of Tongchuan, Northwest China. *Expo. Health* **2019**, *11*, 95–107. [CrossRef]
15. Yao, R.; Yan, Y.; Wei, C.; Luo, M.; Xiao, Y.; Zhang, Y. Hydrochemical Characteristics and Groundwater Quality Assessment Using an Integrated Approach of the PCA, SOM, and Fuzzy c-Means Clustering: A Case Study in the Northern Sichuan Basin. *Front. Environ. Sci.* **2022**, *10*, 907872. [CrossRef]
16. Li, S.; Su, H.; Li, Z. Hydrochemical characteristics and groundwater quality in the thick loess deposits of China. *Environ. Sci. Pollut. Res.* **2022**, *29*, 8831–8850. [CrossRef]
17. Li, C.; Gao, X.; Li, S.; Bundschuh, J. A review of the distribution, sources, genesis, and environmental concerns of salinity in groundwater. *Environ. Sci. Pollut. Res.* **2020**, *27*, 41157–41174. [CrossRef]
18. Keesari, T.; Roy, A.; Pant, D.; Sinha, U.K.; Kumar, P.V.N.; Rao, L.V. Major ion, trace metal and environmental isotope characterization of groundwater in selected parts of Uddanam coastal region, Andhra Pradesh, India. *J. Earth Syst. Sci.* **2020**, *129*, 205. [CrossRef]
19. Subba Rao, N.; Gugulothu, S.; Das, R. Deciphering artificial groundwater recharge suitability zones in the agricultural area of a river basin in Andhra Pradesh, India using geospatial techniques and analytical hierarchical process method. *Catena* **2022**, *212*, 106085. [CrossRef]
20. Xiao, Y.; Liu, K.; Hao, Q.; Li, Y.; Xiao, D.; Zhang, Y. Occurrence, controlling factors and health hazards of fluoride-enriched groundwater in the lower flood plain of Yellow River, Northern China. *Expo. Health* **2022**, *14*, 345–358. [CrossRef]
21. Hao, Q.; Lu, C.; Zhu, Y.; Xiao, Y.; Gu, X. Numerical investigation into the evolution of groundwater flow and solute transport in the Eastern Qaidam Basin since the Last Glacial Period. *Geofluids* **2018**, *2018*, 9260604. [CrossRef]
22. Nag, S.; Roy, M.B.; Roy, P.K. Study on the functionality of land use land cover over the evaluation of groundwater potential zone: A fuzzy AHP based approach. *J. Earth Syst. Sci.* **2022**, *131*, 146. [CrossRef]
23. Ntona, M.M.; Busico, G.; Mastrocicco, M.; Kazakis, N. Modeling groundwater and surface water interaction: An overview of current status and future challenges. *Sci. Total Environ.* **2022**, *846*, 157355. [CrossRef]
24. Khorchani, H.; Kamel, S. Modeling of groundwater yield by the electrical method case of the Triassic sandstone aquifer (Tataouine South-Eastern Tunisia). *J. Water Supply Res. Technol.-Aqua* **2021**, *70*, 600–618. [CrossRef]
25. Marinović, V.; Stevanović, Z. Karst groundwater quantity assessment and sustainability: The approach appropriate for river basin management plans. *Environ. Earth Sci.* **2019**, *78*, 362. [CrossRef]

26. Yin, S.; Xiao, Y.; Gu, X.; Hao, Q.; Liu, H.; Hao, Z.; Meng, G.; Pan, X.; Pei, Q. Geostatistical analysis of hydrochemical variations and nitrate pollution causes of groundwater in an alluvial fan plain. *Acta Geophys.* **2019**, *67*, 1191–1203. [CrossRef]
27. Adimalla, N.; Manne, R.; Zhang, Y.; Xu, P.; Qian, H. Evaluation of groundwater quality and its suitability for drinking purposes in semi-arid region of Southern India: An application of GIS. *Geocarto Int.* **2022**, 1–12. [CrossRef]
28. Li, Y.; Li, P.; Cui, X.; He, S. Groundwater quality, health risk, and major influencing factors in the lower Beiluo River watershed of northwest China. *Hum. Ecol. Risk Assess.* **2021**, *27*, 1987–2013. [CrossRef]
29. Luo, Y.; Xiao, Y.; Hao, Q.; Zhang, Y.; Zhao, Z.; Wang, S.; Dong, G. Groundwater geochemical signatures and implication for sustainable development in a typical endorheic watershed on Tibetan plateau. *Environ. Sci. Pollut. Res.* **2021**, *28*, 48312–48329. [CrossRef]
30. Ray, R.K.; Syed, T.H.; Saha, D.; Sarkar, B.C.; Reddy, D.V. Recharge mechanism and processes controlling groundwater chemistry in a Precambrian sedimentary terrain: A case study from Central India. *Environ. Earth Sci.* **2017**, *76*, 136. [CrossRef]
31. Rajmohan, N.; Masoud, M.H.Z.; Niyazi, B.A.M. Impact of evaporation on groundwater salinity in the arid coastal aquifer, Western Saudi Arabia. *Catena* **2021**, *196*, 104864. [CrossRef]
32. Venkatramanan, S.; Chung, S.Y.; Selvam, S.; Lee, S.Y.; Elzain, H.E. Factors controlling groundwater quality in the Yeonjegu District of Busan City, Korea, using the hydrogeochemical processes and fuzzy GIS. *Environ. Sci. Pollut. Res.* **2017**, *24*, 23679–23693. [CrossRef]
33. Liu, R.-p.; Zhu, H.; Liu, F.; Dong, Y.; El-Wardany, R.M. Current situation and human health risk assessment of fluoride enrichment in groundwater in the Loess Plateau: A case study of Dali County, Shaanxi Province, China. *China Geol.* **2021**, *4*, 487–497. [CrossRef]
34. Xiao, Y.; Shao, J.; Frape, S.; Cui, Y.; Dang, X.; Wang, S.; Ji, Y. Groundwater origin, flow regime and geochemical evolution in arid endorheic watersheds: A case study from the Qaidam Basin, northwestern China. *Hydrol. Earth Syst. Sci.* **2018**, *22*, 4381–4400. [CrossRef]
35. Wang, S.; Xie, Z.; Wang, F.; Zhang, Y.; Wang, W.; Liu, K.; Qi, Z.; Zhao, F.; Zhang, G.; Xiao, Y. Geochemical Characteristics and Quality Appraisal of Groundwater From Huatugou of the Qaidam Basin on the Tibetan Plateau. *Front. Earth Sci.* **2022**, *10*, 874881. [CrossRef]
36. Ma, R.; Zhang, B.; Zhou, X. The effects of climate change and groundwater exploitation on the spatial and temporal variations of heavy metal content in maize in the Luan River catchment of China. *Environ. Sci. Pollut. Res.* **2019**, *27*, 1035–1052. [CrossRef]
37. Manikandan, E.; Rajmohan, N.; Anbazhagan, S. Monsoon impact on groundwater chemistry and geochemical processes in the shallow hard rock aquifer. *Catena* **2020**, *195*, 104766. [CrossRef]
38. Mthembu, P.P.; Elumalai, V.; Brindha, K.; Li, P. Hydrogeochemical Processes and Trace Metal Contamination in Groundwater: Impact on Human Health in the Maputaland Coastal Aquifer, South Africa. *Expo. Health* **2020**, *12*, 403–426. [CrossRef]
39. Yin, S.; Xiao, Y.; Han, P.; Hao, Q.; Gu, X.; Men, B.; Huang, L. Investigation of Groundwater Contamination and Health Implications in a Typical Semiarid Basin of North China. *Water* **2020**, *12*, 1137. [CrossRef]
40. Li, H.-x.; Han, S.-b.; Wu, X.; Wang, S.; Liu, W.-p.; Ma, T.; Zhang, M.-n.; Wei, Y.-t.; Yuan, F.-q.; Yuan, L.; et al. Distribution, characteristics and influencing factors of fresh groundwater resources in the Loess Plateau, China. *China Geol.* **2021**, *4*, 509–526. [CrossRef]
41. Gu, X.; Xiao, Y.; Yin, S.; Liu, H.; Men, B.; Hao, Z.; Qian, P.; Yan, H.; Hao, Q.; Niu, Y.; et al. Impact of Long-Term Reclaimed Water Irrigation on the Distribution of Potentially Toxic Elements in Soil: An In-Situ Experiment Study in the North China Plain. *Int. J. Environ. Res. Public Health* **2019**, *16*, 649. [CrossRef]
42. Xiao, Y.; Hao, Q.; Zhang, Y.; Zhu, Y.; Yin, S.; Qin, L.; Li, X. Investigating sources, driving forces and potential health risks of nitrate and fluoride in groundwater of a typical alluvial fan plain. *Sci. Total Environ.* **2022**, *802*, 149909. [CrossRef]
43. Jiang, W.; Liu, H.; Sheng, Y.; Ma, Z.; Zhang, J.; Liu, F.; Chen, S.; Meng, Q.; Bai, Y. Distribution, Source Apportionment, and Health Risk Assessment of Heavy Metals in Groundwater in a Multi-mineral Resource Area, North China. *Expo. Health* **2022**, *14*, 807–827. [CrossRef]
44. Jiang, W.; Sheng, Y.; Liu, H.; Ma, Z.; Song, Y.; Liu, F.; Chen, S. Groundwater quality assessment and hydrogeochemical processes in typical watersheds in Zhangjiakou region, northern China. *Environ. Sci. Pollut. Res.* **2022**, *29*, 3521–3539. [CrossRef]
45. Jin, X.; He, C.; Zhang, L.; Zhang, B. A Modified Groundwater Module in SWAT for Improved Streamflow Simulation in a Large, Arid Endorheic River Watershed in Northwest China. *Chin. Geogr. Sci.* **2018**, *28*, 47–60. [CrossRef]
46. Xiao, Y.; Liu, K.; Yan, H.; Zhou, B.; Huang, X.; Hao, Q.; Zhang, Y.; Zhang, Y.; Liao, X.; Yin, S. Hydrogeochemical constraints on groundwater resource sustainable development in the arid Golmud alluvial fan plain on Tibetan plateau. *Environ. Earth Sci.* **2021**, *80*, 750. [CrossRef]
47. Xiao, Y.; Shao, J.; Cui, Y.; Zhang, G.; Zhang, Q. Groundwater circulation and hydrogeochemical evolution in Nomhon of Qaidam Basin, northwest China. *J. Earth Syst. Sci.* **2017**, *126*, 26. [CrossRef]
48. Zhu, G.; Zhang, Y.; Ma, H.; Wan, Q.; Zhang, Z.; Sang, L.; Liu, Y.; Xu, Y. Effects of a chain of reservoirs on temporal and spatial variation in water chemistry within an endorheic basin. *Ecol. Indic.* **2021**, *125*, 107523. [CrossRef]
49. Wu, C.; Wu, X.; Qian, C.; Zhu, G. Hydrogeochemistry and groundwater quality assessment of high fluoride levels in the Yanchi endorheic region, northwest China. *Appl. Geochem.* **2018**, *98*, 404–417. [CrossRef]
50. Grizard, P.; Schmitt, J.-M.; Goblet, P. Hydrogeology of an arid endorheic basin (Tsagaan Els, Dornogobi, Mongolia): Field data and conceptualization, three-dimensional groundwater modeling, and water budget. *Hydrogeol. J.* **2019**, *27*, 145–160. [CrossRef]

51. Huang, F.; Ochoa, C.G.; Chen, X.; Zhang, D. Modeling oasis dynamics driven by ecological water diversion and implications for oasis restoration in arid endorheic basins. *J. Hydrol.* **2021**, *593*, 125774. [CrossRef]
52. Gao, G.; Shen, Q.; Zhang, Y.; Pan, N.; Ma, Y.; Jiang, X.; Fu, B. Determining spatio-temporal variations of ecological water consumption by natural oases for sustainable water resources allocation in a hyper-arid endorheic basin. *J. Clean. Prod.* **2018**, *185*, 1–13. [CrossRef]
53. Ndehedehe, C.E.; Ferreira, V.G.; Onojeghuo, A.O.; Agutu, N.O.; Emengini, E.; Getirana, A. Influence of global climate on freshwater changes in Africa's largest endorheic basin using multi-scaled indicators. *Sci. Total Environ.* **2020**, *737*, 139643. [CrossRef]
54. *GB/T 14848-2017*; Standards for Groundwater Quality. General Administration of Quality Supervision (GAQS): Beijing, China, 2017.
55. WHO. *Guidelines for Drinking-Water Quality*, 4th ed.; World Health Organization: Geneva, Switzerland, 2017.
56. Xiao, Y.; Yin, S.; Hao, Q.; Gu, X.; Pei, Q.; Zhang, Y. Hydrogeochemical appraisal of groundwater quality and health risk in a near-suburb area of North China. *J. Water Supply Res. Technol.-Aqua* **2020**, *69*, 55–69. [CrossRef]
57. Qu, S.; Wang, G.; Shi, Z.; Xu, Q.; Guo, Y.; Ma, L.; Sheng, Y. Using stable isotopes (δD, δ18O, δ34S and 87Sr/86Sr) to identify sources of water in abandoned mines in the Fengfeng coal mining district, northern China. *Hydrogeol. J.* **2018**, *26*, 1443–1453. [CrossRef]
58. Zhang, Y.; Li, X.; Luo, M.; Wei, C.; Huang, X.; Xiao, Y.; Qin, L.; Pei, Q. Hydrochemistry and Entropy-Based Groundwater Quality Assessment in the Suining Area, Southwestern China. *J. Chem.* **2021**, *2021*, 5591892. [CrossRef]
59. Zhang, Y.; Dai, Y.; Wang, Y.; Huang, X.; Xiao, Y.; Pei, Q. Hydrochemistry, quality and potential health risk appraisal of nitrate enriched groundwater in the Nanchong area, southwestern China. *Sci. Total Environ.* **2021**, *784*, 147186. [CrossRef]
60. Xiao, Y.; Liu, K.; Hao, Q.; Xiao, D.; Zhu, Y.; Yin, S.; Zhang, Y. Hydrogeochemical insights into the signatures, genesis and sustainable perspective of nitrate enriched groundwater in the piedmont of Hutuo watershed, China. *Catena* **2022**, *212*, 106020. [CrossRef]
61. Zhang, Y.; He, Z.; Tian, H.; Huang, X.; Zhang, Z.; Liu, Y.; Xiao, Y.; Li, R. Hydrochemistry appraisal, quality assessment and health risk evaluation of shallow groundwater in the Mianyang area of Sichuan Basin, southwestern China. *Environ. Earth Sci.* **2021**, *80*, 576. [CrossRef]
62. Qu, S.; Duan, L.; Shi, Z.; Liang, X.; Lv, S.; Wang, G.; Liu, T.; Yu, R. Hydrochemical assessments and driving forces of groundwater quality and potential health risks of sulfate in a coalfield, northern Ordos Basin, China. *Sci. Total Environ.* **2022**, *835*, 155519. [CrossRef]
63. Mahammad, S.; Islam, A.; Shit, P.K. Geospatial assessment of groundwater quality using entropy-based irrigation water quality index and heavy metal pollution indices. *Environ. Sci. Pollut. Res.* **2022**. [CrossRef]
64. USSL. *Diagnosis and Improvement of Saline and Alkali Soils*; Agriculture Handbook; United States Department of Agriculture: Washington, DC, USA, 1954.
65. Qu, S.; Shi, Z.; Liang, X.; Wang, G.; Han, J. Multiple factors control groundwater chemistry and quality of multi-layer groundwater system in Northwest China coalfield—Using self-organizing maps (SOM). *J. Geochem. Explor.* **2021**, *227*, 106795. [CrossRef]

water

MDPI

Article

Spatiotemporal Oasis Land Use/Cover Changes and Impacts on Groundwater Resources in the Central Plain of the Shiyang River Basin

Lifang Wang [1,2], Zhenlong Nie [1,*], Qinlong Yuan [3,*], Min Liu [1], Le Cao [1], Pucheng Zhu [1], Huixiong Lu [4] and Bo Feng [4]

1 Institute of Hydrogeology and Environmental Geology, Chinese Academy of Geological Sciences, Shijiazhuang 050061, China
2 Key Laboratory of Groundwater Sciences and Engineering, Ministry of Natural Resources, Shijiazhuang 050061, China
3 Chongqing Jiangshan Hydropower Construction Engineering Survey and Design Consulting Co., Ltd., Chongqing 400000, China
4 Airborne Survey and Remote Sensing Center of Nuclear Industry, Shijiazhuang 050002, China
* Correspondence: niezhenlong@mail.cgs.gov.cn (Z.N.); yuanqinlong54@163.com (Q.Y.)

Abstract: The impacts of land use/cover changes (LUCCs) on groundwater resources are a global issue. The Shiyang River Basin of China is a typical, ecologically fragile area. Focusing on the Wuwei sub-basin of the central plain, this study analyzed typical remote sensing image data for 17 specific dates since 1970. Before the Comprehensive Treatment Program in 2007, the area of natural oases decreased at a rate of 16.25 km^2/year, while the area of farmland expanded at a rate of 13.85 km^2/year. The farmland expansion preferentially occurred in low-vegetation-coverage oases, where the groundwater depth increased from 4 to 20 m. The consumption of groundwater increased from 7319.5 $\times$ 10^4 m^3/year to 12,943.2 $\times$ 10^4 m^3/year. During the period 2008–2018, the areas of both the natural oases and farmland decreased at rates of 2.57 km^2/year and 8.99 km^2/year, respectively. The groundwater level rose significantly in the south and west, as well as near the main river channel. Groundwater consumption has been restored to 7270.4 $\times$ 10^4 m^3/year. Only 0.12 km^2 of every 1.17 km^2 of the original natural oases were restored through the natural farmland–natural oases conversion process. Groundwater depth increased significantly with the continuous expansion of farmland. Since the farmland area was effectively controlled, the trend of groundwater-level decline was significantly improved. These findings provide scientific support for the ecological restoration and reconstruction of oases, as well as an efficient and balanced development of river basin water resources.

Keywords: Shiyang River basin; natural oasis; land use/cover changes; groundwater level; groundwater resources

1. Introduction

Groundwater is the world's largest and most accessible source of unfrozen freshwater [1,2]; it supports food and water security, economic development, drought-risk alleviation, and other important ecosystem functions [3]. About 2.5 billion people worldwide depend on groundwater supplies [4]. However, as the world's population continues to grow, more people must rely on groundwater resources [5,6], particularly in arid and semiarid areas [7,8]. Groundwater is, as yet, less sensitive to climate variability and change compared to surface water, which makes groundwater more reliable for consumption. However, human activity affects the groundwater's distribution, quantity, and quality [9,10]. Therefore, assessing the impacts of human activities on the groundwater system is a major scientific challenge [11]. Land-use change is one of the most important human interventions that alter groundwater flow systems [9], one that will continue in the future to impact groundwater recharge dynamics [12]. Human-induced land use/cover changes (LUCCs) not only impact groundwater recharge but also have a significant effect on groundwater flow dynamics [5,13]. Therefore, in the LUCC plans that have been established by the International Geosphere–Biosphere Program (IGBP) and the International Human Dimensions Program (IHDP), one core problem is in understanding the impacts of regional land-use and land-cover changes on hydrological processes and water resources [14]. The impact of LUCC-induced changes in regional vegetation ecosystems on the regional hydrological cycle processes [15–17] has been an important research topic in hydrological science for more than 20 years and has become an issue of global importance [16,18]. Groundwater is usually the most important resource in arid regions, especially for socio-economic development, and is the best choice for a reliable water supply. Furthermore, groundwater is the most important factor in maintaining the health of an arid oasis ecosystem [19–21]. In the event of a reduction in water supply, the balance of an oasis ecosystem will be disrupted, and the oasis will degrade rapidly [22]. Globally, groundwater resources are irreplaceable when resolving water scarcity. Therefore, precise evaluation and effective management are important guarantees and necessary preconditions for exploring groundwater resources in an arid region. In this regard, an accurate estimation of the impact of human activities on the groundwater system is critical for developing a reasonable utilization program for regional groundwater resources [23,24]. Previous research on the impact of human activities on the groundwater system has mainly focused on the aspects of the intensity and reasonability of groundwater utilization while ignoring the impact of land-use changes on the groundwater system in the basin [25,26]. In fact, as an important link in the regional hydrological cycle, the groundwater dynamics in a basin (especially the groundwater depth and groundwater balance state) respond strongly to LUCCs [27].

Because of the stresses of an arid climate, desertification, water shortages, and other factors, the Shiyang River Basin is a typical ecologically fragile and climate-sensitive region [28]. Frequent droughts and drastic environmental changes are the main factors limiting sustainable development [28]. Since the 1980s, natural oases have been continuously declining, and their ecosystems have nearly collapsed due to the over-exploitation of soil and water resources [29] and the rapid yet uncontrolled expansion of farmland as a result of the intensification of human activities and extensive economic development [30]. In 2007, the Chinese central government launched the Comprehensive Treatment Program of the Shiyang River Basin (CTSRB) [31,32], the main measures of which included implementing the Grain-for-Green Program, protecting and restoring natural oases, increasing the coverage of forest/grass oases, increasing the surface water supply (i.e., increasing runoff and implementing inter-basin water-diversion projects), and controlling the over-exploitation of groundwater (i.e., limiting the per capita water consumption and closing some pumping wells). At a basin scale, LUCCs lead to changes in the patterns of the water supply and its use, and, in turn, cause inevitable changes in groundwater recharge/discharge processes, dynamics, and water cycle patterns, thereby significantly affecting the ecology, environment, economic development, and many other aspects of the basin [33].

Many studies have been conducted to investigate the land-use-driven changes in the water resource pattern of the Shiyang River Basin [34,35]. In addition, previous research has also focused on water resource management and allocation [36], groundwater level dynamics [32], the quantities of groundwater resources [37], policy mechanisms [38], and changes in ecosystem vulnerability [39] in this basin since the implementation of the CTSRB. However, only a very limited number of studies have been conducted to evaluate spatiotemporal changes in the vegetation cover as a measure of the spatiotemporal LUCCs since the implementation of the CTSRB in 2007 and to examine the response of the groundwater resource systems to these changes. Focusing on the Wuwei subbasin, i.e., the central plain of the Shiyang River basin, the goals of this study were: (1) to explore and compare the changes in the spatiotemporal pattern of the oasis coverage before and after 2007; (2) to analyze the effect of the spatiotemporal pattern of land use on the groundwater system, and the response characteristics of the groundwater system to the LUCCs. The findings of this study have implications for the scientific assessment of basin management effectiveness and the future management of water resources; they also provide scientific support for the ecological restoration and reconstruction of oases.

2. Materials and Methods

2.1. Study Area

The Shiyang River Basin (101°41′–104°16′ E, 36°29′–39°27′ N), which is surrounded by the Tengger Desert and the Badain Jaran Desert, is one of the three major inland river basins in the Gansu Hexi Corridor. It is located in the eastern part of the corridor, to the west of the Wushaoling Mountains, and is at the northern foot of the Qilian Mountains, with altitudes of 1300–4800 m. It is well known as a sandy wasteland, characterized by the predominant presence of sandy lands and the strong irrigation dependence of agricultural plantations. In addition to the oasis agroecosystem on the plain, the vast areas of the basin are dominated by a desert vegetation landscape (Figure 1). The basin has a typical inland arid climate, with an annual mean precipitation of 165.4 mm and a high annual mean evaporation of 2100 mm for the whole region, reaching 4000 mm in the deserts.

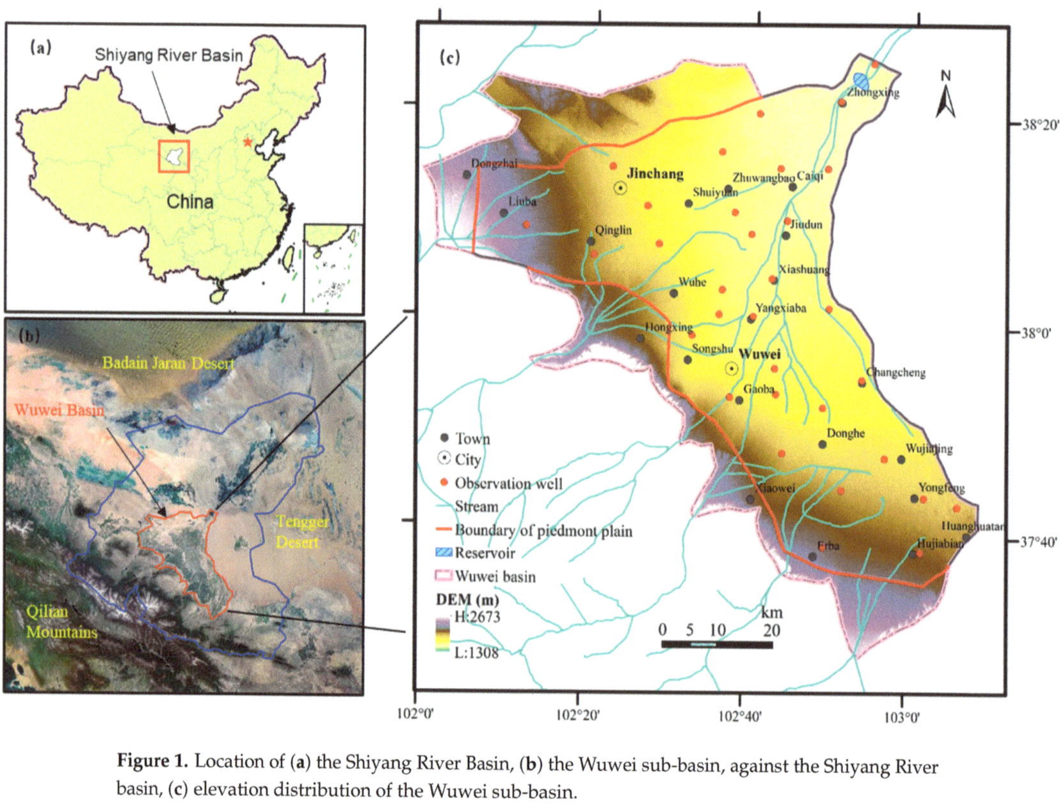

Figure 1. Location of (**a**) the Shiyang River Basin, (**b**) the Wuwei sub-basin, against the Shiyang River basin, (**c**) elevation distribution of the Wuwei sub-basin.

The groundwater aquifers are mainly composed of sand and gravel, deposited in the Upper and Middle Pleistocene, followed by slightly cemented and semi-cemented mud conglomerate, sandstone, and sandy loam, deposited in the Lower Pleistocene. The distribution and burial pattern of the aquifers are controlled by neotectonic movement. Due to the cutting of the buried Quaternary piedmont fault, the base of the fault zone is raised in the south, leading to the phreatic groundwater level being close to the ground surface, i.e., very shallow. In contrast, the aquifers thicken dramatically in the north, where they are mostly phreatic aquifers in the fault terrace and phreatic/confined aquifers in the plain.

The groundwater mainly exists in loose rock pores in the quaternary aquifer. The tertiary mudstone, sandy mudstone, and muddy sand conglomerate underneath are extremely weak in terms of water richness and hydraulic conductivity, and there is no water exchange. The recharging of the Quaternary aquifer is mainly from piedmont runoff recharge from the Qilian Mountains, atmospheric precipitation, surface water infiltration during conveyance, irrigation ditch infiltration, and the infiltration of irrigation groundwater. The groundwater eventually converges in the vicinity of Caiqi Township, Minqin County, where it partly overflows to the surface in the form of springs that flow into the Shiyang River and then on into the Hongyashan Reservoir. Part of this water also flows into the territory in Minqin County, in the form of lateral runoff. In addition, a large volume of groundwater is extracted for farmland irrigation and a small volume evaporates, which, together, constitute another important pathway of regional groundwater discharge. The amount of annual average recharge in the study area was as follows: piedmont runoff, precipitation and reservoir infiltration, surface water infiltration during conveyance, irrigation ditch infiltration, and the infiltration of irrigation groundwater were 18,526.1 × 10^4 m^3/year, 2879.8 × 10^4 m^3/year, 15,219.6 × 10^4 m^3/year, 3694.5 × 10^4 m^3/year, and 14,298.6 × 10^4 m^3/year, accounting for 33.9%, 5.3%, 27.9%, 6.8%, and 26.2% of the total recharge, respectively. The amount of annual average discharge was as follows: lateral outflow, groundwater exploitation, spring overflow, and evaporation were 603.0 × 10^4 m^3/year, 55,044.3 × 10^4 m^3/year, 5596.7 × 10^4 m^3/year, and 1030.9 × 10^4 m^3/year, accounting for 1.0%, 88.4%, 9.0%, and 1.7% of the total discharge, respectively [40,41].

The groundwater generally flows northward from the southern piedmont and eastward from the west, turning towards the northeast after reaching the main channel of the Shiyang River and entering the Minqin Basin through the Hongya-Alagu Mountain Fault. The groundwater is 100–120 m deep in the west and south; it gradually becomes shallower along the direction of groundwater runoff until it reaches the northeastern region, where the groundwater-level depth is as shallow as 5 m in some locations (Figure 2).

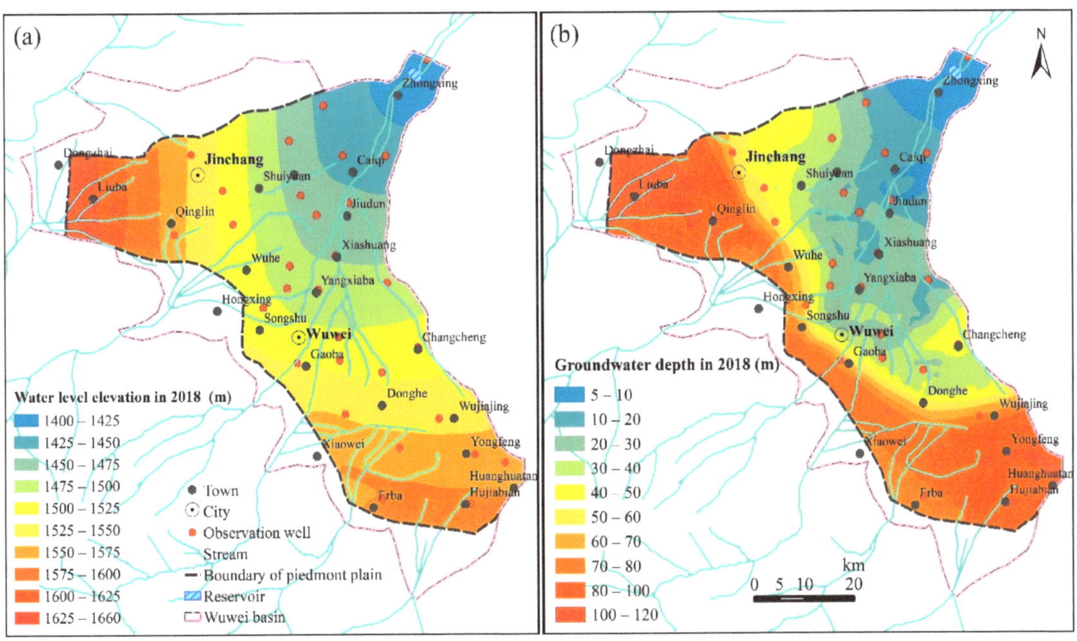

Figure 2. Distribution of (**a**) water level elevation in 2018, (**b**) groundwater depth in 2018.

The vegetation types in the study area include artificial vegetation and natural vegetation. The artificial vegetation is mainly *Haloxylon ammodendron*, which plays an important role in creating windbreaks and for sand fixation. Shrubs are the dominant component of natural vegetation in the area, including *Nitraria* spp., *Tamarix* spp., *Haloxylon ammodendron*, *Kalidium foliatum*, *Reaumuria soongorica*, and *Artemisia arenaria*. The trees include *Populus euphratica*, *Populus gansuensis*, *Elaeagnus angustifolia*, and *Salix matsudana*. The herbs are mainly *Phragmites australis*, *Achnatherum splendens*, *Sophora alopecuroides* L., *Bassia dasyphylla*, and *Agriophyllum squarrosum* [42,43].

2.2. Data Sources and Methodology

2.2.1. Land-Use Types

Raw remote sensing images (https://earthexplorer.usgs.gov/, path/row coordinates: 131–132, 33–34) of the study area, acquired in 1970, 1984, 1990, 1995, 2000, and 2007–2018, were derived from the thematic mapper (TM), enhanced thematic mapper (ETM), and operational land imager (OLI) data obtained by the US Landsat satellites. These data were subjected to atmospheric correction, radiometric correction, geometric correction, image fusion, color synthesis, and cropping [44,45].

In this study, an object-oriented approach was adopted for image segmentation by combining image texture features with actual crop growth features via a bottom-up approach. The nearest-neighbor classification method and the affiliation function classification method were combined to achieve information extraction for the decoding target. Meanwhile, to ensure the correctness and reliability of the results, the interpretation content was calibrated and validated by land-cover type calibration, field spectrum testing, and land-cover type validation.

The normalized difference vegetation index (NDVI) is a commonly used indicator for evaluating vegetation development. In general, NDVI values range between −1 and 1. A high NDVI value indicates that the vegetation is well-developed, whereas a low NDVI value indicates that the vegetation is in poor condition.

The vegetation fractional coverage is estimated using NDVI and the following formula:

$$\mathrm{VFC} = \frac{\mathrm{NDVI} - \mathrm{NDVImin}}{\mathrm{NDVImax} - \mathrm{NDVImin}} \times 100\% \qquad (1)$$

where VFC is the vegetation fractional coverage and $\mathrm{NDVI}_{\mathrm{min}}$ and $\mathrm{NDVI}_{\mathrm{max}}$ are NDVI values with cumulative probabilities of 5% and 95%, respectively [46].

The vegetation coverage was classified according to the vegetation coverage classification criteria for sandy land. The land-use types were classified as artificial oases (i.e., farmland), natural oases, and other land-use types (OL). Given that the remote sensing spectra of artificial oases are significantly affected by the seasons and crop harvest periods, in this study, artificial oases were identified through the visual interpretation of image pixels using characteristic colors, shapes, and textures as interpretation features. The natural oases were classified as high-vegetation-coverage oases (HCO, vegetation fractional coverage (VFC) $\geq$ 30%), medium-vegetation-coverage oases (MCO, 10% $\leq$ VFC < 30%), low-vegetation-coverage oases (LCO, 4% $\leq$ VFC < 10%), and desert oases (DO, VFC < 4%, including sandy land, saline land, shoals, marshland, and bare land). The OL category included urban land, rural residential land, industrial and mining land, rivers/canals, and reservoirs [47]. The regional land-use types on the 17 dates, and particularly the LUCCs during 1984, 2007, and 2018 were analyzed via object-oriented human–computer interactive interpretation according to the established interpretation features of the different land-use types; ArcGIS 10.3 (ESRI, Redlands, CA, USA) was used as the data processing and information extraction platform [44,45].

Geographic information system (GIS) intersection analysis was performed on the 1984, 2007, and 2018 LULC maps by overlaying the various maps, which yielded LUCC maps for 1984–2007 and 2007–2018. Two LULC transition matrices corresponding to these two periods were obtained by adding the map fields with the corresponding classes into the attribute table and then performing statistical analysis of the map units, with each matrix providing information about the direction of transition between two land-use types (i.e., gain vs. loss) and the land area involved in the transition [48]. The LULC transition matrices were used to examine the quantitative changes and the changes in the spatial pattern of the land-use types.

2.2.2. Groundwater Depth

The groundwater depth was measured on site in December 1984, December 2007, and December 2018 via a total of 32 observation boreholes, which all belong to the same unified phreatic aquifer. The changes in groundwater depth from 1984 to 2007 and from 2007 to 2018 were estimated separately by calculating the differences between the measured values and then performing kriging interpolation in a total of 23,105 cells (each interpolation cell had the dimensions of 400 m $\times$ 400 m and an area of 0.16 km^2).

Based on intersection analysis, the changes in the groundwater depth during each period were spatially overlaid on the corresponding land-use change map to visualize and calculate the area of the land-use changes associated with the alterations in groundwater depth.

2.2.3. Groundwater Balance Analysis

A groundwater system is a complex system consisting of inputs, outputs, and geological entities, and the various influencing factors of the system are interlinked and interact with one another. In this study, multivariate regression with a significance test was performed to establish a linear regression equation for identifying the influencing factors that significantly contributed to the changes in the groundwater storage volume.

The groundwater recharge in the study area mainly comes from upstream lateral inflow, surface water infiltration, irrigation ditch infiltration, the infiltration of irrigation groundwater, atmospheric precipitation, and reservoir water infiltration. The groundwater discharge mainly includes groundwater extraction, spring overflow, phreatic evaporation, and downstream lateral outflow. The upstream section of the Wuwei subbasin is bounded by a large fault in front of the Qilian Mountains, and the fault steps lead to a difference of nearly 100 m in the groundwater depth on the two sides of the fault, thereby leading to little inter-annual variation in the recharge. In contrast, atmospheric precipitation and reservoir water infiltration (with both acting as groundwater sources), spring overflow, phreatic evaporation, and downstream lateral outflow (all as groundwater sinks) accounted for a small proportion of the total recharge/discharge of the groundwater system, with no significant inter-annual variations. The surface water conveyance volume affected both the surface water infiltration during conveyance and the irrigation ditch infiltration, while the groundwater extraction volume determined the volume of infiltrated irrigation groundwater. A multivariate regression model was established, using the groundwater storage volume as the dependent variable and the farmland area, surface water conveyance volume, and volume of groundwater extraction as the three independent variables. This model was used to predict and conduct statistical analyses of the historical data. The model can be expressed as follows:

$$\Delta Q = \beta + a\ S_c + b\ Q_s + c\ Q_g$$

where ΔQ is the groundwater storage volume ($\times 10^4$ m^3), S_c is the farmland area (km^2), Q_s is the surface water conveyance volume ($\times 10^4$ m^3), Q_g is the groundwater extraction volume ($\times 10^4$ m^3), and β, a, b, and c are the regression coefficients.

Since agricultural and ecological irrigation was the main water consumption sector in the study area, the surface water conveyance volume and groundwater extraction volume in the model only present the amount of groundwater used in this particular sector.

3. Results and Discussion

3.1. Spatiotemporal LUCCs over the Last 30 Years

3.1.1. Land-Use Intensity Analysis

The land-use category level mainly focuses on the gross gain intensity, gross loss intensity, and uniform intensity of annual change for each land use over different time intervals [49,50]. From 1984 to 2007, the average annual increase in farmland was much bigger than the increases in the other categories (Figure 3). The two categories with the largest average annual reduction were the LCO and the DO. The uniform intensity of annual change for each land use was 1.47. The intensity of change for both farmland and OL was the following: uniform intensity > gross gain intensity > gross loss intensity. This indicated that the increase in farmland and OL was in a relatively stable state. The intensity of change for the DO and LCO was consistent with gross loss intensity > uniform intensity > gross gain intensity. This suggested that the gross losses in the DO and LCO were relatively active. From 2008 to 2018, the LCO had the largest average annual increase, followed by the OL. Farmland decreased by the largest average annual area, and uniform intensity > gross loss intensity > gross gain intensity, indicating that the decrease in farmland was in a relatively stable state. The intensity changes of the DO and the MCO were characterized by gross loss intensity > uniform intensity > gross gain intensity, implying that the gross losses were relatively active. The intensity of changes for the OL and farmland use types were featured as gross gain intensity > uniform intensity > gross loss intensity, implying that the gross gains were relatively active. The uniform intensity of annual change for each land-use type was 1.91, illustrating that the land-use category varied more dramatically in this period. It was presumed that the expansion of OL played a critical role.

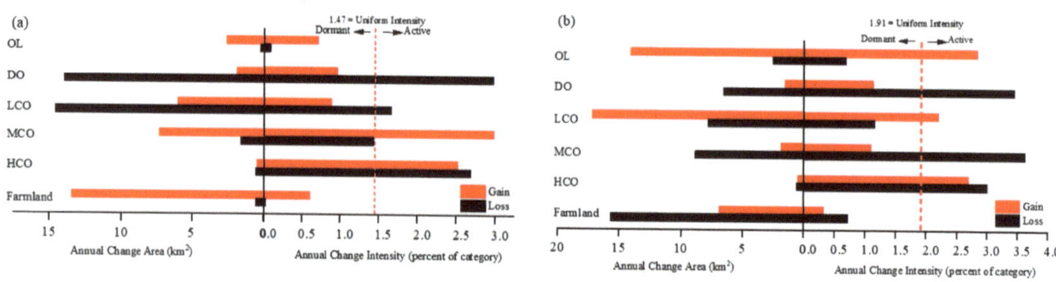

Figure 3. Category level intensity analysis for (**a**) 1984–2007; (**b**) 2008–2018.

The transition level is mainly used to analyze the increase or decrease in specific land-use types at different time intervals, the transition intensity of other land-use types into specific land-use types, and specific land-use types into other land-use types, and to compare the transition intensity of specific land-use types with the uniform intensity of annual change [49,50]. In this study, the intensity of the transitional level of farmland was analyzed (Figure 4). All increases in farmland were at the expense of the loss of natural oases. During 1984–2007, increases in farmland tended to occupy the LCO and HCO. The average intensity of the annual transition was greatest in the LCO, followed by the HCO, and avoided conversion in the MCO and DO. From 2008–2018, increases in farmland also tended to occupy the LCO, and decreases in farmland tended more frequently to convert to OL and avoid conversion to DO, LCO, and MCO.

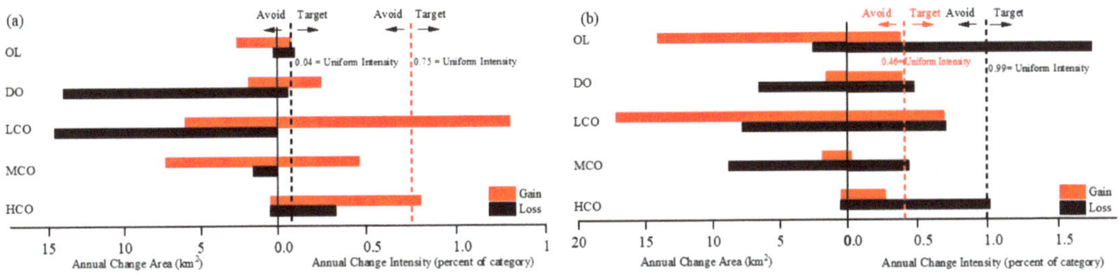

Figure 4. Transition-level intensity analysis of farmland for (**a**) 1984–2007; (**b**) 2008–2018.

3.1.2. Spatiotemporal LUCCs during 1984–2007

The farmland in the study area has continuously expanded with the increase in population during the last 30 years, leading to the shrinkage of natural oases [51,52]. As shown in Figure 5 and Table 1, the farmland, DO, LCO, and MCO had large gain and loss areas from 1984 to 2007.

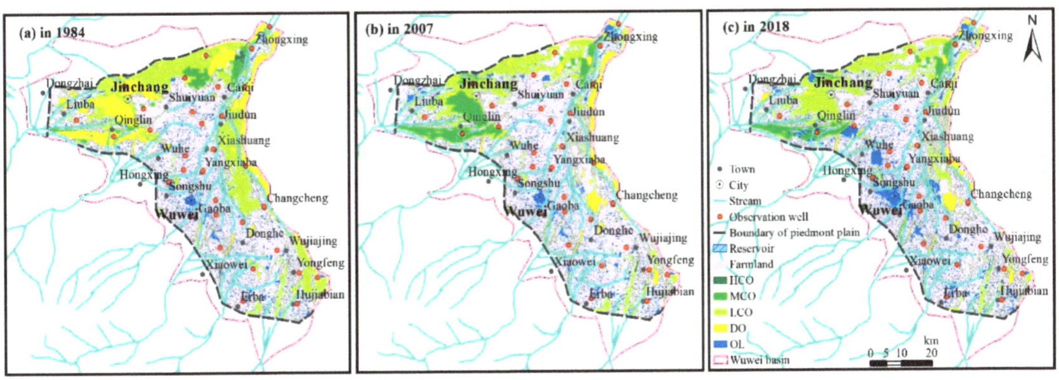

Figure 5. Distributions of land use/cover types, (**a**) in 1984, (**b**) in 2007, and (**c**) in 2018.

Table 1. Transition matrix of the land-cover types (km^2).

Year	Type	Farmland	HCO	MCO	LCO	DO	OL	Total
1984~2007	Farmland	1886.04	1.40	0.28	0.53	2.62	8.56	1899.43
	HCO	3.75	7.72	3.76	1.92	2.81	0.29	20.26
	MCO	12.11	0.00	76.15	20.21	1.74	3.67	113.88
	LCO	260.88	5.29	11.91	536.98	35.06	22.46	872.58
	DO	26.74	3.89	151.76	114.82	147.86	24.74	469.81
	OL	5.43	0.08	0.09	0.04	0.50	307.86	314.01
	Total	2194.95	18.39	243.95	674.49	190.59	367.59	3689.96
2008~2018	Farmland	2022.16	1.96	8.12	60.08	7.18	95.46	2194.95
	HCO	0.55	12.30	2.54	1.01	0.32	1.67	18.39
	MCO	0.77	0.00	146.65	69.86	0.00	26.68	243.95
	LCO	50.99	2.52	4.10	588.57	2.25	26.06	674.49
	DO	8.14	0.15	1.33	57.74	118.28	4.94	190.59
	OL	15.20	0.57	4.07	0.67	6.97	340.10	367.59
	Total	2097.81	17.50	166.81	777.93	135.00	494.92	3689.96

The gain (i.e., conversion of other land use types to farmland) and loss (i.e., conversion of farmland to other land use types) areas of the farmland were 308.91 km^2 and 13.39 km^2, respectively, leading to a net increase of 295.52 km^2 (15.6%). The new farmland was primarily created from the conversion of the LCO (260.88 km^2), which was distributed in Yongfeng and Liuba, as well as Changcheng, Xiashuang, and Jiudun near the river.

The gain and loss areas of the MCO were 167.80 km^2 and 37.73 km^2, respectively, leading to a net increase of 130.07 km^2 (114.2%). The new MCO was primarily from DO (151.76 km^2), which was mostly distributed in nearby Qinglin.

The gain and loss areas of the DO were 42.73 km^2 and 321.95 km^2, respectively, leading to a net reduction of 279.22 km^2 (59.4%). The loss areas, distributed in Qinglin, Liuba, and Zhongxing, were mainly converted into MCO (151.76 km^2) and LCO (114.82 km^2), accounting for 47.1% and 35.7% of the total loss areas, respectively.

The gain and loss areas of the LCO were 137.51 km^2 and 335.60 km^2, respectively, leading to a net reduction of 198.09 km^2 (22.7%). The loss areas were mainly converted into farmland (260.88 km^2), accounting for 77.7% of the total loss areas. The new LCO areas were primarily from DO (114.82 km^2), accounting for 83.5% of the total gain areas, which were mostly distributed in Yongfeng, Xiashuang, and Jiudun.

3.1.3. Spatiotemporal LUCCs during 2008–2018

As shown in Figure 5 and Table 1, the gain and loss areas of the OL, LCO, farmland, MCO, and DO from 2008 to 2018 were relatively large.

The gain and loss areas of the OL were 154.82 km^2 and 27.49 km^2, respectively, leading to a net increase of 127.33 km^2 (34.6%). The gain areas were primarily from farmland (95.46 km^2), accounting for 61.7% of the total gain areas. The new gain areas were distributed near the Wuwei urban area, and the expansion was also significant in Wuhe, Qinglin, and Gaoba.

The gain and loss areas of the LCO were 189.36 km^2 and 85.93 km^2, respectively, leading to a net increase of 103.44 km^2 (15.3%). The new gain areas were primarily from the MCO (69.86 km^2), farmland (60.08 km^2) and DO (57.74 km^2), accounting for 36.9%, 31.7%, and 30.5% of the total gain areas, which were mostly distributed in Jinchang and Qinglin.

The gain and loss areas of the farmland were 75.65 km^2 and 172.79 km^2, respectively, leading to a net reduction of 97.14 km^2 (4.4%). The loss areas were mainly converted into OL (95.46 km^2) and LCO (60.08 km^2), accounting for 55.2% and 34.8% of the total loss areas, respectively; these were mostly distributed in the Wuwei urban area, Wuhe, and Xiashuang areas.

The gain and loss areas of the MCO were 20.16 km^2 and 97.30 km^2, respectively, leading to a net reduction of 77.14 km^2 (31.6%). The loss areas were mainly converted into LCO (69.86 km^2), accounting for 71.8% of the total loss areas, which were mostly distributed in Qinglin and Jinchang.

The gain and loss areas of the DO were 16.72 km^2 and 72.31 km^2, respectively, leading to a net reduction of 55.58 km^2 (29.2%). The loss areas, mostly distributed in Jiudun, were mainly converted into LCO areas (57.74 km^2), accounting for 79.9% of the total loss areas.

3.1.4. LUCCs during 1984–2018

The areas of farmland, natural oases, and OL since 1970 are presented in Figure 6. From 1970 to 1984, human activities were stable, and there were no significant LUCCs. From 1984 to 2007, the farmland and OL expanded at rates of 13.85 km^2/year and 2.42 km^2/year, respectively, while the natural oases decreased at a rate of 16.25 km^2/year. From 2008 to 2018, the OL expanded at a higher rate of 11.52 km^2/year, while farmland and natural oases decreased at rates of 8.99 km^2/year and 2.57 km^2/year, respectively.

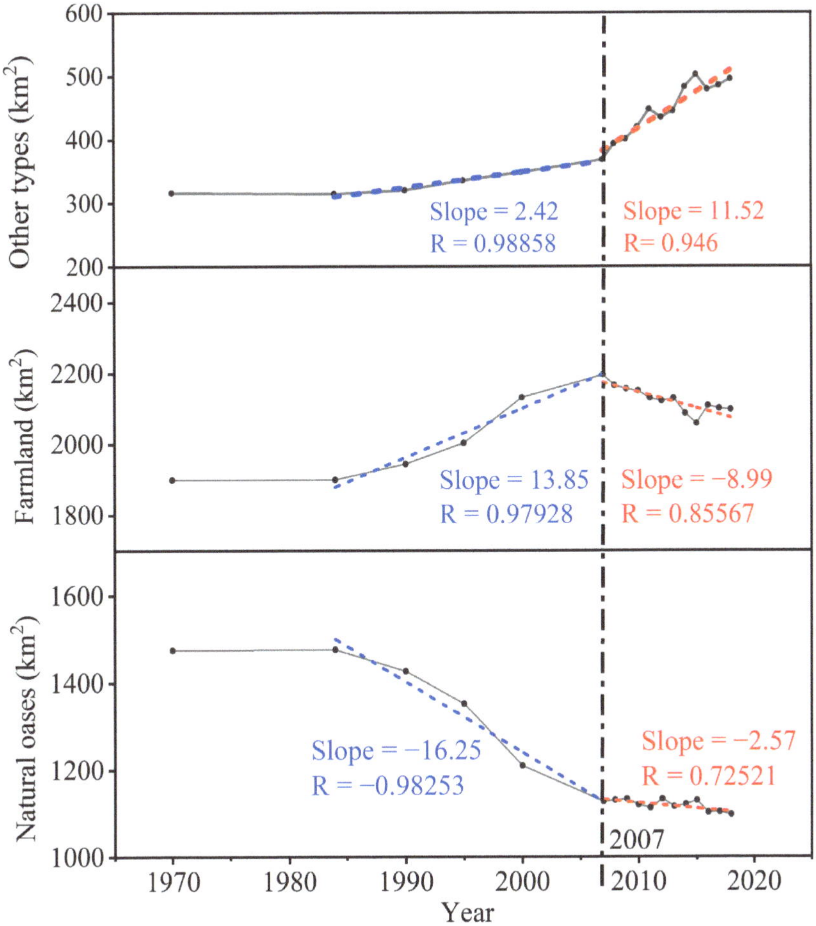

Figure 6. Changes in the areas taken up by the various land use/cover types.

A correlation analysis between the farmland area and the natural oasis area (Figure 7) revealed that the natural oasis area decreased by 1.17 km^2 for every 1 km^2 of increase in farmland area before 2007. After 2007, the natural oases recovered by only 0.12 km^2 for every 1 km^2 of increase in farmland area. That is, only 0.12 km^2 of every 1.17 km^2 of the original natural oases were restored after the natural oasis–farmland–return to natural oasis conversion process was implemented. This indicates that, to some extent, relatively speaking, the natural oases lacked vegetation species diversity and the ecosystem structure was simple and extremely fragile, i.e., the ecosystem stability was low, and it would be very difficult to achieve system recovery in the event of ecosystem damage [22].

During 1984–2018, the farmland and OL increased by 198.38 km^2 and 180.91 km^2, respectively, while the natural oases decreased by 379.29 km^2. In particular, the MCO increased by 52.92 km^2; while the DO, LCO, and HCO decreased by 334.81 km^2, 94.65 km^2, and 2.76 km^2, respectively. In 1984, the HCO, MCO, LCO, and DO accounted for 1.4%, 7.7%, 59.1%, and 31.8% of the natural oasis area, respectively. In 2018, these four vegetation cover types accounted for 1.6%, 15.2%, 70.9%, and 12.3%, respectively, meaning that the first three types increased by 0.2%, 7.5%, and 11.8%, while the DO decreased by 19.5%. This indicates that, to some extent, local human activities provided a certain amount of water supply to ensure the good growth of vegetation species in the natural oases.

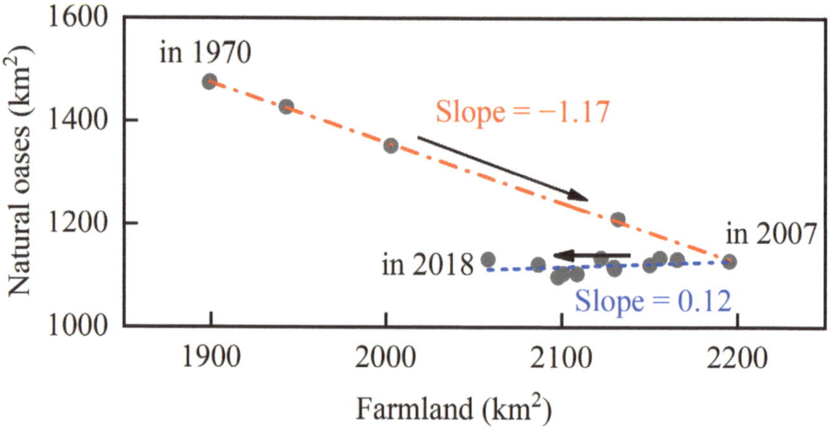

Figure 7. Relationship between the farmland area and the natural oasis area.

3.2. *Impact of LUCCs on Groundwater Depth*

The groundwater depth, as an important indicator of the ecological environment [32,53], is the dominant factor influencing the distribution, growth, population succession, and survival of vegetation species in natural and DO areas [54,55]; it is also an essential variable for measuring and evaluating basin water resource management. A reasonable groundwater depth is beneficial for both agricultural production and ecologically sustainable development [56,57].

3.2.1. Impact of LUCCs on Groundwater Depth from 1984 to 2007

As shown in Figure 8, the changes in groundwater depth from 1984 to 2007 exhibited a trend of significant increase–gradual slowdown–small rebound from west to east and from south to north. The groundwater level declined by more than 40 m in the western and southern parts of the region and by 4–20 m in the central part, while it varied little in the northeast, with a small rebound at some locations.

Figures 5 and 6 jointly illustrate that in the regions where the groundwater depth changed from 1984–2007, the farmland and MCO increased, while the DO and LCO decreased. Significant LUCCs occurred in the western and southeastern parts of the study area, especially near Jinchang. The total area of land-use conversion was 508.96 km^2, accounting for 13.8% of the study area.

Table 2 shows that where the groundwater depth changed from 1984–2007, the area of DO decreased by 11.52–57.44 km^2. The LCO only underwent positive changes in some parts of the study area, with a maximum increase of 11.84 km^2 and a maximum decrease of 51.68 km^2. The farmland underwent positive changes throughout the study area, with a maximum increase of 102.08 km^2. The MCO generally underwent positive changes, with a maximum increase of 55.84 km^2.

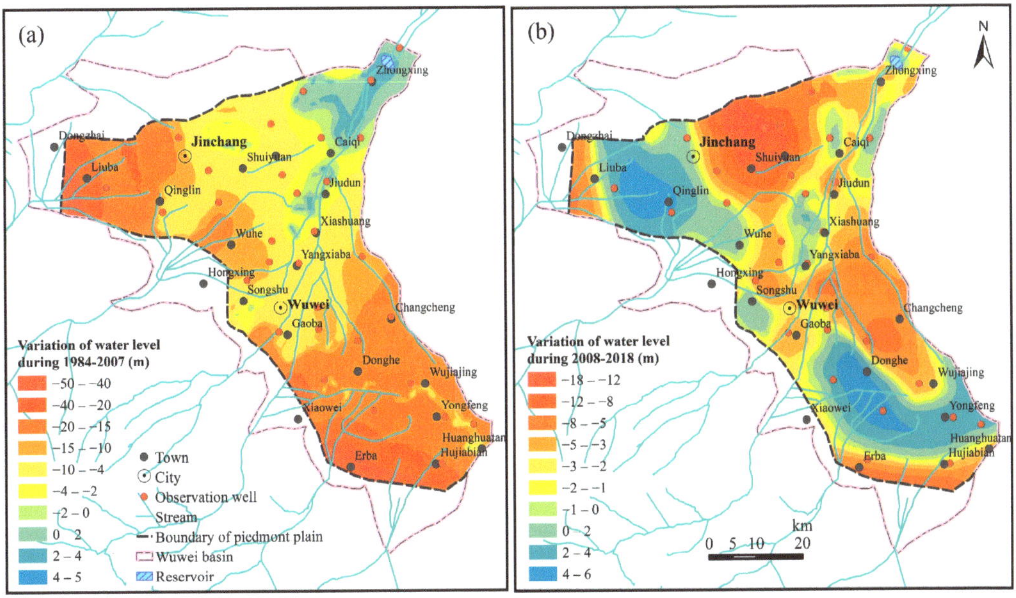

Figure 8. Groundwater depth changes in the plain during (**a**) 1984–2007, and (**b**) 2008–2018 (negative values denote water-level decline).

Table 2. Changes in the areas of the various land-use types over set periods in subregions corresponding to a different range of changes in groundwater depth (km^2).

Year	Water Level Variations (m)	Farmland	HCO	MCO	LCO	DO	OL
1984~2007	−50 to −40	0.8	0	11.52	0.32	−19.04	6.4
	−40 to −20	2.56	0	21.12	11.84	−40.48	4.96
	−20 to −15	50.56	−0.48	55.84	−51.68	−57.44	3.2
	−15 to −10	34.4	0.32	21.76	−41.44	−16.8	1.76
	−10 to −4	102.08	0.96	14.08	−85.6	−45.28	13.76
	−4 to −2	35.04	2.56	−6.4	−24.32	−13.6	6.72
	−2 to 0	12.64	−0.16	−2.08	0	−11.52	1.12
	0 to 2	7.36	−0.96	1.28	3.2	−12.96	2.08
	2 to 4	16.16	1.92	4.16	−4.96	−30.08	12.8
	4 to 5	33.6	−6.4	9.28	−5.6	−31.68	0.8
2008~2018	−18 to −12	3.52	0	0	−5.28	−0.32	2.08
	−12 to −8	−0.8	0.48	0.96	−6.24	1.92	3.68
	−8 to −5	−43.04	−0.48	4.16	27.2	−7.68	19.84
	−5 to −3	−25.92	−0.8	3.04	4.48	−3.84	23.04
	−3 to −2	−14.72	0	1.28	3.04	−5.76	16.16
	−2 to −1	−6.72	1.12	1.76	6.72	−12	9.12
	−1 to 0	−2.24	1.12	−2.24	6.72	−9.28	5.92
	0 to 2	−3.04	0	−23.68	25.92	−19.36	20.16
	2 to 4	−1.92	−0.96	−40.48	35.2	0.8	7.36
	4 to 6	−2.24	−0.48	−21.44	4.8	−0.32	19.68

The regions where the groundwater depth changed from −20 m to −4 m had a total area of 2153.60 km^2, accounting for 58.3% of the total study area; these areas (referred to as the −20 m/−4 m subregion for simplicity) were distributed in the central part of the study area. From 1984 to 2007, the farmland in the −20 m/−4 m subregion increased by

187.04 km^2, accounting for 63.4% of the total increase in farmland in the study area. The MCO increased by 91.68 km^2, accounting for 70.2% of the total increase in the MCO in the study area. The LCO decreased by 178.72 km^2, accounting for 90.2% of the total reduction in the LCO in the study area. The DO decreased by 119.52 km^2, accounting for 42.9% of the total reduction in the DO in the study area. In addition, the total area of land-use conversion in the −20 m/−4 m subregion was 298.72 km^2, accounting for 13.9% of the total area of this subregion and 58.7% of the total area of land-use conversion in the study area, indicating that this subregion was the region wherein the LUCCs were the most active and concentrated. This pattern was created because the groundwater depth was naturally 10–40 m in the −20 m/−4 m subregion, which allowed groundwater extraction wells to be drilled at low cost and the groundwater to be discharged with little difficulty. In addition, the −20 m/−4 m subregion received runoff recharge from the south all year round (especially during the irrigation period), and thus, the groundwater storage was abundant, thereby making this subregion a favorable place for LUCCs.

Under natural conditions, the groundwater in the western and southern parts of the study area was generally more than 50 m below the ground surface, and, thus, its ecological function in the growth of natural oases was negligible. The expansion of farmland was preferentially conducted in the LCO because in 1984, the areas of the different land-use types, listed in decreasing order, were: farmland > LCO > DO > OL > MCO > HCO. The LCO accounted for 872.58 km^2 of land use, which was 1.86 times larger than in DO. In addition, as opposed to the DO, the LCO was covered by vegetation to some extent, indicating that the locations of the LCO were advantageous in terms of having access to relatively more groundwater to support crop growth. After the conversion of LCO to farmland, agricultural irrigation activities were conducted continuously, mainly utilizing surface water conveyance and groundwater irrigation. In addition, the study area displayed a high-slope terrain, with a difference in elevation of 1400 m. During the irrigation period, a large volume of runoff was generated from the cropland, which allowed the nearby natural oasis areas, especially the DO areas, to effectively receive a water supply. Consequently, the plants in these nearby areas could grow well, gradually converting the areas into LCO and, further, into MCO. For example, the groundwater level was deep (>70 m in 2018) in the areas where the groundwater depth changed from −50 to −40 m (referred to as the −50 m/−40 m subregion); thus, the groundwater could not supply water for natural vegetation growth at all. In the −50 m/−40 m subregion, the DO decreased by 19.04 km^2, while the MCO increased by 11.52 km^2, and there were almost no changes in any of the other land-use types investigated in this study. This indicates that the gain in MCO was derived from DO, due to the year-round human irrigation of the nearby farmland.

3.2.2. Impact of LUCCs on Groundwater Depth from 2008 to 2018

As shown in Figure 8, the groundwater level generally declined in the eastern and northern parts of the Wuwei Plain from 2008 to 2018, with a drop of 5–18 m; the maximum drop occurred around Shuiyuan. The groundwater level rose by 2–6 m in the south and west, and the groundwater level also rose at multiple sites in the vicinity of the main river channel.

Figure 5 and Table 2 jointly show that where the groundwater depth changed during 2008–2018, the farmland, MCO, and DO all decreased, while the LCO and OL expanded. The total converted land area was 261.28 km^2, accounting for 7.1% of the total study area.

The areas where the groundwater level changed from −8 to −2 m from 2008 to 2018 spanned 1457.92 km^2, accounting for 39.4% of the total study area. In these areas (referred to as the −8 m/−2 m subregion), the farmland decreased by 83.68 km^2, accounting for 86.2% of the total reduction in farmland in the study area. The DO decreased by 17.28 km^2, accounting for 30.9% of the total reduction in the DO in the study area. The OL increased by 59.04 km^2, accounting for 46.5% of the total increase in the OL in the study area. The LCO increased by 34.72 km^2, accounting for 33.9% of the total increase in the LCO in the study area. The total converted land area in the −8 m/−2 m subregion was 102.24 km^2, accounting for 7.0% of the total area of this subregion and 39.1% of the total converted land area of the study area.

The areas where the groundwater level changed from 0 to 4 m from 2008 to 2018 spanned 804.48 km^2, accounting for 21.8% of the total study area. In these areas (referred to as the 0 m/4 m subregion), the MCO decreased by 64.16 km^2, accounting for 83.7% of the total reduction in MCO in the study area. The DO decreased by 18.56 km^2, accounting for 33.2% of the total reduction in DO in the study area. The LCO increased by 61.12 km^2, accounting for 59.6% of the total increase in LCO in the study area. The OL increased by 27.52 km^2, accounting for 21.7% of the total increase in OL in the study area. The total converted land area in the 0 m/4 m subregion was 89.44 km^2, accounting for 11.1% of the total area of this subregion and 34.2% of the total converted land area in the study area.

After 2007, the farmland shrinkage was concentrated in those areas where the groundwater level changed from −8 to 2 m. In the −8 m/2 m subregion, the farmland was mainly converted to OL near Wuhe, Songshu, and Wuwei. Part of the farmland was converted to the LCO to the east of Yangxiaba and Xiashuang.

The shrinkage in the MCO was due to conversion in the 0 m/4 m subregion, where this land use type was mainly converted to the LCO near Qinglin. Due to groundwater pumping, irrigation, and weakening surface runoff in the 0 m/4 m subregion, the groundwater level rose by 2–6 m; and by 2018, the groundwater level was about 80 m deep, but this depth still prevented the groundwater from being effectively used by natural vegetation species, thereby leading to a gradual decline in the vegetation coverage over the years.

The DO generally decreased in each subregion where the groundwater depth changed over the years, gradually changing into the LCO as farmland irrigation and ecological irrigation practices increased in the nearby areas. The maximum decrease (19.36 km^2) in the DO occurred to the east of Caiqi where the groundwater depth changed from 0 to 2 m.

3.3. Causes of Groundwater Depth Variation

Since the 1980s, with the expansion of farmland, the water demand for farmland irrigation has increased. Surface water was not sufficient for farmland irrigation. At the same time, the utilization rate of surface water was not high for a number of reasons, such as the poor impermeability of the canal system, and the water delivery timing not fully matching the exact moment of farmland irrigation. In addition, the canal water still needed to be diverted to farmland via ditches, with serious losses along the way. Water was still mainly based on extracted groundwater, and wells were growing rapidly. According to the statistics, there were 4294 wells in the Wuwei Basin in 2005, with a distribution density of 0.86 wells/km^2. Due to the wide distribution of wells and the convenience of water extraction and supply, groundwater extraction rose from 5.0×10^8 m^3/year in 1980 to 6.4×10^8 m^3/year in 2006, with an average annual extraction of 5.9×10^8 m^3/year (Figure 9). With the sharp increase in groundwater exploitation, the groundwater level dropped rapidly, and the spring overflow declined significantly. The spring overflow in the mainstream of the Shiyang River decreased from 4.4×10^8 m^3/year in the 1950s to 0.43×10^8 m^3/year in 2010. The pre-mountain region of the Wuwei Basin was the runoff area. Under groundwater dynamics, the groundwater ran east from Liuba, and north from Erba and Huanghuatan. The groundwater depth in these areas was more than 100 m, which meant that groundwater could not provide physiological water for natural vegetation and no longer served its ecological function. To ensure farmland irrigation, the combination of groundwater and surface water supply was adopted. However, some factors, such as the construction of reservoirs and the anthropogenic interception of flow in the upper reaches of the basin, resulted in a weakening of the piedmont runoff recharge from the Qilian Mountains. In addition, the topographic slope of the pre-mountain plain was relatively steep, which tended to generate runoff to the middle of the basin during farmland irrigation. The amount of piedmont runoff recharge from the Qilian Mountains was less than the sum of groundwater extraction and discharge to adjacent areas, resulting in a greater decline in groundwater level. The central and northern parts of the basin received lateral runoff recharge from adjacent areas, coupled with the relatively shallow groundwater depth and weak groundwater exploitation, so that the groundwater decline was smaller.

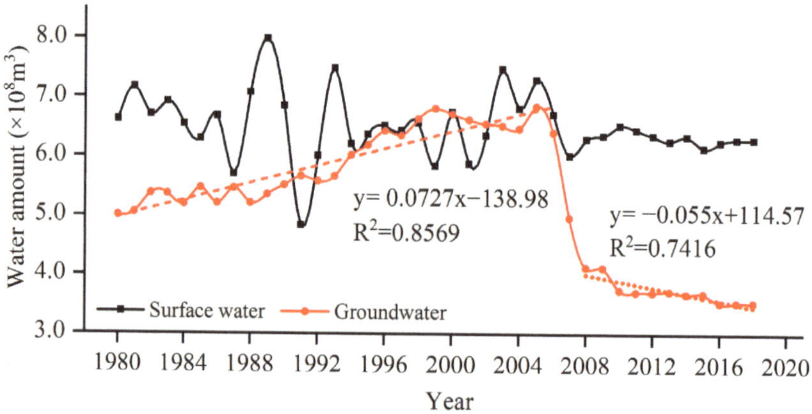

Figure 9. Amount of surface water conveyance and groundwater extraction since 1980.

Since the implementation of CTSRB, the farmland area decreased and groundwater extraction became limited, with an average annual extraction of 4.6 × 10^8 m^3/year, which was about 6.8% lower than the amount before CTSRB [41]. The amount of groundwater recharged by irrigation also changed, including surface water infiltration during conveyance, irrigation ditch infiltration, and the infiltration of irrigation groundwater. In 2007, the farmland area was 2629.9 km^2, and the surface water conveyance was 7.5 × 10^8 m^3. It was calculated that the surface water infiltration during conveyance and irrigation ditch infiltration was 2.0 × 10^8 m^3 in total. However, the groundwater extraction was 6.9 × 10^8 m^3, and the infiltration of irrigation groundwater was 1.8 × 10^8 m^3. Therefore, the groundwater consumption was 3.1 × 10^8 m^3. As the farmland area decreased, the amount of groundwater resource loss also decreased (Figure 10). By 2018, the farmland decreased to 2510 km^2, the surface water conveyance was 6.4 × 10^8 m^3, and the surface water infiltration during conveyance and irrigation ditch infiltration totaled 1.7 × 10^8 m^3. While the groundwater extraction was 4.9 × 10^8 m^3, the infiltration of irrigation groundwater was 1.3 × 10^8 m^3. Thus, the groundwater consumption was 1.8 × 10^8 m^3. With an increased awareness of efficient water use, and the development of high-standard lining and repair projects for canal systems, the losses during the process of surface water conveyance were reduced, and the utilization rate of canal system water increased accordingly. In the areas where the groundwater level had dropped seriously, such as Liuba and Qinlin in the west, groundwater extraction was prohibited. In the areas of Erba and Huanghuatan in the south, the farming area was restricted, and groundwater extraction was limited under the premise of maximizing the utilization of surface water. At the same time, the piedmont runoff recharge capacity from the Qilian Mountains was subsequently enhanced by water conservation in the upper reaches of the basin. All these factors largely alleviated the groundwater resource situation in the piedmont plain, and the groundwater level gradually rose. The central part was irrigated only with limited groundwater, so the groundwater level continued to decline, especially in northern Shuiyuang and eastern Changcheng. The water depth was more than 10 m and the groundwater no longer performed its ecological function in these places. The conversion of DO and LCO to farmland increased the demand for groundwater near Changcheng in the east. The growth of the OL near Shuiyuang in the north also raised water demand, due to urban construction and industrial use. As a result, these two areas became the areas with the greatest drops in water level.

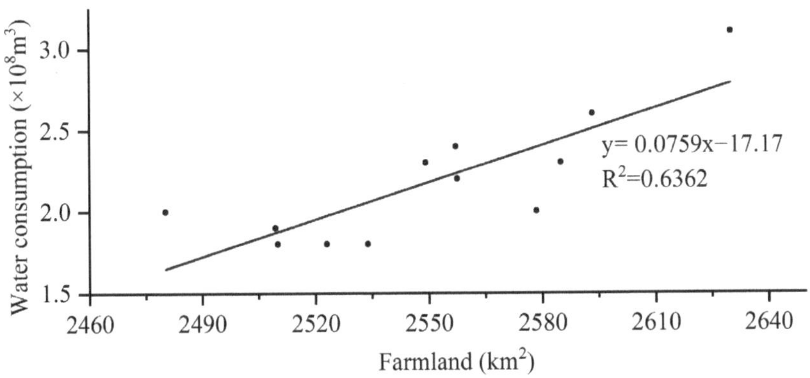

Figure 10. Relationship between farmland area and water consumption in 2007–2018.

3.4. Response of Groundwater Depth to LUCC

The whole region was divided into groundwater depth-increasing areas and groundwater depth-decreasing areas, and its land-use changes were calculated separately to analyze the response of groundwater depth to the land-use changes. During 1984–2007, the area of groundwater depth decreasing accounted for 94.3% of the total, and the average decrease in groundwater depth was 24.2 m (Table 3). Among them, the farmland area increased by 14.2%, yet the LCO and DO decreased by 23.6% and 57.5%, respectively. This indicated that the conversion to farmland reduced groundwater levels significantly. From 2008–2018, the farmland in the area of groundwater depth increasing decreased by 1.1%, and the water level rose by an average of 3.1 m. The farmland in the area of groundwater depth decrease was reduced by 5.9% and the water level dropped by an average of 5.1 m. The DO decreased while the LCO increased. This further indicated that the irrigation behavior would bring some water to the nearby natural oases, which could promote the healthy growth of natural vegetation.

Table 3. Land use changes in various groundwater depth zones from 1984 to 2018.

Water Level Variation Area	Land-Use Types						
	Year	Farmland	HCO	MCO	LCO	DO	OL
Decline area (−24.2 m)	1984 (km^2)	1843.04	6.08	87.84	815.84	430.24	303.36
	2007 (km^2)	2104.64	10.24	209.12	623.2	183.04	356.16
	Rate (%)	14.2	68.4	138.1	−23.6	−57.5	17.4
Rise area (+1.0 m)	1984 (km^2)	63.2	13.76	25.28	58.4	39.36	9.92
	2007 (km^2)	96.8	7.36	34.56	52.8	7.68	10.72
	Rate (%)	53.2	−46.5	36.7	−9.6	−80.5	8.1
Decline area (−5.1 m)	2008 (km^2)	1532.8	13.92	80.48	449.12	144.96	246.72
	2018 (km^2)	1442.88	15.36	89.44	485.76	108	326.56
	Rate (%)	−5.9	10.3	11.1	8.2	−25.5	32.4
Rise area (+3.1 m)	2008 (km^2)	669.12	3.68	163.2	226.88	45.76	120.16
	2018 (km^2)	661.92	2.24	77.6	292.8	26.88	167.36
	Rate (%)	−1.1	−39.1	−52.5	29.1	−41.3	39.3

3.5. Groundwater Balance Analysis

Recent studies have shown that the groundwater in the study area remained in a negative balance all year round, with the rate of change in the groundwater storage volume shifting from $-12{,}943.2 \times 10^4$ m^3/year in 2007 to -7270.4×10^4 m^3/year in 2018, due to the gradual decline in the groundwater resource consumption as the CTSRB was implemented.

According to the data published in the report on the research and demonstration of the rational extraction and utilization of groundwater and the protection of ecological functions in the Shiyang River Basin, a linear regression equation for the changes in the groundwater storage volume as a function of the farmland area, surface water conveyance volume, and groundwater extraction volume was established using Excel:

$$\Delta Q = 25094.58 - 19.0297S_c + 0.5214Q_s - 0.5229Q_g$$

where $R^2 = 0.9434$, indicating a good fit of the linear regression equation to the groundwater storage volume. The significance test (Table 4) showed that the changes in the groundwater storage volume were significantly correlated with the farmland area ($p < 0.05$) and were extremely significantly correlated with the surface water conveyance volume and groundwater extraction volume (both $p < 0.01$). That is, the linear regression equation fitted the raw data well and was reliable.

Table 4. Regression equation parameters for the changes in the groundwater storage volume.

Type	Coefficients	*p*-Value
β	25,094.5822	0.49830
a	−19.0297	0.03714
b	0.5214	0.00085
c	−0.5229	0.00105

After 1984, the farmland underwent erratic expansion, followed by the implementation of the Grain-for-Green Project. In addition, the groundwater storage volume also underwent a rapid depletion to slow recovery process. The missing early data for the groundwater storage volume can be predicted using the established linear regression equation. The farmland area was 1899.43 km^2 in 1984, and it increased by 295.52 km^2 in 1984. Assuming that the surface water conveyance volume and groundwater extraction volume in 1984 were the same as in 2007, the rate of change of the groundwater storage volume in 1984 was estimated to be -7319.5×10^4 m^3/year. With the implementation of the CTSRB, the farmland decreased to 2097.8 km^2 in 2018, and the rate of change of the groundwater storage volume was -7270.4×10^4 m^3/year in 2018. The groundwater in the study area was in a negative balance all year round and was extracted in large volumes. To ensure water resource management in the basin and to achieve sustainable development of the basin, it is necessary to implement and improve integrated basin management initiatives sustainably and efficiently.

The changes in groundwater levels in arid and semi-arid regions were influenced by a combination of factors. Based on the above analysis, it could be seen that the groundwater depth was sensitive to changes in land use types because of human activities. This indicated that groundwater resources in the region were closely related to land use types. Groundwater depth increased significantly with the continuous expansion of farmland. Since the implementation of CTSRB, the farmland area has been effectively controlled, and the trend of groundwater level decline has been significantly improved.

The groundwater depth also changed to different degrees due to other human activities, such as the surface soil nature, the type of crops cultivated, and changes in irrigation systems. Especially in the pre-mountain plain area, where the aquifer particles were relatively coarse, the variations in groundwater depth would be larger. These factors were not considered in this study, and the results of the analyses are still limited and uncertain, to some extent. For these reasons, to propose countermeasures and measures for the rational use of water and land resources in the development of water resources planning programs, the impact of human activities, including a variety of factors such as land use changes, should be considered comprehensively when water-resources planning is carried out.

4. Conclusions

The main conclusions of this study are as follows.

Before the CTSRB in 2007, the area of natural oases decreased at a rate of 16.25 km^2/year, while the area of farmland expanded at a rate of 13.85 km^2/year. The farmland expansion preferentially occurred in LCO and MCO, where the groundwater depth increased from 4 to 20 m. The consumption of groundwater increased from 7319.5 $\times$ 10^4 m^3/year to 12,943.2 $\times$ 10^4 m^3/year.

During 2008–2018, the areas of both the natural oases and farmland decreased at the rates of 2.57 km^2/year and 8.99 km^2/year, respectively. The groundwater level rose significantly in the south, in the west, and near the main river channel. The negative groundwater balance was alleviated, and the groundwater resource consumption was restored to 7270.4 $\times$ 10^4 m^3/year. Through the natural oasis–farmland–natural oasis conversion process, only 0.12 km^2 of every 1.17 km^2 of the original natural oases were restored, indicating that it would be very difficult to achieve system recovery in the event of ecosystem damage. Groundwater depth increased significantly with the continuous expansion of farmland. Since the farmland area was effectively controlled, the trend of groundwater level decline was significantly improved.

Land-use changes inevitably led to changes in the water supply and use patterns, groundwater depth, and groundwater balance status. To fulfill the goals of water resource management, the ecological restoration and reconstruction of oases, and the sustainable development of the basin, the following measures should be continued in the future:

(1) Adjusting the local agricultural planting structure, developing special agriculture with high output value, high efficiency, and good benefits, and advocating and promoting the concept of water planning and water conservation to reduce wasteful behavior regarding water resources.

(2) According to the basin water allocation plan, planners should decompose and implement the indicators of groundwater extraction and reduction. At the same time, groundwater extraction should be reduced in a planned manner, to gradually achieve a balance between groundwater extraction and recharge, and then achieve a greater amount of recharge than extraction.

(3) Minimizing the area of farmland, especially by reducing the reclamation of the upper reaches of the basin. This is because the upper part, i.e., the upper and middle alluvial fans, is the source of the recharge of the groundwater system.

(4) In fact, the groundwater depth in most areas of the basin is far greater than the required depth (less than 10 m) for the normal growth of natural vegetation species [57]. Therefore, inter-basin water transfer projects, for example, can be combined to provide orderly artificial groundwater recharge to the area.

(5) Gradually restoring groundwater levels and implementing ecological water conveyance to maintain the normal growth of natural oases. Due to the slow recovery of vegetated ecosystems, the relationship between vegetation recovery, groundwater, and land use changes over time can be studied further in subsequent studies.

Author Contributions: Conceptualization, L.W. and Z.N.; methodology, L.W., Q.Y. and Z.N.; formal analysis, L.W., H.L. and B.F.; investigation, M.L., L.C. and Q.Y.; data curation, L.W., P.Z. and M.L., writing—original draft preparation, L.W.; writing—review and editing, P.Z. and Q.Y.; supervision, M.L. and L.C.; funding acquisition, Z.N., M.L. and L.C. All authors have read and agreed to the published version of the manuscript.

Funding: This research was funded by the National Natural Science Foundation of China (grant Nos. 41902262 and 41807214), the National Key Research and Development Program of China (grant No. 2017YFC0406103), the Natural Science Foundation of Hebei Province, China (grant No. D2022504014), the Fundamental Research Funds for Central Public Welfare Research Institutes, CAGS (grant No. SK202011), and the China Geological Survey Project (grant Nos. DD20221752 and DD20190349). The authors declare that they have no financial conflicts of interest related to this work.

Data Availability Statement: The data presented in this study are available on request from the corresponding author.

Acknowledgments: The authors would like to sincerely thank Zhang Tongze, who has provided abundant data for this study. We also thank the editors and anonymous reviewers who provided valuable comments.

Conflicts of Interest: The authors declare no conflict of interest.

References

1. Gleeson, T.; Cuthbert, M.; Ferguson, G.; Perrone, D. Global Groundwater Sustainability, Resources, and Systems in the Anthropocene. *Annu. Rev. Earth Planet. Sci.* **2020**, *48*, 431–463. [CrossRef]
2. Lall, U.; Josset, L.; Russo, T. A Snapshot of the World's Groundwater Challenges. *Annu. Rev. Environ. Resour.* **2020**, *45*, 171–194. [CrossRef]
3. Buhay Bucton, B.G.; Shrestha, S.; KC, S.; Mohanasundaram, S.; Virdis, S.G.P.; Chaowiwat, W. Impacts of climate and land use change on groundwater recharge under shared socioeconomic pathways: A case of Siem Reap, Cambodia. *Environ. Res.* **2022**, *211*, 113070. [CrossRef] [PubMed]
4. Grönwall, J.; Danert, K. Regarding Groundwater and Drinking Water Access through A Human Rights Lens: Self-Supply as A Norm. *Water* **2020**, *12*, 419. [CrossRef]
5. Sajjad, M.M.; Wang, J.; Abbas, H.; Ullah, I.; Khan, R.; Ali, F. Impact of climate and land-use change on groundwater resources, study of Faisalabad district, Pakistan. *Atmosphere* **2022**, *13*, 1097. [CrossRef]
6. Ridwansyah, I.; Yulianti, M.; Apip; Onodera, S.; Shimizu, Y.; Wibowo, H.; Fakhrudin, M. The impact of land use and climate change on surface runoff and groundwater in Cimanuk watershed, Indonesia. *Limnology* **2020**, *21*, 487–498. [CrossRef]
7. Nasiri, S.; Ansari, H.; Ziaei, A.N. Determination of water balance equation components in irrigated agricultural watersheds using SWAT and MODFLOW models: A case study of Samalqan plain in Iran. *J. Groundw. Sci. Eng.* **2022**, *10*, 44–56. [CrossRef]
8. Kafando, M.B.; Koïta, M.; Zouré, C.O.; Yonaba, R.; Niang, D. Quantification of Soil Deep Drainage and Aquifer Recharge Dynamics according to Land Use and Land Cover in the Basement Zone of Burkina Faso in West Africa. *Sustainability* **2022**, *14*, 14687. [CrossRef]
9. Zomlot, Z.; Verbeiren, B.; Huysmans, M.; Batelaan, O. Trajectory analysis of land use and land cover maps to improve spatial–temporal patterns, and impact assessment on groundwater recharge. *J. Hydrol.* **2017**, *554*, 558–569. [CrossRef]
10. Zheng, Z.; Liu, L.; Cui, X. Source identification of methane in groundwater in shale gas development areas: A critical review of the state of the art, prospects, and future challenges. *J. Groundw. Sci. Eng.* **2021**, *9*, 245–255. [CrossRef]
11. Tang, Z.; Engel, B.A.; Pijanowski, B.C.; Lim, K.J. Forecasting land use change and its environmental impact at a watershed scale. *J. Env. Manag.* **2005**, *76*, 35–45. [CrossRef]
12. Kim, J.H.; Jackson, R.B. A Global Analysis of Groundwater Recharge for Vegetation, Climate, and Soils. *Vadose Zone J.* **2012**, *11*. [CrossRef]
13. Van Huijgevoort, M.H.J.; Voortman, B.R.; Rijpkema, S.; Nijhuis, K.H.S.; Witte, J.M. Influence of climate and land use change on the groundwater system of the Veluwe, the Netherlands: A historical and future perspective. *Water* **2020**, *12*, 2866. [CrossRef]
14. Huang, M.; Zhang, L. Hydrological responses to conservation practices in a catchment of the Loess Plateau, China. *Hydrol. Process.* **2004**, *18*, 1885–1898. [CrossRef]
15. Jinno, K.; Tsutsumi, A.; Alkaeed, O.; Saita, S.; Berndtsson, R. Effects of land-use change on groundwater recharge model parameters. *Hydrol. Sci. J.* **2009**, *54*, 300–315. [CrossRef]
16. Guerrero-Morales, J.; Fonseca, C.R.; Goméz-Albores, M.A.; Sampedro-Rosas, M.L.; Silva-Gómez, S.E. Proportional variation of potential groundwater recharge as a result of climate change and land-use: A study case in Mexico. *Land* **2020**, *9*, 364. [CrossRef]
17. Zhang, L.; Dawes, W.R.; Walker, G.R. Response of mean annual evapotranspiration to vegetation changes at catchment scale. *Water Resour. Res.* **2001**, *37*, 701–708. [CrossRef]
18. Ghimire, U.; Shrestha, S.; Neupane, S.; Mohanasundaram, S.; Lorphensri, O. Climate and land-use change impacts on spatiotemporal variations in groundwater recharge: A case study of the Bangkok Area, Thailand. *Sci. Total Environ.* **2021**, *792*, 148370. [CrossRef]
19. Calow, R.C.; Robins, N.S.; Macdonald, A.M.; Macdonald, D.M.J.; Gibbs, B.R.; Orpen, W.R.G.; Mtembezeka, P.; Andrews, A.J.; Appiah, S.O. Groundwater Management in Drought-prone Areas of Africa. *Int. J. Water Resour. D* **1997**, *13*, 241–262. [CrossRef]
20. Alley, W.M.; Reilly, T.E.; Franke, O.L. *Sustainability of Ground-Water Resources*; Circular 1186; US Geological Survey: Reston, VI, USA, 1999. [CrossRef]
21. Li, X.; Xiao, D. Dynamics of water resources and land use in oases in middle and lower reaches of Shiyang River watershed, Northwest China. *Adv. Water Sci.* **2005**, *16*, 643–648. (In Chinese) [CrossRef]
22. Li, H. Oasis Water Resource Utilization Scenarios Simulation and Oasis Ecological Security—A Case Study in Wuwei and Minqin Oasis of Shiyang River Stream. Ph.D. Thesis, Peking University, Beijing, China, 2008.
23. Schwartz, W.F.; Zhang, H. *Fundamentals of Groundwater*; John Wiley & Sons: New York, NY, USA, 2003; p. 583.
24. Sato, K.; Iwasa, Y. *Groundwater Hydraulic*; Springer: Tokyo, Japan, 2003; p. 204.

25. Hu, J. Analysis on space-time distribution of groundwater resources based on gis technology in Shiyanghe River Basin. *J. Desert Res.* **2008**, *28*, 781–786.
26. Wang, G.; Yang, L.; Chen, L.; Jumpei, K. Impacts of land use changes on groundwater resources in the Heihe River Basin. *J. Geogr Sci* **2005**, *15*, 405–414. [CrossRef]
27. Shu, L.; Wang, Z.; Yuan, Y.; Zhang, F.; Liu, W.; Lu, C. Land use change and its impact on groundwater in the typical district of Sanjiang Plain during the past 40 years. *J. Hydraul. Eng.* **2021**, *52*, 896–906. [CrossRef]
28. Hao, Y.; Xie, Y.; Ma, J.; Zhang, W. The critical role of local policy effects in arid watershed groundwater resources sustainability: A case study in the Minqin oasis, China. *Sci. Total Environ.* **2017**, *601–602*, 1084–1096. [CrossRef] [PubMed]
29. Hu, H.; Ding, H.; He, B. Dynamic variation of groundwater level in the middle-lower reaches of Shiyanghe River Basin for nearly 40 years. *Northwestern Geol.* **2016**, *49*, 164–174. [CrossRef]
30. Li, Z. Study on Water Crisis in Arid Inland River Basins and Its Countermeasures: A Case Study of Shiyang River Basin in Gansu Province. Ph.D. Thesis, Cold and Arid Regions Environmental and Engineering Research Institute, Chinese Academy of Sciences, Beijing, China, 2007.
31. Fan, Y.; Wang, C.; Nan, Z. Comparative evaluation of crop water use efficiency, economic analysis and net household profit simulation in arid Northwest China. *Agric. Water Manag.* **2014**, *146*, 335–345. [CrossRef]
32. Li, S.; Zhang, Y.; Ma, Y.; Wang, L.; Mu, H. Analysis of groundwater dynamic changes in Shiyang River Basin. *J. Arid. Land Resour. Environ.* **2018**, *32*, 145–151. [CrossRef]
33. Zhou, J.; Lei, L.; Shi, P.; Wang, L.; Wei, W.; Liu, H. Response of runoff to the climate and land use pattern changes in Shiyang River Basin. *Acta Ecol. Sin.* **2015**, *35*, 3788–3796. [CrossRef]
34. Liu, J. Study on Water Resource Distribution Change of Modality Driven with Land Use and Land Coverage in Hexi Inland River Basin of Northwest China. Ph.D. Thesis, Cold and Arid Regions Environmental and Engineering Research Institute, Chinese Academy of Sciences, Beijing, China, 2007.
35. Ma, X.; Li, B.; Wu, C.; Peng, H.; Guo, Y. Predicting of temporal-spatial change of groundwater table resulted from current land use in Minqin oasis. *Adv. Water Sci.* **2003**, *14*, 85–90. [CrossRef]
36. Li, Z. Study on Ecological Flow Analysis and Water Resources Optimization Allocation in Shiyang River Basin. Master Thesis, Chang'an University, Xi'an, China, 2020.
37. Zhang, L. Study on the Effectiveness of Integrated Management of Shiyang River Basin in Terms of Water Resources. Master Thesis, Lanzhou University, Lanzhou, China, 2017.
38. Shi, Y.; Zhang, R.; Dong, P.; Hao, Y.; Shi, P.; Cheng, Z. Effect analysis of efficient utilization of water resources and ecological management measures in Shiyang River Basin: A case study of water resources management in Minqin. *Water Resour. Plan. Des.* **2017**, 53–55. [CrossRef]
39. Zhang, J.; Li, J.; Yang, X.; Yin, S.; Chen, J. Rural social-ecological systems vulnerability evolution and spatial-temporal heterogeneity in arid environmental change region: A case study of Minqin Oasis, northwestern China. *Appl. Geogr.* **2022**, *145*, 102747. [CrossRef]
40. Nie, Z.; Liu, M.; Liu, P.; Wang, L.; Meng, L. *Research and Demonstration of Rational Extraction and Utilization of Groundwater and Protection of Ecological Functions in Shiyang River Basin*; Institute of Hydrogeology and Environmental Geology, CAGS: Shijiazhuang, China, 2021.
41. Wang, L.; Nie, Z.; Liu, M.; Cao, L.; Zhu, P.; Yuan, Q. Rational Allocation of Water Resources in the Arid Area of Northwestern China Based on Numerical Simulations. *Sustainability* **2023**, *15*, 55. [CrossRef]
42. Liu, H.; Liu, S.; Li, Y. Response of riparian vegetation to the change of groundwater level at middle and lower reaches of the Shiyang River. *Arid Zone Res.* **2012**, *29*, 335–341.
43. Liu, S. Change of Riparian Vegetation and Its Driving Factors at the Middle and Lower Reaches of Shiyang River Basin. Ph.D. Thesis, Beijing Forestry University, Beijing, China, 2010.
44. Feng, B.; Nie, Z.; Wang, J.; Lu, H. Long-time series remote sensing dynamic monitoring of the oasis in Shiyang River basin. *Geospat. Inf.* **2020**, *18*, 6. [CrossRef]
45. Lu, H.; Nie, Z.; Liu, M.; Feng, B.; Cheng, X.; Wang, J.; Wang, Q.; Cui, H.; Fan, F. Research on land cover changes in Shiyang River Basin in recent 50 years based on RS and GIS. *Geogr. Res.* **2020**, *29*, 8.
46. Li, M.; Wu, B.; Yan, C.; Zhou, W. Estimation of vegetation fraction in the upper basin of Miyun Reservoir by remote sensing. *Resour. Sci.* **2004**, *26*, 153–159.
47. Gao, Z.; Li, Z.; Wei, H.; Ding, F. Quantitative monitoring of vegetation cover change by using remotely sensed data over Minqin Oasis, Gansu. *Geogr. Res.* **2006**, *25*, 587–595. [CrossRef]
48. Rongqun, Z.; Sun, W.; Ai, D.; Wang, S. Analysis on land cover change based on visual method of monitoring information. *Trans. Chin. Soc. Agric. Eng.* **2015**, *31*, 8. [CrossRef]
49. Aldwaik, S.Z.; Pontius, R.G. Intensity analysis to unify measurements of size and stationarity of land changes by interval, category, and transition. *Landsc. Urban Plan* **2012**, *106*, 103–114. [CrossRef]
50. Huang, B. Processes in Land Use Change and their Environmental Consequences in Coastal Area of Fujian Province from the Perspective of Land-Sea Integration. Ph.D. Thesis, Xiamen University, Xiamen, China, 2019.
51. Wen, X.; Duan, H.; Liao, J.; Xue, X. Rs monitoring of the spatial-temporal variation of vegetation index and desertification in Shiyang River Basin. *J. Jilin Univ. (Earth Sci. Ed.)* **2012**, *42*, 415–422. [CrossRef]

52. Wei, J. The Hexi Area Oasis Extraction Method and Spatial Distribution Based on the Remote Sensing. Master Thesis, Lanzhou University, Lanzhou, China, 2016.
53. Wang, J.; Zhang, G.; Cui, H.; Wang, Q.; Dong, H.; Hao, J. System index attribute and application of groundwater function zoning in northwest inland area of China. *J. Hydraul. Eng.* **2020**, *51*, 796–804. [CrossRef]
54. Ye, H.; Chen, S.; Sheng, F.; Chen, H. Research on dynamic changes of land cover and its correlation with groundwater in the Shule River Basin. *J. Hydraul. Eng.* **2013**, *44*, 83–90. [CrossRef]
55. Yan, D.; Wang, H.; Qin, D.; Wang, J.; Li, Y. The ecological evolution driven by water in the lower reaches of Heihe Basin. *China Environ. Sci.* **2005**, *25*, 38–42.
56. Ren, D.; Xu, X.; Huang, Q.; Huo, Z.; Xiong, Y.; Huang, G. Analyzing the role of shallow groundwater systems in the water use of different land-use types in arid irrigated regions. *Water* **2018**, *10*, 634. [CrossRef]
57. Liu, M.; Nie, Z.; Cao, L. Comprehensive evaluation on the ecological function of groundwater in the Shiyang River watershed. *J. Groundw. Sci. Eng.* **2021**, *9*, 326–340. [CrossRef]

water

MDPI

Article

Hydrochemical and Isotopic Characteristics and the Spatiotemporal Differences of Surface Water and Groundwater in the Qaidam Basin, China

Haijiao Yang [1,2,3,4], Jiahua Wei [1,2,3,4,5,*] and Kaifang Shi [2,3,4]

1 School of Civil Engineering and Water Resources, Qinghai University, Xining 810016, China; 2014990031@qhu.edu.cn

2 State Key Laboratory of Plateau Ecology and Agriculture, Qinghai University, Xining 810016, China; shikaifang2021@163.com

3 Laboratory of Ecological Protection and High Quality Development in the Upper Yellow River, Xining 810016, China

4 Key Laboratory of Water Ecology Remediation and Protection at Headwater Regions of Big Rivers, Ministry of Water Resources, Xining 810016, China

5 State Key Laboratory of Hydroscience and Engineering, Tsinghua University, Beijing 100084, China

* Correspondence: weijiahua@tsinghua.edu.cn

Abstract: In the context of climate change, precipitation and runoff in the arid inland basins of northwest China have undergone significant changes. The Qaidam Basin (QB) is a typical highland arid inland area. Understanding the spatial and temporal variations in surface water and groundwater chemistry and isotopes, as well as their causes, is crucial for future water resource management and ecological protection. Samples of river, lake, and groundwater, as well as others, were collected and tested in five typical watersheds in the summer and winter. The hydrochemistry and isotopic spatiotemporal differences in various water bodies were studied using the significant difference method, water vapor flux models, hydrochemistry, isotopes, and other methods for cause analyses. The results indicate the following: (1) There are differences in hydrochemistry between the southern and northern basins because the southern basin is more influenced by the dissolution of salt rocks and evaporation, whereas the northern basin is mainly affected by carbonate weathering. (2) The enrichment of δD and δ^{18}O in the northern basin gradually increases from west to east, while in the southern basin, it is the opposite. This is because the southern basin receives a larger contribution of water vapor from the mid-latitude westerlies, while the northern basin primarily relies on local evaporation as its water vapor source. (3) Significant differences are observed in the total dissolved solids (TDS) and hydrochemical types of river water and groundwater between the summer and winter due to higher rates of rock weathering and evaporation in the summer. (4) The more pronounced seasonal differences in hydrogen and oxygen stable isotopes in the southern basin are due to higher rates of internal water vapor circulation in the summer. (5) The similarity in characteristics between river water and groundwater is the result of strong exchanges between river water and groundwater from piedmonts to terminals. The spatiotemporal heterogeneity of terminal lakes is attributed to the accumulation of salts and groundwater replenishment from other sources.

Keywords: Qaidam Basin; hydrochemistry; H−O isotopes; water vapor source; surface and groundwater interactions; salt source

1. Introduction

The inland arid basins in northwest China face a scarcity of freshwater resources with respect to uneven spatial and temporal distributions, severe desertification, and fragile ecological environments. Since the 21st century, climate warming has led to increased precipitation in the northwestern arid regions, resulting in varying changes in terrestrial water storage [1]. This has amplified hydrological fluctuations and uncertainties in water resources [2,3].

The Qaidam Basin (QB) is a typical inland plateau basin, and due to its arid climate, the river–lake system is highly susceptible to seasonal changes and precipitation. Climate warming has brought about changes in regional hydrological cycles, resulting in increased precipitation, river discharge and lake areas [4], and expanded land water storage [1,5], while also causing the freshening of resource-based lakes [6]. The spatiotemporal variations in hydrochemistry and isotopes are primarily constrained by factors such as geological structure, lithogenic origins, topography, hydrology, and meteorology, yet the mechanisms underlying their influences remain unclear. Understanding the differences in the hydrochemistry and isotopes of surface water and groundwater is of significant importance for the future management of water resources and the development of salt lakes in the basin.

Due to its unique ecological functions and the value of its salt lake resources, QB has received considerable attention from scholars. They have extensively researched the spatial and temporal distribution of basin runoff changes, river–lake variations, driving mechanisms, and the hydrochemical characteristics of river–lake systems from various perspectives [7–9]. In summary, QB has multiple small rivers with short courses and considerable fluctuations in runoff within a year. Except for the rivers that are primarily replenished by groundwater, most rivers exhibit seasonal variations. Originating from the surrounding mountainous areas, these rivers either undergo frequent transformations with groundwater, disappear in the desert, or overflow onto the surface before finally draining into a terminal lake [10]. Therefore, the central part of the basin is predominantly an area without runoffs.

The relationship between surface water and groundwater transformation is mainly controlled by factors such as geological structures and spatial changes with respect to lithology. For instance, the Bayin River infiltrates the aquifer in the alluvial fan plain and undergoes repeated exchanges with groundwater in the alluvial–lacustrine plain [11]. The shallow groundwater of Nomhon River in the mountainous region is obstructed by uplifting structures, causing a shift in flow towards the northwest, while high-pressure artesian water is formed at the axis of the uplifting structure [12]. The groundwater supply from Nalengele River in the front uplift zone differs during wet and dry seasons, resulting in variations in the salinity of the terminal lake [13].

Over the past 20 years, the total area of lakes in the QB has expanded by 82.63%. The most significant changes in lake area occur in the central part of the basin, where some large shallow lakes are disappearing, while the water storage capacity of small and deep lakes is increasing [14]. The driving mechanisms behind these changes in the lake area vary. For instance, the expansion of Sugan Lake is attributed to glacial and snow melt, while fluctuations in the water levels of Xiao Qaidam Lake and Tuosu Lake are caused by increased precipitation [6].

The chemical characteristics of salt lake water in the QB are closely related to their material sources, which primarily include four types: (1) It is believed that the salt content in lakes is a result of long-term weathering, leaching, transportation, dissolution, and precipitation processes caused by continuous contributions from surrounding weathering and leaching, as well as from groundwater and surface water [15]. For example, the source of potassium is attributed to the widespread distribution of potassium-containing rocks around the basin [16]. The brine and salt deposits in the northern and northeastern parts of the Qarhan Salt Lake are influenced by the supply of chloride-type brine springs [17]. (2) In addition to surface chemical weathering materials supplied by river input, there are other sources, such as the Qarhan Salt Lake in the central part of the basin, primarily receiving contributions from oilfield water in the Tertiary strata [18,19] and remnants from the ancient Qaidam Lake [16,20]. (3) The formation of lithium, boron, and potassium deposits mainly results from volcanic–geothermal water [21], as observed in lakes like Qarhan Salt Lake in the central basin, Da Qaidam Lake and Xiao Qaidam Lake in the east, and West Taijnar

Lakes and Mahai Lake. (4) In the lakes formed by a series of mountain basins distributed to the north, east, and south of the basin, lake water supplies from the surrounding ancient lake areas or capture are the primary sources of salts [22].

The aforementioned research has mainly explored the changes in basin water resources and the hydrochemical characteristics of salt lakes from the perspectives of water resource transformation, structural control factors, and material sources. Howerer, there is no systematic integration of the inland river–lake system. Given the vast topography of QB, there are spatial and seasonal differences in surface water and groundwater. Various internal and external factors, such as hydrogeology, geology, water–rock interactions, weathering and mineral dissolution, localized precipitation, evaporation rates, and water vapor sources [23], have significantly influenced hydrochemistry and isotopes. Further research is needed to identify these differences and their attributions. This article selected five typical watersheds in the basin; collected and tested precipitation, river water, groundwater, lake water, and brine water samples; and analyzed the spatial and seasonal hydrochemical and H−O isotopic characteristic differences of rivers and lakes. The main reasons for these differences are explored from the perspectives of aquifer lithology, water vapor sources, the mechanisms of hydrochemical influence, the transformation relationship between river water and groundwater, and the sources of salinity in terminal lakes.

2. Materials and Methods

2.1. Regional Overview

The QB is situated in the northeastern part of the Qinghai–Tibet Plateau (QTP), with geographic coordinates approximately ranging from 90°00′ to 99°20′ E and 34°40′ to 39°20′ N. It has low temperatures, significant temperature variations, sparse precipitation, high evaporation rates, and a dry and cold climate. The climate is characterized by dryness and coldness, with significant differences in meteorological elements between the summer and winter seasons. According to the data from the Golmud Meteorological Station, the average temperature from June to August is 15.4 °C, accompanied by a monthly precipitation of 103.5 mm and an evaporation of 3242.1 mm. From December to February of the following year, the average temperature dropped to −8.2 °C, while experiencing monthly precipitation of merely 6.74 mm alongside evaporation of 530.1 mm.

The evolution of the basin is closely related to the formation and development of the QTP. This is a Cenozoic rift basin, and its overall structure is characterized by the compression and collision of the East Kunlun Mountains and Qilian Mountains, with the Altun Mountains thrusting southeastward. The uplifted peripheral mountains and subsiding basins have formed a distinct structural pattern. The surrounding mountain areas are dominated by Paleogene strata, while the basin and its margins are mainly composed of Paleogene to Neogene strata, with Quaternary deposits primarily distributed in the central basin and wide intermontane valleys. From the southern and northern mountain areas to the center of the basin, the topography transitions from high to low. The southern and northern landforms exhibit symmetrical patterns, and the aquifers display zonation. The area in front of the mountains has a single unconfined aquifer, gradually transitioning to multi-layer confined aquifers as it approaches the terminal lakes. The grain size of sediments changes from coarse gravel to fine sand, and there are significant geological differences between the south and north of the QB.

The basin has a total of 79 rivers, most of which are seasonal rivers with abundant water during the warm season that dry up during the cold season. The hydrological pattern is characterized by rivers flowing from the surrounding mountains towards the center of the basin, where central lakes with varying salinity levels are developed. Based on geological conditions and watershed factors, the research area can be divided into southern and northern basin watersheds. The southern basin mainly includes the Nalengle River, Wutumeiren, Golmud River, and Qaidam River (including the Nomhon River, Chahanwusu River, Xiangride River, etc.), while the northern basin primarily includes Temurik

River, Yukha River, Haleteng River, Tataling River, Bayin River, and other watersheds (see Figure 1).

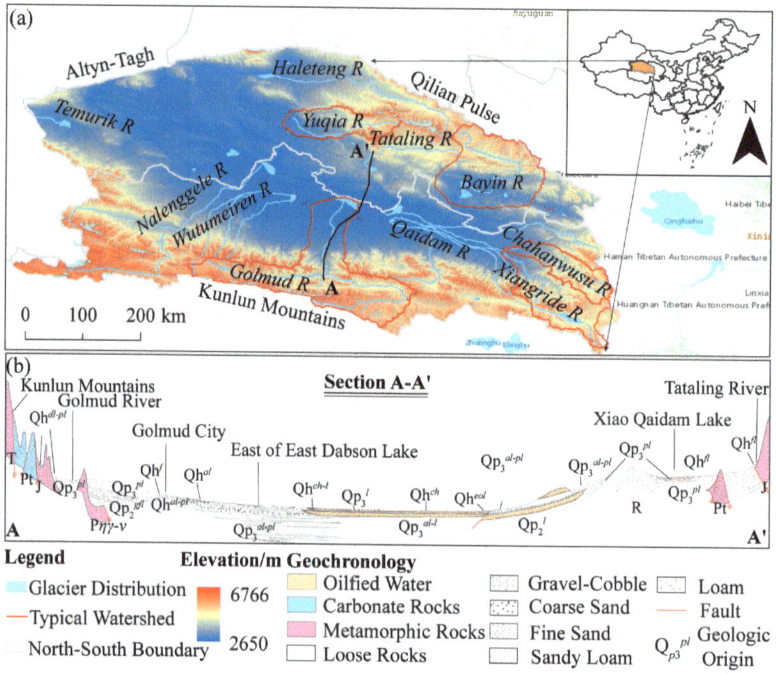

Figure 1. Overview map of QB: (**a**) digital elevation map; (**b**) comprehensive hydrogeological profile (revised according to reference [10]): A−A′ profile represents the cross-section from the front of the Kunlun Mountains to the Tataling River.

2.2. Research Methods

Based on the comprehensive watershed scale (area, river length, annual runoff, etc.), supply characteristics, and downstream water use types, and using the hierarchical clustering method [24], 44 rivers with a watershed area greater than 200 km^2 were classified into five categories. The typical rivers from different orientations in the basin have been selected for comparative studies [25].

The analysis of physicochemical parameters and the hydrogen and oxygen isotope (H−O isotope) characteristics of different water bodies in different basins and seasons was conducted using significant difference testing methods, such as analysis of variance (ANOVA) [26] and non-parametric testing methods [27]. The Piper ternary diagram [28] is used to compare hydrochemical types. The differences in internal and external factors within the basin were analyzed spatially, examining the geological lithology and genesis relative to an understanding of the reasons for differences in water chemistry among the watersheds. Based on the ERA-5 reanalysis dataset and utilizing the Water Accounting Model (WAM)-2Layer water vapor flux statistical model [29], the analysis of deuterium excess (d-excess) [30] was combined to examine the water vapor sources for precipitation in the southern and northern basin watersheds, revealing spatial differences. The Gibbs model [31] was used to analyze the influence mechanisms of water chemistry in summer and winter seasons. The water vapor flux statistical model results are employed to analyze the internal recycling rate of precipitation to identify the reasons for seasonal differences. Finally, hydrochemistry and isotopic methods were employed to investigate the surface water–groundwater interaction and interpret the sources of salinity in the terminal lake; the objective was to analyze the reasons for the differentiation among river water, groundwater, and lake water along the course.

The Water Accounting Model (WAM) uses an area mark in a given atmospheric Euler field to track or backtrack moisture movement, and then calculate the water vapor sources of precipitation (WSoP) or reprecipitation of evapotranspiration (RPoET). It can also determine the regional recycling ratio, which is the proportion of locally evaporated water vapor that condenses to form precipitation to the total precipitation [32,33]. The WAM-2Layer model is an improved version designed to reduce errors caused by vertical wind shear. Unlike the Lagrangian method [34], the WAM tracking method can pinpoint the specific source of water vapor evaporation. In arid regions with low precipitation frequency, a few major precipitation events may contribute significantly to the total precipitation. There may be substantial differences in water vapor transport pathways during precipitation and non-precipitation periods, resulting in different water vapor sources [35]. Therefore, we utilize the WAM-2Layer model to calculate the differences in water vapor sources between the southern and northern parts of the QB, and also consider the water vapor recycling ratio to compare seasonal variations.

2.3. Sampling and Testing

Sample collection was conducted during the summer (July–August 2019) and winter (December 2019–January 2020) seasons. In summer, 63 river water samples, 3 lake water samples, 119 groundwater samples, 2 precipitation samples, and 4 brine samples were collected at the locations indicated in Figure 2. In winter, samples were collected from the Bayin River and Golmud River basins, totaling 48 samples (including 25 surface water, 18 groundwater, 3 lake water, 1 brine, and 1 precipitation sample).

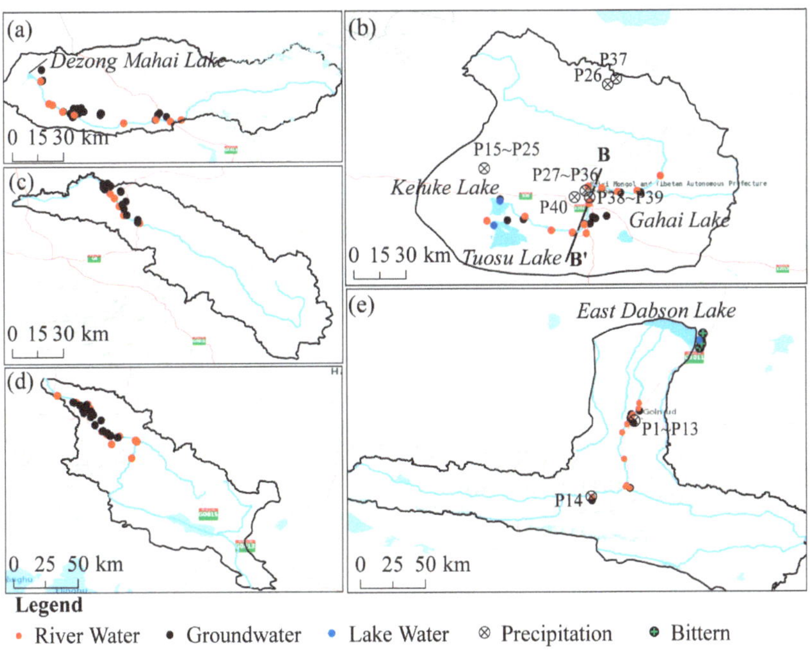

Figure 2. Sampling point distribution map: (**a**) Yuqia River; (**b**) Bayin River; (**c**) Chahanwusu River; (**d**) Xiangride River; and (**e**) Golmud River: B−B′ profile represents the cross-section from the outlet of Bayin River to the terminal lake area.

Prior to sampling, a multi-parameter water quality monitor (EXO SY1 USA) was used for the on-site measurement of water temperature (WT), salinity (SAL), total dissolved solids (TDS), and hydrogen ion concentration index (pH). SAL was calculated based on conductivity and temperature measurements, with a measurement range of 0 to 70 ppt and a resolution of 0.01 ppt. Anions (Cl^- ($\leq$100 mg·L^{-1}), NO_3^-, SO_4^{2-}, Br^- and F^-) were determined using an ion chromatograph (ICS5000+, Thermo Fisher Scientific, Waltham, MA, USA), while cations (Mg^{2+}, Ca^{2+}, Na^+, K^+, Al^{3+}, and Fe^{2+}) were measured using an inductively coupled plasma mass spectrometer (ICAP6300, Thermo Fisher Scientific, Waltham, MA, USA). CO_3^{2-}, HCO_3^-, and Cl^- (>100 mg·L^{-1}) were determined using acid titration methods. H−O isotopes (δD and δ^{18}O) were analyzed using a liquid water isotope analyzer (GLA431-TIWA, ABB, Quebec, QC, Canada). For brine, lake water, and river water with high salinity, as well as groundwater, an LI-2100 (LICA United Technology Limited, Beijing, China) fully automated vacuum extraction system was used for pre-extraction before conducting the tests.

To investigate the sources of atmospheric moisture, we downloaded the ERA-5 re-analysis dataset from the European Centre for Medium-Range Weather Forecasts (https://www.ecmwf.int/en/forecasts/datasets/reanalysis-datasets/era5, accessed on 1 August 2022). Precipitation H−O isotope data were collected with supplementation from the literature, as shown in Appendix A (Table A1).

3. Results

3.1. Summer Indicators

The physicochemical, hydrochemical and isotopic characteristics of surface water and groundwater in the QB during the summer are presented in Appendix A Table A2 and Figure 3.

In summer, the water temperature of water bodies varies from 4.0 °C to 25.6 °C with the mean value (±SD) of 12.8 ± 4.0 °C. Surface water and groundwater tend towards being alkaline; the pH value is 8.4 ± 0.5, SAL is 2.0 ± 19.1 ppt, and TDS is 2324.4 ± 20,630.6 mg/L, with the main ionic components being Cl^-, SO_4^{2-}, HCO_3^-, Na, Ca, and Mg. The δD value is −64.2 ± 11.1‰, and the δ^{18}O value is −9.2 ± 2.0‰. When comparing the five watersheds, the TDS levels are arranged from highest to lowest as follows: Golmud River, Chahanwusu River, Bayin River, Xiangride River, and Yuqia River, with Golmud River primarily having high ion concentrations due to lake water influence.

Based on significant difference analysis [26] (Figure 3), it can be observed that for most physical and chemical parameters, there are no significant differences within the $\alpha \leq 0.05$ range. However, significant differences exist in pH values between the Golmud River, Chahanwusu River, and Xiangride River. HCO_3^- levels show significant differences between the Yuqia River and the Bayin River, Xiangride River, and Golmud River. The high concentration of SO_4^{2-} in groundwater in parts of the Bayin and Golmud River basins is due to the greater impacts of human activities in these areas. H−O isotopes show significant differences: δD values exhibit significant differences between the Bayin River and the Xiangride River, as well as the Yuqia River and Chahanwusu River. The δ^{18}O values show significant differences between the Yuqia River, Chahanwusu River, and the other three rivers.

Spatially, significant variations are observed only in pH, HCO_3^-, and H−O isotopes among the various watersheds.

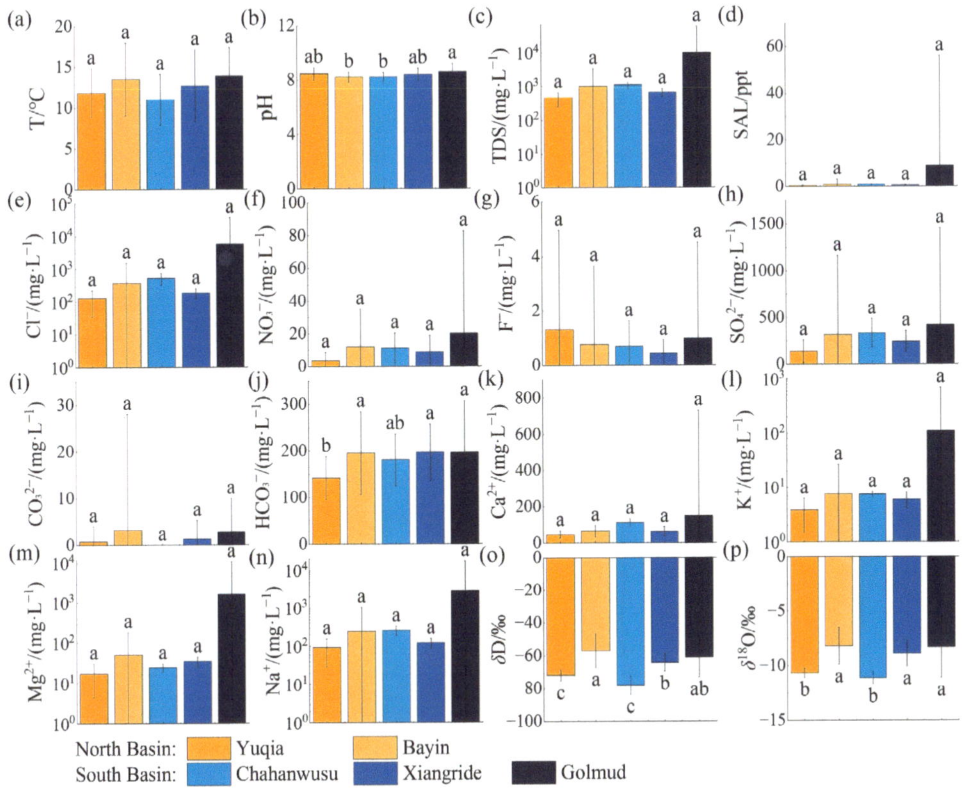

Figure 3. Physical and chemical parameters and the H–O isotope of water bodies in summer: (**a**) water temperature; (**b**) pH; (**c**) TDS; (**d**) SAL; (**e**) Cl^-; (**f**) NO_3^-; (**g**) F^-; (**h**) SO_4^{2-}; (**i**) CO_3^{2-}; (**j**) HCO_3^-; (**k**) Ca^{2+}; (**l**) K^+; (**m**) Mg^{2+}; (**n**) Na^+; (**o**) δD; (**p**) $\delta^{18}O$. (Letters Symbols a, b, and c represent significant differences between each pair of data in the five groups. Identical letters indicate no significant difference, while different letters indicate a significant difference, with a significance level of 0.05).

3.2. Winter Indicators

Based on the analysis above, it is evident that there exist significant disparities in water characteristics between the southern and northern basins, demonstrating a latitudinal zonality. Therefore, the watersheds of the Golmud River in the south and the Bayin River in the north are selected to analyze seasonal differences. Appendix A Table A3 presents the physicochemical characteristics and hydrochemical and isotopic statistics of winter water bodies. Four physicochemical characteristics and H–O stable isotopes were selected for the analysis of significant difference. Because the data do not conform to a normal distribution, the Mann–Whitney U non-parametric test method [27] was employed for the analysis of significant differences (Figure 4).

In winter, the water temperature in Golmud River ranges from −5.5 to 9.0 °C (with an average of 2.1 ± 3.6 °C), and in Bayin River, it ranges from −0.2 to 8.0 °C (with an average of 11.0 ± 7.2 °C), which is more stable compared to summer. In the Golmud River basin during winter, the TDS ranges from 479.2 to 195,635.7 mg/L (with an average of 10,966.5 ± 44,720.2 mg/L), and the salinity is 9.8 ± 40.2 ppt, which is higher than in summer. In the Bayin River basin, the TDS ranges from 344.1 to 5508.9 mg/L (with an average of 888.2 ± 1087.8 mg/L), and the salinity is 0.6 ± 1.0 ppt, which is lower than in summer. When comparing different water types, it is observed that the pH of lake water in the Golmud River basin (ranging from 7.2 to 7.9) is lower than that of river water and groundwater, while the pH of lake water in the Bayin River is higher (ranging from 9.1 to 9.4 for Tuosu Lake and 8.5 to 8.6 for Keluke Lake, with an average of 8.3 for river and lake water in both summer and winter). The TDS in all three lake water samples is high, especially in East Dabson Lake, where the TDS reaches as high as 1.2×10^5 mg/L. Lake water has high TDS and a wide range of variation, with the enrichment of Na, K, and Mg, and the dominant anions are Cl^- and SO_4^{2-}.

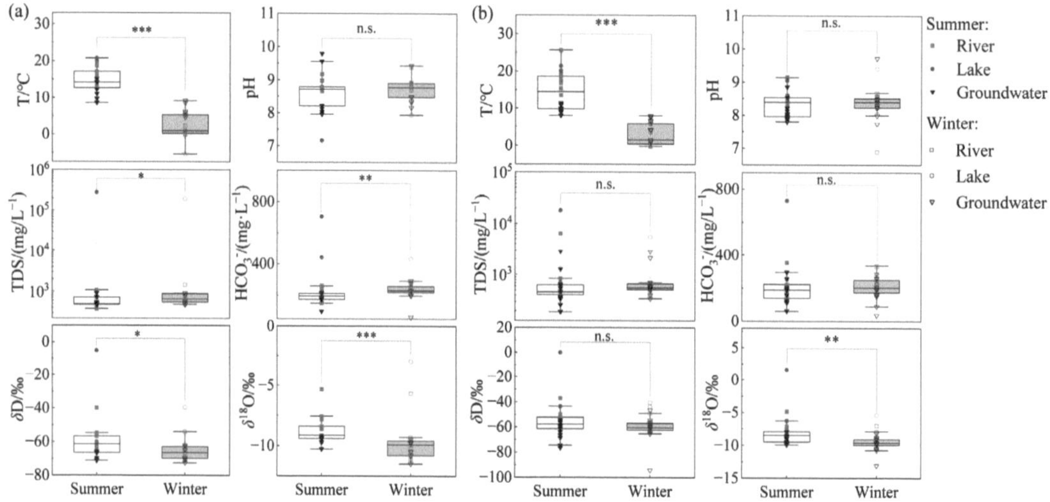

Figure 4. Differences in physicochemical parameters and HO isotopes between summer and winter: (**a**) Golmud River; (**b**) Bayin River. (The significance level is $\alpha = 0.05$, with "***" representing $p \leq 0.001$, "**" representing $p \leq 0.01$, "*" representing $p \leq 0.05$, and "n.s." representing no significant difference).

During the summer and winter seasons, significant differences are observed in the TDS of the Golmud River ($p \leq 0.05$), as well as in SAL and HCO_3^- concentrations ($p \leq 0.01$). The differences in δD are also significant ($p \leq 0.05$), while the differences in δ^{18}O are extremely significant. On the other hand, there are no significant differences in TDS, HCO_3^- concentration, SAL, and δD in the Bayin River. In terms of isotope comparison, significant differences are observed in δD, while extremely significant differences are observed in δ^{18}O in the Golmud River. In the Bayin River, significant differences are observed in δ^{18}O ($p \leq 0.01$), while no significant differences are found in δD.

In summary, the magnitude of the differences in characteristics for different regions during the summer and winter seasons varied. The main differences are observed in water temperature, δ^{18}O, TDS and HCO_3^- in the Golmud River watershed.

4. Discussion

Based on the above analysis, it is evident that there are differences in the water characteristics of the southern and northern parts of the QB during the summer and winter seasons. Surface water and groundwater exhibit spatiotemporal variations, which are related to different underlying surface conditions and external influencing factors. The composition of water sources, surface water to groundwater conversion along the flow path, and the different supply sources of terminal lakes are all related. The following sections will discuss these factors separately.

4.1. Analysis of Spatial Differences

4.1.1. Differences in Hydrochemical Types and Their Causes

Classifying hydrochemical types can provide a more intuitive analysis of the chemical characteristics of water bodies in different basins. By using the Piper ternary diagram and the Shukarev classification method [36], comparisons of the main ion concentrations in the five watersheds were conducted, leading to the categorization of hydrochemical types (Figure 5 and Table 1). It can be observed that the proportions of each ion's concentration in river water and groundwater are similar, while the concentration of Cl^- is higher in lake water. In the northern basin (Figure 5a,b), the concentrations of Ca^{2+} and HCO_3^- are higher than in the southern basin (Figure 5c–e), while Na^+, K^+, and Cl^- concentrations are lower. Table 1 shows that in the northern basin, rivers and groundwater are mainly of the HCO_3·Cl·SO_4-Na·Ca type, whereas in the southern basin, there are various types of rivers and groundwater, but overall, Cl^- concentrations are higher. The Tuosu Lake in the northern basin is of the Cl·SO_4-Na type, while the East Dabson Lake in the southern basin is of the ClNa·Mg type.

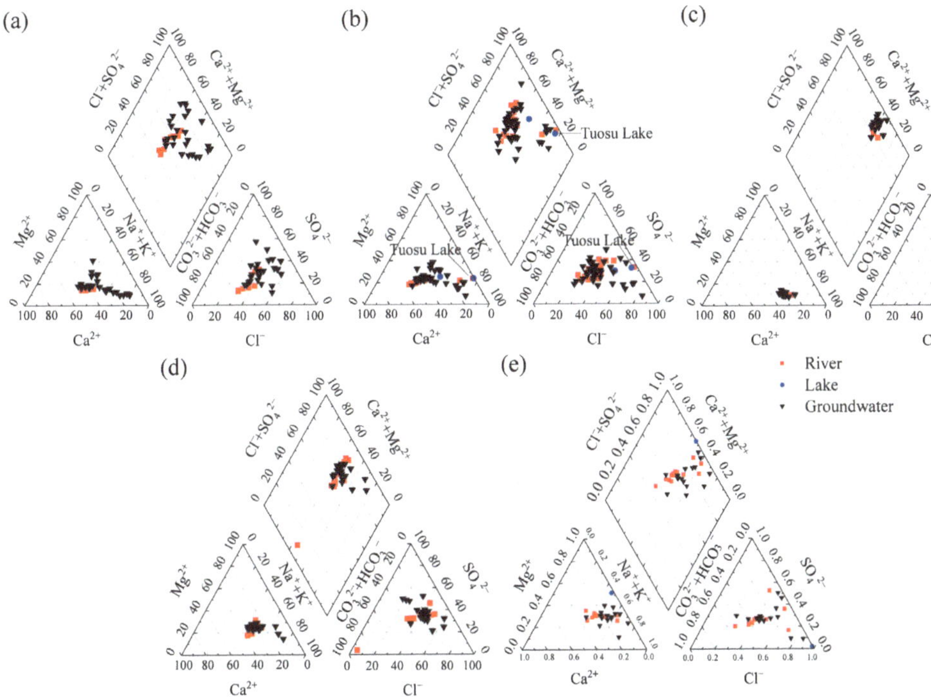

Figure 5. Piper ternary diagram of water bodies in different basins: (**a**) Yuqia River; (**b**) Bayin River; (**c**) Chahanwusu River; (**d**) Xiangride River; (**e**) Golmud River.

Table 1. Hydrochemical types of the main water bodies in different watersheds.

Number	Watershed	River and Groundwater	Lake
(a)	Yuqia River	HCO_3·Cl·SO_4-Na·Ca	/
(b)	Bayin River	HCO_3·Cl·SO_4-Na·Ca	Cl·SO_4-Na
(c)	Chahanwusu River	Cl·SO_4-Na·Ca	/
(d)	Xiangride River	Cl·HCO_3·SO_4-Na·Ca	/
(e)	Golmud River	Cl·HCO_3-Na·Ca	ClNa·Mg

Consequently, there are significant differences in the hydrochemical types between the southern and northern basin watersheds. Additionally, the chemical characteristics of river water and groundwater are similar, whereas lake water differs markedly. The main reason for the differences in the chemical composition of water is the diverse underlying conditions, which can be analyzed from the perspectives of regional geological origin and aquifer lithology. Based on the comprehensive Kunlun Mountains–Xiao Qaidam Lake hydrogeological profile (Figure 1b), it is evident that the surrounding mountains of the basin mainly consist of Paleozoic formations, while the basin and its margin are dominated by Mesozoic–Cenozoic formations. In the southern mountainous regions, there are distributions of Triassic (T, shallow marine deposition), Permian (P), Ordovician (O), Paleoproterozoic (Pt), Jurassic (J), and Devonian (D) formations. The Quaternary system (Q) is mainly distributed within the basin and the wide valley areas between the mountains. The genesis types of the Holocene (Qh) include debris flow accumulation ($Qh^{dl\text{-}pl}$), alluvial fan accumulation ($Qh^{al\text{-}l}$), and marsh sedimentation (Qh^{fl}). In the river channels, the primary accumulations are the Holocene alluvial layers (Qh^{al}), while the periphery of the terminal lake is primarily characterized by chemical deposits–lacustrine layers ($Qh^{ch\text{-}l}$), and the lower part mainly consists of late Pleistocene floodplain deposits ($Qp_3^{al\text{-}pl}$). In the northern part of Lake Dabson (the largest lake in the Qarhan Salt Lake area), the upper part is mainly covered by late Pleistocene lacustrine layers (Qp_3^{l}) and Holocene chemical deposits (Qh^{ch}), while the lower part consists of late Pleistocene lacustrine floodplain deposits ($Qp_3^{al\text{-}l}$). In the southern part of Xiao Qaidam Lake, there is a distribution of continental red clastic sedimentation belonging to the Paleogene–Neogene system (R). The periphery of the lake is composed of late Pleistocene floodplain deposits (Qp_3^{pl}), and the central part of the basin is characterized by marsh sedimentation (Qh^{fl}).

From the perspective of the aquifer system, the surrounding bedrock in mountainous areas in the basin is composed of metamorphic rocks and carbonate rocks. Among these rock formations, numerous springs emerge, transferring the chemical composition of rocks to surface water. In the Kunlun Mountain area, the aquifer formations of metamorphic rocks primarily consist of schist, gneiss, quartz schist, and slate. The carbonate rock fractures are well developed, resulting in abundant spring water flow with excellent water quality. In the Qilian Mountain area, the aquifer formations of metamorphic rocks include sandstone, shale, slate, and gneiss, while the carbonate rocks mainly exhibit thick and massive structures with well-developed joint fissures, creating favorable conditions for the storage and migration of groundwater [10]. This also explains the higher concentration of HCO_3^- in the water bodies of mountainous regions, as documented in reference [25].

Loose rocks are distributed in the intermountain valleys and alluvial fan areas of each watershed. Vertically, the granular composition transitions from coarse to fine particles, starting from the coarse pebble layer and transitioning to coarse sand, fine sand, and silt towards the basin's center from the front-tilted plain. Vertically, the transition comprises a single pebble layer to a multilayer structure with overlapping rock types. Horizontally, parallel to the basin's edge mountains, the main river channels form an independent aquifer system in the front faces of mountainous areas [10].

A significant difference between the southern and northern basins is the varying connectivity of the aquifers in the lacustrine plain. In the southern basin, the confined aquifer in the lacustrine plain is interconnected with the thick aquifer of the alluvial fan, directly receiving groundwater recharge. In the northern basin, the transverse continuity of the confined aquifer is inferior to that of the southern basin. Some areas are isolated by mountains and are independent units. However, vertically, they generally connect with the unconfined aquifer of the alluvial fan primarily through vertical top-supported cross-flow groundwater recharge, and then the groundwater is discharged via groundwater evaporation [10]. The supply of oilfield water in the southern part of the basin (as shown in Figure 1b) is also a leading cause of higher Cl^- content in the water bodies of the southern basin. According to the research findings in [25], it is known that Na^+, Cl^-, and SO_4^{2-}

dominate the water bodies in the southern basin, while Ca^{2+} and HCO_3^- dominate in the northern basin.

Thus, the differences in hydrochemical composition are primarily due to variations in the aquifer's lithology: (1) in the southern region, the influence of salt rock dissolution and oilfield water is more significant, while in the northern mountainous areas, the development of carbonate rock joints and fissures is more pronounced, indicating a more significant impact due to carbonate weathering; (2) the connectivity of the aquifers in the floodplain areas results in different groundwater recharge mechanisms, with direct upstream groundwater recharge in the south and vertical top-supported cross-flow recharge in the north, resulting in relatively lower potentials for evaporation.

4.1.2. Differences in H-O Isotopes and Their Causes

The surface water in the QB exhibits significant spatial variations in H-O isotopes (Figure 6a,b). In the northern basin, the H-O isotope values increase from west to east, while in the southern basin, there is an overall trend of increasing values from east to west. The less pronounced pattern in the Xiangride River basin is attributed to the fact that the river water is primarily supplied by local precipitation in the plains, whereas in other basins, the river water is mainly replenished by atmospheric precipitation from high-altitude mountain areas [37].

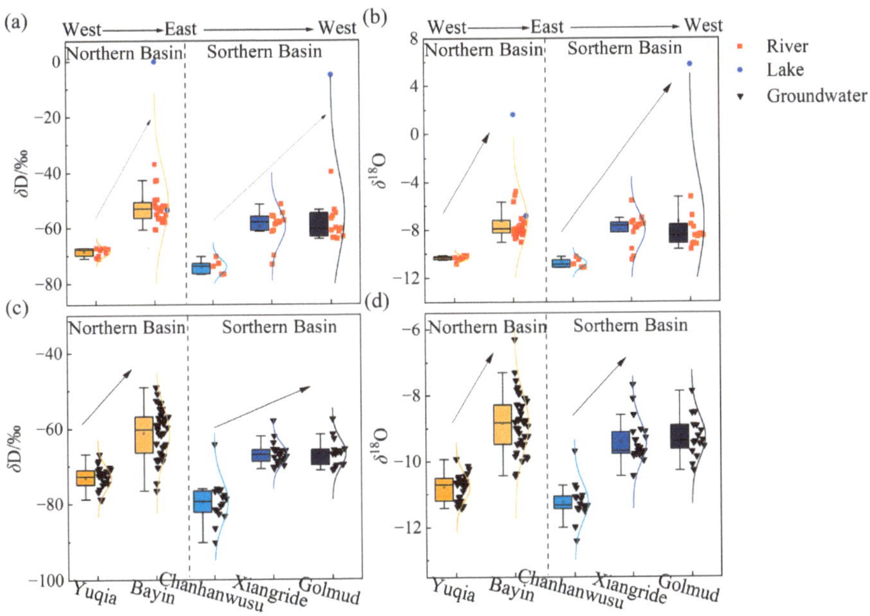

Figure 6. Spatial variation of H-O isotopes in the surface water and groundwater of the QB: (**a**) δD of surface water; (**b**) $\delta^{18}O$ of surface water; (**c**) δD of groundwater; (**d**) $\delta^{18}O$ of groundwater: The arrows represent the spatial trend of H-O.

In comparison to surface water, the H-O isotope values in groundwater are lower (Figure 6c,d), as the isotopic fractionation dynamics caused by evaporation and mixing are weaker than in surface water [38]. The spatial variability in groundwater is not as pronounced as in surface water, but follows almost the same distribution pattern within the basin. This suggests that close hydraulic interactions result in similar spatial variations in the H-O isotope composition of both surface water and groundwater.

In summary, the isotopic composition of watersheds in the southern basin contrasts with that of watersheds in the northern basin. In the northern basin, δD and $\delta^{18}O$ gradually enrich from west to east, whereas in the southern basin, the opposite trend is observed, indicating a significant influence of local meteorological conditions such as precipitation and evaporation. The following analysis will use precipitation H-O isotopes, deuterium excess (d-excess) and atmospheric water vapor circulation to explore the reasons behind these variations.

Deuterium excess is defined as d-excess = $\delta D - 8\delta^{18}O$, and its value is related to the relative humidity of the ambient air when the water vapor at the source of precipitation evaporates [39]. This value remains constant during transport and contains important information about the source of water vapor. Generally, precipitation from high-latitude inland areas has a d-excess value of >10‰, while precipitation from low-latitude oceanic evaporation has a d-excess value of <10‰ [40]. In arid regions, the d-excess values decrease during the evaporation process and increase during the local water vapor recycling process associated with intense evaporation [41]. In their study of the source of summer water vapor in the northern Tibetan Plateau, Yang et al. [42] suggested that when the $\delta^{18}O$ in precipitation is less than −20‰, the source of water vapor is marine air masses, and when $\delta^{18}O$ in precipitation is greater than −13‰, the source of water vapor is local evaporation.

Based on the collected H-O isotope values in precipitation and the river water data collected in this study, the d-excess values for the Golmud and Delingha regions were calculated (Table 2). It can be observed that the $\delta^{18}O$ values for precipitation at both locations are higher than −13‰. This is mainly due to the arid climate and intense evaporation in the QB, where local evaporation contributes significantly to the water vapor. In comparison, the $\delta^{18}O$ value of precipitation in the Golmud region averages −8.80‰, slightly lower than that of river water (−8.36‰), indicating that in addition to local river water evaporation, precipitation has other sources. The higher d-excess values (average of 10.61‰) indicate relatively dry climate conditions in the moisture source region. Therefore, it can be inferred that some of the water vapor comes from the Central Asian inland moisture transported by the westerlies (the average $\delta^{18}O$ value of precipitation in the Central Asian region is −9.79‰ [43]). Precipitation in the Delingha region has a higher d-excess (average of 12.01‰) and a $\delta^{18}O$ value (average of −6.96‰) higher than that of river water (average of −7.87‰), indicating that the water vapor is primarily derived from local evaporation.

Table 2. Statistical table of $\delta^{18}O$ and d-excess.

Sampling Location	Type	$\delta^{18}O$/‰				d-Excess/‰			
		Min	Median	Max	Mean	Min	Median	Max	Mean
Golmud	Rain	−23.44	−8.67	1.89	−8.80	−27.86	11.68	38.70	10.61
	River	−9.65	−8.49	−5.31	−8.36	2.65	7.80	16.17	8.39
Delingha	Rain	−34.14	−5.16	1.09	−6.96	−4.82	11.66	31.45	12.01
	River	−9.03	−8.18	−4.83	−7.87	1.70	9.15	14.38	9.11

Using the southern and northern basins as marked regions, precipitation source tracking calculations were conducted using WAM-2Layer. Additionally, regional evaporation and water vapor recycling rates were calculated [29]. The contribution rates of the average precipitation water vapor sources from 2019 to 2020 are shown in Figure 7.

After calculation, the annual average evaporation in the southern basin was deemed to be 404.9 mm, while in the northern basin, it is 216.8 mm. The average annual precipitation recycling rate (water vapor recycling) in the southern basin is 6.7%, whereas in the northern basin, it is 5.2%. Therefore, the southern basin has a stronger local evaporation capacity, leading to a higher precipitation recycling rate compared to the northern basin [35]. According to Figure 7, the westerly circulation and local water vapor cycle are the main contributors to precipitation in the basin [38,43], but their contributions are not entirely the same for the southern and northern basins. The areas that provide moisture for precipitation in the southern basin are mainly concentrated in the central QTP. Among them, the Mushroom Peak on the southwestern edge of the QB contributes the most moisture, with an annual contribution of around 50 mm. There is also a contribution from the southwestern part of the QTP, which may come from the evaporation of snow on high mountains. The moisture sources for the northern basin are mainly in the northeast of the basin, with an annual contribution of around 20 mm. There is less water vapor contribution from the southwest to the northern basin than to the southern basin.

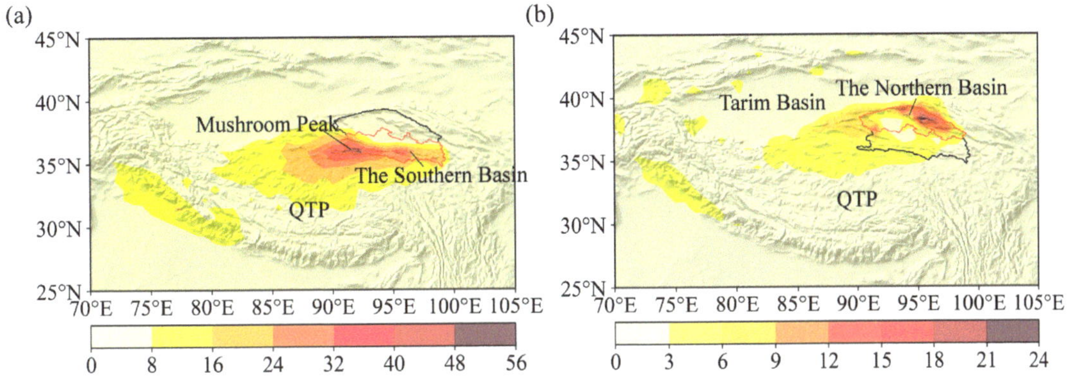

Figure 7. Water vapor source of precipitation in the north and south of the QB (mm/y): (**a**) southern basin; (**b**) northern basin.

The contribution rates of precipitation water vapor sources and the values of $\delta^{18}O$ and d-excess indicate that the sources of water vapor in different regions and the transportation intensity within the basin are not entirely the same. In the case of the southern basin, the contribution of water vapor transported by the westerly winds is relatively significant [44], whereas in the northern basin, the primary source of water vapor is local evaporation [45].

4.2. Analysis of Seasonal Differences

4.2.1. Differences in Hydrochemical Types and Their Causes

By using the Piper ternary diagram to illustrate the differences in major ion concentrations and hydrochemical types for the two watersheds during summer and winter (Figure 8), it can be observed that there are significant differences in the ion concentrations of river water and groundwater for the two watersheds. In the Piper ternary diagram, the summer sample points are positioned above the winter sample points. The seasonal variation trends in ion concentration proportions are similar. During the summer, Ca^{2+} and Cl^-, as well as $SO_4{}^{2-}$, are depleted compared to the winter season, while $HCO_3{}^-$ and Mg^{2+}, along with Na^+, are enriched compared to the summer season.

In the Golmud River watershed, the primary hydrochemical type of river and groundwater is $HCO_3 \cdot SO_4 \cdot Cl\text{-}Na \cdot Mg \cdot Ca$ in summer and $HCO_3 \cdot SO_4\text{-}Na \cdot Mg$ in winter. In the Bayin River watershed, the primary hydrochemical type is $SO_4 \cdot HCO_3 \cdot Cl\text{-}Na \cdot Ca \cdot Mg$ in summer and $HCO_3 \cdot SO_4 \cdot Cl\text{-}Na \cdot Ca \cdot Mg$ in winter. The East Dabson Lake in both seasons is of the $Cl\text{-}Na \cdot Mg$ type, with Ca^{2+} and Mg^{2+} being more enriched during the winter and Na^+ being depleted. The differences in anion concentrations are small, leading to minimal impacts on the hydrochemical type. Tuosu Lake maintains the same hydrochemical type in both seasons, being of the $Cl \cdot SO_4\text{-}Na$ type. Keluke Lake exhibits different hydrochemical types in different seasons. In summer, it is primarily of the $HCO_3 \cdot SO_4 \cdot Cl-Na \cdot Mg \cdot Ca$ type, while in winter seasons, it is mainly the $HCO_3 \cdot SO_4 \cdot Cl-Na \cdot Ca \cdot Mg$ type.

Therefore, the types of river water and groundwater differ in different seasons, while the hydrochemical types of terminal lakes are not influenced by the seasons. It is evident that the differences in hydrochemical composition and types are related to water temperature, precipitation, the leaching effect of groundwater, and the evaporative precipitation effect.

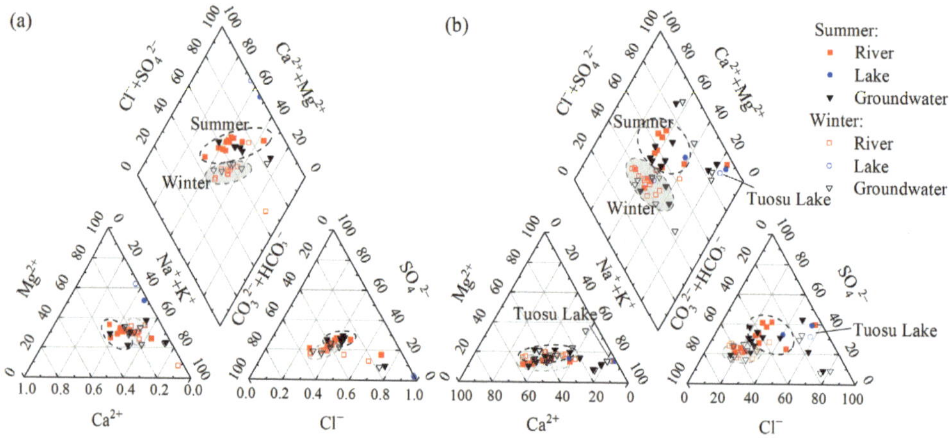

Figure 8. Piper ternary diagram for different kinds of water in summer and winter: (**a**) Golmud River; (**b**) Bayin River.

The Gibbs models are employed to characterize influencing mechanisms, such as atmospheric input, water–rock interaction, and evaporation–crystallization effects [31,46] (Figure 9).

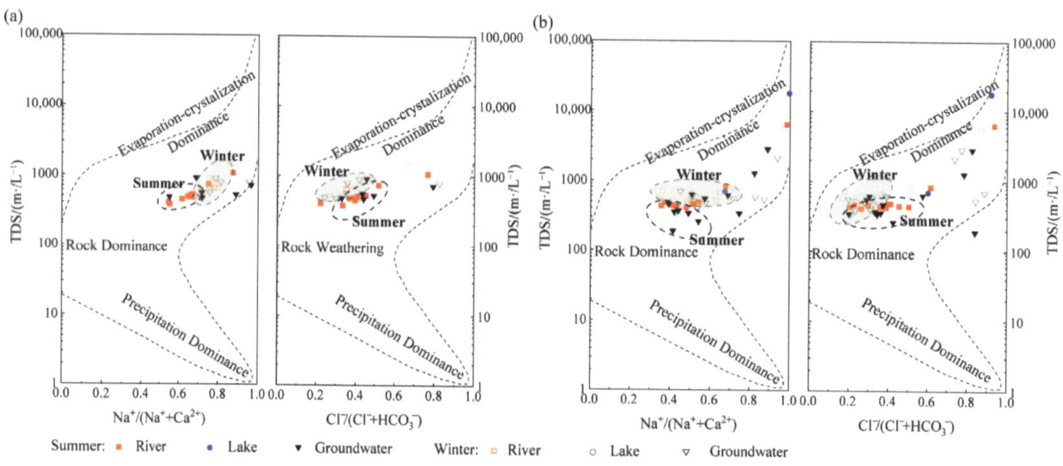

Figure 9. Gibbs plots of water bodies in the summer and winter: (**a**) Golmud River; (**b**) Bayin River.

Generally, the $Na^+/(Na^+ + Ca^{2+})$ and $Cl^-/(Cl^- + HCO_3^-)$ ratios of water samples from both watersheds in different seasons are distributed towards the middle–right position in the Gibbs diagram, indicating that the hydrochemical composition of the study area's water bodies deviates as a result of atmospheric precipitation, and is mainly controlled by rock weathering and evaporation crystallization. The absence of lake water samples in the Gibbs diagram of the Golmud River is due to a TDS that exceeds 10^5 mg/L, indicating that the lake water is influenced by other factors. In the case of the Bayin River, one water sample in the summer groundwater falls outside the graph in the $Cl^-/(Cl^- + HCO_3^-)$ ratio diagram, suggesting that hydrochemical characteristics may be influenced by other factors, such as human activities [36].

Comparing the hydrochemical mechanisms of the two watersheds in the summer and winter seasons, it can be observed that there are similar seasonal differences. In the

$Na^+/(Na^+ + Ca^{2+})$ ratio diagram, the summer samples are located below and to the left of the winter samples, indicating a greater influence of rock weathering in the summer, which is likely due to the higher water temperature and stronger water–rock interaction. In the $Cl^-/(Cl^- + HCO_3^-)$ ratio diagram, the winter samples are positioned above the summer samples, suggesting a higher TDS in winter, which also implies a greater influence of evaporation and precipitation in the winter, as the region is drier and experiences faster water surface evaporation during this season.

Therefore, water bodies are more influenced by water–rock interaction in summer than in winter seasons; this is not only related to water temperatures, but also to factors such as different seasonal atmospheric precipitation and evaporation.

4.2.2. Differences in H-O Isotopes and Their Causes

Evaporation lines with specific slopes represent the isotopic composition characteristics of water samples from the same source, while different positions on the same evaporation line reflect differences in the degree of unbalanced evaporation they have undergone. The δD-$\delta^{18}O$ relationships of precipitation isotopes collected from the literature with precipitation, river water, lake water, and groundwater sampled in this study are illustrated in Figure 10. By comparing with the atmospheric precipitation lines from Golmud and Delingha regions (LWML, $\delta D = 7.840\ \delta^{18}O - 4.566$, $\delta D = 7.833\ \delta^{18}O + 8.606$) [45], it is evident that precipitation sampling points in the Golmud area deviate significantly from the LWML, indicating pronounced H-O isotope fractionation due to intense evaporation and minimal rainfall in Golmud. In contrast, precipitation sampling points in Delingha are generally located on the LWML atmospheric precipitation line. Both river water and groundwater in the Golmud River are situated on the evaporation line, suggesting that these three water sources share the same primary supply. However, lake water sampling points deviate from the evaporation line, especially during winter points, located to the upper left of the evaporation line, indicating additional supply sources (detailed analysis in Section 4.3.2). In the Bayin River, several groundwater sampling points are far from the evaporation line (black points in the lower left of Figure 10b) because they were collected from deep water layers (depth of 90~110 m) in the front edge of alluvial fans and lacustrine plains. The water of Keluke Lake aligns with the evaporation line, sharing the same source as river water and groundwater. On the other hand, sampling points for Tuosu Lake are distant from other water bodies. This is attributed to continuous evaporation in Tuosu Lake and the possibility of local groundwater supply (detailed analysis in Section 4.3.2).

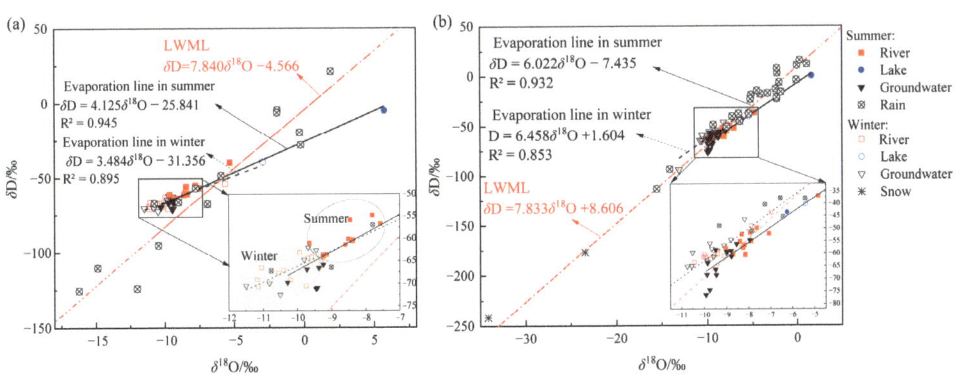

Figure 10. Comparison of the $\delta^{18}O$–δD relationship of water bodies in summer and winter: (**a**) Golmud River; (**b**) Bayin River.

Comparing the evaporation lines with the LWML, it is evident that the slopes ($\delta D/\delta^{18}O$ values) and intercepts of the evaporation lines for river water, lake water, and groundwater in the Golmud River are significantly smaller than the local precipitation line's slope (7.840).

Moreover, the winter $\delta D/\delta^{18}O$ values and intercepts (−31.356 and 3.484) are smaller than those in summer (−25.841 and 4.125), indicating that both summer evaporation and winter water freezing lead to the enrichment of heavy isotopes. Additionally, the impact is more pronounced in winter, suggesting a drier winter compared to summer (consistent with Section 4.2.1). In the Bayin River, the slope and intercept of the evaporation line are closer to the local precipitation line. The summer $\delta D/\delta^{18}O$ value (6.022) is slightly lower than in winter (6.458), which is attributed to the relatively lower humidity in summer.

Comparing the distribution of summer and winter evaporation lines and sampling points on the δD-$\delta^{18}O$ plot, it is evident that the water bodies in the Golmud River exhibit significant seasonality, while this is not the case for the Bayin River. This observation is related to local climate conditions. Based on the results of precipitation source backtracking, the average water vapor source and evaporation factors for the winter and summer seasons from 2019 to 2021 are calculated, as shown in Table 3.

Table 3. Evaporation, precipitation, and water vapor internal circulation rates in summer and winter.

Area	Season	Total Evaporation	Regional Total Precipitation	Regional Total Evaporation	Regional Actual Total Evaporation	Effective Evaporation Rate	Water Vapor Loss Rate	Regional Water Vapor Circulation Rate
		mm/a	mm/a	mm/a	mm/a	%	%	%
Southern basin	Summer	289.26	296.35	21.98	219.83	10.02	2.39	7.42
	Winter	33.86	30.75	1.16	21.86	5.23	−9.61	3.79
Northern basin	Summer	113.65	112.35	7.55	126.09	5.90	7.06	6.05
	Winter	16.09	16.06	0.45	11.88	3.53	2.41	2.56

It can be observed that in the southern basin, the actual total evaporation in summer reaches 219.83 mm, with an effective evaporation rate of 10.02% and an internal cycling rate of 7.24%. In contrast, the effective evaporation rate in winter is only 5.23%, with an internal cycling rate of 3.79%. The evaporation, precipitation, and water vapor internal cycling rates in the northern basin are all lower than those in the southern basin, and the difference in effective evaporation rates between summer and winter is also relatively small.

4.3. Surface Water–Groundwater Interactions and Salt Sources for Terminal Lake

4.3.1. Surface Water–Groundwater Interaction from Mountainous Areas to Terminal Lakes

In arid inland river basins, there are frequent interactions between surface water and groundwater, and the upstream–downstream relationship along the course can also affect the hydrochemical and isotopic characteristics of water bodies. The relatively stable $\delta^{18}O$ and Cl^- are selected as tracers to analyze the surface water–groundwater interaction (taking Bayin River as an example, Figure 11).

It can be observed that in the Bayin River watershed, the frequency of surface water–groundwater interaction is intense due to the control of bedrock barriers, lithology, and terrain. The intermountain valley has larger rock and soil particles. The longitudinal groundwater $\delta^{18}O$ first becomes depleted and then enriched, while surface water first becomes enriched and then depleted. This indicates that groundwater is influenced by the bedrock barrier and surfaces in the region. The Quaternary unconsolidated layer is in non-conformity contact with the bedrock, characterized by the presence of foreland uplifts or aquifer structures. Groundwater is obstructed by the bedrock and emerges at the surface. Surface water and groundwater undergo two (or even three or four) conversions in the intermountain valleys before entering the main basin [25]. In the expansive foothill plain, characterized by large sediment particles, a significant amount of surface water replenishes groundwater. From the Delingha urban area to Yikeshu (alluvial plain to lacustrine plain), the terrain gradually becomes gentle, and the rock and soil particles become smaller. This transition from a single unconfined aquifer to multi-layer confined aquifers leads to relatively stable $\delta^{18}O$ and Cl^- levels in surface water. In contrast, fluctuations in $\delta^{18}O$ and Cl^- are observed in the groundwater, suggesting frequent surface water–groundwater transformations within this region. In the terminal lake area, there is a sharp enrichment of

$\delta^{18}O$ and Cl^- in both groundwater and surface water, indicating that groundwater and surface water are closely related. Simultaneously, this area exhibits strong evaporation and concentration effects. This can be attributed to the large evaporative surface area of the lake and the shallow groundwater depth (<5 m), leading to significant water consumption.

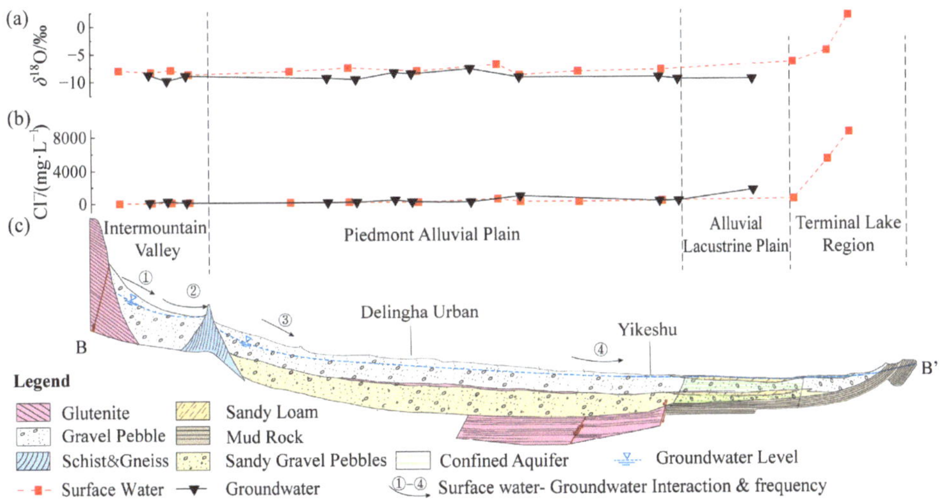

Figure 11. Surface water–groundwater interactions in the Bayin River: (**a**) $\delta^{18}O$; (**b**) Cl^-; (**c**) hydrogeological profile (Figure 2b B−B′. Revised according to reference [47]).

Therefore, the interaction between surface water and groundwater in the Bayin River can be characterized as follows: In the mountainous regions, river water infiltrates into the ground, and subsequently, groundwater emerges at the surface due to the obstruction of bedrock. After exiting the mountains, groundwater continues to seep downward, with increased seepage in the alluvial plain. Controlled by lithological changes and terrain, it overflows the surface before the front edge of an alluvial fan, replenishing the lakes. This phenomenon explains the similarities between the hydrochemical and isotopic characteristics of river water and groundwater.

4.3.2. Salt Sources for Terminal Lake

The fact that the hydrochemical types of the two terminal lakes do not vary with the seasons while exhibiting significant differences in salt content suggests the possibility of additional sources of replenishment. First, the supply of salt content to terminal lakes from the upstream water bodies is analyzed via an analysis of the hydrochemical distribution characteristics along the course, using the Bayin River watershed as an example (Figure 12). Subsequently, hydrochemical and isotopic methods are utilized to identify potential alternative sources.

In Figure 12, we see that all ions except Ca^{2+} and HCO_3^- increase to about the same magnitude. The decrease in Ca^{2+} and the smaller increase in HCO_3^- may be due to the precipitation of calcite. It is evident that in the mountainous region of the Bayin River watershed, the changing trends of Cl^-, SO_4^{2-}, and HCO_3^- anions in river water exhibit opposite results. This indicates ion exchange during the leaching process, where carbonate rocks are replaced by rock salts. In the intermountain valley, there is a sudden increase in ion concentrations in the groundwater, while the Cl^- and SO_4^{2-} contents of surface water initially increase and then decrease, which is controlled by the bedrock barrier, causing changes in ions after surface water is replenished by groundwater. As the water flow slows upon entering the piedmont alluvial plain area, the transport capacity gradually decreases, leading to the accumulation of fine particulate matter. This forms a single layer of sand

and gravel stones in the northern part, gradually transitioning to a multi-layer structure of gravel, sand, silt, and clay (Figure 11c). The infiltration of surface water at the alluvial plain leads to a decreasing and then increasing trend with respect to ion concentrations in the groundwater. At the junction of the alluvial fan and lacustrine plain, there is a sudden increase followed by a decrease in Cl^- and Na^+ in groundwater, and surface water shows a similar trend but with a smaller magnitude of change compared to groundwater. This is because the confined water flows upward, carrying substances from the lower gravel and sandstone into the unconfined aquifer, which then replenishes the surface water. In the alluvial lacustrine plain, the deposition of fine sand, medium sand, silt, and clay begins. The type of groundwater transitions from unconfined water to confined water, and the ion contents of both surface water and groundwater begin to increase sharply. Apart from Mg^{2+}, all other ions in surface water increase to 10^3~10^4 mg/L, with Cl^-, SO_4^{2-}, and Na^+ showing the most noticeable increase rates in groundwater, while HCO_3^- shows a peak before decreasing. The low content of Mg^{2+} and Ca^{2+} is due to their precipitation after saturation, where Na^+ replaces Ca^{2+} and Mg^{2+}. The surrounding area of terminal lakes has a more complex lithology, including black silt with fine sand, clay, and silt layers, yellow-brown fine sand, grayish-blue silt, and silt loam [22], which contributes to the increased ion concentration in the lake area. This also further explains the surface water–groundwater transformation relationship described in Section 4.3.1. Additionally, the intense evaporation in the lake area has led to the significant precipitation of ions such as Cl^- in the evaporite rocks of the region [48].

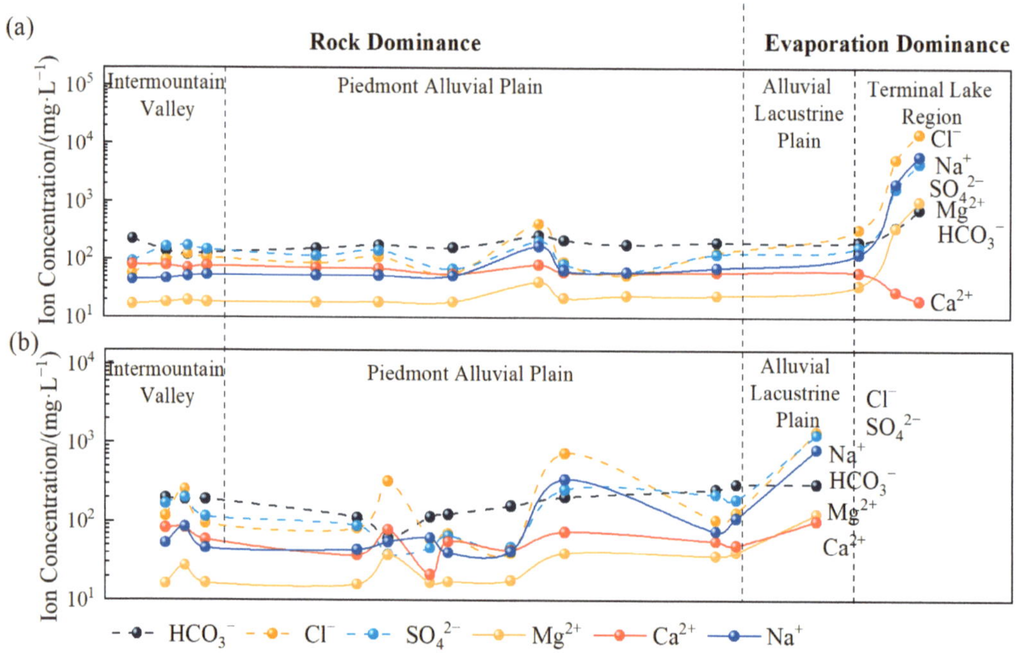

Figure 12. Variation curves of main ions along the Bayin River from the mountain to terminal lake: (**a**) surface water; (**b**) groundwater.

Various types of rocks in mountainous regions serve as the primary sources of salt. Over time, these rocks have undergone weathering and transported salt via surface water–groundwater interactions. As water moves from the mountains to the alluvial plain, the dissolution and precipitation of various substances in groundwater alter the chemical composition of surface water. The strong evaporation in lake areas further concentrates ions, particularly chloride ions. The chemical characteristics of water in the mountainous and alluvial plain regions are mainly influenced by lithological changes, with lower ion concentrations in surface water and groundwater. Downstream in the plains, groundwater carries the filtration process to the surface, making it a dominant factor in the area characterized by groundwater–surface water interactions. In the lake area, intense evaporation leads to significant changes in the chemical properties of both surface water and groundwater, resulting in a substantial increase in ion concentrations, dominated by the effects of evaporation. Therefore, weathering in mountainous areas, precipitation leaching, water–rock interactions during surface water–groundwater conversion processes, and the exchange of chemical components play vital roles in providing salt sources to terminal lakes. This is the reason for the differences in the chemical types of water in Keluke Lake during the summer and winter seasons.

It is evident that the salt content in East Dabson Lake and Lake Tuosu does not solely originate from surrounding freshwater sources. In the central and western parts of the QB, there is a widespread occurrence of oilfield brine in the Jurassic–Cretaceous strata at depths ranging from 500 to 1500 m [49]. Research indicates that chloride-type waters from Qarhan Salt Lake are distributed in the northern and northeastern regions, and the northern edge of the salt lake is marked by a fault distribution (Figure 1b), most likely providing a pathway for deep subsurface water [50]. To analyze the potential sources of replenishment around East Dabson Lake, deep brine water pumped from the northeastern region was collected. The brine water is primarily of the $CaCl_2$ type with a TDS concentration of 281.5~417.2 g/L (Appendix A Tables A2 and A3). In contrast, water recharged by rivers is rich in HCO_3^-, SO_4^{2-}. Therefore, a comparison is made by plotting the Ca^{2+}-SO_4^{2-}-HCO_3^- ternary diagram for different water bodies in the lake area (Figure 13a). It shows that whether in summer or winter, the sampling points of the lake water are closer to those of the brine, suggesting that the lake water receives deep-seated brine. Another compelling piece of evidence is that bromide ions were not detected in any of the river water samples (detection limit: 0.04 mg/L). This would result in all water samples with a Cl^- concentration greater than 40 mg/L exhibiting a Cl/Br ratio exceeding 1000, which may be indicative of the dissolution halite [51,52]. In the brine and lake, the Br ranges from 1.13 to 12.9 mg/L, with a Cl/Br ratio exceeding 10,000, indicating that the lake is supplied with brine.

Further analyses were conducted using the δD-$\delta^{18}O$ relationship (Figure 13b). Water in Zone 1 corresponds to surface and groundwater in the Golmud River watershed and falls along the meteoric water line, indicating that it primarily originates from precipitation in the Kunlun Mountains. Zone 2 represents brine water and winter lake water, both of which fall along the evaporation line, indicating minimal precipitation influence. Summer lake water is distant from Zone 2, suggesting significant evaporation during the summer. The contribution of deep-seated fissure water leads to similar chemical types in East Dabson Lake during both summer and winter seasons. The simultaneous mixing of river water and groundwater leads to the precipitation of $MgSO_4$ evaporites and low sulfate content [53], and its desulfurization process results in a slightly acidic pH.

To analyze the salt sources of Tuosu Lake, the chemical test results of surface water, groundwater, and water from the upstream connected stream, Keluke Lake and rivers are analyzed separately (Table 4). It is observed that the concentrations of F^- and CO_3^{2-} in Lake Tuosu are higher than those in the upstream water bodies, while the concentration of Ca^{2+} is lower, indicating the substitution of F^- with $CaCO_3$. Additionally, during the summer, there is a sudden increase in concentrations of Mg^{2+}, Na^+, Al^{3+}, and Fe^{2+}, suggesting the presence of other rocks in the vicinity of Lake Tuosu that undergo dissolution and precipitation due to water–rock interactions under high-temperature conditions. Previous research indicates that low hills and hillsides with mud volcanoes are distributed on the northeast side of Tuosu Lake, and the formation of Keluke Lake is influenced by the movement of the Himalayas. With the intrusion of magma and the effects of reverse thrust and overthrust structures, significant uplift was seen to occur in the Tertiary strata, leading to the division of Lian Lake into two areas [54]. Therefore, the sources of salt in Tuosu Lake may not only be related to surrounding freshwater replenishment, but also to the presence of lava when the lake formed.

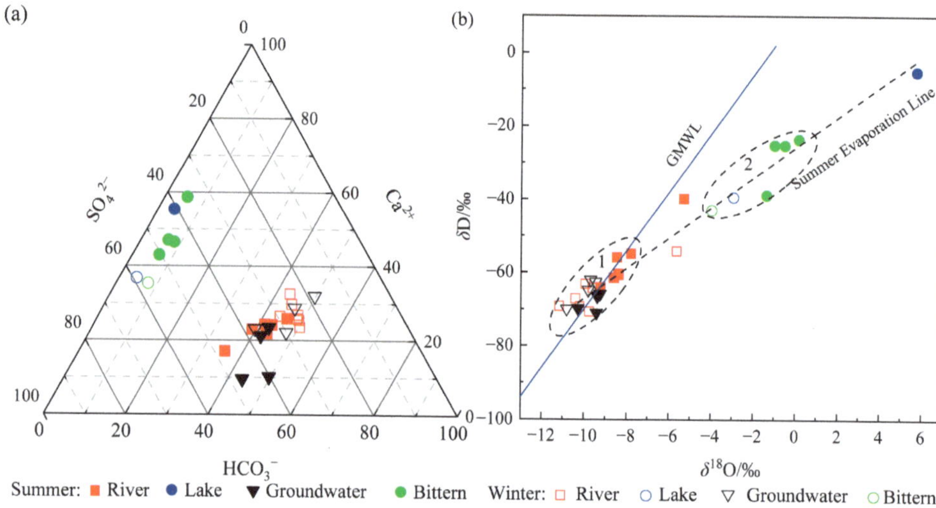

Figure 13. Analysis of different water sources in East Dabson Lake: (**a**) $Ca^{2+}-SO_4^{2-}-HCO_3^-$ ternary diagram; (**b**) $\delta D-\delta^{18}O$ relationship: 1 represents the area where river water and groundwater sampling points are located, and 2 represents the area where lake water and brine sampling points are located.

Table 4. Ion concentrations of different water bodies in the Bayin River watershed. Unit: mg/L.

Sample Name/Type	Season	F^-	CO_3^{2-}	Ca^{2+}	Mg^{2+}	Na^+	Al^{3+}	Fe^{2+}
Tuosu Lake	Summer	0.65	200.02	19.60	1019.00	6008	7.32	4×10^{-3}
	Winter	1.25	68.44	47.34	435.30	1709	0	0
Connected stream	Summer	0.31	0	27.73	465.60	2048	0	7×10^{-4}
	Winter	0.85	20.96	69.06	73.20	177.30	0	0
Keluke Lake	Summer	0.37	0	59.81	51.57	123.50	0	0
	Winter	0.79	0	88.11	50.64	100.30	0	0
River water (mean)	Summer	0.23	0	68.40	33.59	64.48	0	7.05×10^{-3}
	Winter	1.01	0	72.57	45.28	70.14	0	4.55×10^{-5}
Groundwater (mean)	Summer	0.28	0	61.42	52.34	147.29	0	6.98×10^{-2}
	Winter	1.46	1.28	60.16	78.32	211.51	0	3.67×10^{-4}

5. Conclusions

Based on the sampling, testing, and research conducted on the hydrochemistry and isotopic spatiotemporal variations in river water, lake water, and groundwater in five typical watersheds of the QB, the following main conclusions have been drawn:

(1) Spatially, there are significant differences in pH and HCO_3^- across various watersheds. The hydrochemical types of river water and groundwater in the northern region are HCO_3·Cl·SO_4-Na·Ca, while those in the southern region are of the Cl·SO_4-Na·Ca type. The terminal lake of Bayin River is of the Cl·SO_4-Na type, and Golmud River is of the Cl-Na·Mg type. These differences arise primarily due to significant variations in the lithology of the aquifers. The southern basin is more strongly influenced by the dissolution and evaporation of saline rocks, while the northern basin is influenced by carbonate weathering. There are significant differences in H-O isotopes, as the water vapor sources in the southern and northern basins are different. The southern basin

receives a greater contribution of water vapor transported by the mid-latitude westerlies, while water vapor in the northern basin mainly results from local evaporation;

(2) Temporally, there are distinct seasonal differences in water temperatures and $\delta^{18}O$ in the two watersheds, as well as TDS, HCO_3^-, and δD in the Golmud River watershed. River water, groundwater, and water from Keluke Lake exhibit significant seasonal differences with respect to hydrochemical types. During the winter, there is a depletion of Ca^{2+}, Cl^-, and SO_4^{2-} and an enrichment of HCO_3^-, Mg^{2+}, and Na^+ compared to the summer. These variations arise from different hydrochemical influences in the summer and winter seasons, with the water body being more affected by water–rock interactions in the summer. There are significant seasonal differences in H-O isotopes in Golmud River due to the substantial seasonal differences in the effective evaporation rate and the water vapor recycling rate in the southern basin;

(3) Similar characteristics of river water and groundwater are attributed to frequent transformations between surface water and groundwater from the river mouth to the terminal lake. Geological structures, strata, and topography are controlling factors, and seasonal hydro-meteorological conditions influence localized transformation relationships. The minimal seasonal impact on the salt content of terminal lakes is mainly due to the effects of salt accumulation, and this is supplemented via other sources of groundwater. The leaching action of upstream river water and groundwater and intense downstream evaporation contribute to a portion of the salt content in the terminal lake. However, East Dabson Lake is also influenced by deep-seated fissure water from the northern oil fields, while the salt source of Tuosu Lake may be related to the surrounding volcanoes present during the formation of the lake.

Author Contributions: Conceptualization, H.Y. and J.W.; methodology, H.Y. and K.S.; formal analysis, H.Y.; investigation, H.Y.; data curation, H.Y. and K.S.; writing—original draft preparation, H.Y.; writing—review and editing, J.W. All authors have read and agreed to the published version of the manuscript.

Funding: This research was funded by the Major Science and Technology Project of Qinghai Province, titled "rediction and Risk Assessment of Key River-Lake Water Changes on the Qinghai-Tibet Plateau (Qinghai)" grant number 2021-SF-A6.

Data Availability Statement: Data are contained within the article and appendix.

Acknowledgments: Thanks to the Hydrological and Water Resources Survey Center and the Natural Resources Museum of Qinghai Province for providing data and information support.

Conflicts of Interest: The authors declare no conflicts of interest.

Appendix A

Table A1. Supplementing the collection of precipitation H-O isotope data.

Sample Number	Location	Longitude	Latitude	Sampling Date	$\delta^{18}O$/‰	δD/‰	Attribution
P1	Golmud	95.091	36.346	1 July 2010	1.89	21.06	
P2	Golmud	95.091	36.346	5 July 2010	−1.92	−4.60	
P3	Golmud	95.091	36.346	5 July 2010	−1.94	−6.28	
P4	Golmud	95.091	36.346	8 July 2010	−7.79	−56.96	
P5	Golmud	95.091	36.346	28 July 2010	−10.48	−95.39	
P6	Golmud	95.091	36.346	2 August 2010	−5.95	−48.53	
P7	Golmud	95.091	36.346	10 August 2010	−0.24	−28.08	[45]
P8	Golmud	95.091	36.346	16 August 2010	−6.93	−67.51	
P9	Golmud	95.091	36.346	15 September 2010	−0.32	−19.69	
P10	Golmud	95.091	36.346	18 September 2010	−14.88	−110.38	
P11	Golmud	95.091	36.346	20 September 2010	−12.02	−124.00	
P12	Golmud	95.091	36.346	24 September 2010	−16.23	−125.42	
P13	Golmud	95.091	36.346	27 September 2010	−9.01	−66.43	

Table A1. *Cont.*

Sample Number	Location	Longitude	Latitude	Sampling Date	$\delta^{18}O$/‰	δD/‰	Attribution
P14	Golmud	94.563	35.874	June to September 2019	−10.77	−67.21	[43]
P15	Delingha	96.82	37.48	26 July 2019	−3.28	−17.54	
P16	Delingha	96.82	37.48	27 July 2019	−2.39	−13.19	
P17	Delingha	96.82	37.48	29 July 2019	1.09	11.74	
P18	Delingha	96.82	37.48	3 August 2019	−0.08	−4.19	
P19	Delingha	96.82	37.48	4 August 2019	−2.00	−6.62	
P20	Delingha	96.82	37.48	5 August 2019	−2.25	−22.85	
P21	Delingha	96.82	37.48	6 August 2019	−5.42	−37.98	
P22	Delingha	96.82	37.48	11 August 2019	−4.12	−18.06	
P23	Delingha	96.82	37.48	12 August 2019	−2.22	2.62	
P24	Delingha	96.82	37.48	27 August 2019	−2.28	−21.45	
P25	Delingha	96.82	37.48	29 August 2019	−6.55	−37.88	
P26	Delingha	97.464	37.902	1 July 2018	−1.81	−16.24	
P27	Delingha	97.345	37.366	1 July 2018	6.43	−45.32	This article
P28	Delingha	97.345	37.366	3 June 2022	0.00	10.45	
P29	Delingha	97.345	37.366	7 September 2022	−2.26	7.31	
P30	Delingha	97.345	37.366	24 August 2022	−14.10	−93.61	
P31	Delingha	97.345	37.366	22 August 2022	−15.55	−113.32	
P32	Delingha	97.345	37.366	20 June 2022	−5.15	−19.22	
P33	Delingha	97.345	37.366	7 July 2022	−5.23	−23.33	
P34	Delingha	97.345	37.366	25 June 2022	0.31	15.60	
P35	Delingha	97.345	37.366	16 July 2022	−4.51	−14.92	
P36	Delingha	97.345	37.366	18 July 2022	−9.32	−48.90	
P37	Delingha	97.506	37.933	26 May 20222022/5/26	−7.26	−40.40	
P38	Delingha	97.370	37.319	1 December, 2018	−34.14	−241.71	
P39	Delingha	97.370	37.319	1 December, 2018	−23.43	−177.03	
P40	Delingha	97.284	37.349	June to September 2010	−7.85	−51.00	[45]

Table A2. Statistical results of water temperature (T), pH, TDS, SAL values, ionic concentrations and isotopic compositions of various water bodies in five typical watersheds of Qaidam Basin in summer (the unit of T is °C, the unit of SAL is ppt, the unit of ionic concentration and TDS is mg/L, the unit of $\delta^{18}O$ and δD is ‰).

Watershed	Value	T/°C	pH	TDS	SAL	F^-	Cl^-	NO_3^-	SO_4^{2-}	CO_3^{2-}	HCO_3^-	Ca^{2+}	K^+	Mg^{2+}	Na^+	δD	$\delta^{18}O$
Yuqia River	Min.	7.9	7.7	162.2	0.1	0.2	39.5	0.0	12.6	0.0	42.0	10.3	1.0	3.1	33.1	−78.7	−11.4
	Max.	20.1	9.9	978.6	0.8	22.2	415.9	26.5	450.2	13.9	309.8	107.2	11.8	69.5	261.6	−66.8	−9.9
	Mean	11.9	8.5	462.1	0.4	1.3	129.8	3.6	137.9	0.8	142.0	47.9	3.9	17.4	92.6	−71.9	−10.7
	SD	3.0	0.4	210.1	0.2	3.6	95.6	5.0	119.9	3.2	45.9	19.5	2.4	13.0	63.9	3.4	0.4
Bayin River	Min.	8.1	7.6	190.0	0.0	0.0	22.7	0.0	37.6	0.0	61.2	19.6	1.8	14.0	38.8	−76.5	−10.4
	Max.	25.6	9.7	18,152.7	17.2	21.1	8306.0	162.4	5956.5	200.0	731.6	251.4	136.0	1019.0	6008.0	0.0	1.6
	Mean	13.5	8.2	1020.9	0.8	0.8	378.3	11.9	315.6	3.2	195.4	64.8	7.5	51.2	246.6	−56.9	−8.2
	SD	4.5	0.4	2366.9	2.2	2.9	1189.4	23.3	848.2	24.8	89.0	30.5	18.8	130.5	783.5	10.2	1.7
Chanhanwusu River	Min.	7.1	7.8	849.0	0.7	0.3	263.4	0.1	10.2	0.0	32.9	83.3	6.4	14.8	189.9	−90.3	−12.4
	Max.	19.6	8.9	1556.8	1.2	4.7	1077.6	34.6	604.4	0.0	258.2	175.0	9.4	36.7	474.1	−64.2	−9.7
	Mean	11.0	8.2	1123.0	0.9	0.7	547.7	11.2	332.5	0.0	180.8	111.4	7.4	24.6	260.2	−78.2	−11.1
	SD	3.1	0.3	201.3	0.2	0.9	220.8	9.1	152.3	0.0	55.6	21.1	0.8	5.6	71.1	5.6	0.6
Xiangride River	Min.	4.0	7.6	163.5	0.1	0.1	6.3	0.0	8.8	0.0	43.0	15.6	4.2	13.7	88.2	−73.2	−10.5
	Max.	23.3	9.3	1202.5	0.9	2.9	356.9	50.9	606.3	13.9	330.4	113.0	13.9	63.4	241.4	−51.6	−5.6
	Mean	12.7	8.4	689.9	0.5	0.4	189.6	8.8	241.9	1.3	196.4	62.9	6.0	34.9	122.6	−64.2	−8.9
	SD	4.4	0.5	189.6	0.2	0.5	68.9	10.2	112.1	3.9	61.0	23.7	2.1	9.9	35.9	5.5	1.2
Golmud River	Min.	7.8	7.2	378.8	0.3	0.1	56.2	0.0	61.1	0.0	39.8	6.6	2.8	20.3	50.3	−71.2	−10.3
	Max.	20.7	9.8	282,277.1	261.8	19.7	173,154.0	256.2	5813.7	31.0	707.8	3290.0	3248.0	48,900.0	81,780.0	−5.1	5.7
	Mean	13.9	8.6	9829.8	9.0	1.0	5839.5	20.2	415.7	2.7	196.0	147.8	110.9	1617.2	2775.7	−61.0	−8.3
	SD	3.5	0.6	50,565.6	46.9	3.5	31,053.0	62.6	1039.7	7.2	111.4	583.7	582.2	8775.4	14,662.9	12.2	2.8
Qaidam Basin (5 watersheds)	Min.	4.0	7.2	162.2	0.0	0.0	6.3	0.0	8.8	0.0	32.9	6.6	1.0	3.1	33.1	−90.3	−12.4
	Max.	25.6	9.9	282,277.1	261.8	22.2	173,154.0	256.2	5956.5	200.0	731.6	3290.0	3248.0	48,900.0	81,780.0	0.0	5.7
	Mean	12.8	8.4	2324.4	2.0	0.8	1215.3	11.1	286.8	2.0	184.1	80.2	23.6	297.0	612.7	−64.2	−9.2
	SD	4.0	0.5	20,630.6	19.1	2.7	12,627.4	29.5	659.5	15.0	80.9	237.5	236.7	3564.5	5969.7	11.1	2.0
Brine	Min.	17.6	6.7	281,468.7	253.2	12.2	180,546.5	22.3	2994.7	0.0	704.6	1289.4	3790.0	32,520.0	46,280.0	−38.8	−1.4
	Max.	20.3	7.8	417,249.5	375.4	17.4	211,549.5	32.6	4409.0	0.0	795.6	2862.0	12,604.0	57,340.0	74,300.0	−23.6	0.2
	Mean	19.0	7.1	369,910.7	332.8	14.3	198,443.8	28.1	3907.2	0.0	741.4	1857.8	7497.5	48,160.0	64,035.0	−28.2	−0.7
	SD	1.2	0.5	62,814.4	56.5	2.4	13,318.2	4.5	628.2	0.0	39.1	691.8	3694.2	11,434.6	12,441.1	7.1	0.7

Table A3. Statistical results of water temperature (T), pH, TDS, SAL values, ionic concentrations and isotopic composition of various water bodies in Bayin River and Golmud River watershed in winter (the unit of T is °C, the unit of SAL is ppt, the unit of ionic concentration and TDS is mg/L, and the unit of $\delta^{18}O$ and δD is ‰).

Watershed	Value	T	pH	TDS	SAL	F^-	Cl^-	NO_3^-	SO_4^{2-}	CO_3^{2-}	HCO_3^-	Ca^{2+}	K^+	Mg^{2+}	Na^+	δD	$\delta^{18}O$
Bayin River	Min.	−0.2	6.9	344.1	0.0	0.7	43.5	0.3	32.0	0.0	36.4	27.1	3.0	18.4	42.5	−94.2	−13.1
	Max.	8.1	9.7	5508.9	4.6	3.1	1848.3	11.1	1027.3	68.4	335.4	120.6	41.2	298.9	1709.0	−40.3	−5.4
	Mean	3.1	8.4	888.2	0.6	1.2	232.9	5.7	151.6	4.0	206.7	66.7	6.5	51.6	204.1	−59.5	−9.5
	SD	3.0	0.5	1087.8	1.0	0.6	396.5	2.7	216.8	14.1	65.4	26.7	7.5	60.4	370.0	9.6	1.4
Golmud River	Min.	−5.5	7.9	479.2	0.4	0.8	68.3	0.5	45.3	0.0	61.4	11.8	4.7	32.5	84.6	−72.5	−11.5
	Max.	9.1	9.4	195,635.7	175.8	4.9	103,499.3	22.8	4820.8	111.0	432.5	1255.0	2562.0	25,500.0	26,990.0	−39.5	−3.0
	Mean	2.1	8.7	10,966.5	9.8	1.4	5598.3	6.5	342.0	17.2	236.6	106.6	192.8	1388.6	1571.8	−64.9	−9.6
	SD	3.6	0.4	44,720.2	40.2	0.9	23,708.0	5.4	1085.0	25.0	66.6	278.3	617.1	5838.9	6155.6	7.5	2.0
Brine	Value	2.2	7.3	274,318.1	253.8	172,651.2	349.5	44.3	3924.9	0.0	671.4	1018.0	3744.0	30,620.0	64,840.0	−43.0	−4.0

References

1. Yin, J.; Wei, J.; Li, Q.; Ayantobo, O.O. Regional Characteristics and Impact Factors of Change in Terrestrial Water Storage in Northwestern China from 2002 to 2020. *IEEE J. Sel. Top. Appl. Earth Obs. Remote Sens.* **2023**, *16*, 386–398. [CrossRef]
2. Chen, Y.; Li, Z.; Xu, J.; Shen, Y.; Xing, X.; Xie, T.; Li, Z.; Yang, L.; Xi, H.; Zhu, C.; et al. Changes and Protection Suggestions in Water Resources and Ecological Environment in Arid Region of Northwest China. *Bull. Chin. Acad. Sci.* **2023**, *38*, 385–393. [CrossRef]
3. Abbas, M.; Zhao, L.; Wang, Y. Perspective Impact on Water Environment and Hydrological Regime Owing to Climate Change: A Review. *Hydrology* **2022**, *9*, 203. [CrossRef]
4. Zhang, X.; Chen, J.; Chen, J.; Ma, F.; Wang, T. Lake Expansion under the Groundwater Contribution in Qaidam Basin, China. *Remote Sens.* **2022**, *14*, 1756. [CrossRef]
5. Wei, L.; Jiang, S.; Ren, L.; Tan, H.; Ta, W.; Liu, Y.; Yang, X.; Zhang, L.; Duan, Z. Spatiotemporal changes of terrestrial water storage and possible causes in the closed Qaidam Basin, China using GRACE and GRACE Follow-On data. *J. Hydrol.* **2021**, *598*, 126274. [CrossRef]
6. Fan, Q.; Cheng, Y.; Chen, T.; Xu, N. Influence of Water Replenishment of Lakes in the Northern Qaidam Basin on Salt Lake Resources and Ecological Environment. *J. Salt Lake Res.* **2022**, *30*, 11–18.
7. Zhao, D.; Wang, G.; Liao, F.; Yang, N.; Jiang, W.; Guo, L.; Liu, C.; Shi, Z. Groundwater-surface water interactions derived by hydrochemical and isotopic (^{222}Rn, deuterium, oxygen-18) tracers in the Nomhon area, Qaidam Basin, NW China. *J. Hydrol.* **2018**, *565*, 650–661. [CrossRef]
8. Lv, A.; Zhang, C. Analysis of the Characteristics and Driving Forces of Changes in Lake Water Volume in Inland Arid Basins in China. *Water* **2022**, *14*, 3141. [CrossRef]
9. Yang, N.; Wang, G.; Shi, Z.; Zhao, D.; Jiang, W.; Guo, L.; Liao, F.; Zhou, P. Application of Multiple Approaches to Investigate the Hydrochemistry Evolution of Groundwater in an Arid Region: Nomhon, Northwestern China. *Water* **2018**, *10*, 1667. [CrossRef]
10. Wang, Y.; Guo, H.; Li, J. Investigation and evaluation of groundwater resources and environmental problems in Qaidam Basin. In *Investigation and Evaluation of Groundwater Resources and Environmental Problems in Qaidam Basin*; Geological Publishing House: Beijing, China, 2008; pp. 49–211.
11. Yang, N.; Zhou, P.; Wang, G.; Zhang, B.; Shi, Z.; Liao, F.; Li, B.; Chen, X.; Guo, L.; Dang, X.; et al. Hydrochemical and isotopic interpretation of interactions between surface water and groundwater in Delingha, Northwest China. *J. Hydrol.* **2021**, *598*, 126243. [CrossRef]
12. Xiao, Y.; Shao, J.; Cui, Y.; Zhang, G.; Zhang, Q. Groundwater circulation and hydrogeochemical evolution in Nomhon of Qaidam Basin, northwest China. *J. Earth Syst. Sci.* **2017**, *126*, 26. [CrossRef]
13. Han, J.; Xu, J.; Yi, L.; Chang, Z.; Wang, J.; Ma, H.; Zhang, B.; Jiang, H. Seasonal Interaction of River Water-Groundwater-Salt Lake Brine and Its Influence on Water-Salt Balance in the Nalenggele River Catchment in Qaidam Basin, NW China. *J. Earth Sci.* **2022**, *33*, 1298–1308. [CrossRef]
14. Zhang, C.; Lv, A.; Jia, S.; Qi, S. Longterm multisource satellite data fusion reveals dynamic expansion of lake water area and storage in a hyperarid basin of China. *J. Hydrol.* **2022**, *610*, 127888. [CrossRef]
15. Li, W.; He, Q. Discussion on the Origins of the Materials in Qarhan Salt Lake. *J. Hebei Coll. Geol.* **1993**, *16*, 254–263.
16. Zhang, X.; Fan, Q.; Li, Q.; Du, Y.; Qin, Z.; Wei, H.; Shan, F. The Source, Distribution, and Sedimentary Pattern of K-Rich Brines in the Qaidam Basin, Western China. *Minerals* **2019**, *9*, 655. [CrossRef]
17. Yi, J.; Fan, Q.; Wei, H.; Wang, M.; Qin, Z.; Yuan, Q.; Shan, F. The Characteristics of the Mineral Assemblage and the Forming Clue in Qarhan Salt Lake. *J. Salt Lake Res.* **2017**, *25*, 47–54.
18. Li, J.; Chen, F.; Ling, Z.; Li, T. Lithium sources in oilfield waters from the Qaidam Basin, Tibetan Plateau: Geochemical and Li isotopic evidence. *Ore Geol. Rev.* **2021**, *139*, 104481. [CrossRef]
19. Chen, L.; Ma, T.; Ma, J.; Du, Y.; Xiao, C. Identification of material source for the salt lakes in the Qaidam Basin. *Hydrogeol. Eng. Geol.* **2015**, *42*, 101–107. [CrossRef]
20. Zhang, X.; Li, Q.; Qin, Z.; Fan, Q.; Du, Y.; Wei, H.; Gao, D.; Shan, F. Boron isotope geochemistry of a brine-carbonate system in the Qaidam Basin, western China. *Sediment. Geol.* **2019**, *383*, 293–302. [CrossRef]
21. Li, Q.; Fan, Q.; Wang, J.; Qin, Z.; Zhang, X.; Wei, H.; Du, Y.; Shan, F. Hydrochemistry, Distribution and Formation of Lithium-Rich Brines in Salt Lakes on the Qinghai-Tibetan Plateau. *Minerals* **2019**, *9*, 528. [CrossRef]
22. Zhu, Y.; Zhong, J.; Li, W. The Neotectonic Movement and the Evolution of Saline Lakes of Qaidam Basin in Northwestern China. In *The Neotectonic Movement and the Evolution of Saline Lakes of Qaidam Basin in Northwestern China*; Geological Publishing House: Beijing, China, 1994; pp. 90–99.
23. Hussainzadeh, J.; Samani, S.; Mahaqi, A. Investigation of the geochemical evolution of groundwater resources in the Zanjan plain, NW Iran. *Environ. Earth Sci* **2023**, *82*, 123. [CrossRef]
24. Kileshye Onema, J.-M.; Taigbenu, A.E.; Ndiritu, J. Classification and flow prediction in a data-scarce watershed of the equatorial Nile region. *Hydrol. Earth Syst. Sci.* **2012**, *16*, 1435–1443. [CrossRef]
25. Yang, H.; Wei, J.; Ren, Q. Interaction between surface water and groundwater and hydrochemical characteristics in the typical watersheds of the Qaidam Basin. *Arid Zone Res.* **2022**, *39*, 1543–1554. [CrossRef]
26. Alomari, A.H.; Carvalho, F.P.; Saleh, M.A.; Al-Malkawi, G.; Hashim, S.; Bilbiesy, E.; Banikhalaf, R.; Al-Sayaheen, A.; Abosalim, F.; Sharat, D.; et al. Seasonal Variation In Major Ions Chemistry and Radionuclide Concentrations In Groundwater: A Case Study In The Amman-Zarqa Basin (Jordan). *Water Air Soil Pollut.* **2023**, *234*, 256. [CrossRef]

27. Caporaso, J.G.; Kuczynski, J.; Stombaugh, J.; Bittinger, K.; Bushman, F.D.; Costello, E.K.; Fierer, N.; Peña, A.G.; Goodrich, J.K.; Gordon, J.I.; et al. QIIME allows analysis of high-throughput community sequencing data. *Nat. Methods* **2010**, *7*, 335–336. [CrossRef] [PubMed]
28. Piper, A.M. A graphic procedure in the geochemical interpretation of water-analyses. *Trans. AGU* **1944**, *25*, 914. [CrossRef]
29. Shi, K.; Li, T.; Zhao, J.; Su, Y.; Gao, J.; Li, J. Atmospheric recycling of agricultural evapotranspiration in the Tarim Basin. *Front. Earth Sci.* **2022**, *10*, 950299. [CrossRef]
30. Zhang, Y.; Xu, J.; Yan, Y.; Su, W.; Luo, P.; Yu, H. Characteristics and Moisture Sources of the StableIsotopes in Precipitation at the Monsoon MarginalRegion of North-Central China. *Pol. J. Environ. Stud.* **2022**, *31*, 1963–1974. [CrossRef]
31. Gibbs, R.J. Mechanisms Controlling World Water Chemistry. *Sci. New Ser.* **1970**, *170*, 1088–1090. [CrossRef]
32. Van Der Ent, R.J.; Savenije, H.H.G.; Schaefli, B.; Steele-Dunne, S.C. Origin and fate of atmospheric moisture over continents. *Water Resour. Res.* **2010**, *46*, 2010WR009127. [CrossRef]
33. Van Der Ent, R.J.; Tuinenburg, O.A.; Knoche, H.-R.; Kunstmann, H.; Savenije, H.H.G. Should we use a simple or complex model for moisture recycling and atmospheric moisture tracking? *Hydrol. Earth Syst. Sci.* **2013**, *17*, 4869–4884. [CrossRef]
34. Shi, K.; Lang, Q.; Huang, Y.; Zhao, J.; Wang, H.; Chen, G.; Zhang, Q.; Yin, H.; Su, Y.; Wang, P. Analysis of the water-vapor sources in rainstorm processes in TIANJIN city based on the trajectory method. *Atmos. Sci. Lett.* **2023**, e1196. [CrossRef]
35. Yang, K.; Tang, Q.; Lu, H. Analysis of precipitation recirculation rate and water vapor sources in Tibetan Plateau. *Sci. Sin. Terrae* **2022**, *52*, 574–578.
36. Miao, Q.; Li, X.; Xu, Y.; Liu, C.; Xie, R.; Lv, Z. Chemical characteristics of groundwater and source identification in a coastal city. *PLoS ONE* **2021**, *16*, e0256360. [CrossRef] [PubMed]
37. Yang, N.; Wang, G. Spatial variation of water stable isotopes of multiple rivers in southeastern Qaidam Basin, northeast Qinghai-Tibetan Plateau: Insights into hydrologic cycle. *J. Hydrol.* **2024**, *628*, 130464. [CrossRef]
38. Zhang, Y.; Tan, H.; Cong, P.; Shi, D.; Rao, W.; Zhang, X. Isotopic variations in surface waters and groundwaters of an extremely arid basin and their responses to climate change. *Hydrol. Earth Syst. Sci.* **2023**, *27*, 4019–4038. [CrossRef]
39. Dansgaard, W. Stable isotopes in precipitation. *Tellus* **1964**, *16*, 436–468. [CrossRef]
40. Gu, W. Isotope Hydrology. In *Isotope Hydrology*; Science Press: Beijing, China, 2011; pp. 150–173.
41. Froehlich, K.; Gibson, J.J. Deuterium excess in precipitation and its climatological significance. In *Study of Environmentalchange Using Isotope Techniques*; C&S Papers Series 13; International Atomic Energy Agency: Vienna, Austria, 2002; pp. 54–65.
42. Yang, M.; Yao, T.; Tian, L.; Lu, A. Analysis of Precipitation From Different Water Vapor Sources in Tibetan Plateau. *Sci. Geogr. Sin.* **2004**, *24*, 426–431.
43. Rao, W.; Li, Y.; Tan, H.; Li, Y.; Wen, C.; Zhang, X. Stable hydrogen-oxygen isotope composition and atmospheric moisture sources of precipitation in an arid-alpine region: A case study of the Golmud River Watershed on the north slope of the Kunlun Mountains. *JHE* **2021**, *52*, 1116–1125. [CrossRef]
44. Yang, N.; Wang, G. Moisture sources and climate evolution during the last 30 kyr in northeastern Tibetan Plateau: Insights from groundwater isotopes (^{2}H, ^{18}O, ^{3}H and ^{14}C) and water vapour trajectories modeling. *Quat. Sci. Rev.* **2020**, *242*, 106426. [CrossRef]
45. Zhu, J.; Chen, H.; Gong, G. Hydrogen and Oxygen Isotopic ompositions of Precipitation and Its Water Vapor Sources in Eastern Qaidam Basin. *Environ. Sci.* **2015**, *36*, 2784–2790. [CrossRef]
46. Kiyani, V.; Esmaili, A.; Alijani, F.; Samani, S.; Vasić, L. Investigation of drainage structures in the karst aquifer system through turbidity anomaly, hydrological, geochemical and stable isotope analysis (Kiyan springs, western Iran). *Environ. Earth Sci.* **2022**, *81*, 517. [CrossRef]
47. Zhao, Z.; Liu, Z.; Xu, W. *Hydrogeological Survey of Alluvial Fan Water Supply in Bayin River, Delingha City, Qinghai Province*; Qinghai Environmental Geology Exploration Bureau: Xining, China, 2017; pp. 35–42.
48. Zhang, B.; Zhao, D.; Zhou, P.; Qu, S.; Liao, F.; Wang, G. Hydrochemical Characteristics of Groundwater and Dominant Water–Rock Interactions in the Delingha Area, Qaidam Basin, Northwest China. *Water* **2020**, *12*, 836. [CrossRef]
49. Fan, Q.; Ma, H.; Tan, H.; Li, T. Geochemistry Characteristics of Sulfur Isotope in Oilfield Brine of the Western Qaidam Basin. *Bull. Mineral. Petrol. Geochem.* **2009**, *28*, 137–142.
50. Duan, Z.; Yuan, J. The Origin of Saline Materials of Qarhan Salt Lake in Qaidam Basin, China. *Geoscience* **1988**, *2*, 420–428.
51. Davis, S.N.; Whittemore, D.O.; Fabryka-Martin, J. Use of ChlorideBromide Ratios in Studies of Potable Water. *Groundwater* **1988**, *36*, 338–350. [CrossRef]
52. Alcalá, F.J.; Custodio, E. Using the Cl/Br ratio as a tracer to identify the origin of salinity in aquifers in Spain and Portugal. *J. Hydrol.* **2008**, *359*, 189–207. [CrossRef]
53. Li, Q.; Fan, Q.; Wei, H.; Qin, Z.; Zhang, X.; Du, Y.; Shan, F. Sulfur isotope constraints on the formation of $MgSO_4$−deficient evaporites in the Qarhan salt Lake, western China. *J. Asian Earth Sci.* **2020**, *189*, 104160. [CrossRef]
54. Yao, S.; Zhang, J.; Li, W.; Li, H. Preliminary Investigation and Study on the Characteristics of Mud Volcanoes and the Cause of Tuosuo and Keluke Lakes in Lian Lake Area, Delingha. *Northwest. Geol.* **2016**, *49*, 155–163.

sustainability

MDPI

Article

Hydrogeochemical Insights into the Sustainable Prospects of Groundwater Resources in an Alpine Irrigation Area on Tibetan Plateau

Shaokang Yang [1,2,3], Zhen Zhao [1,2,3,*], Shengbin Wang [1,2,3], Shanhu Xiao [1,2,3], Yong Xiao [4,5,6,*], Jie Wang [4,7], Jianhui Wang [1,2,3], Youjin Yuan [1,2,3], Ruishou Ba [1,2,3], Ning Wang [8], Yuqing Zhang [4,7], Liwei Wang [4,7] and Hongjie Yang [4,7]

1 Bureau of Qinghai Environmental Geological Prospecting, Xi'ning 810007, China
2 Key Lab of Geo-Environment of Qinghai Province, Xi'ning 810007, China
3 Qinghai 906 Engineering Survey and Design Institute Co., Ltd., Xi'ning 810007, China
4 Faculty of Geosciences and Engineering, Southwest Jiaotong University, Chengdu 611756, China
5 Fujian Provincial Key Laboratory of Water Cycling and Eco-Geological Processes, Xiamen 361021, China
6 MOE Key Laboratory of Groundwater Circulation and Environmental Evolution, China University of Geosciences (Beijing), Beijing 100083, China
7 Sichuan Province Engineering Technology Research Center of Ecological Mitigation of Geohazards in Tibet Plateau Transportation Corridors, Chengdu 611756, China
8 School of Water and Environment, Chang'an University, Xi'an 710054, China
* Correspondence: zhaozhen_906@163.com (Z.Z.); xiaoyong@swjtu.edu.cn (Y.X.)

Abstract: The Tibetan Plateau is the "Asia Water Tower" and is pivotal for Asia and the whole world. Groundwater is essential for sustainable development in its alpine regions, yet its chemical quality increasingly limits its usability. The present research examines the hydrochemical characteristics and origins of phreatic groundwater in alpine irrigation areas. The study probes the chemical signatures, quality, and regulatory mechanisms of phreatic groundwater in a representative alpine irrigation area of the Tibetan Plateau. The findings indicate that the phreatic groundwater maintains a slightly alkaline and fresh status, with pH values ranging from 7.07 to 8.06 and Total Dissolved Solids (TDS) between 300.25 and 638.38 mg/L. The hydrochemical composition of phreatic groundwater is mainly HCO_3-Ca type, with a minority of HCO_3-Na·Ca types, closely mirroring the profile of river water. Nitrogen contaminants, including NO_3^-, NO_2^-, and NH_4^+, exhibit considerable concentration fluctuations within the phreatic aquifer. Approximately 9.09% of the sampled groundwaters exceed the NO_2^- threshold of 0.02 mg/L, and 28.57% surpass the NH_4^+ limit of 0.2 mg/L for potable water standards. All sampled groundwaters are below the permissible limit of NO_3^- (50 mg/L). Phreatic groundwater exhibits relatively good potability, as assessed by the entropy-weighted water quality index (EWQI), with 95.24% of groundwaters having an EWQI value below 100. However, the potential health risks associated with elevated NO_3^- levels, rather than NO_2^- and NH_4^+, merit attention when such water is consumed by minors at certain sporadic sampling locations. Phreatic groundwater does not present sodium hazards or soil permeability damage, yet salinity hazards require attention. The hydrochemical makeup of phreatic groundwater is primarily dictated by rock–water interactions, such as silicate weathering and cation exchange reactions, with occasional influences from the dissolution of evaporites and carbonates, as well as reverse cation-exchange processes. While agricultural activities have not caused a notable rise in salinity, they are the main contributors to nitrogen pollution in the study area's phreatic groundwater. Agricultural-derived nitrogen pollutants require vigilant monitoring to avert extensive deterioration of groundwater quality and to ensure the sustainable management of groundwater resources in alpine areas.

Keywords: hydrochemistry; groundwater quality; nitrogen contamination; agricultural pollution; alpine region

1. Introduction

Alpine regions, namely high mountain regions, are of great importance to the whole world as they are the source of water for almost all watersheds [1–3]. It is estimated that about 25% of the world's population will rely on the water resources of alpine regions by 2050 [4,5]. Although water resources are abundant in the world's alpine regions, the majority of water resources are in the form of solid water (such as snow and ice) [6] and are difficult for human society to use directly. The only water resource that is easily accessible to humans is liquid freshwater, including surface water (river and lake water, etc.) [7] and groundwater [8]. Meanwhile, the surface water of rivers and lakes is generally unstable [9] and difficult to use as a sustainable water supply for the human community [10]. Groundwater is the ideal source of water for human society because it is a good buffer against external changes [11–13]. Thus, a deep understanding of the fresh liquid water beneath the ground surface is crucial for a sustainable water supply for human society in alpine regions.

Groundwater availability is constrained by both water quantity and water quality. The alpine region is usually rich in water resources due to abundant precipitation and snow/glacier melt water, especially in the basins or valleys [14]. Thus, water quantity is not the most dominant factor influencing the availability of groundwater in alpine regions. On the contrary, water quality is becoming an increasingly important factor limiting groundwater availability [15]. Naturally, groundwater quality in alpine regions is determined by many geogenic toxic elements (such as fluoride and arsenic) [16,17] or potential high salinity [18,19]. In addition, climate change is the most important global concern in all aspects, including the hydrosphere and related spheres [20–24]. The alpine zone is experiencing significant warming and increased humidity, which directly affects groundwater quality and the hydrological cycle [18,24–27]. The melting of glaciers and the degradation of permafrost have resulted in enhanced weathering and dissolution of minerals within mountainous regions and permafrost layers. Consequently, the hydraulic connectivity among various water bodies has been progressively augmented [28–30]. Numerous studies have demonstrated that the hydrological cycle's mechanisms have become increasingly complex due to these influences, leading to alterations in the hydrochemical characteristics and the patterns of hydrochemical quality within groundwater systems [28,31,32].

Furthermore, amidst rapid socio-economic development and population expansion, human activities have emerged as a significant factor affecting water ecosystems [15]. It has been documented that anthropogenic substances are present in numerous aquifers globally [33–36]. Groundwater contamination by toxic substances is an escalating environmental concern due to their potential detrimental effects on human health and aquatic ecosystems [37–39]. Agricultural and livestock activities in the alpine zone have experienced notable expansion [40]. The excessive application of chemical fertilizers and pesticides, coupled with pollutants from livestock and poultry farming, may pose a significant risk to the chemical integrity of groundwater [41–43]. Considering the dual pressures from dramatic shifts in the natural environment and human activities, it is imperative to focus on the hydrochemical characteristics and their formation mechanisms, as well as the sustainability of groundwater resources in alpine areas.

The Tibetan Plateau, often referred to as the "Asia Water Tower", is a pivotal area for ecological security and water resource management in Asia. As a quintessential alpine region, it exhibits heightened sensitivity to environmental fluctuations [44]. The current study examines the Tongde Basin in the northeastern Tibetan Plateau as a case study to elucidate the hydrochemical formation patterns and the availability of groundwater resources in the basin's eastern sector, an area characterized by relatively intensive agricultural and livestock activities. Specifically, the study aims to (1) delineate the hydrochemical characteristics of groundwater; (2) uncover the mechanisms behind groundwater's hydrochemical formation; and (3) assess the overall water quality of groundwater, including its suitability for sustainable potable and irrigation usages. This research is poised to offer scientific guidance on hydrochemical aspects, thereby enhancing water ecosystem security and ensuring water use sustainability in the alpine region of the Tibetan Plateau and other similar regions worldwide.

2. Materials and Methods

2.1. Study Area Description

The study area, situated in the central region of the Tongde Basin on the northeastern Tibetan Plateau, is defined by longitudes from 100°43′22.79″ E to 101°5′16.80″ E and latitudes from 35°6′7.21″ N to 35°21′14.43″ N, encompassing an area of 929.46 km^2 (Figure 1). The topography of the study area is characterized by higher elevations in the southeast and lower elevations in the northwest. The northwestern flank of the study area is typified by a piedmont sloping plain and a valley belt plain, whereas the southeastern side is predominantly composed of tectonically uplifted mid- to high-altitude mountainous terrain. The region's elevation varies between 3219 m and 4057 m above sea level. The hydrological system of the study area is predominantly composed of three rivers. The Baqv River, one of these, courses from east to west across the northern plain. The Cihawuqv River and the Ningxiuqv River, which are tributaries of the Baqv River, originate from the southern mountainous region, flow towards the northern plain, and eventually converge with the Ba River. These rivers bring the surface water of mountainous areas to the plain and recharge aquifers through infiltration in the piedmont. The study area is characterized by a continental plateau climate, with extended winters and abbreviated summers, frequent winds, and scant precipitation. The annual mean temperature fluctuates between −0.77 °C and 1.69 °C, presenting generally low temperatures. Annual precipitation varies between 544 mm and 837 mm, with a mean of 688 mm. Over 80% of the annual rainfall occurs from May through September. This region has a potential evaporation rate of 790 mm, demonstrating a minimal disparity with precipitation.

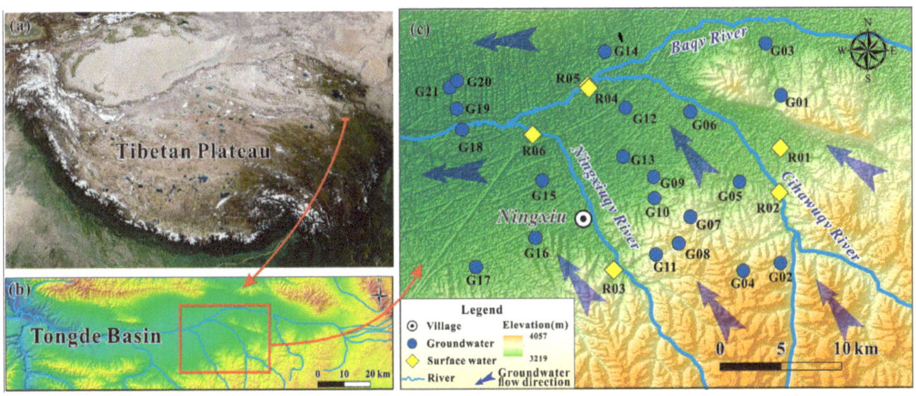

Figure 1. Spatial location of (**a**) the Tongde Basin on the Tibetan Plateau, (**b**) study area in Tongde Basin, and (**c**) the sampling sites of surface water and groundwater.

Geologically, the Quaternary sediments are predominantly found in the piedmont sloping plain and valley belt plain of the northwestern part of the study area. These sediments primarily consist of alluvial deposits and constitute the principal aquifer system within the region. From the southeast to the northwest, the aquifer type transitions from unconfined to confined. Groundwater migrates from the southeastern piedmont towards the northwestern valley, subsequently veering westward (Figure 1c). The aquifer exhibits a substantial variability in well yields, with individual wells producing less than 100 m^3/d to 5000 m^3/d, and as one moves closer to the river valley from both sides of the river, the yield gradually increases. The mid- to high-altitude mountainous strata in the southeastern part of the study area are predominantly composed of Triassic sandstone, and the aquifer type is predominantly characterized as a bedrock fracture system.

Groundwater in the study area primarily originates from the infiltration of atmospheric precipitation and the seasonal melt of snow. Its genesis, occurrence, and distribution are collectively influenced by a suite of interrelated factors. In comparison to surface water, groundwater typically exhibits superior water quality. The higher yields of individual wells and the shallower depth of groundwater in the river valley area facilitate easy extraction of this resource. For the irrigation area, groundwater serves as the principal source of water supply. It is not only indispensable for agricultural and pastoral production but also serves as a vital resource for sustaining the livelihoods of the local population. Consequently, it is essential to assess the water quality of groundwater and comprehend its formation mechanisms to establish a scientific foundation for the conservation of local groundwater resources.

2.2. Water Sampling and Physicochemical Analyses

A total of 27 water samples were collected across the study area in October 2023. Among them, 21 samples are groundwater, while 6 are river water. Among the groundwater samples, four were spring water samples, all of which were collected from the ground surface. The remaining groundwater samples were from phreatic aquifers, with sampling depths ranging from 10 to 30 m. All water samples were evenly distributed throughout the entire study area and had strong representativeness. Except for the spring water samples, all samples were pumped for more than 15 min before collection. The impact of accumulated water in the boreholes/wells on the samples was eliminated. The in situ physicochemical parameters, such as the pH value and electrical conductivity (EC), of the samples were measured on site. The sampling operation was carried out after all in situ physicochemical parameters were stable. The water samples were collected in 500 mL high-density polyethylene bottles after the target water was rinsed 2–3 times. Duplicate sampling was performed at each site to mitigate the risk of potential sampling errors. The samples were sent to the laboratory for hydrochemical analysis within 48 h under a storage temperature of 4 °C.

The on-site measurement device for in situ physicochemical parameters was a portable multi-parameter device (HACH HQ4D, Loveland, CO, USA). The main cations were determined using an inductively coupled plasma mass spectrometer (Agilent 7500ce ICP-MS, Tokyo, Japan). The main anions and NH_4^+ were determined using ion chromatography (Shimadzu LC-10ADVP, Kyoto, Japan). HCO_3^- and TDS were determined using acid–base titration and gravimetric methods, respectively. All samples were tested three times and interspersed with standard controls and blank controls to ensure the accuracy of the test results. The ion charge balance errors of the test results were all within 5, indicating that the test results were reliable.

2.3. Entropy-Weighted Water Quality Index

The entropy-weighted water quality index (EWQI) is an advanced water quality assessment based on the traditional water quality index [45]. This approach introduces the concept of entropy to avoid the influence of human subjectivity to the greatest extent and assigns corresponding weights to each hydrochemical indicator [46,47]. Generally, the evaluation of EWQI goes through the following four steps.

Firstly, the initial eigenvalue matrix **X** for the hydrochemical indicators of water samples is established. The formula for matrix **X** is as follows:

$$X = \begin{bmatrix} x_{11} & x_{12} & \cdots & x_{1n} \\ x_{21} & x_{22} & \cdots & x_{2n} \\ \vdots & \vdots & \ddots & \vdots \\ x_{m1} & x_{m2} & \cdots & x_{mn} \end{bmatrix} \tag{1}$$

where x_{mn} represents the concentration of the nth hydrochemical index of the mth water sample.

Secondly, due to the difference in the dimensions of different hydrochemical indicators, the matrix **X** is normalized to obtain the standardized matrix **Y**. The normalization method and the matrix **Y** are shown in Formulas (2) and (3), respectively.

$$y_{ij} = \frac{x_{ij} - \min(x_{ij})}{\max(x_{ij}) - \min(x_{ij})} \tag{2}$$

$$Y = \begin{bmatrix} y_{11} & y_{12} & \cdots & y_{1n} \\ y_{21} & y_{22} & \cdots & y_{2n} \\ \vdots & \vdots & \ddots & \vdots \\ y_{m1} & y_{m2} & \cdots & y_{mn} \end{bmatrix} \tag{3}$$

where y_{mn} represents the result of normalized x_{mn}.

Next, the entropy weighting method is used to allocate weights to each hydrochemical indicator, which would make the results more formal and objective. The allocation criteria are based on the following formulas:

$$p_{ij} = \frac{y_{ij}}{\sum_{i=1}^{m} y_{ij}} \tag{4}$$

$$e_j = -\frac{1}{\ln m} \sum_{i=1}^{m} p_{ij} \ln(p_{ij}) \tag{5}$$

$$\omega_j = \frac{1 - e_j}{\sum_{j=1}^{n} (1 - e_j)} \tag{6}$$

where p_{ij} represents the proportion of the ith sample in the jth hydrochemical parameter; e_j is the information entropy of each hydrochemical parameter; ω_j represents the entropy weight of each hydrochemical parameter.

Finally, using Formulas (7) and (8), the quantitative grading q_i of each water sample is determined and combined with ω_j to obtain the EWQI of each water sample.

$$q_{ij} = \frac{x_{ij}}{S_j} \times 100 \tag{7}$$

$$EWQI = \sum_{j=1}^{n} \omega_j q_{ij} \tag{8}$$

where x_{ij} is the measured value of the jth hydrochemical parameter of the ith water sample; S_j is the acceptable limit value of the hydrochemical index j recommended by the World Health Organization or Chinese guideline.

2.4. Human Health Risk Assessment

Excessive amounts of a certain toxic ion in drinking water can potentially harm people's health. For different populations, the tolerance to toxic substances usually varies [48]. For non-carcinogenic ions, the health risks of ingestion usually dominate. Therefore, this article focuses on assessing the health risks of oral contact [49].

Firstly, the calculation method for the estimated daily intake (EDI) of toxic elements is based on Formula (9).

$$CDI = \frac{C_i \times IR \times EF \times ED}{BW \times AT} \tag{9}$$

where C_i is the concentration of different toxic ions; IR is the daily water intake; EF and ED represent the exposure frequency and exposure duration, respectively; BW refers to the average weight of different populations; AT is the average time of exposure occurrence.

Afterwards, the non-carcinogenic risk (HQ) for specific toxic substances is calculated using Formula (10).

$$HQ_i = \frac{CDI}{RfD_i} \quad (10)$$

where RfD is the reference value corresponding to the intake dose of a specific pollutant. The overall non-carcinogenic risk (HI) is the sum of the non-carcinogenic risks caused by all toxic substances, and the calculation formula is:

$$HI = \sum_{i=1}^{n} HQ_i \quad (11)$$

2.5. Irrigation Water Quality Assessment

Irrigation water for agriculture is one of the main ways groundwater is consumed in the study area. Unsuitable groundwater can cause crop reduction and even damage soil functions [50]. The assessment of the irrigation adaptability of groundwater has significance for local agricultural development. The irrigation water quality index (IWQI) can objectively evaluate the quality of agricultural water through multiple parameters [51]. For this research, considering the actual situation of the study area, the sodium adsorption ratio (SAR), soluble sodium percentage (%Na), and permeability index (PI) are selected to evaluate the irrigation adaptability of groundwater [52].

SAR measures the cation exchange that occurs in the soil due to irrigation water. It reflects the degree of Na^+ adsorption in the soil. The larger the SAR value, the higher the Na^+ content in the soil and the stronger the soil alkalization [53]. The calculation for SAR is shown in Formula (12).

$$SAR = \frac{Na^+}{\sqrt{\frac{Mg^{2+} + Ca^{2+}}{2}}} \times 100\% \quad (12)$$

%Na is an important indicator reflecting the potential Na hazard in irrigation water. It reflects the proportion of Na^+ in the cations of irrigation water. Higher Na^+ can damage soil structure and reduce its permeability [49]. The calculation for %Na is shown in Formula (13).

$$\%Na = \frac{Na^+ + K^+}{Na^+ + K^+ + Mg^{2+} + Ca^{2+}} \times 100\% \quad (13)$$

PI is another important indicator reflecting irrigation adaptability. It is used to reflect the impact of Na^+, HCO_3^-, Ca^{2+}, and Mg^{2+} in irrigation water on soil permeability. The calculation for PI is shown in Formula (14).

$$PI = \frac{Na^+ + \sqrt{HCO_3^-}}{Ca^{2+} + Mg^{2+} + Na^+} \times 100\% \quad (14)$$

3. Results and Discussion

3.1. General Hydrochemical Characteristics

To preliminarily realize the overall hydrochemical characteristics of surface water and groundwater in the study area, the physicochemical data of collected river waters and groundwaters are summarily presented in Tables 1 and 2. The corresponding drinking guidelines of various indexes recommended by the Chinese Guideline [54] and World Health Organization [55] are also included in the Tables for comparison.

The pH value of collected river waters was 7.76–7.97 with a mean of 7.85, and that of collected groundwaters was 7.07–8.06 with an average of 7.70. Both river water and groundwater in the study area had pH values slightly greater than 7, demonstrating a neutral to a little alkaline nature. River water and groundwater TDS was in the range of 297.39–368.57 mg/L (averaging 328.90 mg/L) and 300.25–638.38 mg/L (averaging 399.86 mg/L), respectively. Notably, groundwater had a higher TDS value than river water in the study area, although both had relatively low TDS. This may be related to the relatively long residence time of groundwater in the aquifer, which could provide more than enough time for groundwater to react with the deposits. Both the slightly alkaline nature and the very fresh characteristic of river water and groundwater imply that water in the study area—regardless if it is above or beneath the ground surface—has relatively good quality and keeps a natural, or approximately natural, status.

Table 1. Summary of river water physicochemical indexes in the study area and corresponding drinking water standard.

Index	Unit	Min	Max	Mean	SD *	Guideline	% of the Sample Exceeding the Guideline
pH	/	7.76	7.97	7.85	0.09	6.5–8.5 **	0%
TDS	mg/L	297.39	368.57	328.90	27.47	1000 **	0%
K^+	mg/L	2.07	2.49	2.26	0.21	/	
Na^+	mg/L	9.41	36.05	22.54	8.95	200 **	0%
Ca^{2+}	mg/L	38.08	102.20	71.48	24.82	75 ***	50%
Mg^{2+}	mg/L	9.72	46.17	21.67	14.93	50 ***	0%
Cl^-	mg/L	10.64	39.00	18.91	10.44	250 **	0%
$SO_4{}^{2-}$	mg/L	14.41	57.64	36.02	18.91	250 **	0%
$HCO_3{}^-$	mg/L	262.39	341.71	302.05	30.81	/	
$NO_3{}^-$	mg/L	8.46	17.23	12.58	3.85	50.0 ***	0%
$NO_2{}^-$	mg/L	0.00	0.00	0.00	0.00	0.02 **	0%
$NH_4{}^+$	mg/L	0.08	0.24	0.16	0.08	0.2 **	33.33%

* Standard deviation; ** Chinese Guideline [54]; *** WHO Guideline [55].

Table 2. Summary of groundwater physicochemical indexes in the study area and corresponding drinking water standard.

Index	Unit	Min	Max	Mean	SD *	Guideline	% of the Sample Exceeding the Guideline
pH	/	7.07	8.06	7.70	0.25	6.5–8.5 **	0%
TDS	mg/L	300.25	638.38	399.86	90.01	1000 **	0%
K^+	mg/L	0.56	54.36	5.99	11.55	/	
Na^+	mg/L	7.62	117.10	40.28	29.30	200 **	0%
Ca^{2+}	mg/L	48.1	112.22	76.82	21.28	75 ***	42.86%
Mg^{2+}	mg/L	8.51	49.82	22.68	10.73	50 ***	0%
Cl^-	mg/L	10.64	106.35	35.28	28.44	250 **	0%
$SO_4{}^{2-}$	mg/L	9.61	220.94	44.83	44.73	250 **	0%
$HCO_3{}^-$	mg/L	250.18	524.77	336.77	62.99	/	
$NO_3{}^-$	mg/L	3.58	42.86	16.09	8.80	50.0 ***	0%
$NO_2{}^-$	mg/L	0.00	0.08	0.01	0.02	0.02 **	9.09%
$NH_4{}^+$	mg/L	0.06	1.42	0.22	0.29	0.2 **	28.57%

* Standard deviation; ** Chinese Guideline [54]; *** WHO Guideline [55].

Among the major cations, Ca^{2+} is the predominant ion for river water, with a concentration in the range of 38.08–102.20 mg/L, followed by Mg^{2+}, Na^+, and K^+. The sampled river water in the study area had Mg^{2+}, Na^+, and K^+ concentrations in the ranges of 9.72–46.17 mg/L, 9.41–36.05 mg/L, and 2.07–2.49 mg/L, respectively. The concentration of major anions in river water was in the order of $HCO_3{}^- > SO_4{}^{2-} > Cl^-$; the concentration of $HCO_3{}^-$ (262.39–341.71 mg/L) is far greater than that of $SO_4{}^{2-}$ (14.41–57.64 mg/L) and Cl^- (10.64–39.00 mg/L). Ca^{2+} is the predominant ion among the major cations of groundwater and ranges from 48.10 mg/L to 112.22 mg/L (mean of 76.82 mg/L). The second most predominant cation for groundwater is distinct from river water. Na^+ ranks second among the major cations of groundwater, with the concentration varying between 7.62 mg/L and 117.10 mg/L (mean of 40.28 mg/L). Mg^{2+} and K^+ rank third and fourth among the major cations of groundwater, respectively, with concentrations in the ranges of

8.51–49.82 mg/L (mean of 22.68 mg/L) and 0.56–54.36 mg/L (mean of 5.99 mg/L). The major anions of groundwater present the same concentration order as river water, i.e., HCO_3^- > SO_4^{2-} > Cl^-. The collected groundwaters in the study area have a HCO_3^- concentration ranging from 250.18 mg/L to 524.77 mg/L. SO_4^{2-} concentration varies from 9.61 to 220.94 mg/L, with a mean of 44.83 mg/L. Cl^- is in the range of 10.64–106.35 mg/L, with an average of 35.28 mg/L. All the major ions (including cations and anions) of river water and groundwater were within the drinking water standards except Ca^{2+} (Tables 1 and 2). About 50.00% and 42.86% of sampled river waters and groundwaters were above the standard of 75 mg/L recommended by WHO for drinking purposes [55]. However, they only slightly exceeded the recommended drinking standard of Ca^{2+}.

Nitrogen contaminants (including NO_3^-, NO_2^-, NH_4^+) were also detected in this research. The concentrations of NO_3^- were in the range of 8.46–17.23 mg/L for river water and 3.58–42.86 mg/L for groundwater, respectively, with a mean of 12.58 mg/L and 16.09 mg/L. All sampled waters—regardless of whether they were above or below the ground surface—were within the NO_3^- the drinking standard concentration recommended by WHO (below 50 mg/L) [55]. The concentration of NO_2^- was very low for river water and within the drinking standard of 0.02 mg/L given by the Chinese Guideline [54]. Groundwater in the study area had higher NO_2^- concentrations than river water, with a maximum of 0.08 mg/L. Approximately 9.09% of the collected groundwater samples had NO_2^- concentrations beyond the drinking standard of 0.02 mg/L. The concentration of NH_4^+ varied from 0.08 mg/L to 0.24 mg/L, with an average of 0.16 mg/L for river water and between 0.06 mg/L and 1.42 mg/L, with an average of 0.22 mg/L for groundwater. Both river water and groundwater had NH_4^+ contents exceeding the recommendations by the Chinese Guideline [54] in some sampling locations (accounting for 33.33% and 28.57% of sampling sites for river water and groundwater, respectively).

3.2. Hydrochemical Types

The hydrochemical types of surface (river) water and groundwater in the present study area were visually identified based on the Piper trilinear diagram. The Piper trilinear diagram is composed of three sub-diagrams, including two triangles below and one diamond above. The two triangles below were used to demonstrate the major cation (Na^++K^+, Mg^{2+}, Ca^{2+}) and ion (Cl^-, SO_4^{2-}, HCO_3^-) compositions, respectively. The diamond above is used to synthetically illustrate the hydrochemical types of water.

As demonstrated in Figure 2, the collected river waters dominantly plotted in A and B of the left triangle, demonstrating that Ca^{2+} was the predominant cation, followed by Mg^{2+}. For groundwater, it can be seen that although most of the collected groundwater samples were situated in A, many present a distributing trend towards C dominance, with some directly plotting in C. Some sporadic groundwater samples were situated in the boundary area of B dominance. All of the above suggests that Ca^{2+} is an overwhelmingly dominant cation in the unconfined aqueous environment, followed by Na^{2+} and Mg^{2+}. This is consistent with the above cation orders based on the average concentration. Overall, the surface water and groundwater in the study area were mainly characterized as [Ca^{2+}] type.

For anions, it can be seen that all collected river water samples were situated in the left corner of D. dominance in the right triangle (Figure 2), implying that river water in the study area is dominated by HCO_3^-. Groundwater samples were also all situated in D, suggesting that HCO_3^- is the predominant anion for groundwater. Notably, groundwater samples also showed a trend from the left corner to the upper and right direction of D, indicating that the groundwater is much saltier than river water. As a whole, river water and groundwater in the study area were characterized as [HCO_3^-] type.

For the synthetical chemical feature, all river waters were situated in the left corner of dominance 1 of the diamond, indicating that river waters in the study area are very fresh in nature. The collected groundwaters were also dominantly in the 1 of the diamond but showed a distributing trend from the left corner to the right direction with one even in the 3. This indicates that groundwater has saltier hydrochemical features than river water. Overall, river waters were characterized by hydrochemical type HCO_3-Ca, and groundwaters were hydrochemical type HCO_3-Ca and HCO_3-Na·Ca.

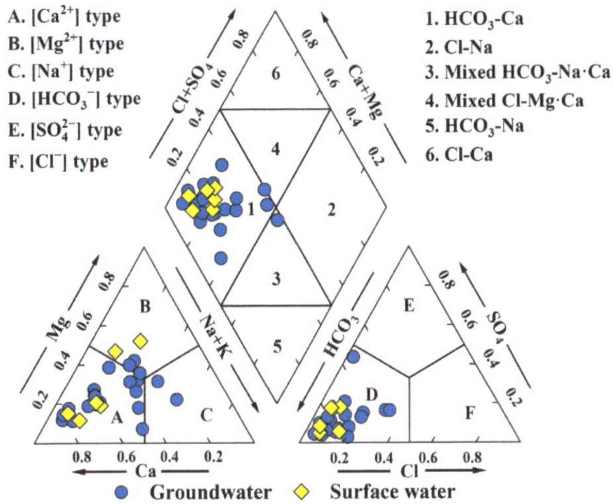

Figure 2. Piper diagram illustrating the hydrochemical composition of river water and groundwater in the study area.

3.3. Formation of Groundwater Chemistry

Natural factors form the basis of the hydrochemical composition of groundwater. In general, the major natural mechanisms potentially governing the hydrochemical composition of groundwater are precipitation, rock–water interactions, and evaporation [56]. "Precipitation" represents the hydrochemical feature of recharge water, which is usually atmospheric water. "Rock–water interactions" signifies all the processes that occur between aquifer media and groundwater from the moment water infiltrates the aquifer until it is extracted or moves on. "Evaporation" refers to the evaporation effects on groundwater, primarily affecting phreatic groundwater that is located at shallow depths. These three major natural mechanisms can be revealed by the relationship of the $Na^+/(Na^+ + Ca^{2+})$ ratio versus TDS and $Cl^-/(Cl^- + HCO_3^-)$ versus the TDS ratio [57].

As depicted in Figure 3, all groundwater samples collected from the phreatic aquifers plotted in the rock dominance, indicating that the hydrochemical composition of phreatic groundwater is predominantly controlled by rock–water interactions in nature. No groundwater samples situated in the precipitation dominance, implying that no sampled phreatic groundwater retained the hydrochemical composition of recharge water. This is because the groundwater flows very slowly, giving it enough time to react with the aquifer media. In addition, the collected phreatic groundwaters were not in the evaporation dominance, suggesting that evaporation effects are also not significant to the hydrochemical composition of phreatic groundwater in the study area.

To further reveal the specifics of the rock–water interaction, the relationships between the Ca^{2+}/Na^+ ratio and Mg^{2+}/Na^+ ratio and between the Ca^{2+}/Na^+ ratio and HCO_3^-/Na^+ ratio were explored. Three major rock/mineral types can be identified by these two relations, including carbonates, silicates, and evaporites. As demonstrated in Figure 4, the collected phreatic groundwater samples were dominantly situated in or around the silicate dominance, indicating that silicate weathering (Formula (15)) is the dominant process contributing to the chemical constituents of phreatic groundwater in the study area. Meanwhile, phreatic groundwater in some sporadic sites is also observed distributing towards the evaporite and carbonate dominance, suggesting that the dissolution of evaporites and carbonates can also contribute chemical constituents to groundwater to some degree.

$$2NaAlSi_3O_8 + 2CO_2 + 11H_2O \rightarrow Al_2Si_2O_5(OH)_4 + 4H_4SiO_4 + 2Na^+ + 2HCO_3^- \quad (15)$$

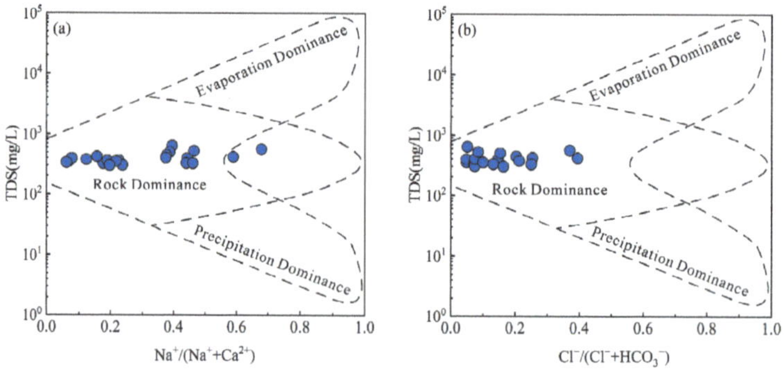

Figure 3. Gibbs diagram of (**a**) $Na^+/(Na^++Ca^{2+})$ versus TDS and (**b**) $Cl^-/(Cl^-+HCO_3^-)$ versus TDS for the sampled groundwaters in the study area.

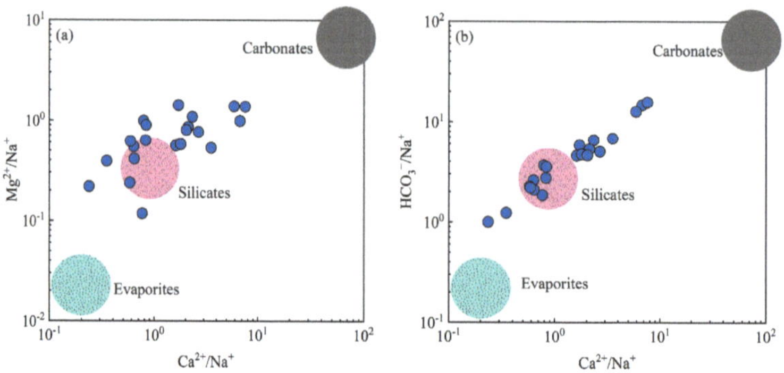

Figure 4. End-member diagram of (**a**) Ca^{2+}/Na^+ versus Mg^{2+}/Na^+ and (**b**) Ca^{2+}/Na^+ versus HCO_3^-/Na^+ illustrating the minerals contributing to groundwater chemistry.

Besides the aforementioned rock weathering and mineral dissolution, ion exchange is also a non-negligible process that occurs in the aquifer, particularly in fine sedimentary aquifers. The chlor-alkali index (CA1-1 and CAI-2) is introduced to gain insights into potential ion exchange processes, which can be calculated by Formulas (16) and (17). As shown in Figure 5a, most of the collected samples have negative values for both CA1-1 and CAI-2, implying the occurrence of the cation-exchange reaction (Formula (18)) in the aquifer. Meanwhile, three samples had positive values for both CA1-1 and CAI-2, indicating existing reverse cation-exchange reactions (Formula (19)) in the aquifer of these sporadic sites. The relation between (Na^++K^+-Cl^-) and (Ca^{2+}+Mg^{2+}-HCO_3^--SO_4^{2-}) is also introduced to verify these ion exchange processes. As depicted in Figure 5b, it is confirmed that the cation-exchange reaction is the dominant ion-exchange process in the phreatic aquifers, and the reverse cation-exchange reaction also occurred at some sporadic sites.

$$CAI - 1 = \frac{Cl^- - (Na^+ + K^+)}{Cl^-} \quad (16)$$

$$CAI-2 = \frac{Cl^- - (Na^+ + K^+)}{HCO_3^- + SO_4^{2-} + CO_3^{2-} + NO_3^-} \tag{17}$$

$$2Na^+ + CaX_2\ (\text{or } MgX_2)\ (\text{solids}) \rightarrow Ca^{2+}\left(\text{or } Mg^{2+}\right) + 2NaX\ (\text{solids}) \tag{18}$$

$$Ca^{2+}\left(\text{or } Mg^{2+}\right) + 2NaX\ (\text{solids}) \rightarrow 2Na^+ + CaX_2\ (\text{or } MgX_2)\ (\text{solids}) \tag{19}$$

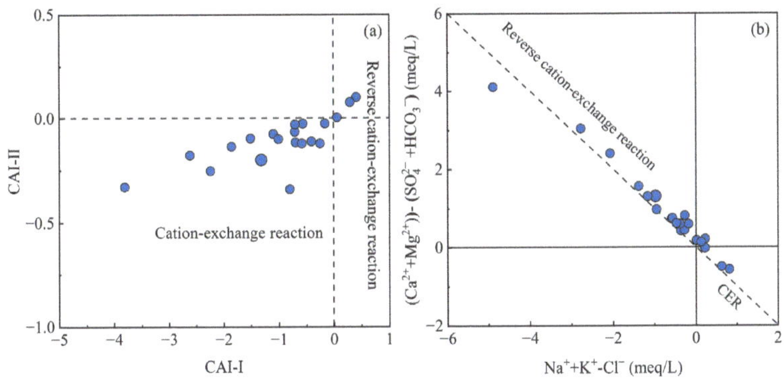

Figure 5. Scatter plots of (**a**) CAI-I versus CAI-II and (**b**) (Na^++K^+-Cl^-) versus (Ca^{2+}+Mg^{2+}-HCO_3^--SO_4^{2-}) of groundwater in the study area. RCER defines the reverse cation-exchange reaction; CER denotes the cation-exchange reaction.

Although all sampled groundwater had NO_3^- concentrations below the drinking standard of 50 mg/L recommended by WHO [55], most groundwater samples exceeded the geological background limit of 10 mg/L [58] (Figure 6a). This implies that phreatic groundwater in the study area has been influenced by human activities in its hydrochemical composition, especially nitrite. The relation of the Cl^-/Na^+ ratio versus NO_3^-/Na^+ ratio was employed to gain insights into the specific source of the nitrate pollutant. As depicted in Figure 6b, the sampled groundwaters dominantly plotted adjacent to agricultural activities, implying that NO_3^- in phreatic groundwater primarily originates from agricultural practices. It can be concluded that agricultural practices have led to widespread inputs of nitrogen pollutants to phreatic aquifers, although the level of pollution is still relatively low.

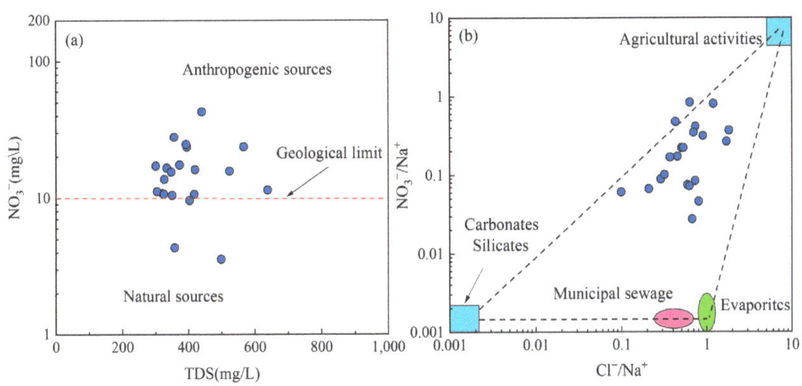

Figure 6. Scatter plots of (**a**) TDS versus NO_3^- and (**b**) Cl^-/Na^+ versus NO_3^-/Na^+ of groundwater in the study area.

The Pearson correlation coefficients are introduced to gain insights into the influence of agricultural practices on groundwater chemical composition besides NO_3^-. As shown in Table 1, the groundwater at some sampling sites had a relatively high content of NH_4^+, and approximately 33.33% of samples exceeded the permissible limit recommended by the Chinese Guideline [54]. Considering the potential contamination sources in the study area, these exceeding NH_4^+ contaminants originated from chemical fertilizers of agricultural practices. In addition, some hydrochemical indicators such as Cl^-, Na^+, K^+, and TDS presented a positive relation with NH_4^+ (Figure 7), indicating that agricultural practices also brought chemical solutes into phreatic groundwaters, especially Cl^-, Na^+, and K^+. Although agricultural practices introduce some solutes (including nitrogen contaminants and chemical solutes) and increase the salinity of groundwater, to some extent, the perturbations are relatively limited for the overall framework of groundwater hydrochemistry (Figure 7).

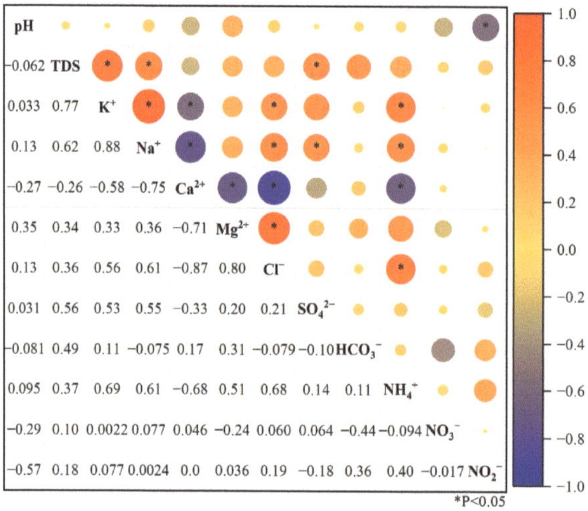

Figure 7. Correlation of various hydrochemical parameters and their correlation coefficients for groundwater in the study area.

3.4. Groundwater Quality Evaluation

3.4.1. Drinking Water Quality

Groundwater is widely used as drinking water in alpine areas like the present study area. The chemical quality suitability of groundwater for drinking purposes was evaluated using the EWQI approach. The results show that the phreatic groundwater in the study area had a relatively large range of EWQI values from 25.7 to 159.9 (Figure 8), suggesting that phreatic groundwater in the present area has a wide variation of hydrochemical quality. As demonstrated in Figure 8, most of the collected phreatic groundwater samples (95.24%) had EWQI values below 100, suggesting it is suitable for direct drinking purposes. Particularly, about 80.95% of the sampled groundwaters had EWQI values below 50, belonging to the excellent water quality rank (rank 1). Only one groundwater sample (4.76%) had an EWQI value beyond 100 and reached 159.88, which is characterized by a poor water quality rank (rank 4). Phreatic groundwater in this sampling site (G09) should be avoided for direct drinking. The main reason for the poor quality of phreatic groundwater at sampling site G09 is the relatively high content of NH_4^+, originating from agricultural activities.

In addition to the overall quality revealed by the EWQI assessment, the potential health hazards of toxic substances, including NH_4^+, NO_2^-, and NO_3^-, were assessed using the HHRA model. The HQ values of NH_4^+ and NO_2^- in phreatic groundwater were all below 1 for all sampling sites (Figure 9a,b), implying that these two nitrogen pollutants would not pose significant health hazards to water consumers through the oral pathway. Meanwhile, the HQ values of NO_3^- showed a relatively large variation for all population groups. The HQ_{NO3} values range from 0.08 to 1.02 for adult males, between 0.09 and 1.08 for adult females, from 0.09 to 1.10 for children, and between 0.16 and 1.87 for infants (Figure 9c). Only one sampling site (G10) had HQ_{NO3} values exceeding the permissible limit of 1 for adult male, adult female, and children, with maximum values of 1.02, 1.08, and 1.10, respectively. Four sampling sites were identified with HQ_{NO3} values exceeding the permissible limit of 1 in infants, G01, G06, G10, and G11, with respective HQ_{NO3} values of 1.03, 1.08, 1.87, and 1.22.

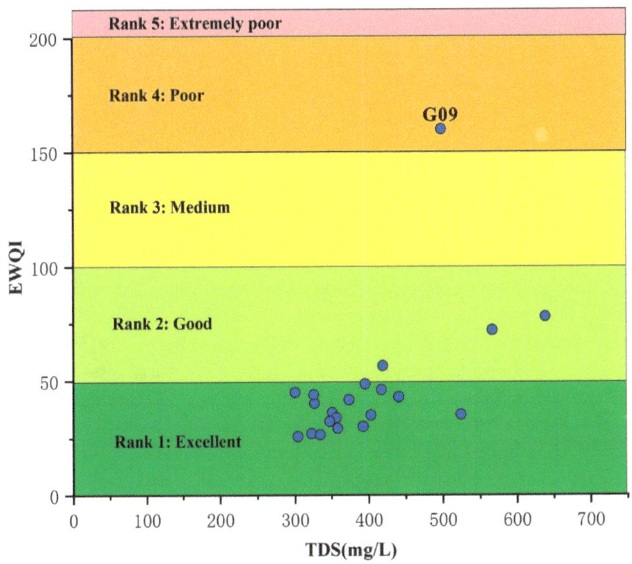

Figure 8. Scatter plots of TDS versus EWQI of groundwater in the study area.

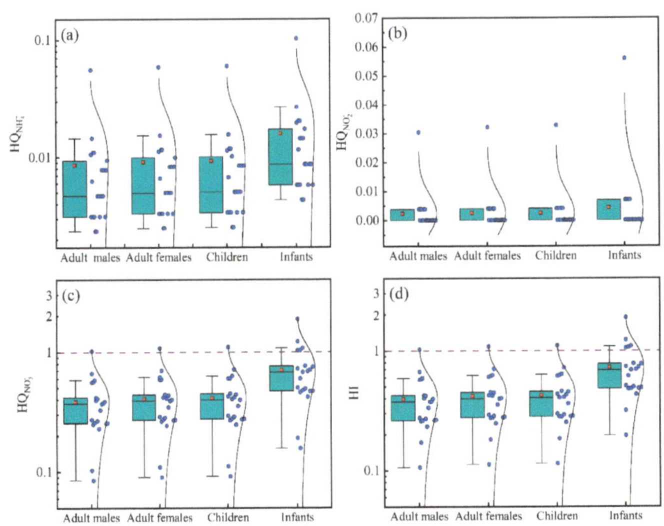

Figure 9. Box plots of HQ values of (**a**) NH_4^+, (**b**) NO_2^-, (**c**) NO_3^-, and (**d**) HI values of groundwater in the study area.

The HI values are in the range of 0.11–1.02 for adult males, 0.11–1.08 for adult females, 0.11–1.10 for children, and 0.20–1.88 for infants, with an average of 0.39, 0.42, 0.42, and 0.72, respectively (Figure 9d). Approximately 4.76%, 4.76%, 4.76%, and 19.05% of collected groundwater samples had HI values beyond the permissible limit of 1 for the aforementioned four population groups, indicating potential health risk. As discussed above, the HQ values of NH_4^+ and NO_2^- for all populations were very small, while those of NO_3^- were larger in variation. Thus, the overall health risks are primarily posed by the nitrate pollutant phreatic in groundwater. However, phreatic groundwaters at most sampling sites had HI values below the safe limit of 1, indicating a low and negligible risk. The only groundwater sample (G10) with potential health risks for adult males and females (HI value just slightly beyond the permissible limit of 1) had values of 1.02 and 1.08, respectively. This suggests that the potential health risk of groundwater at sampling site G10 is very limited for adults. Although four sampling sites had HI values exceeding the permissible limit of 1 for infants, two were just slightly beyond 1. Only two sampling sites had relatively large HI values, with 1.88 for G10 and 1.23 for G11. For children, the only sampling site with a potential health risk was with an HI value of 1.10. Therefore, approximately 4.76% (1 sampling site, G10) and 9.52% (2 sampling sites, G10 and G11) had slightly significant health risks for water consumption in children and infants, respectively.

3.4.2. Irrigation Water Quality

Groundwater, especially in the phreatic aquifers, is primarily utilized for agricultural irrigation in alpine areas. The suitability of water quality for irrigation depends on the content of hydrochemical ions in water and how they affect the soils and plants [59]. EC is a comprehensive parameter demonstrating the overall content of chemical ions in water. Generally, water with EC values less than 250 μS/cm and in the range of 250–750 μS/cm is regarded as excellent and good quality, respectively, and suitable for irrigation. Water with EC values in the range of 750–2250 μS/cm is considered a doubtful quality for irrigation, and that beyond 2250 μS/cm is unsuitable for irrigation. As demonstrated in Table 3, the EC values of phreatic groundwater samples vary from 509.73 μS/cm to 944.06 μS/cm, averaging at 670.22 μS/cm. Approximately 76.19% of sampled phreatic groundwaters had EC values in the range of 250–750 μS/cm, implying they are good for irrigation. Meanwhile, about 23.81% of the samples had EC values in the range of 750-2250 μS/cm, belonging to the doubtful quality category.

Table 3. Assessment results of groundwater quality for irrigation in the study area.

Index	Unit	Range	Class	Percentage
EC	μS/cm	<250	Excellent	/
		250–750	Good	76.19%
		750–2250	Doubtful	23.81%
		>2250	Unsuitable	/
SAR	meq/L	<10	Excellent	100%
		10–8	Good	/
		18–26	Doubtful	/
		>26	Unsuitable	/
%Na	%	<20	Excellent	52.38%
		20–40	Good	38.10%
		40–60	Permissible	9.52%
		60–80	Doubtful	/
		>80	Unsuitable	/
PI	%	<25	Unsuitable	/
		25–75	Moderately suitable	95.24%
		>75	Suitable	4.76%

The excessive Na^+ in irrigation water compared with ions of Mg^{2+} and Ca^{2+} would damage the permeability of the soil, which can be assessed by the SAR. The calculated results show that the SAR values of sampled phreatic groundwaters in the study area range between 0.19 and 3.34 and are all below 10, indicating excellent quality (Table 3). The Wilcox diagram is introduced to assess the potential overall hazards for agricultural irrigation. It can be seen from Figure 10 that about 76.19% and 23.81% of the collected phreatic groundwater samples belong to the C2S1 and C3S1 categories, respectively, indicating medium and high salinity hazards but a low sodium hazard for all. It can be seen that most of the study area has groundwater of medium salinity hazard and low sodium hazard (C2S1) rather than high salinity hazard and low sodium hazard (C3S1) (Figure 11a). Groundwaters with high salinity hazards and low sodium hazards (C3S1) are scattered throughout the research area (Figure 11a).

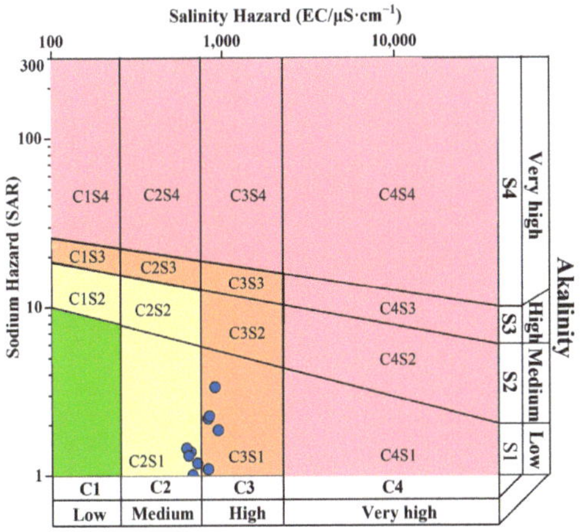

Figure 10. Wilcox diagram demonstrating irrigation quality of groundwater in the study area.

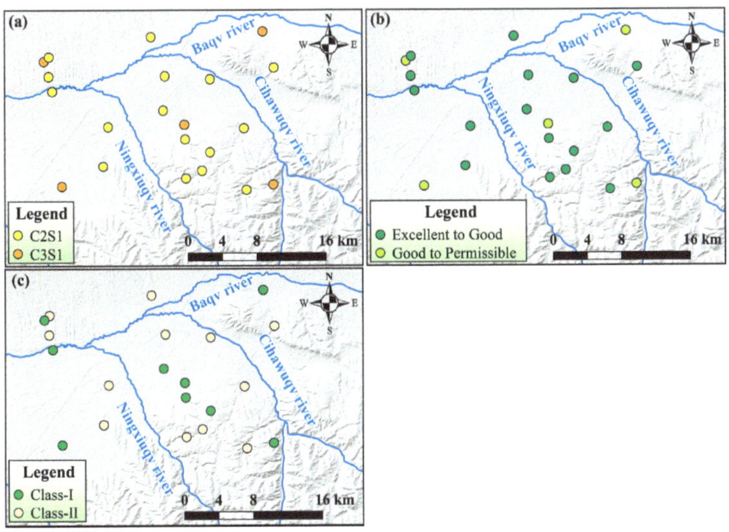

Figure 11. Spatial distribution of groundwater quality for irrigation demonstrated by (a) Wilcox diagram classification, (b) USSL diagram classification, and (c) Doneen diagram classification.

The %Na was further used to assess the potential effect of Na^+ on soil permeability. As seen in the calculated results, it ranges from 5.26 to 50.76, with a mean of 21.47. It can be seen from Table 3 that about 52.38% of the samples are divided into the excellent quality category for irrigation in terms of %Na. Approximately 38.10% and 9.52% of the groundwater samples belong to the good and permissible irrigation quality, respectively. In addition, the USSL diagram, which considers the EC and %Na at the same time, is introduced to evaluate the integrated irrigation quality of phreatic groundwater. As shown in Figure 12, the majority (76.19%) of collected groundwater samples are situated in the category of excellent to good quality, and the rest of the samples (23.81%) belong to the category of good to permissible quality. Thus, all phreatic groundwaters are suitable for irrigation in terms of the %Na. The distribution of groundwater quality classified by the USSL diagram (Figure 11b) is similar to that classified by the Wilcox diagram (Figure 11a). Most of the study area's groundwater quality is excellent to good. Groundwaters with good to permissible quality are scattered throughout the research area (Figure 11b).

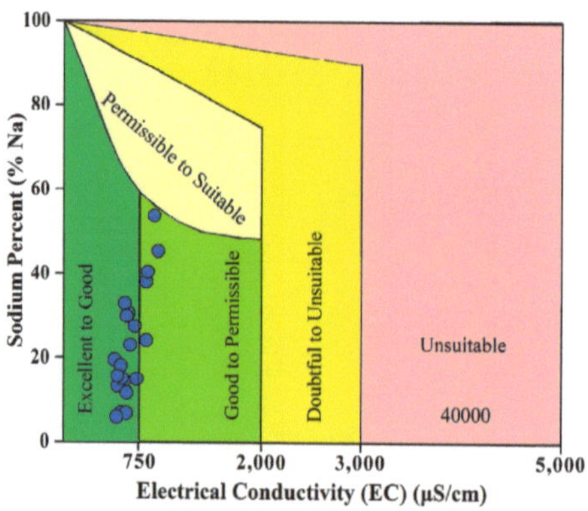

Figure 12. USSL diagram contracted by the relation between EC and sodium percent of groundwater in the study area.

The PI value is employed to gain insights into the potential effects of dissolved chemical ions in irrigation water on soil. The calculated PI values are in the range of 39.51–75.59, with a mean of 53.82. Approximately 95.24% and 4.76% of phreatic groundwater samples belong to the moderately suitable and suitable categories for irrigation in terms of PI assessment results (Table 3). An integrated diagram constructed by PI and the total concentration of chemical ions is employed. As depicted in Figure 13, about 42.86% of sampled groundwaters are situated in the Class-I category, indicating good quality for irrigation in terms of permeability. Other samples (57.14%) belong to the Class-II category, demonstrating suitable quality for irrigation purposes. Thus, there is no potential threat to soil permeability when using phreatic groundwater for irrigation in terms of the assessment of PI. Groundwaters that have the irrigation quality of Class-I are mainly distributed away from the river channel, and groundwaters adjacent to rivers are irrigation quality Class-II (Figure 11c).

Overall, phreatic groundwater in the study area is of relatively good quality for irrigation purposes and would not pose a sodium hazard or cause potential permeability damage to soil. Although groundwater at most sampling sites (76.19%) is suitable for irrigation, a small portion of sampling sites (23.81%) have a potential salinity hazard for the phreatic groundwater and should be considered during the long-term sustainable irrigation practice.

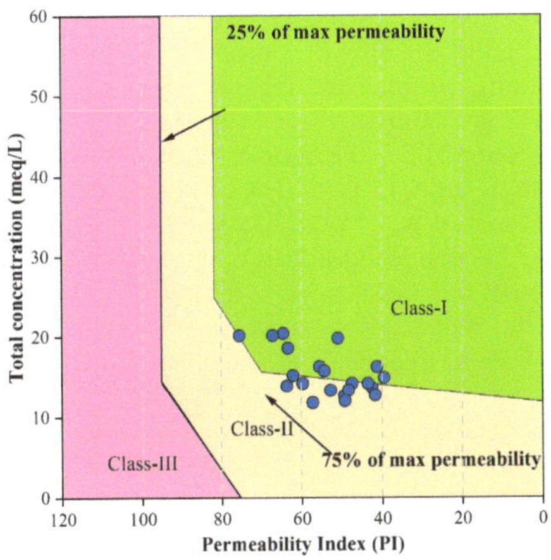

Figure 13. Doneen diagram illustrating the relation between permeability index and total concentration of groundwater in the study area.

3.5. Implication for Sustainable Development of Groundwater Resources in Alpine Regions

Groundwater in alpine regions is significant for the local development and the water circulation of the hydrosphere as it is in the headwater region. At the same time, it is vulnerable to external disturbances, including global climate change and human activities. Generally, groundwater in the alpine regions is of a natural status, which is usually of good quality. However, with the strengthening of human activities and the intensification of climate change, groundwater resources are facing unprecedented challenges. The present research takes a typical irrigation alpine region of the Tibetan Plateau to gain insights into the hydrochemical status, quality, and formation of groundwater in alpine regions with dense human irrigation practices.

Phreatic groundwater in present irrigation alpine regions retains the natural, slightly alkaline, and fresh features but is slightly saltier than recharged river water. All forms of nitrogen contaminants (NO_3^-, NO_2^-, and NH_4^+) had higher concentrations in phreatic groundwater than in river water and originated from agricultural practices like chemical fertilizer application. This demonstrates that phreatic groundwater in the present alpine plain has deteriorated by agricultural practices [60,61]. Thus, agricultural contaminants should be considered and managed. It is recommended to transition from chemical fertilizers to organic alternatives, such as animal manure, for their eco-friendly benefits and sustainable environmental development.

Groundwater quality has been influenced by the aforementioned agricultural nitrogen contaminants and their availability for direct drinking usage at some sporadic sites (G09). The relatively high content of NO_3^- would potentially threaten the health of local humans, especially for minors. Groundwater serves as a viable resource for sustainable long-term irrigation without adversely affecting soil permeability. However, the potential for salinization at some sites must be carefully managed to ensure the sustainability of agricultural practices if groundwater is used for long-term irrigation. As global climate change and human activities escalate, it is anticipated that groundwater resources in alpine regions will be increasingly threatened by pollution and are at risk of natural degradation, such as salinization. Thus, more attention should be paid to the groundwater resources of these headwater regions.

4. Conclusions

The Tibetan Plateau is the "Third Pole" of the earth and the "Asia Water Tower", and its alpine regions are sensitive and fragile to changes in the external environment. Groundwater resources are essential for the sustainable development of alpine regions on the Tibetan Plateau. The present research takes a typical area on the Tibetan Plateau to obtain insights into the hydrochemistry and suitability of phreatic groundwater for sustainable usage in the alpine agricultural area. The main findings are below.

Phreatic groundwater keeps a slightly alkaline and fresh nature in the present alpine irrigation area with a pH varying from 7.07 to 8.06 and TDS in the range of 300.25–638.38 mg/L, which is very similar to the hydrochemical features of river water but slightly saltier. Groundwater is characterized by a hydrochemical type of HCO_3-Ca and HCO_3-Na·Ca. This composition is also slightly more variable compared to that of river water, which typically exhibits a HCO_3-Ca type. The majority of major ions are in relatively low concentrations and within the desirable limit for drinking purposes, except for Ca^{2+}. All forms of nitrogen contaminants have higher concentrations in phreatic groundwater than in river water. All phreatic groundwaters have NO_3^- content in the range of 3.58–42.86 mg/L with a mean of 16.09 mg/L, which are all within the drinking standard of 50 mg/L. The maximum concentration of NO_2^- and NH_4^+ in groundwater reaches 0.08 mg/L and 1.42 mg/L, respectively. Approximately 9.09% and 28.57% of collected groundwaters are beyond the permissible limit of NO_2^- (0.02 mg/L) and NH_4^+ (0.2 mg/L) for drinking purposes.

Groundwater has relatively good hydrochemical quality for domestic usage based on the entropy-weighted water quality index assessment. Although the EWQI value of phreatic groundwater is in a large range of 25.7 to 159.9, most groundwaters (95.24%) have an EWQI value below 100 and are suitable for direct drinking. At one of the sampling sites, designated as G09, the EWQI exceeds 100, reaching up to 159.9. This high value indicates that the water is classified within the poor water quality category and ascribed to the relatively high content of NH_4^+ (1.42 mg/L). The high concentration of NO_3^- in groundwater would potentially threaten the health of minors at some sporadic sites in the study area, with a maximum HI value of 1.88 for infants and 1.10 for children. The SAR and %Na value of phreatic groundwater vary between 0.19 and 3.34 and from 5.26 to 50.76, respectively. The Na^+ content in groundwater is acceptable for irrigation and would not damage the soil permeability. The PI value of groundwater ranges from 39.51 to 75.59 and also presents no threat to the soil permeability. While the EC value of groundwater has a relatively large variation from 250–750 μS/cm, 23.81% of collected groundwaters had doubtful quality for irrigation and may pose a potential salinity hazard during long-term irrigation practices.

The hydrochemical composition of groundwater is primarily governed by rock–water interactions rather than the original recharge water (precipitation) and evaporation. Silicate weathering is the predominant process that contributes dissolved chemicals to groundwater. The dissolution of evaporites and carbonates can also contribute some chemical constituents to phreatic groundwater but to a very limited degree. The cation-exchange reaction is the dominant ion-exchange process in phreatic aquifers and influences the hydrochemical composition of groundwater in most sampling sites. Reverse cation-exchange reactions also occur in the aquifer but only at some sporadic sites. Agricultural practice has resulted in the elevation of nitrogen contaminants, which are dominant in the form of NO_3^- and NH_4^+ rather than NO_2^-. The major chemicals are not significantly brought into the phreatic groundwater during this process. It is crucial to address agricultural nitrogen pollutants to safeguard invaluable groundwater resources and foster the sustainable development of these water assets in alpine irrigation regions.

Author Contributions: S.Y.: field investigation, methodology, formal analysis, writing–original draft; Z.Z.: conceptualization, formal analysis, writing–reviewing and editing; S.W.: conceptualization, formal analysis; S.X.: field investigation, formal analysis; Y.X.: conceptualization, formal analysis, writing—reviewing and editing; J.W. (Jie Wang): field investigation, data curation, methodology, formal analysis; J.W. (Jianhui Wang): field investigation, methodology, formal analysis; Y.Y.: formal analysis, data curation, methodology; R.B.: formal analysis, data curation, methodology; N.W.: methodology, writing–reviewing and editing; Y.Z.: field investigation, methodology, formal analysis; L.W.: methodology, data curation, formal analysis; H.Y.: field investigation, formal analysis. All authors have read and agreed to the published version of the manuscript.

Funding: This research was funded by the Key Lab of Geo-environment of Qinghai Province (Grant No. 2023-KJ-02; Grant No. 2023-KJ-15); the Kunlun Talents—Science and Technology Leading Talents Program of Qinghai Province (Grant to Zhen Zhao); the Applied Basic Research Project of Science and Technology Program of Qinghai Province (Grant No. 2024-ZJ-771); the National Natural Science Foundation of China (Grant No. 42477059; Grant No. 42107067); and the Student Research Training Program of Southwest Jiaotong University (Grant No. 241925); Innovative Practice Bases of Geological Engineering and Surveying Engineering of Southwest Jiaotong University (YJG-2022-JD04).

Data Availability Statement: The data used to support the findings of the present study will be provided upon request by the corresponding author.

Conflicts of Interest: Authors Shaokang Yang, Zhen Zhao, Shengbin Wang, Shanhu Xiao, Jianhui Wang, Youjin Yuan and Ruishou Ba were employed by Qinghai 906 Engineering Survey and Design Institute Co., Ltd. The remaining authors declare that the research was conducted in the absence of any commercial or financial relationships that could be construed as a potential conflict of interest.

References

1. Immerzeel, W.W.; Lutz, A.F.; Andrade, M.; Bahl, A.; Biemans, H.; Bolch, T.; Hyde, S.; Brumby, S.; Davies, B.J.; Elmore, A.C.; et al. Importance and vulnerability of the world's water towers. *Nature* **2020**, *577*, 364–369. [CrossRef] [PubMed]
2. Crawford, J.T.; Hinckley, E.-L.S.; Litaor, M.I.; Brahney, J.; Neff, J.C. Evidence for accelerated weathering and sulfate export in high alpine environments. *Environ. Res. Lett.* **2019**, *14*, 124092. [CrossRef]
3. Xiao, Y.; Zhang, Y.; Yang, H.; Wang, L.; Han, J.; Hao, Q.; Wang, J.; Zhao, Z.; Hu, W.; Wang, S.; et al. Interaction regimes of surface water and groundwater in a hyper-arid endorheic watershed on Tibetan Plateau: Insights from multi-proxy data. *J. Hydrol.* **2024**, *644*, 132020. [CrossRef]
4. Viviroli, D.; Kummu, M.; Meybeck, M.; Kallio, M.; Wada, Y. Increasing dependence of lowland populations on mountain water resources. *Nat. Sustain.* **2020**, *3*, 917–928. [CrossRef]
5. Tian, L.; Guo, S.; Feng, J.; He, C. Quantifying the altitudinal response of water yield capacity to climate change in an alpine basin on the Tibetan Plateau through integrating the WRF-Hydro and Budyko framework. *CATENA* **2024**, *242*, 108087. [CrossRef]
6. Halloran, L.J.S.; Millwater, J.; Hunkeler, D.; Arnoux, M. Climate change impacts on groundwater discharge-dependent streamflow in an alpine headwater catchment. *Sci. Total Environ.* **2023**, *902*, 166009. [CrossRef]
7. Xu, N.; Lu, H.; Li, W.Y.; Gong, P. Natural lakes dominate global water storage variability. *Sci. Bull.* **2024**, *69*, 1016–1019. [CrossRef]
8. Guo, X.; Feng, Q.; Yin, Z.; Si, J.; Xi, H.; Zhao, Y. Critical role of groundwater discharge in sustaining streamflow in a glaciated alpine watershed, northeastern Tibetan Plateau. *Sci. Total Environ.* **2022**, *822*, 153578. [CrossRef]
9. Schädler, B.; Weingartner, R. Impact of Climate Change on Water Resources in the Alpine Regions of Switzerland. In *Alpine Waters*; Bundi, U., Ed.; Springer: Berlin/Heidelberg, Germany, 2010; pp. 59–69. [CrossRef]
10. Xiao, Y.; Liu, K.; Hao, Q.; Xiao, D.; Zhu, Y.; Yin, S.; Zhang, Y. Hydrogeochemical insights into the signatures, genesis and sustainable perspective of nitrate enriched groundwater in the piedmont of Hutuo watershed, China. *Catena* **2022**, *212*, 106020. [CrossRef]
11. He, Q.; Kuang, X.; Ma, E.; Chen, J.; Feng, Y.; Zheng, C. Evolution of runoff components and groundwater discharge under rapid climate warming: Lhasa river basin, Tibetan Plateau. *J. Hydrol.* **2024**, *628*, 130556. [CrossRef]
12. Evans, S.G.; Ge, S.; Voss, C.I.; Molotch, N.P. The Role of Frozen Soil in Groundwater Discharge Predictions for Warming Alpine Watersheds. *Water Resour. Res.* **2018**, *54*, 1599–1615. [CrossRef]
13. Somers, L.D.; McKenzie, J.M.; Mark, B.G.; Lagos, P.; Ng, G.-H.C.; Wickert, A.D.; Yarleque, C.; Baraër, M.; Silva, Y. Groundwater Buffers Decreasing Glacier Melt in an Andean Watershed—But Not Forever. *Geophys. Res. Lett.* **2019**, *46*, 13016–13026. [CrossRef]
14. Meng, Y.; Liu, G.; Li, M. Tracing the Sources and Processes of Groundwater in an Alpine Glacierized Region in Southwest China: Evidence from Environmental Isotopes. *Water* **2015**, *7*, 2673–2690. [CrossRef]
15. Shen, H.; Rao, W.; Tan, H.; Guo, H.; Ta, W.; Zhang, X. Controlling factors and health risks of groundwater chemistry in a typical alpine watershed based on machine learning methods. *Sci. Total Environ.* **2023**, *854*, 158737. [CrossRef] [PubMed]
16. Peña Reyes, F.A.; Crosta, G.B.; Frattini, P.; Basiricò, S.; Della Pergola, R. Hydrogeochemical overview and natural arsenic occurrence in groundwater from alpine springs (upper Valtellina, Northern Italy). *J. Hydrol.* **2015**, *529*, 1530–1549. [CrossRef]
17. Giménez-Forcada, E.; Luque-Espinar, J.A.; López-Bahut, M.T.; Grima-Olmedo, J.; Jiménez-Sánchez, J.; Ontiveros-Beltranena, C.; Díaz-Muñoz, J.Á.; Elster, D.; Skopljak, F.; Voutchkova, D.; et al. Analysis of the geological control on the spatial distribution of potentially toxic concentrations of As and F- in groundwater on a Pan-European scale. *Ecotoxicol. Environ. Saf.* **2022**, *247*, 114161. [CrossRef]
18. Gomez, L.; Alvarez, A.; D'Ambrosio, S.; Zalazar, G.; Aravena, R. Use of isotopes techniques to reveal the origin of water salinity in an arid region of Central-Western Argentina. *Sci. Total Environ.* **2021**, *763*, 142935. [CrossRef]
19. Qu, S.; Zhao, Y.; Zhang, K.; Wang, J.; Li, M.; Yang, X.; Ren, X.; Hao, Y.; Yu, R. Multi-isotopes (δD, δ18Owater, 87Sr/86Sr, δ34S and δ18Osulfate) as indicators for groundwater salinization genesis and evolution of a large agricultural drainage lake basin in Inner Mongolia, Northwest China. *Sci. Total Environ.* **2024**, *946*, 174181. [CrossRef]

20. Zhao, Y.; Zhang, Y.; Yan, Y.; Wen, Y.; Zhang, D. Geographic distribution and impacts of climate change on the suitable habitats of two alpine Rhododendron in Southwest China. *Glob. Ecol. Conserv.* **2024**, *54*, e03176. [CrossRef]
21. Kumar, S.; Chatterjee, U.; David Raj, A.; Sooryamol, K.R. Global Warming and Climate Crisis/Extreme Events. In *Climate Crisis: Adaptive Approaches and Sustainability*; Chatterjee, U., Shaw, R., Kumar, S., Raj, A.D., Das, S., Eds.; Springer Nature: Cham, Switzerland, 2023; pp. 3–18. [CrossRef]
22. Fu, C.-c.; Li, X.-q.; Cheng, X. Unraveling the mechanisms underlying lake expansion from 2001 to 2020 and its impact on the ecological environment in a typical alpine basin on the Tibetan Plateau. *China Geol.* **2023**, *6*, 216–227. [CrossRef]
23. Liu, Y.; Wei, L.; Huang, A.; Peng, B.; Shu, Q. Spatial and temporal evolution of soil water and its response to the environment in the Yangtze River source area under climate change. *Hydrogeol. Eng. Geol.* **2023**, *50*, 39–52.
24. Vörösmarty, C.J.; Green, P.; Salisbury, J.; Lammers, R.B. Global water resources:: Vulnerability from climate change and population growth. *Science* **2000**, *289*, 284–288. [CrossRef] [PubMed]
25. Wang, Z.; Cao, S.; Cao, G.; Hou, Y.; Wang, Y.; Kang, L. Implications for water management in alpine inland river basins: Evidence from stable isotopes and remote sensing. *Ecol. Indic.* **2023**, *154*, 110580. [CrossRef]
26. Zhang, Q.; Shen, Z.; Pokhrel, Y.; Farinotti, D.; Singh, V.P.; Xu, C.-Y.; Wu, W.; Wang, G. Oceanic climate changes threaten the sustainability of Asia's water tower. *Nature* **2023**, *615*, 87–93. [CrossRef]
27. Qin, Y.; Jin, X.; Jin, Y.; Mao, X.; Du, K. Simulation and prediction of runoff in cold alpine basins under climate change conditions. *J. Salt Lake Res.* **2024**, *32*, 29–38.
28. Han, F.; Liu, T.; Huang, Y.; Zan, C.; Pan, X.; Xu, Z. Response of water quality to climate warming and atmospheric deposition in an alpine lake of Tianshan Mountains, Central Asia. *Ecol. Indic.* **2023**, *147*, 109949. [CrossRef]
29. Yi, B.; Liu, J.; He, W.; Lu, X.; Cao, X.; Chen, X.; Zeng, X.; Zhang, Y. Optical variations of dissolved organic matter due to surface water-groundwater interaction in alpine and arid Datonghe watershed. *Sci. Total Environ.* **2023**, *864*, 161036. [CrossRef] [PubMed]
30. Ronchetti, F.; Deiana, M.; Lugli, S.; Sabattini, M.; Critelli, V.; Aguzzoli, A.; Mussi, M. Water isotope analyses and flow measurements for understanding the stream and meteoric recharge contributions to the Poiano evaporite karst spring in the North Apennines, Italy. *Hydrogeol. J.* **2023**, *31*, 601–619. [CrossRef]
31. Kumar, R.; Saika, M.; Vishwakarma, D.K.; Al-Ansari, N.; Nand Lal, K.; Elbeltagi, A.; Kallem, S.; Prasad, V.; Kuriqi, A. Assessment of Climate Change Impact on Snowmelt Runoff in Himalayan Region. *Sustainability* **2022**, *14*, 1150. [CrossRef]
32. Singh, R.; Kayastha, S.P.; Shrestha, S.M.; Sapkota, R.P. Hydro-geochemical conditions under projected climate change scenarios of Marshyangdi River, Nepal. *Theor. Appl. Climatol.* **2024**, *155*, 5375–5387. [CrossRef]
33. Fida, M.; Li, P.; Wang, Y.; Alam, S.M.K.; Nsabimana, A. Water Contamination and Human Health Risks in Pakistan: A Review. *Expo. Health* **2023**, *15*, 619–639. [CrossRef]
34. Taucare, M.; Viguier, B.; Figueroa, R.; Daniele, L. The alarming state of Central Chile's groundwater resources: A paradigmatic case of a lasting overexploitation. *Sci. Total Environ.* **2024**, *906*, 167723. [CrossRef] [PubMed]
35. Wang, M.; Su, C.; Wang, X.; Jiang, J.; Ren, F.; Liu, H. Spatial pattern, hydrogeochemical controlling processes and non-carcinogenic risks of fluoride-enriched groundwater in the North Henan Plain, Northern China. *Appl. Geochem.* **2024**, *163*, 105934. [CrossRef]
36. Deiana, M.; Mussi, M.; Pennisi, M.; Boccolari, M.; Corsini, A.; Ronchetti, F. Contribution of water geochemistry and isotopes (δ18O, δ2H, 3H, 87Sr/86Sr and δ11B) to the study of groundwater flow properties and underlying bedrock structures of a deep landslide. *Environ. Earth Sci.* **2019**, *79*, 30. [CrossRef]
37. Morin-Crini, N.; Lichtfouse, E.; Liu, G.; Balaram, V.; Ribeiro, A.R.L.; Lu, Z.; Stock, F.; Carmona, E.; Teixeira, M.R.; Picos-Corrales, L.A.; et al. Worldwide cases of water pollution by emerging contaminants: A review. *Environ. Chem. Lett.* **2022**, *20*, 2311–2338. [CrossRef]
38. Richards, L.A.; Kumari, R.; White, D.; Parashar, N.; Kumar, A.; Ghosh, A.; Kumar, S.; Chakravorty, B.; Lu, C.; Civil, W.; et al. Emerging organic contaminants in groundwater under a rapidly developing city (Patna) in northern India dominated by high concentrations of lifestyle chemicals. *Environ. Pollut.* **2021**, *268*, 115765. [CrossRef]
39. Qu, S.; Duan, L.; Shi, Z.; Mao, H.; Wang, G.; Liu, T.; Yu, R.; Peng, X. Identifying the spatial pattern, driving factors and potential human health risks of nitrate and fluoride enriched groundwater of Ordos Basin, Northwest China. *J. Clean. Prod.* **2022**, *376*, 134289. [CrossRef]
40. Deléglise, C.; François, H.; Loucougaray, G.; Crouzat, E. Facing drought: Exposure, vulnerability and adaptation options of extensive livestock systems in the French Pre-Alps. *Clim. Risk Manag.* **2023**, *42*, 100568. [CrossRef]
41. Gugulothu, S.; Subbarao, N.; Das, R.; Dhakate, R. Geochemical evaluation of groundwater and suitability of groundwater quality for irrigation purpose in an agricultural region of South India. *Appl. Water Sci.* **2022**, *12*, 142. [CrossRef]
42. Alam, S.M.K.; Li, P.; Fida, M. Groundwater Nitrate Pollution Due to Excessive Use of N-Fertilizers in Rural Areas of Bangladesh: Pollution Status, Health Risk, Source Contribution, and Future Impacts. *Expo. Health* **2024**, *16*, 159–182. [CrossRef]
43. Jiang, W.; Meng, L.; Liu, F.; Sheng, Y.; Chen, S.; Yang, J.; Mao, H.; Zhang, J.; Zhang, Z.; Ning, H. Distribution, source investigation, and risk assessment of topsoil heavy metals in areas with intensive anthropogenic activities using the positive matrix factorization (PMF) model coupled with self-organizing map (SOM). *Environ. Geochem. Health* **2023**, *45*, 6353–6370. [CrossRef] [PubMed]
44. Ma, W.; Wu, T.; Wu, X.; Yang, S.; Li, R.; Zhou, S.; Li, X.; Zhu, X.; Hao, J.; Ni, J.; et al. Carbon budgets and environmental controls in alpine ecosystems on the Qinghai-Tibet Plateau. *Catena* **2023**, *229*, 107224. [CrossRef]

45. Abidi, J.H.; Elzain, H.E.; Sabarathinam, C.; Selmane, T.; Selvam, S.; Farhat, B.; Ben Mammou, A.; Senapathi, V. Evaluation of groundwater quality indices using multi-criteria decision-making techniques and a fuzzy logic model in an irrigated area. *Groundw. Sustain. Dev.* **2024**, *25*, 101122. [CrossRef]
46. Siddique, M.; Islam, A.; Hasanuzzaman, M.; Khan, R.; Akbor, M.A.; Hossain, M.; Sajid, M.; Mia, M.; Mallick, J.; Rahman, M.; et al. Multivariate statistics and entropy theory for irrigation water quality and entropy-weighted index development in a subtropical urban river, Bangladesh. *Environ. Sci. Pollut. Res.* **2022**, *29*, 8577–8596. [CrossRef]
47. Rahman, M.A.; Siddique, M.A.B.; Khan, R.; Reza, A.H.M.S.; Khan, A.H.A.N.; Akbor, M.A.; Islam, M.S.; Hasan, A.B.; Hasan, M.I.; Elius, I.B. Mechanism of arsenic enrichment and mobilization in groundwater from southeastern Bangladesh: Water quality and preliminary health risks assessment. *Chemosphere* **2022**, *294*, 133556. [CrossRef]
48. Haji, M.; Karuppannan, S.; Qin, D.; Shube, H.; Kawo, N.S. Potential Human Health Risks Due to Groundwater Fluoride Contamination: A Case Study Using Multi-techniques Approaches (GWQI, FPI, GIS, HHRA) in Bilate River Basin of Southern Main Ethiopian Rift, Ethiopia. *Arch. Environ. Contam. Toxicol.* **2021**, *80*, 277–293. [CrossRef]
49. Chen, F.; Yao, L.; Mei, G.; Shang, Y.; Xiong, F.; Ding, Z. Groundwater Quality and Potential Human Health Risk Assessment for Drinking and Irrigation Purposes: A Case Study in the Semiarid Region of North China. *Water* **2021**, *13*, 783. [CrossRef]
50. Khan, S.; Ullah, Q.; Khan, A.A.; Hassan, S.S.; Shakoor, A.; Ijaz, M. Geostatistical investigation of groundwater quality zones for its applications in irrigated agriculture areas of Punjab (Pakistan). *Environ. Earth Sci.* **2022**, *81*, 91. [CrossRef]
51. Li, R.; Yan, Y.; Xu, J.; Yang, C.; Chen, S.; Wang, Y.; Zhang, Y. Evaluate the groundwater quality and human health risks for sustainable drinking and irrigation purposes in mountainous region of Chongqing, Southwest China. *J. Contam. Hydrol.* **2024**, *264*, 104344. [CrossRef] [PubMed]
52. Mukate, S.; Wagh, V.; Panaskar, D.; Jacobs, J.A.; Sawant, A. Development of new integrated water quality index (IWQI) model to evaluate the drinking suitability of water. *Ecol. Indic.* **2019**, *101*, 348–354. [CrossRef]
53. Benaafi, M.; Pradipta, A.; Tawabini, B.; Al-Areeq, A.M.; Bafaqeer, A.; Humphrey, J.D.; Nazal, M.K.; Aljundi, I.H. Suitability of treated wastewater for irrigation and its impact on groundwater resources in arid coastal regions: Insights for water resources sustainability. *Heliyon* **2024**, *10*, e29320. [CrossRef] [PubMed]
54. *GB/T 14848-2017*; Standards for Groundwater Quality. General Administration of Quality Supervision: Beijing, China, 2017.
55. WHO. *Guidelines for Drinking-Water Quality*, 4th ed.; Incorporating the First Addendum; World Health Organization: Geneva, Switzerland, 2017.
56. Zhang, Y.; Xiao, Y.; Yang, H.; Wang, S.; Wang, L.; Qi, Z.; Han, J.; Hao, Q.; Hu, W.; Wang, J. Hydrogeochemical and isotopic insights into the genesis and mixing behaviors of geothermal water in a faults-controlled geothermal field on Tibetan Plateau. *J. Clean. Prod.* **2024**, *442*, 140980. [CrossRef]
57. Gibbs, R.J. Mechanisms controlling world water chemistry. *Science* **1970**, *170*, 1088–1090. [CrossRef] [PubMed]
58. Xiao, Y.; Hao, Q.; Zhang, Y.; Zhu, Y.; Yin, S.; Qin, L.; Li, X. Investigating sources, driving forces and potential health risks of nitrate and fluoride in groundwater of a typical alluvial fan plain. *Sci. Total Environ.* **2022**, *802*, 149909. [CrossRef]
59. Francis, V.; Krishnaraj, S.; Kumar, S.; Andiyappan, R.K.; Govindan, P. GIS-based groundwater potential zonation and assessment of groundwater quality and suitability for drinking and irrigation purposes in the Shanmughanadhi river basin, south India. *Kuwait J. Sci.* **2024**, *51*, 100243. [CrossRef]
60. Han, R.; Liu, W.; Zhang, J.; Zhao, T.; Sun, H.; Xu, Z. Hydrogeochemical characteristics and recharge sources identification based on isotopic tracing of alpine rivers in the Tibetan Plateau. *Environ. Res.* **2023**, *229*, 115981. [CrossRef]
61. Kang, X.B.; Xu, M.; Li, K.; Liu, X. Exploration and utilisation of groundwater resources in extreme arid regions of the Tibetan Plateau: A case study on Ali area, China. *J. Earth Syst. Sci.* **2020**, *129*, 208. [CrossRef]

water

MDPI

Article

Investigation on the Hydrogeochemical Characteristics and Controlling Mechanisms of Groundwater in the Coastal Aquifer

Guangnan An [1,2], Hao Kang [2], Rongbing Fu [1,*], Damao Xu [1] and Jia Li [2]

[1] College of Environment Science and Engineering, Tongji University, Shanghai 200092, China
[2] Appraisal Center for Environment & Engineering, Ministry of Ecology and Environment, Beijing 100041, China
* Correspondence: furongbing@tongji.edu.cn

Abstract: Groundwater contamination in coastal areas has attracted widespread attention. However, studies on the hydrogeochemical characteristics and controlling mechanisms in coastal aquifers are still lacking. In this study, 71 sets of groundwater samples were collected during the dry and wet seasons in a coastal city, Shandong Province. Correlation and principal component analyses were used to identify pollution sources. Meanwhile, Piper diagrams, Gibbs plots, ion ratios, and saturation indices were employed to investigate the hydrogeochemical controlling mechanisms. The results revealed that pollution components included Na^+, $NH_4{}^+$, Cl^-, $SO_4{}^{2-}$, $NO_3{}^-$, $NO_2{}^-$, Pb, As, Se, TDS, TH, F^-, and Mn. Pollution compositions in the study area were primarily derived from natural processes and anthropogenic activities. The contamination of nitrogen resulted primarily from agricultural activities. The exceedance of $SO_4{}^{2-}$ was mainly due to the leaching of waste by rainfall. High Na^+, Cl^-, and F^- were related to sea intrusion. Pb and Se might have been caused by anthropogenic activities. The exceedance of As was caused by anthropogenic inputs and natural factors. The poor seepage conditions and anoxic conditions promoted the enrichment of Mn. The concentration of most components in the dry season was larger than that in the wet season. There were no significant differences in water chemistry type during the wet season and dry season. Groundwater chemical compositions were dominated by the dissolution of halite, gypsum, and anhydrite, as well as the cation exchange reaction. The influence of seawater intrusion on groundwater was not serious.

Keywords: seawater intrusion; pollution source; dry and wet season; groundwater

1. Introduction

Groundwater is significant resource for industrial utilization, agricultural irrigation, and domestic drinking in the world [1–3]. However, the quality of groundwater is under serious threat, including agricultural activities, heavy-metal contamination, and seawater intrusion [4–6]. Groundwater quality is suffering deterioration in urban areas, especially in coastal regions [7]. Once an aquifer system is severely polluted, it is challenging to remediate it to the initial environment [8–10]. With approximately 40% of the world's population living within 100 km of a coastal area, freshwater deposited in coastal aquifers has been an important resource to support the development of coastal areas, providing drinking water for almost one billion people in coastal areas worldwide [11,12]. Due to the high intensity of human activities, water pollution, eutrophication, and seawater intrusion in coastal areas are becoming increasingly serious, threatening human health [13–15]. Many coastal areas, including the Netherlands, Italy, and China's Jiaodong Peninsula are suffering from a serious lack of water [16–19]. Therefore, a comprehensive knowledge of the hydrogeochemical processes in the coastal aquifer is critical for these valuable water resources.

Seawater intrusion is considered to be a widespread environmental hazard that affects groundwater in many coastal areas [20,21]. Even 1% mixing of seawater will make groundwater undrinkable and cause a series of problems, including soil salinization and ecosystem degradation [11]. In coastal areas, it is important to study groundwater chemistry due to the different degrees of mixing interaction between seawater and groundwater [22,23]. The hydrogeochemical processes happening in groundwater systems and the interactions with the media influence the quality of groundwater [24]. Moreover, analysis of groundwater chemical characteristics can trace the origin and evolution of groundwater [25]. The primary processes influencing the hydrogeochemistry and groundwater quality are water–rock interactions, particularly dissolution and weathering processes, carbonate balance, oxidation–reduction, and adsorption–desorption, in addition to human activities [22,26]. Therefore, prevention of groundwater pollution in coastal aquifers necessitates understanding the processes of geochemical evolution and dynamics of flow systems [27].

Several studies have been implemented to investigate the hydrogeochemical characteristics of coastal areas and the effects of seawater intrusion. Multivariate statistical analyses and clustering methods, including principal component analysis (PCA), factor analysis (FA), and hierarchical cluster analysis (HCA), have been employed to investigate the groundwater quality and pollution source [28–33]. Meanwhile, Isotopic methods are also often an important tool for studying seawater intrusion and groundwater evolution [16,27,34]. Hydrogen and oxygen isotopes can reveal the cause of groundwater salinization and the source of recharge. Strontium is widely used to study groundwater–rock interactions, mixing processes, and the discharge of seawater to groundwater. Furthermore, Piper diagrams, Gibbs plots, ion ratios, and saturation indices play a significant role in studying the hydrogeochemical controlling mechanisms [35–40]. In addition, numerical simulation methods can be used to evaluate the exchange of groundwater with seawater and the transport characteristics of nutrients [41–44]. Water quality indices (WQIs), machine learning, and the partial least squares regression model (PLSR) are typically used to evaluate the pollution level in groundwater [45–47].

Groundwater in the coastal cities adjacent to the Yellow Sea are at risk of seawater intrusion. Considering the uses of groundwater, it can be potentially contaminated by anthropogenic factors. The contamination of coastal aquifers involves complex hydrogeochemical processes which control the groundwater environment in different ways. Therefore, it is desirable to identify the pollution source and investigate the controlling processes of water chemistry. Although scholars have reported the influence scope of seawater intrusion, the hydrogeochemistry and controlling mechanisms of groundwater in the dry and wet seasons are still poorly characterized. The objectives of this work were to (1) identify the pollution source in the study area, (2) investigate the hydrogeochemical characteristics and controlling mechanisms of groundwater, and (3) determine the spatiotemporal variation of groundwater hydrochemistry.

2. Study Area

The study area was located in Shandong Province (Figure 1), with geographical coordinates of 119°19′14″ to 119°39′38″ E longitude and 35°18′05″ to 35°36′27″ N latitude. The general topography of the study area is high in the northwest, sloping slightly to the southeast, with mostly coastal plains in the east and south. The study area is located in the central part of the Qinling–Dabie–Sulu orogenic belt, and the exposed strata include the Paleozoic Jingshan Group and the Cenozoic Quaternary (Table S1); a regional fracture structure is present, mostly in the northeast direction. At the same time, magmatic rocks are widely distributed in the study area (Figure S1).

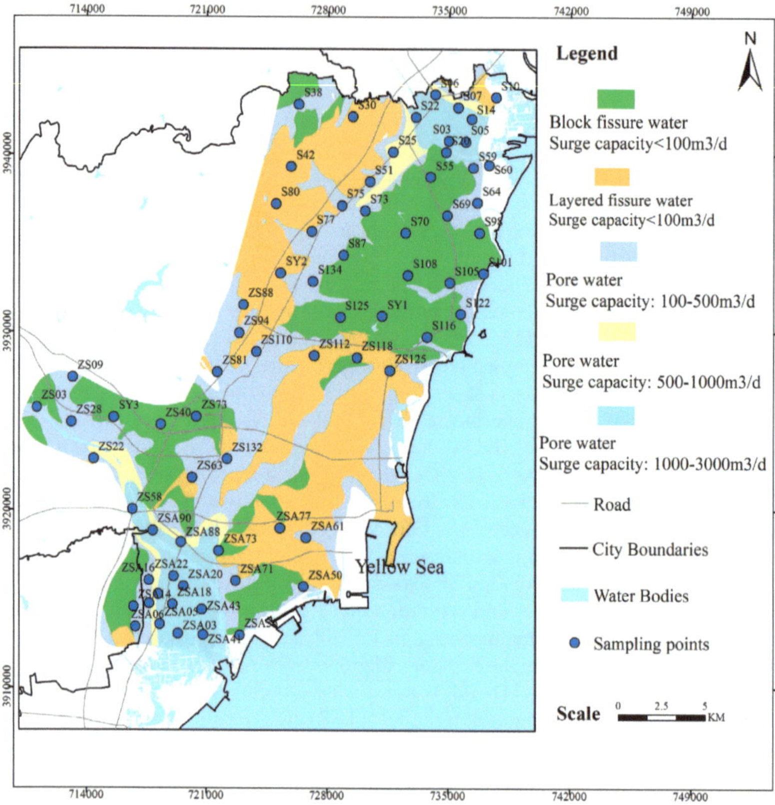

Figure 1. Location map of the study area and the distribution of sampling wells.

The climate of the study area is temperate maritime monsoon climate. The average humidity is about 72%. The mean annual temperature is about 12 °C, and the mean annual rainfall reaches 826.8 mm. The mean annual evaporation reaches 1299.0 mm. The water system in the study area features the Yellow Sea in the southeast, and the water system is dendritic in shape. The river is mainly recharged from natural precipitation, and the river flows into the Yellow Sea from west to east. According to the conditions of groundwater deposit, groundwater can be classified into four types: loose rock-like pore water, clastic rock-like fracture water, carbonate karst water, and bedrock fracture water. Near the coastal area, the groundwater type is primarily loose rock pore water and bedrock fracture water. Groundwater is recharged through atmospheric precipitation, river infiltration, and irrigation infiltration. On the whole, the hydraulic gradient of the groundwater is small, and the groundwater flows from northwest to southeast, before finally flowing into the Yellow Sea (Figures S3 and S4). Groundwater is discharged through artificial exploitation, evaporation, and runoff.

3. Materials and Methods

3.1. Samples Collection and Analysis

In order to investigate the hydrogeochemistry of the study area during the dry season and wet season, we collected 71 sets of groundwater samples in April and July 2022. The distributions of the sampling wells are shown in Figure 1. A peristaltic pump (AML919) was used to collect groundwater samples. Then, the samples were filtered through 0.45 μm membrane into polypropylene bottles. Samples for total Fe and Mn analysis were acidified using a nitric acid reagent. In the field, all samples were stored in a thermostat with ice packs. Samples were stored in a refrigerator after transport to the laboratory. The major anions were analyzed by ion chromatography (ICS-2000). Cations were determined by inductively coupled plasma mass spectrometry (ICP-5000). Sample testing was completed at the Eighth Geological Brigade of Shandong Province Geology and Mining Bureau. The sample collection, storage and test methods were in strict compliance with the standards of the Ministry of Natural Resources of China. A quality assurance (QA)/quality control (QC) program was established to guarantee the precision of the measurements. Parallel samples were utilized to control the accuracy of the test. The charge balance error was guaranteed to be within 5%.

3.2. *Saturation Index*

The saturation index (*SI*) is significant for assessing the equilibrium and reactivity between minerals and groundwater [48]. The ratio of the activity product to solubility product is employed to evaluate the hydrogeochemical state and reaction in groundwater. A saturation index larger than 0 means that the mineral is supersaturated and will precipitate. A saturation index less than 0 indicates that the mineral is unsaturated and will dissolve. In this work, PHREEQC version 2.8 was applied to determine the *SI* value of different minerals.

$$SI = \log_{10}\frac{IAP}{LP},$$

where *IAP* represents the ratio of activity product, and *LP* represents the solubility product.

3.3. *Multivariate Statistical Analyses*

The groundwater controlling mechanism can be determined using graphical and multivariate statistical methods, such as Piper plots, Gibbs plots, and ion ratios. Multivariate statistical methods play an important role in identifying pollution sources, such as correlation analysis and PCA, using GIS and Origin.

4. Results and Discussion

4.1. *Groundwater Chemistry Characteristics*

Statistical descriptions of the hydrogeochemical data in the dry and wet seasons are shown in the Table 1. Each component exhibited different concentration variations. The range of pH from 7.24 to 8.52 in the dry season and the range of pH from 7.69 to 8.89 in the wet season indicated that the groundwater was predominantly alkaline. The value of TDS ranged from 209.22 mg/L to 8880.98 mg/L, with a mean of 675.87 mg/L, in the dry season, and ranged from 161.32 mg/L to 4064.75 mg/L, with a mean of 571.49 mg/L, in the wet season, indicating that the seasonal variation of TDS was significant. For most components except NO_2^-, Ni, Se, F^-, and Fe, the mean concentration in the dry season was larger than that in the wet season. The mean concentration of major ions during the dry season was in the order $Na^+ > Ca^{2+} > Mg^{2+} > K^+$ and $HCO_3^- > Cl^- > SO_4^{2-} > NO_3^-$. The sequence changed to $Ca^{2+} > Na^+ > Mg^{2+} > K^+$ and $HCO_3^- > Cl^- > SO_4^{2-} > NO_3^-$ during the wet season. The results of CV showed that Na^+, K^+, Mg^{2+}, NH_4^+, Cl^-, NO_3^-, NO_2^-, Pb, As, Br, TDS, and Mn were sensitive to environmental and anthropogenic activities in the dry season. The spatial distribution of Na^+, K^+, NH_4^+, Cl^-, NO_3^-, NO_2^-, Pb, As, Br, and Mn was highly heterogeneous.

Table 1. Statistic descriptions of main ionic parameters in groundwater.

Component	Dry Season					Wet Season				
	Max	Min	Mean	SD	CV	Max	Min	Mean	SD	CV
Na^+ (mg/L)	2962.00	12.00	96.58	346.41	358.69	936.00	6.73	59.77	112.09	187.53
K^+ (mg/L)	69.50	0.29	7.13	12.29	172.39	65.00	0.54	5.82	9.88	169.71
Ca^{2+} (mg/L)	240.00	27.20	94.06	41.73	44.36	223.00	31.70	88.47	39.34	44.46
Mg^{2+} (mg/L)	266.00	6.72	30.56	31.65	103.57	80.60	3.53	22.93	13.60	59.32
NH_4^+ (mg/L)	5.00	0.00	0.12	0.60	490.19	1.50	0.00	0.03	0.18	524.34
Cl^- (mg/L)	4565.69	1.97	132.99	539.24	405.46	2316.95	5.08	104.85	277.82	264.97
SO_4^{2-} (mg/L)	761.98	15.11	108.56	94.43	86.98	445.29	14.09	93.23	63.45	68.06
HCO_3^- (mg/L)	878.00	26.91	207.84	131.38	63.21	528.14	20.18	198.14	99.35	50.14
NO_3^- (mg/L)	648.90	1.27	97.53	110.81	113.62	652.17	1.50	88.12	109.19	123.91
NO_2^- (mg/L)	22.27	0.00	0.64	2.69	423.75	13.91	0.00	0.80	2.37	297.38
Pb (ug/L)	57.32	0.36	1.78	6.91	387.06	20.66	0.36	1.49	2.93	196.34
Ni (ug/L)	11.62	0.67	3.17	2.29	72.11	39.50	1.84	6.27	4.66	74.39
As (ug/L)	27.92	0.48	2.57	3.96	153.98	16.65	0.48	2.40	2.76	115.14
Se (ug/L)	15.02	1.64	1.93	1.63	84.54	16.87	1.64	2.53	2.21	87.23
Br (mg/L)	13.01	0.02	1.02	1.61	158.24	6.85	0.02	0.71	1.18	165.34
TDS (mg/L)	8880.98	209.22	675.87	1015.47	150.25	4064.75	161.32	571.49	489.41	85.64
TH (mg/L)	1388.03	141.71	361.17	192.22	53.22	853.52	107.46	331.93	142.85	43.04
F^- (mg/L)	2.28	0.01	0.66	0.32	47.83	1.19	0.01	0.70	0.19	27.59
Fe (mg/L)	0.11	0.02	0.02	0.01	55.74	0.02	0.02	0.02	0.00	0.00
Mn (mg/L)	2.97	0.00	0.28	0.58	207.78	2.10	0.00	0.10	0.34	319.66
pH	8.52	7.24	7.94	0.30	358.69	8.89	7.69	8.25	0.19	2.25

4.2. Source Identification of Pollution Components

According to Class III groundwater quality standards, the contamination components included Na^+, NH_4^+, Cl^-, SO_4^{2-}, NO_3^-, NO_2^-, Pb, As, Se, TDS, TH, F^-, and Mn in groundwater. In order to identify the source for each pollution composition, the correlation coefficients between the components were plotted in the dry season and in the wet season (Figure 2). In the dry season, there was strong positive correlation between Na^+ and Mg^{2+}, Cl^-, SO_4^{2-}, Se, Br, and TDS. There was strong positive correlation between NH_4^+ and NO_2^-. Meanwhile, Cl^- showed strong positive correlation with SO_4^{2-}, Se, Br, and TDS. There was a significant correlation between NO_3^- and Ca^{2+}. As, Se, and TDS showed strong positive correlation. Moreover, there was significant correlation between As and Na^+, K^+, Mg^{2+}, Cl^-, and SO_4^{2-}. However, there was no correlation between Pb and other components.

In the wet season, the correlation of the different components showed some changes with the dry season. There was strong positive correlation between Na^+ and Cl^-, Se, and TDS. There was a significant correlation between NO_3^- and Ca^{2+}, TH. Likewise, there was no correlation between Pb and other components. As and K^+ showed a significant correlation. In the dry season, the concentration of Ni exhibited significant differences between dry season (concentration < 0.05 mg/L) and wet season (concentration > 0.05 mg/L). It could be seen that Ni exhibited a low correlation with Ca^{2+} and SO_4^{2-}.

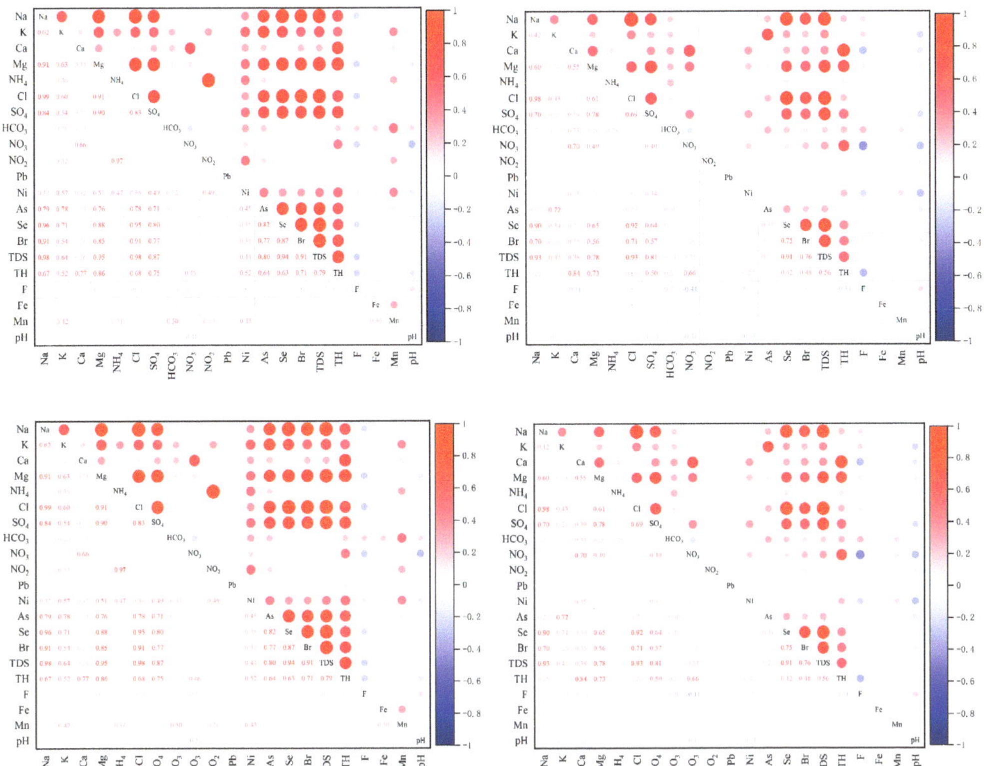

Figure 2. Correlation matrix of major ions in groundwater during the dry season (**left**) and wet season (**right**). The size of the solid circles represents the correlation degree.

Correlation analysis and principal component analysis were utilized to identify sources of pollution components in groundwater. The variance contribution of PC1 reached 41.2% in the dry season (Figure 3). The variance contribution of PC1 reached 33.0% in the wet season. Four principal constituents were extracted for dry season and wet season, contributing to 73% and 63.6% of total variance, respectively.

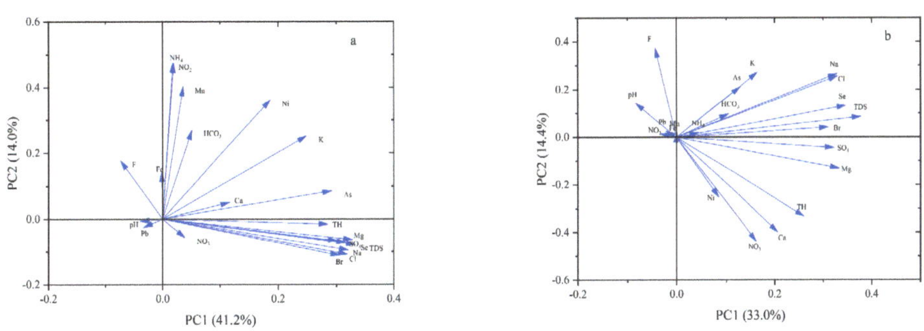

Figure 3. Principal component analysis of different components in the dry season (**a**) and the wet season (**b**).

The strong positive relationships among these components supported these same sources. In the dry season, according to the distribution characteristics of Na^+ and Cl^-, it could be seen that the high concentration of Na^+ and Cl^- was mainly distributed along the Yellow Sea (Figures 4 and 5), but the concentration of Na^+ at only one sampling point (ZSA41) exceeded the standard for Class III groundwater. The correlation coefficient between Na^+ and Cl^-, Se, and TDS reached more than 95%, indicating that these components came from same pollution source. Seawater intrusion could play a significant role in the contamination process. Consequently, aquifers close to the coastal zone are at risk of potential seawater intrusion. The SO_4^{2-} exceeded the standard at sampling points S134 and ZSA41. The hydrogeological investigation revealed the presence of some solid wastes in the study area, and the components in the wastes could enter the groundwater under rainfall leaching. The sampling site (ZSA41) was distributed in an industrial agglomeration area, and the water table at this site was significantly lower than the surrounding water table; therefore, seawater intrusion was very serious. At the same time, the industrial activities led to the exceedance of SO_4^{2-}. The concentration of As at two sampling sites (ZSA41 and S10) in the study area exceeded 10 ug/L. On one hand, the input of anthropogenic contaminants caused the exceedance of arsenic. On the other hand, the high pH promoted the desorption of As from the sediment into the groundwater at S10. The concentration of Se exceeded 10 ug/L at only one sampling site (ZSA41), which could have been the result of human activities. The correlation coefficient of TDS with Na^+ and Cl^- reached 98%, suggesting that seawater intrusion contributed to the increase in TDS. Furthermore, the correlation coefficient between TDS and SO_4^{2-} reached 87%, indicating that anthropogenic pollution also contributed to the high TDS. Therefore, the pollution source of elements in PC1 were explained by sea intrusion and anthropogenic activities. Mn, NH_4^+, and NO_2^- exhibited high loadings in PC2. The seepage condition in the study area was poor, and the sediments were under a relatively hypoxic environment, which offered a beneficial condition for the enrichment of Mn. NH_4^+ and NO_2^- were mainly from anthropogenic input. Ca and NO_3^- exhibited high loadings in PC3. The exceedance rate of NO_3^- reached 86%, and the correlation with other pollution components was relatively weak. The discharge of sewage and the usage of fertilizers resulted in excessive NO_3^- concentration. The main source of Ca was the dissolution of gypsum and anhydrite. Therefore, components in PC3 were mainly derived from natural sources and anthropogenic input. HCO_3^- was the main component in PC4, and HCO_3^- came mainly from natural sources. The concentration of Pb exceeded 10 ug/L at only one sampling site (ZSA77), which could have been caused by human activities. According to the distribution characteristics of F^-, the sampling points with a high concentration of F^- were distributed near the sea, and the concentration at ZSA43, S98, and S69 exceeded the limit, suggesting that seawater intrusion played a significant role in the pollution process. The high TH was mainly contributed by Ca^{2+} and Mg^{2+}. Meanwhile, the correlation coefficient with TDS reached 79%, indicating a similar source to TDS. Thus, groundwater in the study area was influenced by both natural processes and human activities in the dry season.

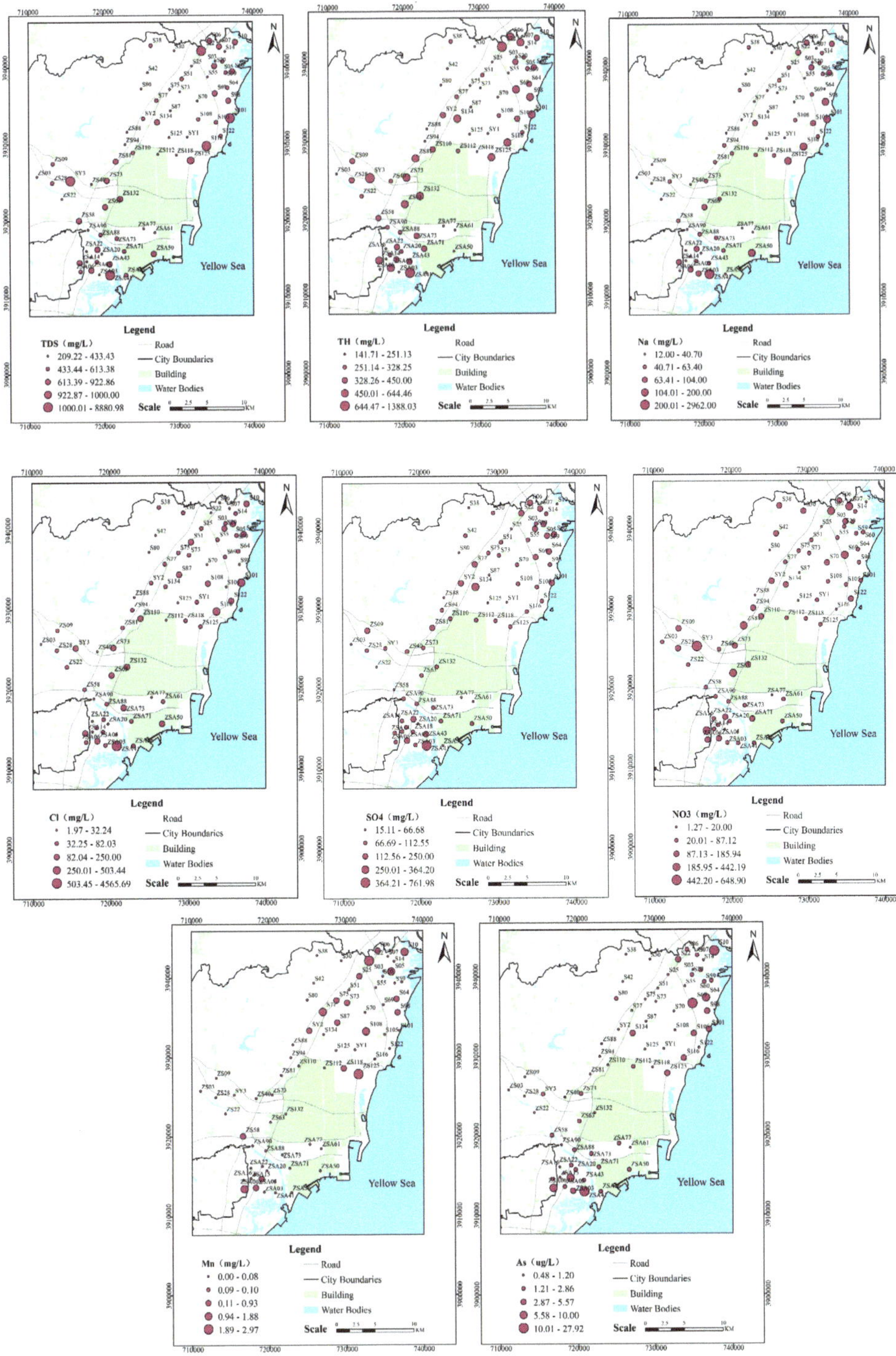

Figure 4. Spatial distribution of pollution components in the dry season.

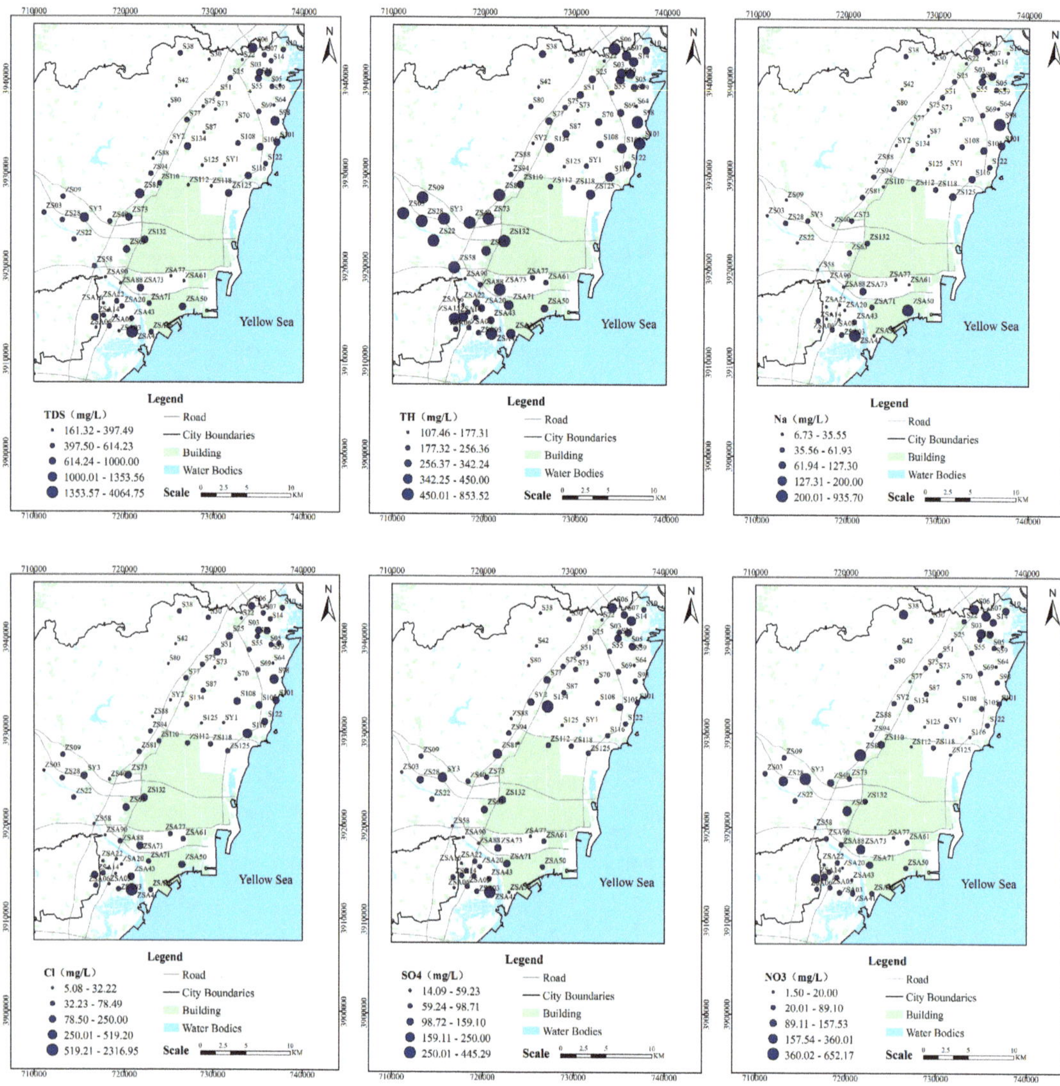

Figure 5. *Cont.*

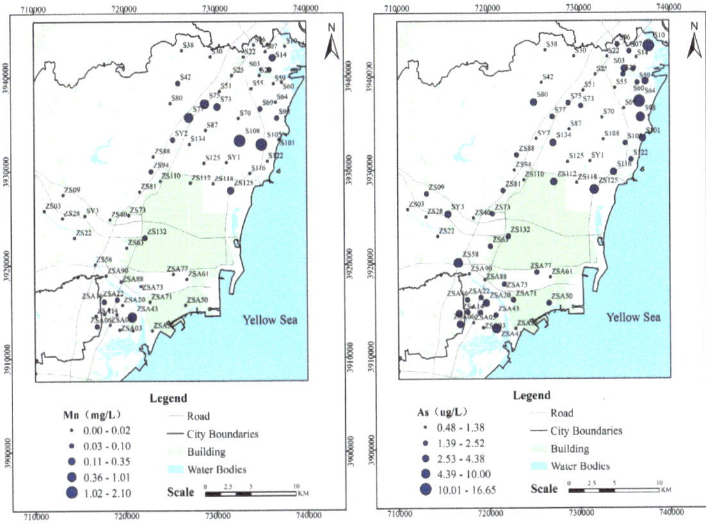

Figure 5. Spatial distribution of pollution components in the wet season.

In the wet season, the components in PC1 were similar to those in the dry season, whereas the components in PC2, PC3, and PC4 were different from those in the dry period. F^- exhibited high loading in PC2. As and K^+ were the main components in PC3. NH_4^+ and HCO_3^- exhibited high loadings in PC4. For most of the pollution components, the concentration in the dry season was greater than that in the wet season, indicating that rainfall reduced the concentration of ions in groundwater.

4.3. Hydrogeochemical Facies

Piper diagrams can be employed to understand the chemical components of groundwater and to determine the hydrogeochemical facies [49]. All sampling points in the dry and wet seasons are plotted in Figure 6. Groundwater in the study area was mainly Ca–HCO_3 type, Ca–Mg–Cl type, and Ca–Cl type, followed by Na–Cl type and Ca–Na–HCO_3 type in the dry season. The hydrogeochemical facies in the wet season were similar to those in the dry season.

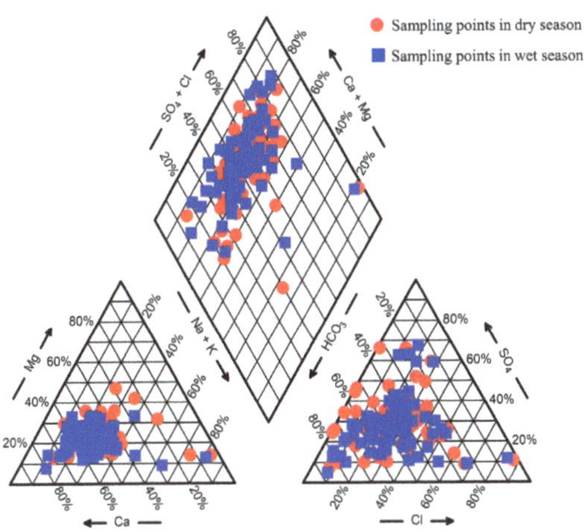

Figure 6. Piper diagrams during the dry and wet seasons.

4.4. Mechanisms Controlling Groundwater Chemistry

Gibbs diagrams are built by drawing the ratio of $Na^+/(Na^+ + Ca^{2+})$ and $Cl^-/(Cl^- + HCO_3^-)$ to TDS and are extensively employed to identify hydrogeochemical processes such as precipitation, water–rock interactions, and evaporation [50]. If TDS is high and the ratio of $Na^+/(Na^+ + Ca^{2+})$ or $Cl^-/(Cl^- + HCO_3^-)$ is close to 1, the governing mechanism is an evaporation effect. If TDS is medium and the ratio of $Na^+/(Na^+ + Ca^{2+})$ or $Cl^-/(Cl^- + HCO_3^-)$ is less than 0.5, the governing mechanism is water–rock interaction. If TDS is low and the ratio of $Na^+/(Na^+ + Ca^{2+})$ or $Cl^-/(Cl^- + HCO_3^-)$ is close to 1, the controlling mechanism is a precipitation effect. As shown in Figure 7, most groundwater samples were distributed in the water–rock-dominant area, suggesting that water chemical components were mainly derived from the water–rock interaction in the region. One sample (ZSA41) was located in the evaporation-dominant area during the dry and wet seasons. The concentration of Na^+ and Cl^- at the sampling site far exceeded the Class III water standard. Therefore, it was not representative of natural interaction processes. This also reflected the impact of seawater intrusion on groundwater.

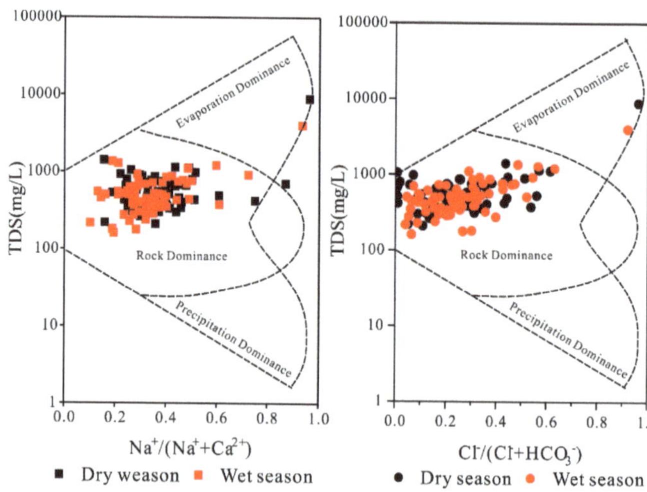

Figure 7. Gibbs diagrams showing the source of water chemistry components.

4.5. The Ratio of Major Ions

In order to investigate the details of water–rock interaction in the region, some ratio maps of major ions were performed. The relationships of Na^+/Cl^- for the dry season and wet season are displayed in Figure 8a. In general, if Na^+ and Cl^- are derived from the dissolution of halite, then the ratio of Na^+/Cl^- should be about equivalent to 1. If the ratio is larger than 1, then Na^+ is primarily derived from silicate weathering. In this study, a ratio approximately equal to 1 was observed for most samples, indicating that the dissolution of halite was one of the sources of Na^+ and Cl^- during the dry and wet seasons. The saturation index showed that halite was always in the dissolved state.

The sources of Ca^{2+} and SO_4^{2-} are commonly the dissolution of gypsum and anhydrite. Figure 8b displays that most of points were located around the 1:1 line. It could be concluded that the dissolution of gypsum and anhydrite was the main source of Ca^{2+} and SO_4^{2-}. Some points were distributed above the Ca^{2+}/SO_4^{2-} ratio 1:1 line, suggesting that the dissolution of aragonite, calcite, and dolomite was a potential source of Ca^{2+} in groundwater. However, the SI of aragonite, calcite, and dolomite for most samples was higher than 0, implying that the contribution of aragonite, calcite, and dolomite to Ca^{2+} was limited. Therefore, the cation exchange process could have contributed to the increase in Ca^{2+}.

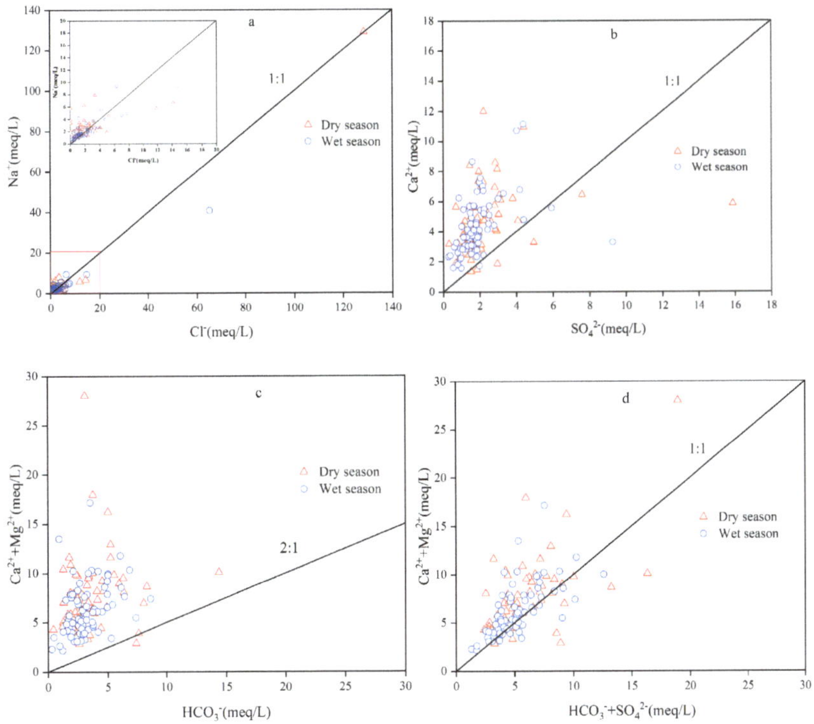

Figure 8. Diagrams showing the correlations of major cations/anions to distinguish the hydrogeochemical processes. (**a**) Na^+ versus Cl^-; (**b**) Ca^{2+} versus $SO_4{}^{2-}$; (**c**) (Ca^{2+} + Mg^{2+}) versus $HCO_3{}^-$; (**d**) (Ca^{2+} + Mg^{2+}) versus ($HCO_3{}^-$ + $SO_4{}^{2-}$).

The relationship between Ca^{2+} + Mg^{2+} and $HCO_3{}^-$ can reflect the source of Ca^{2+} and Mg^{2+}. In this study, the ratio of almost all samples was greater than 0.5, indicating the presence of extra sources of Ca^{2+} and Mg^{2+}, with a lower contribution of carbonate dissolution to Ca^{2+} and Mg^{2+}. This could be confirmed by the SI of aragonite, calcite, and dolomite (Figure 9).

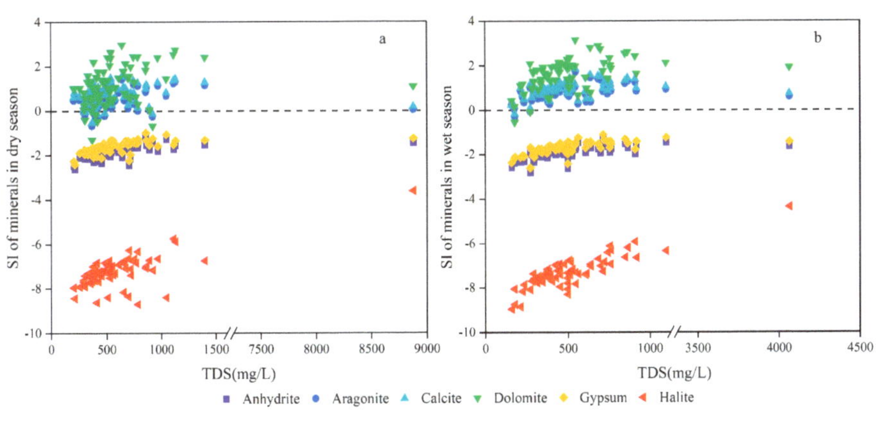

Figure 9. Saturation index diagram against TDS: (**a**) SI of minerals in dry season; (**b**) SI of minerals in wet season.

The $(Ca^{2+} + Mg^{2+})/(HCO_3^- + SO_4^{2-})$ relationship was plotted to characterize the mineralization processes. The plot showed that most points were distributed around the 1:1 line. As mentioned above, the contribution of carbonate dissolution was small; therefore, the dissolution of sulfate minerals was the primary source. Therefore, it could be concluded that the dissolution of gypsum and anhydrite was a significant source of major ions.

The $(Na^+/Cl^-)/Cl^-$ and $(SO_4^{2-}/Cl^-)/Cl^-$ ratios can be used to study the degree of mixing of groundwater with seawater and the dissolution of minerals (Figure 10). Most sampling points were located above the standard seawater ratio line ($Na^+/Cl^- = 0.85$), suggesting a contribution of mineral dissolution to the whole solute loading. Most sampling points were located above the standard seawater ratio line ($Na^+/Cl^- = 0.11$), showing some enrichment of sulfate.

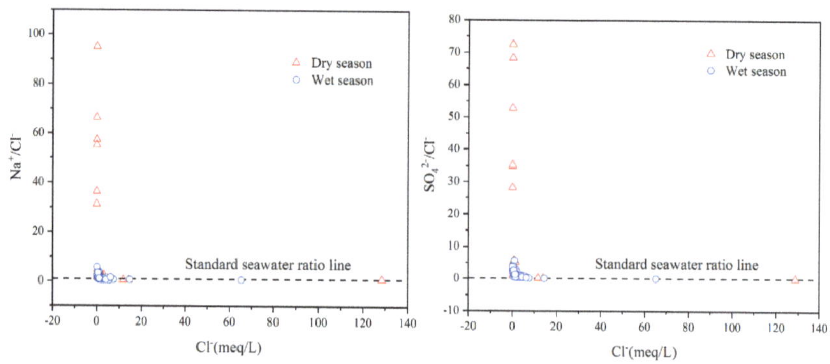

Figure 10. Evolution of hydrogeochemical ratios with rising salinity.

4.6. Seawater Fraction in Groundwater

In order to better characterize the process of seawater mixing into groundwater, the contribution of seawater needs to be calculated. Since Cl^- does not react with other ions or minerals, the contribution of seawater in groundwater can be evaluated by using Cl^- as a tracer. The following formula is employed to calculate the proportion of seawater [7]:

$$f_{sea} = \frac{m_{Cl(sample)} - m_{Cl(freshwater)}}{m_{Cl(seawater)} - m_{Cl(freshwater)}},$$

where $m_{Cl(sample)}$ indicates the level of the sample, $m_{Cl(seawater)}$ indicates the level of Cl^- in seawater, and $m_{Cl(freshwater)}$ indicates the level of Cl^- in freshwater.

The findings demonstrated that the contributions of seawater to groundwater were minimal. During the dry season, the contribution of seawater was greater than 1% at only three sampling sites: S101 (2.2%), S116 (2.6%), and ZSA41 (23.6%). The contribution of seawater to groundwater was less than 1% for the remaining sampling sites. During the wet season, the contribution of seawater was greater than 1% at seven sampling sites: S98 (2.7%), S101 (1.3%), S105 (1.1%), S116 (1.4%), SY3 (1.0%), ZSA41 (12.0%), and ZSA50 (1.2%). The contribution of seawater to groundwater was less than 1% for the remaining sampling sites. It could be concluded that the influence of seawater intrusion on groundwater was not serious.

5. Practical Implications and Perspectives

Groundwater contamination in coastal aquifers is a serious problem around the world. The pollution of coastal aquifers is not only from seawater intrusion, but usually includes pollution from human activities. Therefore, when managing groundwater in coastal aquifers, it is important to not only control groundwater extraction, but also to focus on pollution sources from human activities. Moreover, the pollution sources usually include natural processes and anthropogenic activities.

6. Conclusions

Groundwater contamination is a widespread problem in coastal aquifers. In this study, the sources of contaminants in the coastal zone were identified. Meanwhile, the control processes of hydrogeochemistry were discussed in detail. The below conclusions could be drawn.

Pollution components included Na^+, $NH_4{}^+$, Cl^-, $SO_4{}^{2-}$, $NO_3{}^-$, $NO_2{}^-$, Pb, As, Se, TDS, TH, F^-, and Mn. The groundwater quality was influenced by both natural processes and anthropogenic activities. Na^+ and Cl^- originated from the dissolution of halite and sea intrusion. Mg^{2+} and Ca^{2+} were derived from the dissolution of gypsum and anhydrite, as well as cation exchange processes. The discharge of sewage and the utilization of fertilizers led to the contamination of nitrogen in groundwater. The high $SO_4{}^{2-}$ was most likely due to the leaching of waste by rainfall. The exceedance of As was caused by anthropogenic inputs and natural factors. The poor seepage conditions and anoxic conditions provided a beneficial condition for the enrichment of Mn. High Pb and Se resulted from anthropogenic activities. The concentration of most components in the dry season was larger than that in the wet season. The seasonal variations were significant due to the influence of precipitation.

Groundwater chemistry was mainly Ca–HCO_3 type, Ca–Mg–Cl type, and Ca–Cl type, followed by Na–Cl type and Ca–Na–HCO_3 type in the dry season. The hydrogeochemical facies in the wet season were similar to those in the dry season. The groundwater chemistry was influenced by natural processes and anthropogenic activities.

The impact of seawater intrusion on groundwater was not serious. The high level of Na^+ and Cl^- at some points was the result of seawater intrusion caused by extensive groundwater exploitation.

In the future, numerical simulations of coastal aquifers will be carried out, which will make it possible both to calculate the exchange of groundwater with seawater and to analyze the migration processes of pollutants.

Supplementary Materials: The following supporting information can be downloaded at: https://www.mdpi.com/article/10.3390/w15091710/s1.

Author Contributions: Conceptualization, G.A. and R.F.; writing—original draft, G.A.; supervision, R.F., H.K., D.X. and J.L.; data analysis, G.A. and H.K. All authors read and agreed to the published version of the manuscript.

Funding: This research received no external funding.

Institutional Review Board Statement: No.

Informed Consent Statement: Informed consent was obtained from all subjects involved in the study.

Data Availability Statement: Data is unavailable.

Acknowledgments: The authors thank the reviewers for their comments.

Conflicts of Interest: The authors declare no conflict of interest.

References

1. Kim, J.-H.; Kim, K.-H.; Thao, N.T.; Batsaikhan, B.; Yun, S.-T. Hydrochemical assessment of freshening saline groundwater using multiple end-members mixing modeling: A study of Red River delta aquifer, Vietnam. *J. Hydrol.* **2017**, *549*, 703–714. [CrossRef]
2. Lee, K.-J.; Yun, S.-T.; Yu, S.; Kim, K.-H.; Lee, J.-H.; Lee, S.-H. The combined use of self-organizing map technique and fuzzy c-means clustering to evaluate urban groundwater quality in Seoul metropolitan city, South Korea. *J. Hydrol.* **2019**, *569*, 685–697. [CrossRef]
3. Haselbeck, V.; Kordilla, J.; Krause, F.; Sauter, M. Self-organizing maps for the identification of groundwater salinity sources based on hydrochemical data. *J. Hydrol.* **2019**, *576*, 610–619. [CrossRef]
4. Custodio, E. Coastal aquifers of Europe: An overview. *Hydrogeol. J.* **2010**, *18*, 269–280. [CrossRef]

5. Qiao, W.; Guo, H.; He, C.; Shi, Q.; Xiu, W.; Zhao, B. Molecular Evidence of Arsenic Mobility Linked to Biodegradable Organic Matter. *Environ. Sci. Technol.* **2020**, *54*, 7280–7290. [CrossRef]
6. Guo, H.; Zhou, Y.; Jia, Y.; Tang, X.; Li, X.; Shen, M.; Lu, H.; Han, S.; Wei, C.; Norra, S.; et al. Sulfur Cycling-Related Biogeochemical Processes of Arsenic Mobilization in the Western Hetao Basin, China: Evidence from Multiple Isotope Approaches. *Environ. Sci. Technol.* **2016**, *50*, 12650–12659. [CrossRef]
7. Korrai, S.; Gangu, K.K.; Prasada Rao, P.V.V.; Jonnalagadda, S.B. A study on assessment of vulnerability of seawater intrusion to groundwater in coastal areas of Visakhapatnam, India. *Environ. Dev. Sustain.* **2021**, *23*, 5937–5955. [CrossRef]
8. Chen, R.; Teng, Y.; Chen, H.; Hu, B.; Yue, W. Groundwater pollution and risk assessment based on source apportionment in a typical cold agricultural region in Northeastern China. *Sci. Total Environ.* **2019**, *696*, 133972. [CrossRef]
9. Yang, Q.; Li, Z.; Lu, X.; Duan, Q.; Huang, L.; Bi, J. A review of soil heavy metal pollution from industrial and agricultural regions in China: Pollution and risk assessment. *Sci. Total Environ.* **2018**, *642*, 690–700. [CrossRef] [PubMed]
10. Ran, H.; Guo, Z.; Yi, L.; Xiao, X.; Zhang, L.; Hu, Z.; Li, C.; Zhang, Y. Pollution characteristics and source identification of soil metal(loid)s at an abandoned arsenic-containing mine, China. *J. Hazard. Mater.* **2021**, *413*, 125382. [CrossRef]
11. Chen, G.; Xiong, G.; Lin, J.; Xu, X.; Yu, H.; Liu, W.; Fu, T.; Su, Q.; Wang, Y.; Dai, Y.; et al. Elucidating the Pollution Sources and Groundwater Evolution in Typical Seawater Intrusion Areas Using Hydrochemical and Environmental Stable Isotope Technique: A Case Study for Shandong Province, China. *Lithosphere* **2021**, *2021*, 4227303. [CrossRef]
12. de Graaf, I.E.M.; Gleeson, T.; van Beek, L.P.H.; Sutanudjaja, E.H.; Bierkens, M.F.P. Environmental flow limits to global groundwater pumping. *Nature* **2019**, *574*, 90–94. [CrossRef] [PubMed]
13. Han, D.; Currell, M.J. Review of drivers and threats to coastal groundwater quality in China. *Sci. Total Environ.* **2022**, *806*, 150913. [CrossRef] [PubMed]
14. Johnson, J.M.; Moore, L.J.; Ells, K.; Murray, A.B.; Adams, P.N.; MacKenzie Iii, R.A.; Jaeger, J.M. Recent shifts in coastline change and shoreline stabilization linked to storm climate change. *Earth Surf. Process. Landf.* **2015**, *40*, 569–585. [CrossRef]
15. Ping, Y.; Xiao-yong, D.; Fei, G.; Mei-na, L.; Sheng-hua, L.; Jian-dong, Q.; Liang-yong, Z. Coastal erosion in Shandong of China: Status and protection challenges. *China Geol.* **2018**, *1*, 512–521. [CrossRef]
16. Balasubramanian, M.; Sridhar, S.G.D.; Ayyamperumal, R.; Karuppannan, S.; Gopalakrishnan, G.; Chakraborty, M.; Huang, X. Isotopic signatures, hydrochemical and multivariate statistical analysis of seawater intrusion in the coastal aquifers of Chennai and Tiruvallur District, Tamil Nadu, India. *Mar. Pollut. Bull.* **2022**, *174*, 113232. [CrossRef]
17. Daniele, L.; Tardani, D.; Schmidlin, D.; Quiroga, I.; Cannatelli, C.; Somma, R. Seawater intrusion and hydrogeochemical processes in the Ischia Island groundwater system. *J. Geochem. Explor.* **2022**, *234*, 106935. [CrossRef]
18. Satheeskumar, V.; Subramani, T.; Lakshumanan, C.; Roy, P.D.; Karunanidhi, D. Groundwater chemistry and demarcation of seawater intrusion zones in the Thamirabarani delta of south India based on geochemical signatures. *Environ. Geochem. Health* **2021**, *43*, 757–770. [CrossRef]
19. Chen, H.; Zhang, Y.; Wang, X.; Ren, Z.; Li, L. Salt-Water Intrusion in the Lower Reaches of the Weihe River, Shandong Province, China. *Hydrogeol. J.* **1997**, *5*, 82–88. [CrossRef]
20. Barker, A.P.; Newton, R.J.; Bottrell, S.H.; Tellam, J.H. Processes affecting groundwater chemistry in a zone of saline intrusion into an urban sandstone aquifer. *Appl. Geochem.* **1998**, *13*, 735–749. [CrossRef]
21. Barlow, P.M.; Reichard, E.G. Saltwater intrusion in coastal regions of North America. *Hydrogeol. J.* **2010**, *18*, 247–260. [CrossRef]
22. Salem, Z.E.; ElNahrawy, A.; Attiah, A.M.; Edokpayi, J.N. Vertical and spatial evaluation of the groundwater chemistry in the Central Nile Delta Quaternary aquifer to assess the effects of human activities and seawater intrusion. *Front. Environ. Sci.* **2022**, *10*. [CrossRef]
23. Argamasilla, M.; Barberá, J.A.; Andreo, B. Factors controlling groundwater salinization and hydrogeochemical processes in coastal aquifers from southern Spain. *Sci. Total Environ.* **2017**, *580*, 50–68. [CrossRef] [PubMed]
24. Mollema, P.N.; Antonellini, M.; Dinelli, E.; Gabbianelli, G.; Greggio, N.; Stuyfzand, P.J. Hydrochemical and physical processes influencing salinization and freshening in Mediterranean low-lying coastal environments. *Appl. Geochem.* **2013**, *34*, 207–221. [CrossRef]
25. Galliari, J.; Santucci, L.; Misseri, L.; Carol, E.; Alvarez, M.d.P. Processes controlling groundwater salinity in coastal wetlands of the southern edge of South America. *Sci. Total Environ.* **2021**, *754*, 141951. [CrossRef] [PubMed]
26. Carreira, P.M.; Marques, J.M.; Nunes, D. Source of groundwater salinity in coastline aquifers based on environmental isotopes (Portugal): Natural vs. human interference. A review and reinterpretation. *Appl. Geochem.* **2014**, *41*, 163–175. [CrossRef]
27. Han, D.M.; Song, X.F.; Currell, M.J.; Yang, J.L.; Xiao, G.Q. Chemical and isotopic constraints on evolution of groundwater salinization in the coastal plain aquifer of Laizhou Bay, China. *J. Hydrol.* **2014**, *508*, 12–27. [CrossRef]
28. Xiao, J.; Wang, L.; Chai, N.; Liu, T.; Jin, Z.; Rinklebe, J. Groundwater hydrochemistry, source identification and pollution assessment in intensive industrial areas, eastern Chinese loess plateau. *Environ. Pollut.* **2021**, *278*, 116930. [CrossRef]
29. Quino Lima, I.; Ramos Ramos, O.; Ormachea Muñoz, M.; Quintanilla Aguirre, J.; Duwig, C.; Maity, J.P.; Sracek, O.; Bhattacharya, P. Spatial dependency of arsenic, antimony, boron and other trace elements in the shallow groundwater systems of the Lower Katari Basin, Bolivian Altiplano. *Sci. Total Environ.* **2020**, *719*, 137505. [CrossRef]
30. Dai, L.; Wang, L.; Li, L.; Liang, T.; Zhang, Y.; Ma, C.; Xing, B. Multivariate geostatistical analysis and source identification of heavy metals in the sediment of Poyang Lake in China. *Sci. Total Environ.* **2018**, *621*, 1433–1444. [CrossRef]

31. Alshehri, F.; Almadani, S.; El-Sorogy, A.S.; Alwaqdani, E.; Alfaifi, H.J.; Alharbi, T. Influence of seawater intrusion and heavy metals contamination on groundwater quality, Red Sea coast, Saudi Arabia. *Mar. Pollut. Bull.* **2021**, *165*, 112094. [CrossRef]
32. Unglert, K.; Radić, V.; Jellinek, A.M. Principal component analysis vs. self-organizing maps combined with hierarchical clustering for pattern recognition in volcano seismic spectra. *J. Volcanol. Geotherm. Res.* **2016**, *320*, 58–74. [CrossRef]
33. Underwood, K.L.; Rizzo, D.M.; Schroth, A.W.; Dewoolkar, M.M. Evaluating Spatial Variability in Sediment and Phosphorus Concentration-Discharge Relationships Using Bayesian Inference and Self-Organizing Maps. *Water Resour. Res.* **2017**, *53*, 10293–10316. [CrossRef]
34. Petelet-Giraud, E.; Négrel, P.; Aunay, B.; Ladouche, B.; Bailly-Comte, V.; Guerrot, C.; Flehoc, C.; Pezard, P.; Lofi, J.; Dörfliger, N. Coastal groundwater salinization: Focus on the vertical vriability in a multi-layered aquifer through a multi-isotope fingerprinting (Roussillon Basin, France). *Sci. Total Environ.* **2016**, *566-567*, 398–415. [CrossRef]
35. Hao, C.; Zhang, W.; Gui, H. Hydrogeochemistry characteristic contrasts between low- and high-antimony in shallow drinkable groundwater at the largest antimony mine in hunan province, China. *Appl. Geochem.* **2020**, *117*, 104584. [CrossRef]
36. Lone, S.A.; Jeelani, G.; Mukherjee, A.; Coomar, P. Geogenic groundwater arsenic in high altitude bedrock aquifers of upper Indus river basin (UIRB), Ladakh. *Appl. Geochem.* **2020**, *113*, 104497. [CrossRef]
37. Chen, W.; Zhang, Y.; Shi, W.; Cui, Y.; Zhang, Q.; Shi, Y.; Liang, Z. Analysis of Hydrogeochemical Characteristics and Origins of Chromium Contamination in Groundwater at a Site in Xinxiang City, Henan Province. *Appl. Sci.* **2021**, *11*, 11683. [CrossRef]
38. Xiao, Y.; Gu, X.; Yin, S.; Pan, X.; Shao, J.; Cui, Y. Investigation of Geochemical Characteristics and Controlling Processes of Groundwater in a Typical Long-Term Reclaimed Water Use Area. *Water* **2017**, *9*, 800. [CrossRef]
39. Tanasković, I.; Golobocanin, D.; Miljević, N. Multivariate statistical analysis of hydrochemical and radiological data of Serbian spa waters. *J. Geochem. Explor.* **2012**, *112*, 226–234. [CrossRef]
40. Kai, Z.; Jing-xian, Q.; Yi, C.; Bai-heng, M.; Li, Y.; Hua-ming, G.; Xin-zhou, W.; Lin-ying, W.; Hai-tao, L. Hydrogeochemical characteristics of groundwater and pore-water and the paleoenvironmental evolution in the past 3.10 Ma in the Xiong'an New Area, North China. *China Geol.* **2021**, *4*, 476–486. [CrossRef]
41. Liu, J.; Tokunaga, T. Future Risks of Tsunami-Induced Seawater Intrusion Into Unconfined Coastal Aquifers: Insights From Numerical Simulations at Niijima Island, Japan. *Water Resour. Res.* **2019**, *55*, 10082–10104. [CrossRef]
42. Xu, Z.; Hu, B.X. Development of a discrete-continuum VDFST-CFP numerical model for simulating seawater intrusion to a coastal karst aquifer with a conduit system. *Water Resour. Res.* **2017**, *53*, 688–711. [CrossRef]
43. Miao, T.; Huang, H.; Guo, J.; Li, G.; Zhang, Y.; Chen, N. Uncertainty Analysis of Numerical Simulation of Seawater Intrusion Using Deep Learning-Based Surrogate Model. *Water* **2022**, *14*, 2933. [CrossRef]
44. Yechieli, Y.; Kafri, U.; Goldman, M.; Voss, C. Factors controlling the configuration of the fresh–saline water interface in the Dead Sea coastal aquifers: Synthesis of TDEM surveys and numerical groundwater modeling. *Hydrogeol. J.* **2001**, *9*, 367–377. [CrossRef]
45. Ibrahim, H.; Yaseen, Z.M.; Scholz, M.; Ali, M.; Gad, M.; Elsayed, S.; Khadr, M.; Hussein, H.; Ibrahim, H.H.; Eid, M.H.; et al. Evaluation and Prediction of Groundwater Quality for Irrigation Using an Integrated Water Quality Indices, Machine Learning Models and GIS Approaches: A Representative Case Study. *Water* **2023**, *15*, 694. [CrossRef]
46. Gad, M.; El Osta, M. Geochemical controlling mechanisms and quality of the groundwater resources in El Fayoum Depression, Egypt. *Arab. J. Geosci.* **2020**, *13*, 861. [CrossRef]
47. El Osta, M.; Masoud, M.; Alqarawy, A.; Elsayed, S.; Gad, M. Groundwater Suitability for Drinking and Irrigation Using Water Quality Indices and Multivariate Modeling in Makkah Al-Mukarramah Province, Saudi Arabia. *Water* **2022**, *14*, 483. [CrossRef]
48. Zhu, G.; Wu, X.; Ge, J.; Liu, F.; Zhao, W.; Wu, C. Influence of mining activities on groundwater hydrochemistry and heavy metal migration using a self-organizing map (SOM). *J. Clean. Prod.* **2020**, *257*, 120664. [CrossRef]
49. Piper, A.M. A graphic procedure in the geochemical interpretation of water-analyses. *Eos Trans. Am. Geophys. Union* **1944**, *25*, 914–928. [CrossRef]
50. Gibbs, R.J. Mechanisms Controlling World Water Chemistry. *Science* **1970**, *170*, 1088–1090. [CrossRef]

water

MDPI

Article

Research on Surface Water Quality Assessment and Its Driving Factors: A Case Study in Taizhou City, China

Sihe Deng [1], Cheng Li [1], Xiaosan Jiang [1,*], Tingting Zhao [2] and Hui Huang [3]

1 School of Resources & Environment Sciences, Nanjing Agricultural University, Nanjing 210095, China
2 Taizhou Bureau of Agriculture and Rural Affairs, Taizhou 225300, China
3 Environmental Monitoring Center of Taizhou Ecological Environment Bureau, Taizhou 225300, China
* Correspondence: gis@njau.edu.cn

Abstract: It is necessary to assess and analyze the factors that influence surface water since they are crucial to human activities such as agriculture, raising livestock, and industry. Previous research has mostly focused on how land use and landscape patterns affect the quality of surface waters; it has seldom addressed the industrial and agricultural production activities that are directly connected to human society. Therefore, the research area's surface water quality was assessed by single factor index (SFI) and composite water quality index (WQI), divided into flood and non-flood periods, and water quality indicators with severe pollution and significant seasonal variations were selected; A total of 28 indicators were selected from three main factors-topography, socio-economic, and land use type-and analyzed using the Spearman correlation coefficient model. (1) SFI data reveal substantial seasonal changes in pH, DO, NH_3-N, TN, and TP water quality indicators. The well-developed agricultural and aquaculture in the studied region is the primary cause of the excess TN and NH3-N concentrations; (2) The sample points' water quality index (WQI) scores range from 50 to 80, with 62% of them having "medium" water quality; (3) The study area's seasonal variation in water quality is primarily caused by human socio-economic activities (GDP, industrial effluent discharge, COD discharge, aquatic product quality, and the proportion of primary, secondary, and tertiary industries), as well as land use type (forest, shrubland, and cropland). Topography has little effect on the study area's surface water quality. This study offers a fresh viewpoint on surface water quality management and driver analysis, and a new framework for managing and safeguarding aquatic ecosystems.

Keywords: surface water quality evaluation; spearman correlation coefficient; driving factors; land use; human activity

1. Introduction

The enormous risk of depletion and pollution facing the world's water resources may be attributed mostly to the misuse of water resources and pollution caused by the rise in human demand [1,2]. The degradation of water quality and the scarcity of water are two signs of the worldwide water deficit [3–5]. The two most significant water sources in the natural world are surface water and groundwater [6]. Both fluoride [7] and nitrate [8] are common pollutants in groundwater, with high nitrate levels and fluoride pollution affecting the vast majority of groundwater [9]. Surface water pollution has a greater potential than groundwater contamination, and the environmental implications of climate change have been a major threat to surface water quality [10], with human activities seen as the most significant cause of the worsening of surface water quality in many nations and areas throughout the world in recent decades [11]. The conflict between people and water resources has gotten worse as a result of the rapid expansion of society and the economy as well as the acceleration of global climate change. As a result, the creation of water management plans, including the creation of precise and logical techniques to evaluate surface water quality, is necessary [12]. Assessing the current state of water quality in the river basin and its driving factors is critical for understanding how geography, human socio-economics, and land use impact water quality changes [13,14]. A new problem is presently being faced in how to precisely identify the important variables that influence water quality due to the numerous water quality monitoring indicators and the intricate pollution process [15].

A surface water quality assessment [16] compares collected data to predetermined criteria for physical, chemical, nutritional, and biological indicators of the river to provide a qualitative or quantitative evaluation of the river's water quality. The Chinese government's method for assessing water quality in reports on water quality monitoring is known as single-factor assessment (SFA) [17]. To establish the water quality grade, the surface water environmental quality standards (e.g., GB3838-2002 (2002)) are used to choose the worst monitoring indications out of all monitoring indicators. The estimate of the water quality risk level is simplified by this technique, which is quick and straightforward yet yields conservative evaluation findings [18]. The development of the water quality index (WQI) approach [19] offers a solution to this issue. The Water Quality Index (WQI) is a mathematical tool that can combine a lot of water quality data into a single value [20], accurately analyze the water quality and eutrophication level of the water body, and offer the general public and decision-makers with a description of the current water quality situation and an evaluation of the water quality trends indicator [21]. The combination of the two approaches enables a new level of assessment of the water quality state of the research area and its geographical and temporal features, even though the two methods have distinct evaluation criteria and provide different findings. In this study, data on river water quality were combined with standards for the environmental quality of China's surface waters using SFA and WQI methods. Subtle differences in water quality were identified, and indicators with notable seasonal differences in water were chosen to investigate the factors that influence changes in water quality.

Numerous academics have investigated the causes of surface water quality change in recent years, including analyzing the effects of climate change and land use on water quality [22]. Precipitation, temperature, and evapotranspiration are a few of the variables that affect water quality due to the climate [5,23,24]. However, human-driven land use change has emerged as a significant driver of both terrestrial and aquatic ecosystems [25]. Rivers play a significant role in determining the types of land uses, natural vegetation is commonly related to the quality of the water [26], and urban and agricultural land contributes to the global rise in nutrients and sediments in freshwater ecosystems [27]. According to Shi's [28] research, the percentage of different land use types in a river basin has a significant impact on the water quality of rivers, as does the pattern of the landscape, size, density, aggregation, and variety of land use types. In Brazil, low-order river basins and riparian zones have been investigated at length by Mello et al. [29]. They discovered that forests have a significant role in preserving the water quality of low-order rivers, whereas agriculture and urban development are the primary causes of water quality deterioration. In the Qingyi River Basin in China, Yang et al. [30] investigated the effects of land use and landscape patterns on water quality. The findings demonstrated that cropland and construction land had negative effects on water quality and positive correlations with NH_4^+-N, TP, and EC. The ability of forest land to filter water pollution was shown to be positively connected with DO.

These studies, however, coupled land-use type changes with human activities and lacked the data necessary to define the variables that affect water quality [22]. Changes in land use are only one aspect of human activity; other contributing elements, such as industrial and agricultural operations, sewage discharge and treatment, etc., also have a role. In other words, the combined influence of climatic and land use changes on water quality should be less than but not equal to 100% [31]. As a result, changes in land use cannot accurately reflect the whole spectrum of human activities. The natural topography and human socio-economic activity have an influence on water quality in addition to climate and land use, which cannot be disregarded. As a result, the terrain [32] (elevation, slope, and aspect) and socioeconomic [33] variables are used in this study (population density, sewage discharge, livestock and poultry production, aquatic production, proportion of primary, secondary and tertiary industries, and the number of fertilizers and pesticides applied). various forms of land use (urban land, wetlands, aquatic bodies, cropland, forest land, shrubland, and grassland) to investigate the effects on watershed water quality.

The hydrological influence of land use, climate change, and human activities has been studied in several nations, and the study methodologies may be broadly classified into three groups [34]: the paired catchment approach, hydrological modeling, and statistical analysis. In small experimental catchments, the paired catchment approach [35,36] is often regarded as a popular technique for mitigating the effects of climate change. Its greatest applicable range is a small watershed area of 100 Km^2, but it is challenging to apply to watersheds outside of this small region. A conceptual foundation for understanding the link between climate and water resources is provided by hydrological modeling [30]. The Soil and Water Assessment Tool (SWAT) model is a commonly used tool for hydrological research all over the world. It is based on physical, basin-scale hydrophysical processes and requires a significant amount of data and parameter inputs, which are frequently unavailable in some regions due to time or financial constraints, especially in some developing countries [37]. The SWAT model also lays greater emphasis on the hydrology of the entire watershed's continuous time-scale process of topography, soil [38], and weather than it does on how human activities impact water quality [39]. An efficient way to assess the effects of human activity on water quality is using statistical analysis [34], which is straightforward and simple to use. Given that river water quality monitoring will provide a sizable and ongoing time series of data, several statistical approaches are utilized for analysis and discrimination, and sample sites may be categorized to demonstrate the link between water quality [40]. Analyzes hydrological, climatic, topographic, and human activity data from monitoring stations situated inside and around the research region to provide valid judgments about the link between hydrological responses and variables influencing water quality. Correlation analysis can be well used to investigate the relationship between water quality variables, and for continuous observation time series data, it is more necessary to use statistical methods to deal with the nature of correlation in random variable series [41].

Taizhou City is located in the central region of Jiangsu Province in China, with copious water resources but a dense population and significant water pollution. Urbanization, agriculturalization, and industrialization have increased freshwater consumption and pollution, altered the geographical and temporal distribution of water quality, and resulted in more complicated repercussions on water resources and hydrological processes, social development, and the economy in Taizhou and even the Yangtze River estuary. Therefore, this study used data from 12 state-controlled monitoring sections in Taizhou from 2020 to 2021, separated into flood season and non-flood season, to evaluate the present situation of river water quality in Taizhou by computing the water quality single factor index (SFI) and water quality index (WQI). The river water quality indicators with substantial seasonal variations were chosen, and the Spearman correlation coefficient model was used to examine the relationship between river water quality and topography, social economy, and land use, with the goal of (1) evaluating Taizhou City's water quality; (2) Identify the major pollutant parameters and influencing variables of natural and human activities in Taizhou City throughout the flood and non-flood seasons, and offer scientific support for the healthy development of water ecology in Taizhou City, Jiangsu Province.

2. Materials and Methods

2.1. Study Area

The research area of this study is Taizhou (119°38′24″ E~120°32′20″ E, 32°01′57″ N~33°10′59″ N, as shown in Figure 1), which has 3 districts under the jurisdiction of Hailing District, Gaogang District, and Jiangyan District, etc., and manages 3 cities at the county level, Xinghua City, Jingjiang City, and Taixing City, with an area of 5787 Km^2, of which 77.85% is land area and 22.15% is water area. The territory's river network is extensive and intertwined, and the rivers are generally bordered by the Tong Yang Highway, with the Huai River basin to the north of the road and the Yangtze River basin to the south.

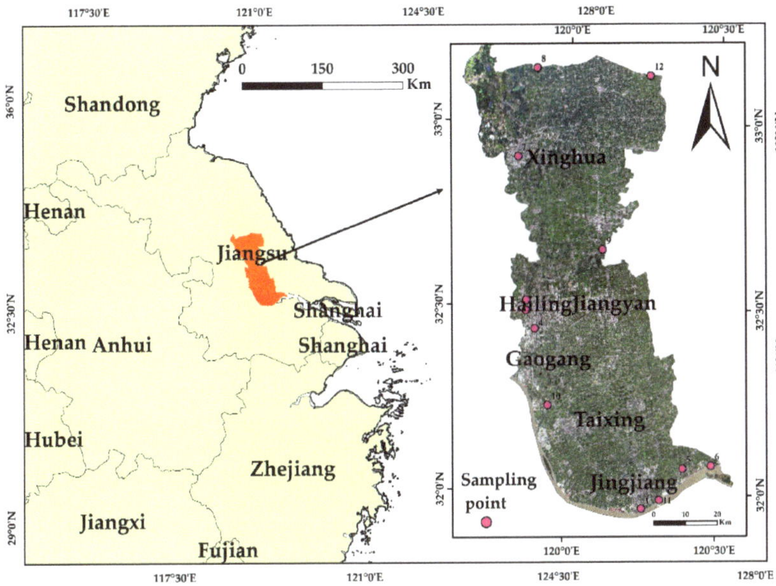

Figure 1. Study area and location of sampling points.

The study location has a flat terrain. Except for an isolated hill in Jingjiang, the remainder is the alluvial plains of the two major river systems, the Jianghuai and Huaihe Rivers. With an elevation of 2–5 m along the river in the south, 5–7 m in the Gaosha region in the middle, and 1.5–5 m in the north Lixiahe area, the terrain is high in the middle and low in the north and south. In the north subtropical humid zone, Taizhou has a subtropical humid monsoon climate with overt monsoon features, including hot and rainy summers and warm, sparsely precipitated winters.

The research area's permanent population was 4.5218 million as of 2021, and its GDP was 602.526 billion yuan. Among these, the primary industry's added value increased by 2.8% to 31.813 billion yuan, while the secondary industry's added value increased by 9.3% to 291.859 billion yuan. The tertiary industry-added value increased by 11.8% to 278.854 trillion yuan. Additionally, The research area has built an industrial system centered on the health sector, with assistance from the manufacturing sector, in the core region of the Yangtze River Delta. The textile industry, the chemical and pharmaceutical industries, the new material industries, and the high-tech shipbuilding industries are among the top industries. Meanwhile, It has established an "8 + N" agricultural industrial chain structure, with the major body consisting of grain and oil, livestock and poultry, river crabs, fruits and vegetables, freshwater fish and shrimp, leisure agriculture, flower gardening, and food seasoning. The agricultural areas of Lixiahe in the north, Gaosha in the middle, and the Yangtze River agricultural area in the south have all been constructed. To create a green development of agriculture, green recycling agriculture, emission reduction, and low carbon are simultaneous goals.

2.2. Data Source and Processing

The main data and its sources are as follows.

(1) The water quality information is sourced from the China Surface Water Quality Automatic Monitoring Real-time Data Release System from 2021 to 2022. (https:

//szzdjc.cnemc.cn:8070/GJZ/Business/Publish/Main.html accessed on 23 May 2022). The China Environmental Monitoring Center is the provider of the system. The information is derived from the newly constructed and fully operational China Surface Water Quality Automatic Monitoring Station. Check the water quality every 4 h and publish the most recent information. Monitoring indexes included water temperature (WT), pH, Dissolved oxygen (DO), Electrical conductivity (EC), Permanganate index (COD_{Mn}), Chemical oxygen demand (COD), Five days biochemical oxygen demand (BOD_5), Ammonia nitrogen (NH_3-N), Total nitrogen (TN), and Total phosphorus (TP).

(2) Topographic data such as elevation, slope and aspect affect rainfall and surface runoff by influencing the hydrothermal distribution and vegetation growth in the catchment, which further affects the generation and transport of non-point source pollution [32], therefore, this paper selects elevation, slope and aspect factors to explore the influence of topography on water quality distribution patterns. Elevation data is 30 m × 30 m spatial resolution DEM data, downloaded form geospatial data cloud (http://www.gscloud.con/ accessed on 15 March 2022);

(3) The immediate reasons for the decline in water quality on the surface include excessive industrial and agricultural effluent discharge and unsustainable industrial structures [42]. Therefore, this paper selects population density, total sewage discharge, chemical oxygen demand (COD) emissions, pork production, poultry production, aquatic product production, the proportion of primary, secondary and tertiary industries, fertilizer and pesticide use, and other factors to explore the socio-economic impact on the distribution pattern of surface water quality. Socio-economic data from Taizhou Statistical Yearbook and National Economic and Social Development Bulletin;

(4) According to studies, the percentage of different land use types in the watershed has a substantial impact on the water quality of rivers, as do landscape patterns, size, density, aggregation, and variety of land use types [28]. As a result, the statistical unit for the percentage of land cover types was chosen as the 12 km buffer zone [33] with a high correlation to the concentration of water quality parameters, and eight land use types—cropland, forest, grassland, shrub land, wetland, water body, bare land, and urban land—were chosen to investigate their effects on the distribution pattern of water quality. The European Space Agency (ESA) released the 2020 Global Land Cover product (https://biewer.esa-worldcover.org/worldcover accessed on 27 April 2022), which is where the data on land use is taken from. The spatial resolution is 10 m × 10 m, and the total classification accuracy is 74.4%.

The selection index of each element is shown in Table 1.

This research looked at water quality during the flood season (FS) and the non-flood season (NFS). According to the quantity of precipitation, May through September in the study area was categorized as the flood season, while all other months were labeled as the non-flood season. The water quality monitoring locations in the research area were distributed in 12 provincially regulated sectors (Figure 1 and Table 2). Water samples were collected in compliance with the "Environmental Quality Standards for Surface Water" established by China (GB3838-2002). DO use portable water quality testing analyzer YSI (yellow springs instrument. Zhejiang Luheng Environmental Technology, Hangzhou, China) for on-site measurement of the water body at 25 cm below the water surface, pH and water temperature using online water quality detector for determination, NH_3-N, TN, TP determination methods for "water quality Determination of ammonia nitrogen salicylic acid spectrophotometric method" (HJ536-2009), "water quality Determination of total nitrogen The methods for the determination of COD, COD_{Mn} and BOD_5 are "Determination of Chemical Oxygen Demand (COD) by Dichromate Method" (HJ828-2017), "Determination of Water Quality (COD) by Dichromate Method" (HJ828-2017) and "Determination of Water Quality (COD) by Ammonium Molybdate" (HJ670-2013). Determination of permanganate index" (GB/T 11892-1989), "Determination of five-day biochemical oxygen demand of water quality Dilution and inoculation method" (HJ505-2009).

Table 1. Indicators of terrain, human society and economy and land use.

Elements	Indicators	Contracted Form	Reference
Terrain factors	Elevation Aspect Slope	Elevation Aspect Slope	Panuska.J.C. et al. [32]
Social economic factors	The population density Gross domestic product Per capita gross domestic product Total sewage discharge Industrial sewage discharge Chemical oxygen demand sewage discharge Urban sewage treatment rate Pork production Poultry production Aquatic production Proportion of primary industry Proportion of second industry Proportion of tertiary industry Fertilizer usage Pesticide uasge	Pop_Den GDP GDP_Per_Cap Tot_Sew_Dis Ind_Sew_Dis COD_Sew_Dis Urb_Sew_Dis Pork_Pro Pou_Pro Aqua_Pro Pri_Ind_Pro Sec_Ind_Pro Ter_Ind_Pro Fert_Use Pest_Use	Chen Ijnyue. et al. [33]
Land-use factors	Proportion of cropland area Proportion of forest area Proportion of grassland area Proportion of shrub area Proportion of wetland area Proportion of water area Proportion of bare land area Proportion of urban land area	Crop_Pro Forest_Pro Grass_Pro Shrub_Pro Wet_Pro Water_Pro Bare_Pro Urban_Pro	Wang X. et al. [42] De Mello, K. et al. [27]

Table 2. Information of water quality automatic monitoring station.

Number	River	River Basin	The Name of Section	Area of Operation
1	Shiwei harbor	Yangtze river basin	Xinshiwei harbor bridge	Jingjiang city
2	Xintongyang river	Huaihe river basin	Taixi	Hailing district
3	Yinjiang river	Huaihe river basin	Hailing bridge	Hailing district
4	Nanguan river	Yangtze river basin	Xianghe bridge	Hailing district
5	Xiashi harbor	Yangtze river basin	Xiashigang bridge	Jingjiang city
6	Xiaqinglong harbor	Yangtze river basin	Xiaqinglong harbor (left shore)	Jingjiang city
7	Luting river	Huaihe river basin	Refrigeration plant south	Xinhua city
8	Zhula river	Huaihe river basin	Jigeng	Xinhua city
9	Fengting river	Huaihe river basin	Laogedong	Jiangyan district
10	Gumagan river	Yangtze river basin	Maxundian west	Gaogang district
11	Pengqi harbor	Yangtze river basin	Pengqi harbor (left shore)	Jingjiang city
12	Gao harbor	Yangtze river basin	Gao barbor (left shore)	Xinhua city

2.3. Water Quality Assessment Methods

2.3.1. Single Factor Pollution Index

Single factor pollution index (SFI) is assessed based on the relationship between water quality metrics and benchmark values [27,28], and various pollution parameter types are computed in different methods [17].

If the value of the index is smaller, the better, as permanganate index and ammonia concentration, then:

$$P_i = \frac{C_i}{S_i} \quad (1)$$

If the value of the index is bigger, the better, as dissolved oxygen, then:

$$P_i = \frac{C_{max} - C_i}{C_{max} - S_i} \quad (2)$$

If the value of the index is closer to a certain value, the better pH value, then:

$$P_i = \frac{C_i - S_{iave}}{S_{max} - S_{iave}} \text{ or } P_i = \frac{C_i - S_{iave}}{S_{iave} - S_{min}} \quad (3)$$

where the pollution index is P_i and C_i is the measured value of water quality parameter i, the maximum value of the evaluation standard is S_i (here we used the boundary value of class II in surface water quality standard ((GB3838-2002), Table 3), C_{max} is the saturated concentration of the evaluation factor, S_{max} and S_{min} are the highest and lowest average value of the evaluation standard respectively, and S_{iave} is the average value of the highest and lowest evaluation criteria of the pollutant.

Table 3. Environment quality standards for surface water.

Risk Levels		Class I (Very Clean)	Class II (Not Risky)	Class III (Moderate)	Class IV (Slightly Risky)	Class V (Risky)
Pollution factors		Concentration (mg/L)				
pH				6~9		
DO	≥	7.5	6	5	3	2
COD_{Mn}	≤	2	4	6	10	15
BOD_5	≤	3	3	4	6	10
NN_3-N	≤	0.15	0.5	1.0	1.5	2.0
TN	≤	0.2	0.5	1.0	1.5	2.0
TP	≤	0.02	0.1	0.2	0.3	0.4

The saturation concentration of DO is:

$$\mathrm{DO} = \frac{468}{31.6 + T} \quad (4)$$

where T is the temperature, and the unit is °C.

2.3.2. Water Quality Index

The water quality measures (DO, EC, COD_{Mn}, BOD_5, NH_3-N, TN, and TP) were chosen as assessment indicators, and the water quality index (WQI) was utilized to assess Taizhou's water quality in 2021–2022. The WQI scale ranges from 0 to 100, and Table 4 [43] lists its evaluation criteria.

Table 4. WQI evaluation criteria.

Range	Class
$10 \leq \mathrm{WQI} < 25$	Very bad
$25 \leq \mathrm{WQI} < 50$	Bad
$50 \leq \mathrm{WQI} < 70$	Medium
$70 \leq \mathrm{WQI} < 90$	Good
$90 \leq \mathrm{WQI} \leq 100$	Excellent

Four stages are taken to determine the WQI [44]:

(1) Selection of key parameters;
(2) Converting important parameters to a common scale was accomplished by applying sub-index curves to convert each parameter to a scale from 0 to 100 [45]. The sub-index curves might be segmented-lined, segmented-nonlinear, linear, nonlinear, etc;
(3) Assigning weights to parameters;
(4) Aggregation of indices to produce a WQI.

The generic equation of WQI [46] is:

$$\mathrm{WQI} = \frac{\sum_{i=1}^{n} H_i \times P_i}{\sum_{i=1}^{n} P_i} \quad (5)$$

where WQI is the comprehensive water quality index, H_i is the standardized value of the ith water quality parameter, P_i is the weight of the ith water quality parameter, and n is the number of water quality parameters involved in the evaluation. P_i values range from 1 to 4, where 4 represents the most important parameter for the survival of aquatic organisms (such as dissolved oxygen) and 1 represents the parameter that affects aquatic life (such as chlorine), as shown in Table 5.

Table 5. Scores and weights of water quality factors in WQI method.

Factors	Weight	Standardization Values										
		100	90	80	70	60	50	40	30	20	10	0
DO	4	≥7.5	>7	>6.5	>6	>5	>4	>3.5	>3	>2	≥1	<1
EC	2	<750	<1000	<1250	<1500	<2000	<2500	<3000	<5000	<8000	≤12,000	>12,000
COD_{Mn}	3	<1	<2	<3	<4	<6	<8	<10	<12	<14	≤15	>15
BOD_5	3	<0.5	<2	<3	<4	<5	<6	<8	<10	<12	≤15	>15
NH_3-N	3	<0.01	<0.05	<0.1	<0.2	<0.3	<0.4	<0.5	<0.75	<1	≤1.25	>1.25
TN	3	<0.1	<1	<1.5	<2	<3	<4	<5	<6	<8	≤10	>10
TP	1	<0.01	<0.02	<0.05	<0.1	<0.15	<0.2	<0.25	<0.3	<0.35	≤0.4	>0.4

2.4. Spearman's Correlation Coefficient

A statistical technique called correlation analysis analyzes the interdependence of data patterns to determine the relationships between variables [47]. Calculating the Spearman correlation coefficient to measure the size of the correlation between two sets of data is a common correlation analysis approach. To evaluate the monotonic connection between two variables, the Spearman correlation coefficient, a nonparametric statistical correlation measure, is utilized [48]. Since the observed indicators of water quality do not follow a normal distribution, the Spearman coefficient is employed to analyze the pattern of correlation between the different factors. The degree and direction of any monotonic link between two rank variables or a rank variable and a measured variable are quantified by the Spearman correlation coefficient r_s, a distribution-free rank statistical measure [49]. A negative (−) correlation happens if one variable declines while the other grows, and vice versa. A positive (+) correlation occurs when two variables increase or decrease equally. The expression for r_s is [50]:

$$r_s = 1 - \frac{6\sum d_i^2}{N\,(N^2 - 1)} \quad (6)$$

where $d_i = X_i - Y_i$ is the difference between each pair of ranking variables, N is the total number of samples, and the value of r_s is between −1 and 1.

3. Results

3.1. Assessment of Water Quality in Taizhou

3.1.1. Assessment of SFI

There are noticeable discrepancies in the water quality parameters in the flood season (FS) and the non-flood season (NFS) in Taizhou from January 2021 to May 2022, according to distribution data (Figure 2) and descriptive statistics (Table 6). There are significant seasonal variations in DO, EC, BOD_5, NH_3-N, and TN. BOD_5 average concentration during the flood season (4.28 mg/L) is 2.63 times higher than the average concentration during the non-flood season (1.63 mg/L); NH_3-N and TN average concentration during the flood season (1.22 mg/L, 1.18 mg/L) was twice as high as average concentration during the non-flood season (0.63 mg/L, 0.65 mg/L); however, average DO concentration during the flood season (6.14 mg/L) was only about 2/3 in non-flood season. Changes in NH_3-N and TN exhibit obvious seasonal variations as well. and the two water quality metrics' average concentrations were much greater during the flood season than they were during the non-flood season. The non-flood season had a wider range of EC variation than the flood season did, as seen by the bigger standard deviation (5.69 vs. 156.60) and coefficient of variation (0.17 vs. 1.41) in the non-flood season.

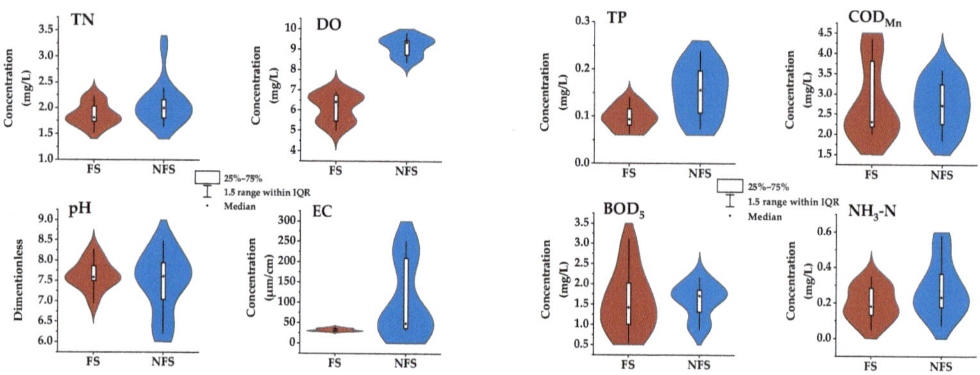

Figure 2. Distribution of water quality monitoring results.

Table 6. Scores and weights of water quality factors in WQI method.

Indicators	Flood Season			Non-Flood Season			*T* Value
	Mean (Std.)	Range	CV	Mean (Std.)	Range	CV	
pH	7.63 (0.56)	6.12–8.79	0.07	7.47 (0.90)	5.19–8.85	0.12	0.710 **
DO/(mg/L)	6.14 (1.17)	3.30–8.00	0.19	9.15 (1.69)	5.10–12.00	0.18	−12.828 **
EC	32.54 (5.69)	23.2–54.4	0.17	110.12 (155.60)	28.2–509.2	1.41	−2.782 ***
COD_{Mn}/(mg/L)	2.80 (1.37)	1.30–7.00	0.49	2.72 (0.82)	1.50–5.40	0.30	0.352 **
BOD_5/ (mg/L)	4.28 (1.95)	0.20–7.25	0.46	1.63 (0.53)	0.20–2.50	0.33	9.016 **
NH_3-N/(mg/L)	1.22 (1.19)	0.04–5.69	0.98	0.63 (1.08)	0.02–6.09	1.70	1.979 **
TN/(mg/L)	1.18 (0.50)	0.19–2.53	0.42	0.65 (0.37)	0.02–1.85	0.57	7.862 **
TP/(mg/L)	0.10 (0.04)	0.05–0.24	0.40	0.15 (0.19)	0.06–1.006	1.22	−3.084 ***

Notes: An independent sample T-test was conducted for water quality indexes in flood season and non-flood season, ** $p < 0.05$, *** $p < 0.01$.

Except for COD_{Mn} and BOD_5, the findings of the single-factor evaluation demonstrate (Figure 3) that the five indicators of pH, DO, NH_3-N, TN, and TP in the research region exhibit obvious hydrological seasonal changes. The biggest seasonal variance was seen in DO, followed by TP, and TN. Among these, the distribution trends of the pH, NH_3-N,

TN, and TP pollution indices values indicated that the flood season was smaller than the non-flood season, although the DO index indicated the contrary. The pollution index was higher during the flood season than during the non-flood season. The primary pollution indicator during the flood season is TN, and the accompanying pollution index is 3.73; in contrast, during the non-flood season, the main pollution indications are TN and TP, and the corresponding pollution indices are 4.16 and 1.53, respectively. As a result, TN and TP indicators with severe pollution in flood and non-flood seasons, as well as DO with large seasonal changes, were chosen to conduct a thorough water quality evaluation and analysis of influencing variables on a macro scale.

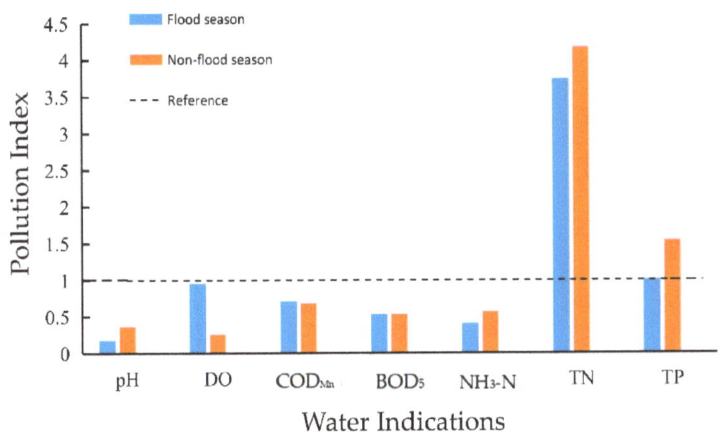

Figure 3. Distribution of single-factor pollution indices for water quality indicators in flood and non-flood periods.

3.1.2. Assessment of WQI

In the research region, Figure 4 depicts the dispersion of the WQI index throughout the flood season and non-flood season. As can be observed, there are no appreciable differences in WQI values in the study region between the flood season and the non-flood season. The most substantial shift in WQI occurs in the southwest of the research region, specifically the west of Hailing District, Gaogang District, Taixing, and Jingjiang. This area's WQI value during flood season is between 70 and 90, whereas its WQI value during the non-flood season is between 50 and 70. Water quality is better during flood season than during the non-flood season. During the flood season, the Study area's average WQI score is 67, which is typically considered to be in the "medium" level. The WQI value of 67% of the sample stations is below 70, and the south has higher water quality than the north. The average WQI value during the non-flood season is 71, which is at the "good" threshold. In the non-flood season, the central region's water quality is somewhat better than that of the other areas, and the WQI value is below 70 at close to 58% of the sample locations. Overall, the study area's water quality score (WQI) falls into the 50–80 range, which is considered to be of "medium" quality.

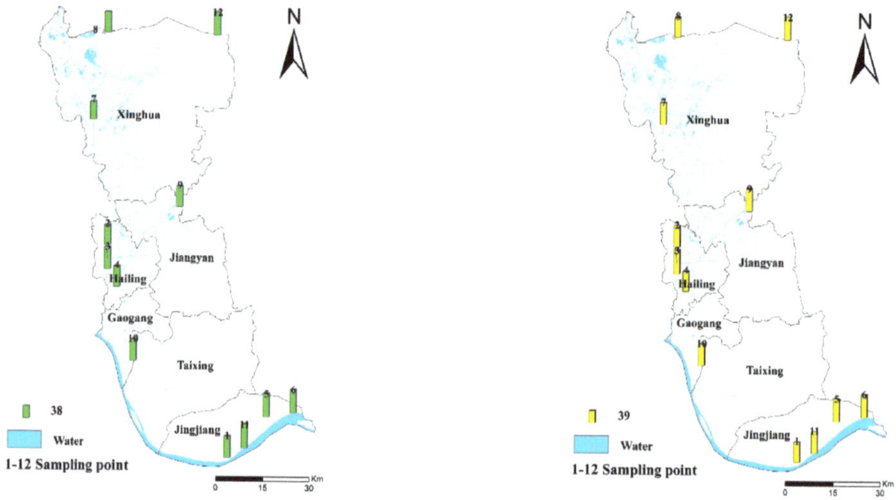

(**a**) Flood–season (**b**) Non–flood season

Figure 4. Spatial distribution of WQI evaluation results.

3.2. Spearman Correlation Analysis

To perform Spearman correlation analysis [42], choose the concentrations of three generally used water quality measures such as TN, TP, and DO, which have notable water quality concerns in both flood season and non-flood season, as well as the WQI comprehensive water quality assessment index and three types of driving variables such as topographical, human socio-economic activity, and land use type. The results are given in Figure 5. To facilitate comparison, the results are summarized in Table 7.

The result shows that during the flood season, the concentration of TN was substantially linked negatively with five indicators, including the proportion of forest land area (r = −0.34, $p < 0.01$) and the proportion of water area (r = −0.44, $p < 0.01$), and considerably positively with seven indicators, including COD discharge (r = 0.34, $p < 0.01$), GDP (r = 0.40, $p < 0.001$), and total sewage discharge (r = 0.48, $p < 0.001$). According to Figure 5b, the primary factor controlling TN concentration during non-flood seasons is the fraction of wetland and forest area, which is inversely connected with wetland area (r = −0.28, $p < 0.05$) and forest area (r = −0.19, $p < 0.05$); had a substantial (r = 0.33, $p < 0.05$) positive correlation with the percentage of the tertiary industry.

During the flood season, the concentration of TP was significantly negatively correlated with four indicators including the proportion of forest area (r = −0.37, $p < 0.01$), the proportion of shrub area (r = −0.63, $p < 0.001$), and the proportion of wetland area (r = −0.27, $p < 0.05$), at the same time, with COD emissions (r = 0.46, $p < 0.001$), aquatic product production (r = 0.36, $p < 0.01$) and GDP ((r = 0.54, $p < 0.01$) < 0.001) and other seven indicators showed a significant positive correlation. The TP concentration in the non-flood season was significantly negatively correlated with the proportion of grassland area (r = −0.31, $p < 0.01$), the average annual water temperature (r = −0.48, $p < 0.001$), and was significantly positively correlated with GDP (r = 0.27, $p < 0.05$), and the urban sewage treatment rate (r = 0.28, $p < 0.05$).

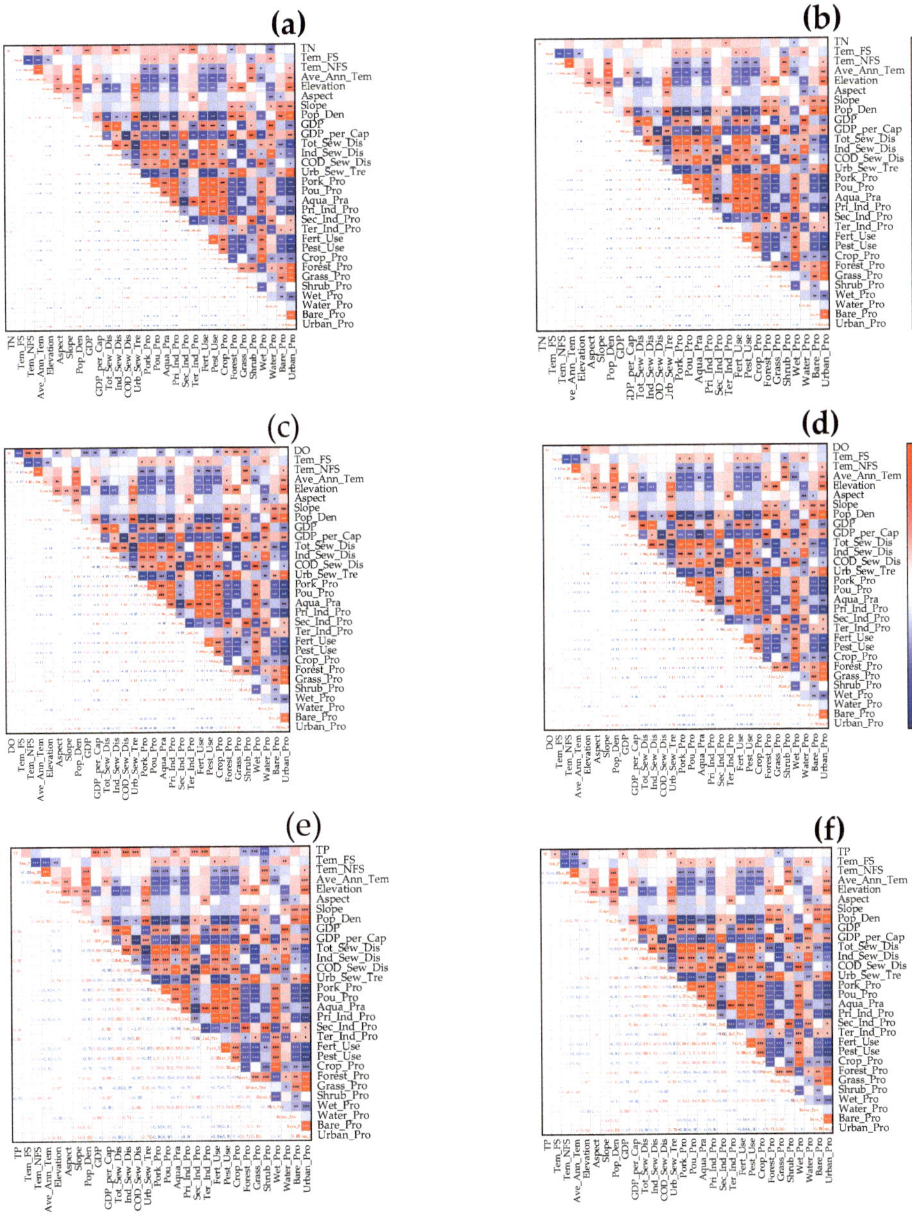

Figure 5. *Cont.*

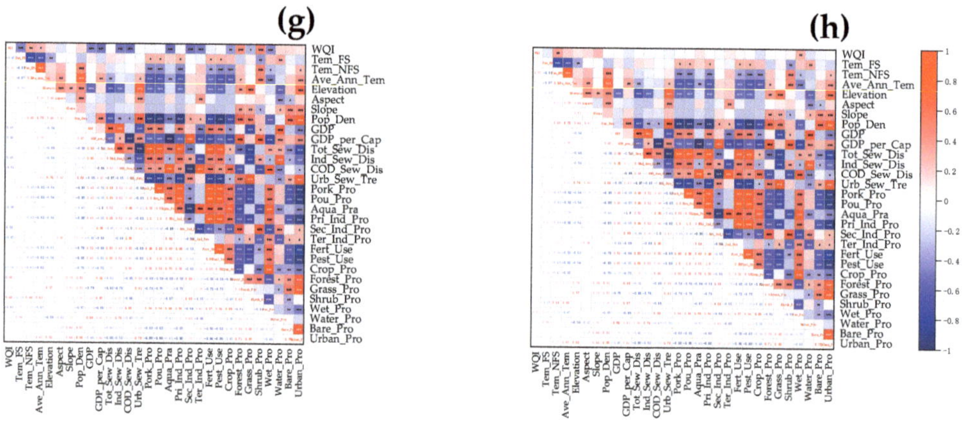

Figure 5. Distribution of single factor pollution index. * $p < 0.10$, ** $p < 0.05$, *** $p < 0.01$. (**a**) TN actuation analysis in flood season; (**b**) TN actuation analysis in non-flood season; (**c**) TP actuation analysis in flood season; (**d**) TP actuation analysis in non-flood season; (**e**) DO actuation analysis in flood season; (**f**) DO actuation analysis in non-flood season; (**g**) WQI actuation analysis in flood season; (**h**) WQI actuation analysis in non-flood season.

Table 7. Spearman correlation analysis summary results.

Driving Factors	TN		DO		TP		WQI	
	HFS	LFS	HFS	LFS	HFS	LFS	HFS	LFS
Elevation	0.20	−0.03	−0.03	0.40 ***	0.02	0.20	0.13	0.14
Aspect	0.39 **	0.06	−0.35 **	0.20	0.04	0	−0.17	0.18
Slope	−0.21	−0.15	0.22	0.24	0.15	−0.11	0.06	0.26
Tem_HFS	−0.08	−0.06	−0.72 ***	−0.16	0.26	0.27 *	−0.45 ***	−0.15
Tem_LFS	0.23 *	0.08	0.57 ***	−0.02	−0.25	−0.59 ***	−0.34 **	0.37 **
Ave_Ann_Tem	0.34 **	0.22	0.44 ***	−0.14	−0.10	−0.48 ***	−0.27 *	0.21
Pop_Den	0.20 *	0.20	0.10	0.10	0.09	−0.02	0.16	0.03
GDP	−0.40 ***	−0.23	−0.41 ***	0.18	−0.54 ***	−0.27 *	−0.46 ***	−0.01
GDP_Per_Cap	−0.25	−0.07	−0.40 **	−0.11	−0.36 **	0.03	−0.50 ***	−0.13
Tot_Sew_Dis	0.19	−0.07	−0.15	0.04	−0.17	0.26	0.02	−0.02
Ind_Sew_Dis	0.48 **	−0.27	−0.45 ***	0.08	−0.59 ***	−0.22	−0.49 ***	−0.06
Urb_Sew_Tre	0.18	−0.01	−0.25 *	0.10	0.15	−0.28 *	−0.04	0.09
COD_Sew_Dis	0.34 **	0.24	−0.30 **	0.02	0.46 ***	−0.10	−0.50 ***	0.07
Pork_Pro	−0.25	−0.24	−0.04	0.05	−0.18	−0.08	−0.08	0.03
Pou_Pro	−0.25	−0.24	−0.04	0.05	−0.18	−0.08	−0.08	0.03
Aqua_Pro	0.25	0.07	−0.40 **	0.11	0.36 **	−0.03	−0.50 ***	0.13
Pri_Ind_Pro	−025	−0.24	−0.04	0.05	−0.18	−0.08	−0.08	0.03
Sec_Ind_Pro	0.34 **	−0.24	0.30 **	−0.02	−0.46 ***	0.10	−0.50 ***	−0.07
Ter_Ind_Pro	0.49 ***	0.33 *	−0.35 **	0.09	0.56 ***	−0.04	−0.47 ***	0.11
Fert_Use	0.17	0.20	−0.08	0.15	−0.13	−0.13	−0.11	0.08
Pest_Use	−0.25	−0.24	−0.04	0.05	−0.18	−0.08	−0.08	0.03
Crop_Pro	−0.09	0.22	−0.34 ***	−0.02	0.18	0.11	−0.38 **	0.17
Forest_Pro	−0.34 **	−0.18 *	0.33 **	0.43 ***	−0.37 **	0.17	0.45 ***	0.01
Grass_Pro	0.27 *	0.20	0.42 ***	−0.05	−0.40 ***	−0.31 *	−0.32 *	−0.04
Shrub_Pro	−0.06	−0.15	0.39 **	−0.06	−0.63 ***	−0.16	0.49 ***	0.08
Wet_Pro	−0.06	−0.28 *	0.29 *	0.11	0.27 *	0.01	−0.43 ***	0.33 **
Water_Pro	−0.44 **	−0.12	−0.23	−0.09	−0.18	−0.03	0.16	−0.17
Bare_Pro	−0.08	0.17	0.18	0.12	0.26	0.01	0.16	−0.07

Note: * $p < 0.10$, ** $p < 0.05$, *** $p < 0.01$.

DO concentrations during the flood season were negatively correlated with ten indicators including flood water temperature (r = −0.72, $p < 0.001$), COD emissions (r = −0.30, $p < 0.01$), GDP (r = −0.41, $p < 0.001$), the proportion of cropland area (r = −0.34, $p < 0.01$) and industrial sewage discharge (r = −0.45, $p < 0.001$). Six indicators were significantly positively correlated including non-flood water temperature (r = 0.57, $p < 0.001$), average annual water temperature (r = 0.44, $p < 0.001$), and the proportion of forest area (r = 0.33, $p < 0.01$). The non-flood season DO concentration driver was relatively simple and significantly positively correlated with two indicators: the proportion of forest area (r = 0.43, $p < 0.001$) and elevation (r = 0.40, $p < 0.01$). Other indicators were not significantly correlated with the non-flood DO concentration.

Flood season WQI index and proportion of cropland (r = −0.38, $p < 0.01$), flood season temperature (r = −0.45, $p < 0.001$), industrial sewage discharge (r = −0.49, $p < 0.001$) and GDP (r = −0.50, $p < 0.001$) and other ten indicators were significantly negatively correlated; at the same time, with the proportion of forest area (r = 0.45, $p < 0.001$), the proportion of shrub area (r = 0.50, $p < 0.001$) and the proportion of wetland area (r = 0.43, $p < 0.001$) were significantly positively correlated. The driving mechanism of WQI in non-flood season is relatively simple, and there is a significant positive correlation between WQI in non-flood season and the proportion of wetland area (r = 0.33, $p < 0.01$) and temperature in non-flood season (r = 0.37, $p < 0.001$), and other indicators were not significantly correlated with WQI concentration in non-flood seasons.

4. Discussion

4.1. Main Factors Influencing Water Quality

The driving force study revealed that land use, particularly the usage of forest, shrub, and agriculture, as well as human socioeconomic activities (GDP, industrial sewage, and COD discharge), are significant driving forces influencing changes in the water quality of Taizhou rivers. The combined influence of natural elements and human activities needs to be taken into account to elucidate the variables that affect water quality.

4.1.1. Impact of Anthropogenic Activities

The primary causes of excessive N and P in rivers are GDP, COD brought on by industrial growth, and non-point source pollution brought on by excessive sewage discharge; however, the presence of forest areas and wetlands can prevent the accumulation of N and P concentrations. A considerable share is accounted for by Taizhou's coal mines, nonferrous metals, papermaking, fertilizer manufacture, and other important water pollutant discharge sectors. Unreasonable industrial construction and sewage discharge will seriously harm the aquatic environment. According to data, Taizhou utilized 73.03% of the total industrial energy in 2020, with heavy industry using the majority of it, followed by light industry. Furthermore, population increase has resulted in the discharge of a huge volume of sewage from residents' homes, as well as the eutrophication of water bodies driven by chemical fertilizers and pesticides used in industrial and agricultural operations [51]. In the same year, 144,930 tons of chemical fertilizers (in pure form) were utilized in Taizhou City, a 1.4% reduction from the year before. Although there has been a yearly decline in Taizhou City's use of chemical fertilizers since 2016, the rate of use is still rather low. Crops are unable to absorb 45–65% of N and P. Some of them reach the water body with the scouring of rains, which raises the chance that groundwater and runoff carrying nitrogen and phosphorus fertilizer nutrients may pollute the river and worsen its richness. According to research by Peng [52] and colleagues, the primary input sources of nitrate in surface water remote from metropolitan areas include nitrate fertilizer, soil ammonium, manure, and septic tank waste. The driving factors of water quality in the watershed were investigated [33], and it was discovered that nitrogen and phosphorus nutrients released from production and living in Guangzhou, as well as human landscape alterations, were the primary causes of non-point source pollution. Furthermore, unrecycled livestock and domestic sewage provide 63% of nitrogen and phosphorus emissions in freshwater [53]. Taizhou is situated in the plain area of the Yangtze River's middle and lower sections. It serves as a significant hub for the production of grain, oil, and freshwater goods in China. Over 31.94% of the population works in agriculture. Rural regions have a thriving freshwater crab farming sector. Rivers, livestock, and poultry breeding [54], particularly the release of manure and urine brought on by pig breeding, cause rivers to contain an excessive amount of N and P, which raises the risk of river pollution.

Land use has been significantly impacted by human activities throughout the world, having an impact on watersheds and their ecosystems [55]. Land use in watersheds can also have an impact on rivers and the overall quality of the ecosystem [56], which in turn modifies the community composition by changing the hydrological, physicochemical, and benthic habitat conditions. Taizhou is located in the Yangtze River Basin and the Huai River Basin. The intricate river network system creates a pattern of land usage in which Taizhou is built by water and born by water. The influence of land use and human activities on river ecosystems and water quality is determined by a variety of factors, including riparian forest buffer zones, the size of the watershed buffer zone, the position of the reach within the watershed, the existence of additional pressures, and the land use measure [57]. The riparian zones allow sediment and pollutants to enter water bodies through surface runoff, groundwater flow, organic inputs, and atmospheric deposition [40], while the existence of built-up land hardens the underlying surface and weakens its retention and absorption of sediment and pollutants, leading to an increase in N and P concentrations in the watershed, and the increase in agricultural and urban land is also directly responsible for the increase in N and P concentrations in the watershed, which are both nutrient concentrations are in turn negatively correlated with increases in the percentage of the forest land area [58]. The percentage of agricultural land in Taizhou has risen from 68.34% to 70.10% during the last ten years. The expansion of agricultural land has resulted in a fall in the proportion of forest land and wetlands. Numerous studies have shown that the extent of agriculture and farmland in the catchment decreases water quality, habitat, and biological assemblages and that land use within the riparian zone and throughout the watershed is particularly successful in predicting TN, nitrate, and orthophosphate.

4.1.2. Impact of Natural Factors

According to Mackie [59] and other research, riparian zones can capture 89% of nitrogen and 80% of phosphorus. The potential of different types of natural vegetation to absorb nitrogen and phosphorus pollution varies significantly, whereas forest land and shrubs have a substantial capacity to do so [26]. The urban greenery coverage rate of 42.57% in Taizhou in 2020 is slightly higher than the average urban greenery coverage rate of 42.1% in China, and the higher urban greenery coverage allows shrubs and forest land to trap some of the pollutants lost to water bodies [60], and it is believed that riparian zone buffers covered by forestland have a positive impact on water quality, significantly reducing sediment loads and nitrogen and phosphorus nutrient concentrations in water, which can stabilize river channels and floodplains and control the flow of sediment, nutrients, and agricultural pollutants from plains and hillsides into rivers [61]. By limiting the geographical distribution and composition of various land uses, topography has an impact on the number of pollutants, the distribution of pollution sources, and even water quality [5]. The research location is in China's Yangtze River Plain, where there is limited topographical undulation but a large number of river networks and significant potential for pollution movement and transformation. Distinct forms of land use play different roles in the paths from nonpoint sources to waterways [62]. Wang et al. [63] found that urban land use can also have adverse effects on rivers and water quality, especially when reaching critical quantities and close to river courses, the main changes associated with the increase in the urban land area include an increase in the number of pollutants in runoff transport, increased water temperature due to loss of riparian vegetation and exposure to surface runoff, and reduced channel and biological habitat structure due to limited sediment input and interactions between riparian and river margins [64]. Taizhou features densely populated cities and a high pace of urbanization. The city's sewage discharge treatment rate was 96.39% in 2020, an 8.69% increase from 2015. The high sewage treatment rate may filter or remove most water contaminants, reducing pollutants released into rivers. The study concluded that there is not a strong association between urban land and river water quality since there are not a lot of urban lands close to the sample location. However, this does not imply that urban land does not affect Taizhou's water quality. To further understand how Taizhou's urban land affects the quality of its water, more sample locations should be established in the future. Furthermore, topography [42] and light [65] influence rainfall and runoff by influencing the distribution of water and heat in the watershed as well as vegetation growth, and the impact on river water quality cannot be overlooked.

4.2. Limitations

In this study, the influence of climate change on water quality was neglected and only the geography, socioeconomics, and land use variables were chosen to evaluate the association of watershed water quality. In actuality, changes in precipitation patterns, global warming, and human responses to climate change will all have a considerable impact on surface water quality [66,67]. Rain causes rivers to redistribute their water, which alters how nutrients are transported through the river. Wolf et al. [68] discovered that increases in annual nitrogen loading in the Iowa watershed, USA, were caused by increases in the river's annual flow and annual base flow as a result of increasing precipitation in the watershed. In a study of seven rivers in the Midwest of the United States, Baeumler and Gupta [69] discovered that climate change, including changes in rainfall and temperature, was primarily to blame for the increase in nitrogen loads in these rivers. Changes in land use and fertilizer application had little impact on nitrogen load changes. Furthermore, the study area's industrial business is developing, and the stainless steel and shipbuilding industries will discharge various chemical elements into the water body. Due to the insufficient study data, this work has not conducted extensive investigation to determine whether this has resulted in excessive heavy metals in the water body [70]. In the future, this is something that has to be taken into consideration.

The evaluation model in this work is constrained by the little amount of data and the short data period, which prevents accurate calibration and verification on an annual scale. Furthermore, changes in water quality are quite complicated, and monitoring the eight indicators alone is insufficient to predict water quality. To undertake a thorough and realistic evaluation of water quality in the future, the breadth and volume of data collecting should be expanded, and the influence of riparian ecosystems should be taken into account [71].

4.3. Water Quality Management

The United Nations has reclassified improving water quality as one of the top concerns on the global agenda [72]. Surface water quality in China has been seriously threatened during the past 40 years by the country's rapid industrialization and urbanization [73]. As a result, a nationwide monitoring network with long-term access to water quality parameters is urgently required to allow dynamic real-time monitoring of water quality and collect reliable water quality data. To regulate the total water quality in China from a macro viewpoint, the government should also apply differentiated management of regional water quality indicators according to various areas and regulate water quality parameters, reference standards, regional variations, and water quality evaluation methodologies [74].

Most of the study area's water pollution control infrastructure construction was finished in recent years, but there are still issues, including ineffective sewage treatment, rain and pollution flowing together, and a mismatch between sewage discharge standards and environmental protection requirements. The research area's sewage discharge and treatment rate was 96.39% in 2020. The next level of water pollution control will put more of an emphasis on management requirements and criteria for water quality classification than it would on building infrastructure. The fecal discharge of tailwater is a critical remedy for the developing aquaculture and animal farming industries in the research region. As a result, the land used for raising livestock and poultry must be carefully planned. Changes to the land cover and crop planting zones should be made in extremely polluted regions, such as growing cash crops in the higher reaches of related rivers, reservoirs, and lakes rather than food crops. To enhance the quality of the water in the watershed, aquiculture zones should be divided into tailwater discharge categories, and Standardized manure disposal practices and standards for livestock and poultry production.

The analysis of the water quality metrics reveals that TN is the primary element influencing the management of water quality in the research region. Fertilizers, pesticides, and insecticides used in agricultural production are the primary sources of TN and NH_3-N [75]. Water quality management focuses on lowering NH_3-N in rivers owing to changes in assessment indicators, while disregarding the influence of nitrate on total water quality in the selection of criteria for WQI calculation. In reality, nitrate has turned into a common water contaminant as a result of nitrogen loss in agriculture and significant nitrate fertilizer use. Monitoring for nitrates shouldn't be disregarded.

In the future, the remote online water quality monitoring system should be improved, the frequency and scope of monitoring water quality and related driving factors should be increased, and the main influencing variables of the water environment in different times and situations should be mined in combination with drones and satellite remote sensing big data, the level of water pollution tracing and emergency management should be improved, and the safety of drinking water should be improved.

5. Conclusions

This study assesses the water quality of Taizhou using SFA and WQI water quality assessment methodologies. It picks water quality indicators with substantial seasonal fluctuations and pollution levels and performs Spearman correlation analysis using topographical, socioeconomic, and land use data to investigate the driving factors of river water quality in Taizhou City. The results demonstrate that:

(1) The concentration of TN (1.18 mg/L) and NH_3-N (1.22 mg/L) surpassing the norm are the primary causes of Taizhou's water pollution, but the presence of TP (0.15 mg/L) over the threshold cannot be disregarded. There are noticeable changes between the hydrological seasons and driving variables of DO, BOD_5, NH_3-N, and TN. While DO concentrations are lower during flood seasons than during non-flood seasons, NH_3-N, TN, and BOD_5 concentrations are greater during flood seasons than during non-flood seasons, indicating that the factors affecting water quality during flood seasons are more complicated than during non-flood seasons. According to the WQI result, Taizhou's water quality is generally good, with the center and northern regions of the city having the most surpassing points. When there is flood season, the water quality is higher than during the non-flood season.

(2) Seasonal variations in river water quality in Taizhou are significantly influenced by changes in land use, particularly changes in forest, shrub, and farmland, as well as increasing human industrial production activities (industrial sewage and COD discharge). The primary causes of the high concentration of TN and NH_3-N and the deterioration of water quality are the vast amounts of nitrogen and phosphorus fertilizers released by human activities and the aggravation of man-made land landscapes.

Reducing the concentration of total nitrogen is a critical issue for water pollution control and management, creating a reasonable evaluation index system will be necessary for managing and controlling river water quality pollution in the future.

Author Contributions: Writing-original draft, S.D.; Writing-review and editing, S.D., C.L. and X.J.; Data curation, S.D.; Methodology, S.D. and C.L.; Software, S.D.; Investigation, S.D.; Visualization, C.L.; Formal acquisition, C.L.; Supervision, C.L., X.J., T.Z. and H.H.; Project administration, X.J., T.Z. and H.H.; Resources, C.L., X.J., T.Z. and H.H. All authors have read and agreed to the published version of the manuscript.

Funding: This work was supported by Development Research and Master Plan of Modern Agricultural Industry in the North Ecological Economic Zone of Taizhou City, China (Grant No.tzyjy100045).

Conflicts of Interest: The authors declare no conflict of interest.

References

1. Alvarez, P.J.J.; Chan, C.K.; Elimelech, M.; Halas, N.J.; Villagran, D. Emerging opportunities for nanotechnology to enhance water security. *Nat. Nanotechnol.* **2018**, *13*, 634–641. [CrossRef] [PubMed]
2. Bakker, K. Water Security Research: Challenges and Opportunities. *Science* **2012**, *337*, 914–915. [CrossRef]
3. Nong, X.; Shao, D.; Zhong, H.; Liang, J. Evaluation of water quality in the South-to-North Water Diversion Project of China using the water quality index (WQI) method. *Water Res.* **2020**, *178*, 115781. [CrossRef]
4. Yu, C.; Yin, X.; Li, H.; Yang, Z. A hybrid water-quality-index and grey water footprint assessment approach for comprehensively evaluating water resources utilization considering multiple pollutants. *J. Clean. Prod.* **2020**, *248*, 119225. [CrossRef]
5. Wu, J.; Lu, J. Spatial scale effects of landscape metrics on stream water quality and their seasonal changes. *Water Res.* **2021**, *191*, 116811. [CrossRef] [PubMed]
6. Carroll, S.; Liu, A.; Dawes, L.; Hargreaves, M.; Goonetilleke, A. Role of Land Use and Seasonal Factors in Water Quality Degradations. *Water Resour. Manag.* **2013**, *27*, 3433–3440. [CrossRef]
7. Xiao, Y.; Liu, K.; Hao, Q.; Li, Y.; Xiao, D.; Zhang, Y. Occurrence, Controlling Factors and Health Hazards of Fluoride-Enriched Groundwater in the Lower Flood Plain of Yellow River, Northern China. *Expo. Health* **2022**, *14*, 345–358. [CrossRef]
8. Xiao, Y.; Liu, K.; Hao, Q.; Xiao, D.; Zhu, Y.; Yin, S.; Zhang, Y. Hydrogeochemical insights into the signatures, genesis and sustainable perspective of nitrate enriched groundwater in the piedmont of Hutuo watershed, China. *Catena* **2022**, *212*, 106020. [CrossRef]
9. Xiao, Y.; Hao, Q.; Zhang, Y.; Zhu, Y.; Yin, S.; Qin, L.; Li, X. Investigating sources, driving forces and potential health risks of nitrate and fluoride in groundwater of a typical alluvial fan plain. *Sci. Total Environ.* **2022**, *802*, 149909. [CrossRef]
10. Uhl, A.; Hahn, H.J.; Jager, A.; Luftensteiner, T.; Siemensmeyer, T.; Doll, P.; Noack, M.; Schwenk, K.; Berkhoff, S.; Weiler, M.; et al. Making waves: Pulling the plug-Climate change effects will turn gaining into losing streams with detrimental effects on groundwater quality. *Water Res.* **2022**, *220*, 118649. [CrossRef]
11. Zheng, L.; Jiang, C.; Chen, X.; Li, Y.; Li, C.; Zheng, L. Combining hydrochemistry and hydrogen and oxygen stable isotopes to reveal the influence of human activities on surface water quality in Chaohu Lake Basin. *J. Environ. Manag.* **2022**, *312*, 114933. [CrossRef] [PubMed]
12. Mararakanye, N.; Le Roux, J.J.; Franke, A.C. Long-term water quality assessments under changing land use in a large semi-arid catchment in South Africa. *Sci. Total Environ.* **2022**, *818*, 151670. [CrossRef] [PubMed]
13. Ahn, S.R.; Kim, S.J. Assessment of integrated watershed health based on the natural environment, hydrology, water quality, and aquatic ecology. *Hydrol. Earth Syst. Sci.* **2017**, *21*, 5583–5602. [CrossRef]
14. Cook, N.A.; Sarver, E.A.; Krometis, L.H.; Huang, J. Habitat and water quality as drivers of ecological system health in Central Appalachia. *Ecol. Eng.* **2015**, *84*, 180–189. [CrossRef]
15. Aboelnga, H.T.; El-Naser, H.; Ribbe, L.; Frechen, F.B. Assessing Water Security in Water-Scarce Cities: Applying the Integrated Urban Water Security Index (IUWSI) in Madaba, Jordan. *Water* **2020**, *12*, 1299. [CrossRef]
16. Tango, P.J.; Batiuk, R.A. Chesapeake Bay recovery and factors affecting trends: Long-term monitoring, indicators, and insights. *Reg. Stud. Mar. Sci.* **2016**, *4*, 12–20. [CrossRef]
17. Hu, Z.; Hu, L.; Cheng, H.; Qi, D. Application of Pollution Index Method Based on Dynamic Combination Weight to Water Quality Evaluation. *Earth Environ. Sci.* **2018**, *153*, 062008. [CrossRef]
18. Xia, R.; Chen, Z. Integrated Water-Quality Assessment of the Huai River Basin in China. *J. Hydrol. Eng.* **2015**, *20*, 05014018. [CrossRef]
19. Yan, F.; Liu, L.; Li, Y.; Zhang, Y.; Chen, M.; Xing, X. A dynamic water quality index model based on functional data analysis. *Ecol. Indic.* **2015**, *57*, 249–258. [CrossRef]
20. Stambuk-Giljanovic, N. *Water Quality Evaluation by Index in Dalmatia*; ELsevier Science: Amsterdam, The Netherlands, 1999.
21. Gazzaz, N.M.; Yusoff, M.K.; Aris, A.Z.; Juahir, H.; Ramli, M.F. Artificial neural network modeling of the water quality index for Kinta River (Malaysia) using water quality variables as predictors. *Mar. Pollut. Bull.* **2012**, *64*, 2409–2420. [CrossRef]
22. Yunyun, L.; Jianjia, C.; Yimin, W.; Wenting, J.; Aijun, G. Spatiotemporal Impacts of Climate, Land Cover Change and Direct Human Activities on Runoff Variations in the Wei River Basin. *Water* **2016**, *8*, 220. [CrossRef]
23. de Paul Obade, V.; Moore, R. Synthesizing water quality indicators from standardized geospatial information to remedy water security challenges: A review. *Environ. Int.* **2018**, *119*, 220–231. [CrossRef] [PubMed]
24. Mian, H.R.; Saleem, S.; Hu, G.; Sadiq, R. Water Distribution Systems: Hydraulics and Quality Modeling. In *Encyclopedia of Water*; CRC press: Boca Raton, FL, USA, 2019; pp. 1–11.
25. Zhang, J.; Guo, Q.; Du, C.; Wei, R. Quantifying the effect of anthropogenic activities on water quality change in the Yangtze River from 1981 to 2019. *J. Clean. Prod.* **2022**, *363*, 132415. [CrossRef]
26. Taniwaki, R.H.; Cassiano, C.C.; Fransozi, A.A.; Vásquez, K.V.; Posada, R.G.; Velásquez, G.V.; Ferraz, S.F.B. Effects of land-use changes on structural characteristics of tropical high-altitude Andean headwater streams. *Limnologica* **2019**, *74*, 1–7. [CrossRef]

27. de Mello, K.; Valente, R.A.; Ribeiro, M.P.; Randhir, T. Effects of forest cover pattern on water quality of low-order streams in an agricultural landscape in the Pirapora river basin, Brazil. *Environ. Monit. Assess.* **2022**, *194*, 189. [CrossRef] [PubMed]
28. Shi, P.; Zhang, Y.; Li, Z.; Li, P.; Xu, G. Influence of land use and land cover patterns on seasonal water quality at multi-spatial scales. *Catena* **2017**, *151*, 182–190. [CrossRef]
29. Mello, K.d.; Valente, R.A.; Randhir, T.O.; dos Santos, A.C.A.; Vettorazzi, C.A. Effects of land use and land cover on water quality of low-order streams in Southeastern Brazil: Watershed versus riparian zone. *Catena* **2018**, *167*, 130–138. [CrossRef]
30. Qiangqiang, Y.; Guanglai, X.; Xiancheng, Y.; Aijuan, L.; Chen, C. Responses of water quality to land use and landscape pattern in the Qingyijiang River watershed. *Acta Ecol. Sin.* **2020**, *40*, 9048–9058. [CrossRef]
31. Swain, S.; Taloor, A.K.; Dhal, L.; Sahoo, S.; Al-Ansari, N. Impact of climate change on groundwater hydrology: A comprehensive review and current status of the Indian hydrogeology. *Appl. Water Sci.* **2022**, *12*, 1–25. [CrossRef]
32. Panuska, J.C.; Moore, I.D.; Kramer, L.A. Terrain analysis-integration into the agricultural non-point source (AGNPS) pollution model. *J. Soil Water Conserv.* **1991**, *46*, 59–64.
33. Chen, J.; Chen, S.; Fu, R.; Yin, X.; Wang, C.; Li, D.; Peng, Y. Analysis of water quality status and driving factors in Guangdong Province. *Acta Ecol. Sin.* **2022**, *42*.
34. Tan, M.L.; Ibrahim, A.L.; Yusop, Z.; Duan, Z.; Ling, L. Impacts of land-use and climate variability on hydrological components in the Johor River basin, Malaysia. *Hydrol. Sci. J.* **2015**, *60*, 873–889. [CrossRef]
35. Lorup, J.K.; Refsgaard, J.C.; Mazvimavi, D. Assessing the effect of land use change on catchment runoff by combined use of statistical tests and hydrological modelling Case studies from Zimbabwe. *J. Hydrol.* **1998**, *205*, 147–163. [CrossRef]
36. Li, Z.; Liu, W.-z.; Zhang, X.-c.; Zheng, F.-l. Impacts of land use change and climate variability on hydrology in an agricultural catchment on the Loess Plateau of China. *J. Hydrol.* **2009**, *377*, 35–42. [CrossRef]
37. Wei, X.; Liu, W.; Zhou, P. Quantifying the Relative Contributions of Forest Change and Climatic Variability to Hydrology in Large Watersheds: A Critical Review of Research Methods. *Water* **2013**, *5*, 728–746. [CrossRef]
38. Yuan, L.; Forshay, K.J. Evaluating Monthly Flow Prediction Based on SWAT and Support Vector Regression Coupled with Discrete Wavelet Transform. *Water* **2022**, *14*, 2649. [CrossRef]
39. Karakuş, C.B. Evaluation of water quality of Kızılırmak River (Sivas/Turkey) using geo-statistical and multivariable statistical approaches. *Environ. Dev. Sustain.* **2019**, *22*, 4735–4769. [CrossRef]
40. Zhang, Z.; Tao, F.; Du, J.; Shi, P.; Yu, D.; Meng, Y.; Sun, Y. Surface water quality and its control in a river with intensive human impacts–a case study of the Xiangjiang River, China. *J. Environ. Manag.* **2010**, *91*, 2483–2490. [CrossRef]
41. Lehmann, A.; Rode, M. Long-term behaviour and cross-correlation water quality analysis of the river Elbe, Germany. *Wat. Res.* **2001**, *35*, 2153–2160. [CrossRef]
42. Wang, X.; Liu, X.; Wang, L.; Yang, J.; Wan, X.; Liang, T. A holistic assessment of spatiotemporal variation, driving factors, and risks influencing river water quality in the northeastern Qinghai-Tibet Plateau. *Sci. Total Environ.* **2022**, *851*, 157942. [CrossRef]
43. Jonnalagadda, S.B.; Mhere, G. Water quality of the Odzi River in the Eastern Highlands of Zimbabwe. *Water Res.* **2001**, *35*, 2371–2376. [CrossRef]
44. Pak, H.Y.; Chuah, C.J.; Tan, M.L.; Yong, E.L.; Snyder, S.A. A framework for assessing the adequacy of Water Quality Index-Quantifying parameter sensitivity and uncertainties in missing values distribution. *Sci. Total Environ.* **2021**, *751*, 141982. [CrossRef]
45. Pesce, F.S.; Wunderlin, A.D. *Use of Water Quality Indices to Verify the Impact of Cordoba City (Argentina) on Suquýa River*; ELsevier Science: Amsterdam, The Netherlands, 1999.
46. Sánchez, E.; Colmenarejo, M.F.; Vicente, J.; Rubio, A.; García, M.G.; Travieso, L.; Borja, R. Use of the water quality index and dissolved oxygen deficit as simple indicators of watersheds pollution. *Ecol. Indic.* **2007**, *7*, 315–328. [CrossRef]
47. Gejingting, X.; Ruiqiong, J.; Wei, W.; Libao, J.; Zhenjun, Y. Correlation analysis and causal analysis in the era of big data. *Mater. Sci. Eng.* **2019**, *563*, 042032. [CrossRef]
48. Zhang, W.-Y.; Wei, Z.-W.; Wang, B.-H.; Han, X.-P. Measuring mixing patterns in complex networks by Spearman rank correlation coefficient. *Phys. A Stat. Mech. Appl.* **2016**, *451*, 440–450. [CrossRef]
49. Xiao, C.; Ye, J.; Esteves, R.M.; Rong, C. Using Spearman's correlation coefficients for exploratory data analysis on big dataset. *Concurr. Comput. Pract. Exp.* **2016**, *28*, 3866–3878. [CrossRef]
50. Krishnaraj, A.; Deka, P.C. Spatial and temporal variations in river water quality of the Middle Ganga Basin using unsupervised machine learning techniques. *Environ. Monit. Assess.* **2020**, *192*, 744. [CrossRef]
51. Tao, J.; Sun, X.H.; Cao, Y.; Ling, M.H. Evaluation of water quality and its driving forces in the Shaying River Basin with the grey relational analysis based on combination weighting. *Environ. Sci. Pollut. Res.* **2022**, *29*, 18103–18115. [CrossRef]
52. Peng, J.; Zheng, B.; Wang, X.; Wang, M. The driving force of basin water resource utilization to lake nutrients. *Hydrol. Sci. J.* **2020**, *65*, 855–866. [CrossRef]
53. Yan, K.; Xu, J.-C.; Gao, W.; Li, M.-J.; Yuan, Z.-W.; Zhang, F.-S.; Elser, J. Human perturbation on phosphorus cycles in one of China's most eutrophicated lakes. *Resour. Environ. Sustain.* **2021**, *4*, 100026. [CrossRef]
54. Zheng, L.; Zhang, Q.; Zhang, A.; Hussain, H.A.; Liu, X.; Yang, Z. Spatiotemporal characteristics of the bearing capacity of cropland based on manure nitrogen and phosphorus load in mainland China. *J. Clean. Prod.* **2019**, *233*, 601–610. [CrossRef]

55. Measho, S.; Chen, B.; Pellikka, P.; Trisurat, Y.; Guo, L.; Sun, S.; Zhang, H. Land Use/Land Cover Changes and Associated Impacts on Water Yield Availability and Variations in the Mereb-Gash River Basin in the Horn of Africa. *J. Geophys. Res. Biogeosciences* **2020**, *125*, e2020JG005632. [CrossRef]
56. Schuetz, T.; Gascuel-Odoux, C.; Durand, P.; Weiler, M. Nitrate sinks and sources as controls of spatio-temporal water quality dynamics in an agricultural headwater catchment. *Hydrol. Earth Syst. Sci.* **2016**, *20*, 843–857. [CrossRef]
57. White, J.Y.; Walsh, C.J. Catchment-scale urbanization diminishes effects of habitat complexity on instream macroinvertebrate assemblages. *Ecol. Appl.* **2020**, *30*, e02199. [CrossRef]
58. Kreiling, R.M.; Bartsch, L.A.; Perner, P.M.; Hlavacek, E.J.; Christensen, V.G. Riparian Forest Cover Modulates Phosphorus Storage and Nitrogen Cycling in Agricultural Stream Sediments. *Environ. Manag.* **2021**, *68*, 279–293. [CrossRef]
59. Mackie, C.; Levison, J.; Binns, A.; O'Halloran, I. Groundwater-surface water interactions and agricultural nutrient transport in a Great Lakes clay plain system. *J. Great Lakes Res.* **2021**, *47*, 145–159. [CrossRef]
60. Carter, L.D.; Dzialowski, A.R. Predicting sediment phosphorus release rates using landuse and water-quality data. *Freshw. Sci.* **2012**, *31*, 1214–1222. [CrossRef]
61. Turunen, J.; Markkula, J.; Rajakallio, M.; Aroviita, J. Riparian forests mitigate harmful ecological effects of agricultural diffuse pollution in medium-sized streams. *Sci. Total Environ.* **2019**, *649*, 495–503. [CrossRef]
62. Yu, S.; Xu, Z.; Wu, W.; Zuo, D. Effect of land use types on stream water quality under seasonal variation and topographic characteristics in the Wei River basin, China. *Ecol. Indic.* **2016**, *60*, 202–212. [CrossRef]
63. WANG, L.; LYONS, J.; KANEHL, P. Impacts of urbanization on stream habitat and fish across multiple spatial scales. *Environ. Manag.* **2001**, *28*, 255–266. [CrossRef] [PubMed]
64. Paul, M.J.; Meyer, J.L. Streams in the urban landscape. *Annu. Rev. Ecol. Syst.* **2001**, *32*, 333–365. [CrossRef]
65. Hamid, A.; Bhat, S.U.; Jehangir, A. Local determinants influencing stream water quality. *Appl. Water Sci.* **2019**, *10*, 1–16. [CrossRef]
66. Ryberg, K.R.; Chanat, J.G. Climate extremes as drivers of surface-water-quality trends in the United States. *Sci. Total Environ.* **2022**, *809*, 152165. [CrossRef]
67. Murdoch, P.S.; Baron, J.S.; Miller, T.L. Potential effects of climate change on surface-water quality in North America. *J. Am. Water Resour. Assoc.* **2000**, *36*, 347–366. [CrossRef]
68. Wolf, K.A.; Gupta, S.C.; Rosen, C.J. Precipitation Drives Nitrogen Load Variability in Three Iowa Rivers. *J. Hydrol. Reg. Stud.* **2020**, *30*, 100705. [CrossRef]
69. Baeumler, N.W.; Gupta, S.C. Precipitation as the Primary Driver of Variability in River Nitrogen Loads in the Midwest United States. *JAWRA J. Am. Water Resour. Assoc.* **2019**, *56*, 113–133. [CrossRef]
70. Rahman, M.A.T.M.T.; Paul, M.; Bhoumik, N.; Hassan, M.; Alam, M.K.; Aktar, Z. Heavy metal pollution assessment in the groundwater of the Meghna Ghat industrial area, Bangladesh, by using water pollution indices approach. *Appl. Water Sci.* **2020**, *10*, 1–15. [CrossRef]
71. Hilary, B.; Chris, B.; North, B.E.; Angelica Maria, A.Z.; Sandra Lucia, A.Z.; Carlos Alberto, Q.G.; Beatriz, L.G.; Rachael, E.; Andrew, W. Riparian buffer length is more influential than width on river water quality: A case study in southern Costa Rica. *J. Environ. Manag.* **2021**, *286*, 112132. [CrossRef]
72. Hanna, D.E.L.; Lehner, B.; Taranu, Z.E.; Solomon, C.T.; Bennett, E.M. The relationship between watershed protection and water quality: The case of Québec, Canada. *Freshw. Sci.* **2021**, *40*, 382–396. [CrossRef]
73. Woo, S.Y.; Kim, S.J.; Lee, J.W.; Kim, S.H.; Kim, Y.W. Evaluating the impact of interbasin water transfer on water quality in the recipient river basin with SWAT. *Sci. Total Environ.* **2021**, *776*, 145984. [CrossRef]
74. Guan, X.; Ren, X.; Tao, Y.; Chang, X.; Li, B. Study of the Water Environment Risk Assessment of the Upper Reaches of the Baiyangdian Lake, China. *Water* **2022**, *14*, 2557. [CrossRef]
75. Li, J.L.; Zuo, Q.T. Forms of Nitrogen and Phosphorus in Suspended Solids: A Case Study of Lihu Lake, China. *Sustainability* **2020**, *12*, 5026. [CrossRef]

water

MDPI

Article

Identification of Hydrochemical Characteristics, Spatial Evolution, and Driving Forces of River Water in Jinjiang Watershed, China

Yuchen Zhu [1,2], Hongjie Yang [3], Yong Xiao [3,*], Qichen Hao [1,2], Yasong Li [1,2], Jiahuan Liu [1,2], Liwei Wang [3], Yuqing Zhang [3], Wenxu Hu [3] and Jie Wang [3]

1 Fujian Provincial Key Laboratory of Water Cycling and Eco-Geological Processes, Xiamen 361021, China; zhuyuchen413@163.com (Y.Z.); haoqichen_iheg@163.com (Q.H.); liyasong712@126.com (Y.L.); liujiahuan1105@163.com (J.L.)

2 Institute of Hydrogeology and Environmental Geology, Chinese Academy of Geological Science, Shijiazhuang 050061, China

3 Faculty of Geosciences and Environmental Engineering, Southwest Jiaotong University, Chengdu 611756, China; yanghj@my.swjtu.edu.cn (H.Y.); leeway@my.swjtu.edu.cn (L.W.); zhangyuq@my.swjtu.edu.cn (Y.Z.); hwxxwh@my.swjtu.edu.cn (W.H.); cfqswangjie@my.swjtu.edu.cn (J.W.)

* Correspondence: xiaoyong@swjtu.edu.cn

Abstract: Rivers are an important source of water in humid regions, but their availability is greatly limited by water chemistry. In order to accurately identify the changes in river water chemical composition, the compositional analysis method (CoDA) is proposed from the perspective of compositional data analysis theory, which considers the geochemical riverine system as a whole and detects the compositional changes of the entire watershed. The basic data analysis is carried out by traditional analysis methods, and the results show that the hydrochemical characteristics of different sections of the basin have significant features. The water chemistry of Dongxi River is of the HCO_3^- Ca type. The water of the Xixi River shows a gradual evolution from the HCO_3-Ca type and high SO_4^{2-} content in the upper reaches to the Cl-Ca type in the lower reaches. The hydrochemistry of river water in the watershed is mainly affected by rock weathering leaching (PC1) and agricultural and domestic pollutant discharge (PC2), with a contribution rate of 48.4% and 19.7%, respectively. Rock weathering, mining, and agricultural pollution are the main factors affecting the chemical composition of river water in different regions. The spatial composition of a single sample at different scales is monitored by the Mahalanobis distance approach in an iterative manner to minimize the influence of a single anomaly on the composition center. The results show that the main reasons for the change in river water chemical composition along the Xixi River are attributed to mine pollution, domestic pollution, and tea plantation and that along the Dongxi River is caused by domestic pollution. The hydrochemical composition changes after the confluence of the Xixi River and Dongxi River are mainly affected by human activities and seawater in urban areas. This research could provide new perspectives and methods for detecting the influences of human and natural factors on the hydrochemistry of river water in humid regions worldwide.

Keywords: hydrochemistry; human activity; component data analysis; Mahalanobis distance; Jinjiang watershed

1. Introduction

The water resource is the most important natural and strategic resource, which plays a key role in maintaining the stability of the natural ecosystem and developing the economy of human society [1]. As one of the main components of freshwater resources, surface water is favored because it is easy to obtain and utilize. However, surface water is also the most vulnerable to be influenced by external factors [2–4]. With the rapid development of the economy, the quality and quantity of surface water were greatly affected because water resource was in huge demand, and domestic sewage and industrial effluent were discharged into the surface water without meeting the discharge requirements [1,5,6]. Hence, it is of immense significance to study the surface water of the basin in depth for better protection and utilization of surface water resources, and it is also beneficial to alleviate the contradiction of water resource shortage [7–9].

Surface water formation is the result of the interaction between the development of rivers and the environment, and it will be affected by groundwater recharge, rock weathering, atmospheric precipitation, human activities, and other effects [4,10]. In most cases around the world, groundwater is one of the main recharge sources for surface water. In mountainous areas, groundwater feeds into surface water in the form of springs or underground rivers, and the mixing of surface water and groundwater will affect the hydrogeochemical composition of surface water [11–13]. Along with the runoff process, the river will exchange materials with the surrounding land, and different types of land use will also affect the materials exchanged with the surface water [14,15]. In recent years, the global temperature change caused by drastic climate change has also directly affected the precipitation recharge and evaporation of surface water [4,16–18]. In addition, in coastal areas, seawater backflow can also alter river hydrochemistry [19].

In addition to natural factors, human activity is also an important part, which affects the physical and hydrochemical properties of surface water [20]. Around the world, rivers are disturbed by human activities to varying degrees [21]. For example, the development of mines may introduce some unique elements into the river [22]. The discharge of untreated sewage can cause the water quality of rivers to decline [23]. Fertilizers applied in agriculture can also find their way into rivers [24]. Human activities have changed the composition of rivers and posed a great threat to surface water resources. Because the physical and hydrochemical properties of the surface water in the basin are the comprehensive embodiment of the hydrological environment conditions and human activities, it contains abundant hydrogeological information.

The analysis of hydrogeochemical characteristics is the basis of studying hydrochemical composition and water quality evolution [25,26]. In order to interpret the abundant information in surface water, scholars have been exploring the hydrochemical characteristics of rivers. On the basis of a large number of practical cases, Gibbs summarized three mechanisms controlling surface hydrochemistry, atmospheric precipitation, rock weathering, and evaporation, respectively, which have been widely used by later generations. The Piper diagram is used to determine the type of hydrochemistry, etc. Some multivariate statistical methods such as hierarchical cluster analysis (HCA) and principal component analysis (PCA) are also used in the analysis of the hydrochemistry [27,28]. However, these river hydrochemistry analyses mainly focus on the statistics and spatial distribution of the main components in the water body, which cannot be comprehensively analyzed as a whole. And they are easy to be affected by component anomalies that are caused by some accidental factors during sampling, thus obtaining some unreliable or wrong conclusions.

In order to eliminate these effects, component data analysis (CoDA) is used to solve the problem. The component data analysis method considers the properties of the component data, combined with the Mahalanobis distance [29]. It considers each hydrochemical component as a whole and obtains the overall characteristics of the multi-component system [30]. Based on component data analysis, the variation characteristics of basin hydrochemistry can be comprehensively evaluated, and the interference of individual regional anomalies can be identified, which has guiding significance for monitoring potential pollution events and local climate-induced changes at the basin scale.

The Jinjiang river, located in the southwest of China's Fujian Province, serves as a major source for local water demand. With the rapid development of urbanization and economy in southwest Fujian, the hydrochemical components of Jinjiang watershed have changed significantly under the combined influence of climate change and human-made pollution [31]. In recent years, scholars have mainly conducted in-depth studies on typical pollutants in the water bodies of Jinjiang watershed, such as nitrogen and phosphorus, organics, and heavy metals. However, the existing hydrochemical analysis is mostly focused on the small watersheds in the downstream urban area, and the hydrochemical components covering the whole Jinjiang watershed have not been comprehensively and systematically analyzed.

This paper in view of the river water hydrochemical characteristics of the Jinjiang watershed, which are strongly influenced by human activities, based on the CoDA method, systematically collected and tested the water samples of the upper and lower reaches and tributaries of Jinjiang watershed. The specific aims are to (1) consider the obtained hydrochemical component data as a whole and analyze the hydrochemical components of the whole basin by using biplot and Mahalanobis distance calculations, based on the CoDA method; (2) identify the composition change and variation rules of hydrochemical components in Jinjiang watershed; (3) explain the influence and contribution of natural conditions and human factors on river hydrochemical components; and (4) provide suggestions for the utilization and protection of water resources in Jinjiang watershed.

2. Study Area

Jinjiang is located in the southeast part of the Fujian province, southeastern China. The Jinjiang watershed is defined within the latitude range of 24°45′ N to 25°35′ N and the longitude range of 117°40′ E to 118°42′ E, with an approximate total area of 5606 km^2. It flows from northeast to southeast with a total length of 404.8 km and runs through Yong Chun County, Anxi County, Nan'an City, Licheng District, and Jinjiang City. This watershed is covered by mountains and hills for 81.4% of the whole area. The plain topography only accounts for approximately 18.6% of the watershed and is distributed in the coastal area in the southeast, which has the largest city—Jinjiang, the most densely populated area in the watershed. This watershed is featured by the typical summer monsoon climate with an average annual rainfall of 1686 mm. But the rainfall presents significant uneven characteristics temporally, with 71.2% occurring from April to September [32].

The Jinjiang watershed includes two major tributaries, namely, Dongxi (West Creek) and Xixi (East Creek). In more detail, it can be divided into five sections, based on the water catchments of the tributaries, which are Xixi tributaries, Xixi mainstream, Dongxi tributaries, Dongxi main stem, and the lower course of Jinjiang, respectively (Figure 1). Dongxi (West Creek) and Xixi (East Creek) are located in the middle-upper area of the watershed. These two tributaries cover an area of 1917 km^2 and 3101 km^2, respectively. Dongxi and Xixi originate from the snowy mountains of Chengxiang in Yongchun and Tiziling, Taozhou in Anxi. Dongxi and Xixi converge at Shuangxi in Nan'an City and finally merge into Quanzhou Bay at Qianpu. The Shanmei Reservoir built in Dongxi is the only water conservancy hub in Jinjiang watershed [33]. The average discharge of the main branch is 5.12 billion m^3 per year, which peaks in the mountainous areas in the upper course and then falls gradually as it flows downstream through lower coastal plains [34]. In contrast, the population distribution and economic development are gradually stronger as the river flows downstream [35].

The strata exposed in Jinjiang watershed are mainly Triassic, Jurassic, and Quaternary, among which the volcanic rocks of the Upper Jurassic Nangyuan Formation are the most widely distributed, and the lithology is mainly tuff. Among the igneous rocks, late Jurassic volcanic rocks and Yanshan-age intrusive rocks are widely exposed. The chemical composition of the exposed strata shows homogeneity [35,36], all of which are dominated by silicate rocks, but the strata contain a variety of metallic and non-metallic ores, such as kaolinite, molybdenum, tungsten, iron, manganese, copper, and zinc, making a wide range of types of mines within Jinjiang watershed. With the rapid development of regional industry and agriculture in recent years, a large number of pollutants (nitrogen and phosphorus nutrient use, iron and manganese mining, etc.) have entered the river, making the quality of the water environment in the basin gradually deteriorate [37–39].

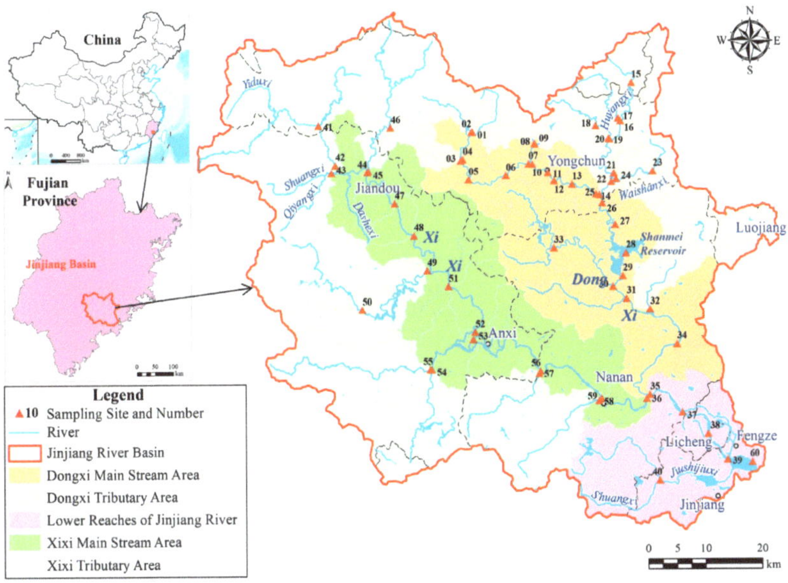

Figure 1. Location of the study area and the river water sampling sites.

Groundwater in Jinjiang watershed is mainly distributed in aquifers of unconsolidated rock pores, weathered eluvial pores and fractures, and bedrock fissures. The porous water in loose rock formations is mainly distributed in the terraces on both sides of the valley, the alluvial fans in front of the mountains, and the marine sedimentary formations in the coastal estuary areas, mainly composed of fine sand, coarse sand, and gravel, with a thickness of about 5–20 m and a specific yield of less than 100 $m^3/d{\cdot}m$. In some areas near the river, the groundwater receives water supply from the river, and the specific yield can exceed 100 $m^3/d{\cdot}m$ [40]. Weathered fissures are mainly distributed at the foot of tablelands and piedmont slopes, with the aquifer composed of clastic rock and conglomerate generally having a thickness of 10–20 m and a relatively low water volume, with a specific yield of 10 $m^3/d{\cdot}m$. Bedrock fissure water is distributed in the fractures and faults of mountainous bedrocks, with the water yield controlled by structural faults. Its water content is extremely uneven, with an overall low water volume and a specific yield of about 10 $m^3/d{\cdot}m$. Groundwater recharge mainly comes from the vertical infiltration of atmospheric precipitation and the lateral infiltration of rivers during the rainy season.

3. Data and Methods

3.1. Sample Collecting and Analytic Testing

Based on data from the Shuttle Radar Topography Mission (SRTM), the basin is divided into 60 drainage areas by using the ArcGIS hydrologic toolset, and at every confluence, we set a sampling spot, which adds up to cover the whole basin.

Water samples were collected from 5 to 20 June 2020 (rainy season). The sampling sites were distributed throughout Jinjiang watershed along the Jinjiang river and its tributaries (Figure 1). The water samples were collected the sample water from 150 cm deep in the middle of the stream to get rid of floating suspended pollutants as much as possible.

Water samples were collected in 2.5 L clean high-density polyethylene bottles that had been thoroughly prerinsed with the water to be sampled. The bottles used for trace element determination were acidified with nitric acid to pH < 2. All samples were stored in potable incubators at the temperature of 4 °C and transported to the Laboratory of Groundwater Sciences and Engineering of the Institute of Hydrogeology and Environmental Geology, Chinese Academy of Geological Sciences for hydrochemical analysis.

Water quality parameters such as total dissolved solids (TDS) and pH were measured and recorded on-site using a MANTA+30 multiparameter instrument (EUREKA, Austin, TX, USA). Major cations (Ca^{2+}, Mg^{2+}, Na^+, K^+) as well as some trace elements and ions (Fe, Mn, Zn, Pb) were determined using a full-spectrum plasma generation spectrometer (ICAP-6300, Tokyo, Japan) with a detection limit of 1 μg/L. $SO_4{}^{2-}$, Cl^- were measured using ion chromatography (Shimadzu LC-10ADvp, Tokyo, Japan). $HCO_3{}^-$ was measured by titration with hydrochloric acid [28]. Total phosphorus and total nitrogen were determined using potassium persulfate oxidation and UV spectrophotometer (UV 2450, Shimadzu, Kyoto, Japan) [41,42]. The consistency and accuracy of the analyzed results had been checked for quality assurance. Triplicate analyses were conducted for each sample to ensure the consistency of analyzed parameters. The accuracy of hydrochemical analysis for each sample was checked using the ionic charge balance error percentage (%CBE), which can be determined by Formula (1). The results showed that the recovery ratio was within ±10%, and the %CBE values were less than ±5% for all samples, implying that the analytical accuracy was reasonably good [43].

$$\%\text{CBE} = \left(\sum_{\text{i}}^{\text{m}} \text{ce}_i^+ - \sum_{\text{i}}^{\text{m}} \text{ce}_i^-\right) / \left(\sum_{\text{j}}^{\text{m}} \text{ce}_j^+ - \sum_{\text{j}}^{\text{m}} \text{ce}_j^-\right) \quad (1)$$

where ce_i^+ and ce_i^- are the charge concentration of a specific cation and anion, respectively, in milliequivalents per lit.

3.2. Component Data Analysis Methods

3.2.1. Log-Ratio Transformation and Biplot

The compositional data are influenced by constant row sum, leading to a "closure effect", causing null-correlation contained in the correlation structure of the data [44]. The multivariate statistical methods cannot be applied to the compositional data [45], and we have to transform the compositional data into a real number space [46]. Currently, the centered log-ratio transformation (CLR) [5] and isometric log-ratio transformation (ILR) [47] are commonly used. The formulas of CLR and ILR are as follows:

$$\text{CLR}(x) = \left(\ln \frac{x_i}{\text{g}(x)}\right) \quad (2)$$

$$\text{ILR}_i(x) = \sqrt{\frac{D-i}{D-i+1}} \ln \frac{\sqrt[D-i]{\prod_{j=i+1}^{D} x_j}}{x_i} \quad (3)$$

As shown in the formula, i represents the number of sample sites, $\text{g}(x)$ represents the geometric mean of the concentration of elements at each sample site, i.e., $\text{g}(x) = \sqrt[D]{x_1 \cdot x_2 \cdots x_D}$; among which, x_i represents the concentration of elements. D refers to dimensions, here it represents the number of variables.

We carried out robust principal component analysis (RPCA) on the transformed data to obtain the loadings and scores. Then we displayed them on the biplot in the form of a 2d multi-dimensional scatterplot [48,49]. The point coordination displays the scores of the principal components for two samples, and the distances between points approximate the correlation of two samples, meaning that the closer they are, the stronger the correlation. The vector length is proportional to the total variance of the variables, representing the component loading [50]. The projection length of the vector on the corresponding axis is the amount of variance described by each component variable captured by the corresponding principal component, i.e., characterizing the contribution of that component to the corresponding principal component of the axis, as shown in Figure 2. The vector length approximates the standard deviation of a column index. The cos(α) represents the correlation between indexes. The projection of the sample point to the vector represents the value of the index at that point. Sample point spacing indicates the correlation between samples.

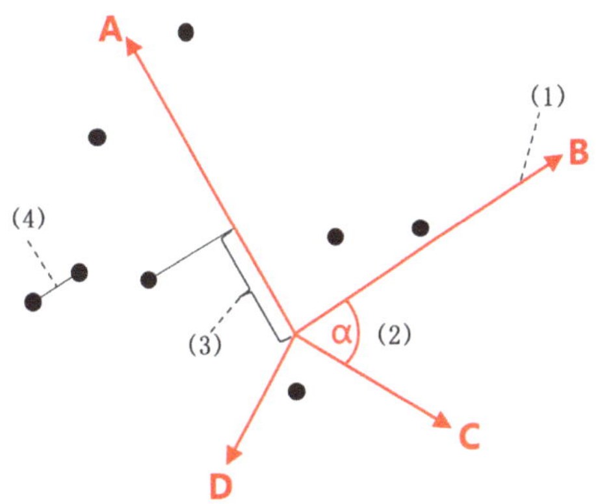

Figure 2. Description of the biplot. The point approximation represents the sample information (row information) of the transformed data matrix and the vectors approximation represent the metric information (column information) of the transformed data matrix. (1) The length of the vector approximates the standard deviation of a column of indicators. (2) The cos(α) represents the correlation between indexes. (3) The projection of a sample point onto a vector represents the value of the indices in that point. (4) The distances between points approximate the correlation of two samples.

3.2.2. Mahalanobis Distance

The Mahalanobis distance (MD) is a measure of the distance between points and centroid, which considers the multivariate covariance [51]. MD is usually used to identify outliers for multivariate data [52,53], and sometimes, it can also be used for the determination of representativity of data extracted from a larger population of samples [54]. When there are changes in the chemical composition of water in the river which are caused by inflows of tributaries with large differences in components and pollutants, it will be reflected as an increase in MD. The formula of MD is as follows:

$$D^2 = (x_i - \overline{x})S^{-1}(x_i - \overline{x})' \tag{4}$$

where, D^2 is MD, x_i is one of the indicators of the samples, which in this paper represents the element concentration; $\overline{x}$ is the average of the indicator, which represents the average value of the ion in the water's chemical composition; S^{-1} is the inverse matrix of the covariance matrix. By using log-ratio transformation, the data will be transformed from simplex space into real number space before MD can be calculated. This paper aims to show the anomaly between samples through the MD of the sample composition.

This paper uses R-package robCompositions to carry out CLR and ILR and uses the R-package FactoMineR, Factoextra to analyze the composition and draw the biplot [55]. The indicators with a detection rate lower than 40% were not involved in the analysis. Finally, the 12 indicators were chosen for analysis: $Na^+ + K^+$, Ca^{2+}, Mg^{2+}, $SO_4{}^{2-}$, $HCO_3{}^-$, Cl^-, Fe, Mn, Zn, Pb, TN, and TP; among which, those below the limit of detection (LoD) were replaced by 2/3 of the LoD. Based on the current land use type and the test results, 13 samples with little interference from human activities are selected (i.e., the samples that have the closest composition to their natural state, sample numbers: 02, 15, 16, 17, 18, 19, 20, 21, 22, 23, 24, 41, and 46) as the "original composition" and then calculate their composition center and the covariance matrix. MD is then calculated using the 60 water samples and the center of "original composition" and the analysis continues to examine the composition variations of the samples with the "original composition". After this, starting from the "original composition", every sample site of the main stem is added from upstream to downstream iteratively. The chemical composition centers of Xixi mainstream and Dongxi mainstream and the MD between every sample and the two composition centers are calculated. Finally, two clustered line charts with MD data of Xixi mainstream and Dongxi mainstream are drawn. The clustered line charts that use the results of iterative calculation of MD can minimize the influence of individual outliers on the whole trend. Except the advantage of visualizing the overall trends, the clustered line charts can also help identify the anomalies in the water's chemical composition.

4. Results and Discussion

4.1. Physicochemical Characteristics of River Water

Due to the significant differences in the natural environment and production and living conditions between the two main tributaries of Dongxi and Xixi, as well as the three downstream areas of Jinjiang, the differences in the characteristics of their respective tributaries and main streams are not significant. Therefore, in order to highlight the overall physical and chemical characteristics of the three areas, the sample data from the tributaries and main streams of Dongxi and Xixi will be combined and analyzed here.

The ranges of chemical compositions of cross-sectional water in Jinjiang watershed are statistically given in Table 1 and Figure 3. (It should be noted that the sample at Point 60, which are affected by seawater, show significant numerical differences from the other points and is not included in the statistical analysis in Table 1). The pH of river water ranges from 7.45 to 9.58, with an average value of 7.87, showing a weak alkalinity. The pH values of the Dongxi watershed are relatively higher than Xixi watershed and lower reaches. The variation range of TDS is 26.97~150.20 mg/L, and the average value is 78.74 mg/L. The downstream of the Jinjiang watershed, affected by seawater near the mouth of the sea, has the highest TDS. The cations in Jinjiang watershed are mainly made up of Ca^{2+} and Na^{+}. For Dongxi and Xixi, the Ca^{2+} is the main cation. The Na^{+} in lower reaches is obviously higher than the other two regions. The anions in Jinjiang watershed are mainly made up of HCO_3^- and SO_4^{2-}. For water bodies containing metal minerals such as Fe, Mn, Zn, and Pb, the order from highest to lowest content is Xixi watershed, lower reaches, and Dongxi watershed, respectively. Total nitrogen (TN) and total phosphorus (TP) in the Dongxi watershed are slightly higher than that in the Xixi watershed. The TN and TP are shown in the middle of the others after flowing downstream. The variation coefficients of Ca^{2+} and K^{+} are relatively small, and their contents are mainly controlled by the water-rock action. The variation coefficients of Zn and Pb are greater than 1 (strong variation), showing high volatility and dispersion at different points, which may be influenced by both environmental changes in groundwater and anthropogenic pollution.

Table 1. Descriptive statistics of river water chemical index in Jinjiang watershed (unit: mg/L).

Location	Parameter	pH	TDS	K^+	Na^+	Ca^{2+}	Mg^{2+}	SO_4^{2-}	HCO_3^-	Cl^-	Pb	Fe	Mn	Zn	TN	TP
Dongxi watershed	Min	7.45	26.97	1.69	1.38	1.62	0.21	2.37	11.87	1.04	0.00	0.01	0.00	0.00	0.58	0.03
	Max	9.58	115.80	5.85	11.88	17.77	3.23	13.67	74.20	12.14	0.01	1.29	0.45	0.17	5.87	0.58
	Mean	7.96	78.74	3.82	6.52	10.87	1.67	7.94	41.25	5.85	0.00	0.46	0.13	0.02	2.43	0.19
	SD	0.52	22.96	0.99	2.71	4.04	0.74	2.79	17.08	2.76	0.00	0.34	0.09	0.03	1.21	0.11
	CV/%	6.57	29.16	26.01	41.51	37.19	44.06	35.14	41.42	47.20	160.43	72.95	69.60	188.20	49.79	59.69
Xixi watershed	Min	7.58	47.66	2.35	2.47	4.57	0.49	3.80	0.59	1.73	0.00	0.47	0.06	0.01	0.22	0.02
	Max	8.65	122.60	4.30	19.37	22.17	6.35	51.47	53.42	9.01	0.01	1.44	1.11	0.21	3.59	0.51
	Mean	7.76	78.59	3.02	5.65	11.53	1.84	16.87	25.61	3.97	0.01	0.71	0.26	0.04	1.81	0.24
	SD	0.22	20.12	0.53	3.55	4.00	1.58	12.81	13.95	1.91	0.00	0.21	0.22	0.05	0.91	0.15
	CV/%	2.88	25.60	17.52	62.81	34.67	86.12	75.91	54.46	48.20	75.29	30.23	83.37	116.03	50.15	63.62
Lower reaches	Min	7.48	81.92	4.07	6.68	11.57	1.77	11.56	36.21	6.24	0.00	0.39	0.15	0.01	1.09	0.06
	Max	7.85	150.20	5.82	23.05	17.40	3.84	16.48	59.36	33.63	0.01	1.23	0.35	0.05	3.28	0.50
	Mean	7.61	116.22	4.93	14.72	13.88	2.59	13.35	46.90	19.50	0.00	0.77	0.24	0.03	2.23	0.23
	SD	0.17	36.22	0.81	9.22	2.72	0.96	2.24	9.58	14.40	0.01	0.35	0.09	0.02	0.90	0.19
	CV/%	2.17	31.16	16.49	62.64	19.59	36.91	16.82	20.44	73.85	200.00	45.35	36.93	69.28	40.46	84.99

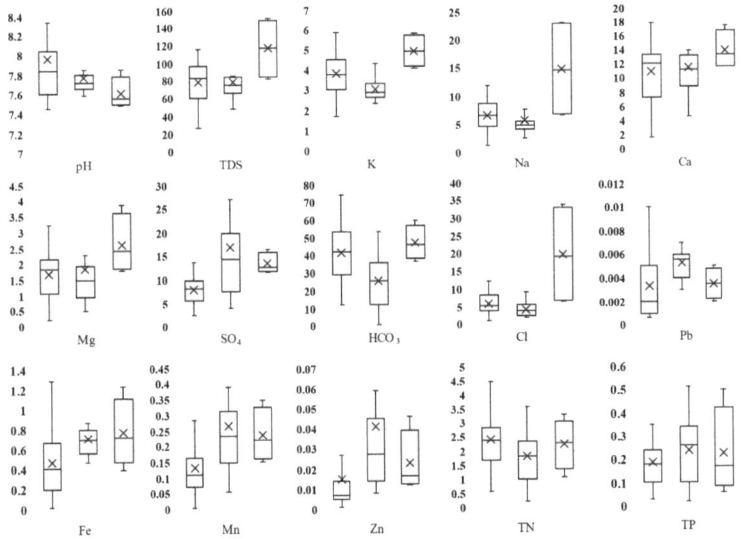

Figure 3. Box plots of river water chemical index in Jinjiang watershed (unit: mg/L). In each plot, the left box represents the Dongxi watershed, the middle box represents the Xixi watershed, and the right box represents the lower reaches.

The Piper trilinear diagram can reflect the composition of ions in water bodies [56]. As shown in Figure 4, the cations are mainly Ca^{2+} and Na^{+}. In the anion triangles, the Dongxi watershed is dominated by HCO_3^{-}, while the composition of the Xixi watershed is mainly made up of SO_4^{2-}. The water chemistry in the Dongxi basin is all HCO_3-Ca type. The water chemistry in the Xixi basin is HCO_3-Ca type with high SO_4^{2-} content in the upper reaches of the basin, and the water chemistry type gradually changes to Cl-Ca type in the lower reaches of Bailai Township. The estuary of Jinjiang watershed, due to the influence of seawater, has a relatively higher proportion of Cl-Na type of water chemistry.

4.2. Evolution of Water Chemical Component Characteristics and Controlling Factors

Hydrochemical components are influenced by both natural factors and human activities. The Gibbs diagram [57] is a way to ensure the dominant natural factors influencing the local hydrochemical components. For study areas, most water samples have a TDS value of around 100 mg/L. So, in Figure S1, the water samples gather in the rock dominance of the Gibbs diagrams. It indicates that hydrochemical components are dominantly governed by rock–water interaction in nature. For the end-member diagram (Figure S2), the samples are located in the silicate-controlling district, which shows that silicate rock weathering has a huge influence [58].

Robust principal component analysis (RPCA) is performed on the CLR-transformed component matrix to obtain component scores and loadings. By using these data, the biplot of compositions analysis and distribution map of the contribution rate of each variable were made (Figures 5 and 6). The first two principal components explain 68.1% of the total variance, with 48.4% by principal component 1 (PC1) and 19.7% by principal component 2 (PC2). The line of SO_4^{2-} is the longest, followed by TP, Ca^{2+}, Mg^{2+}, Na^{+} + K^{+}, TN, Fe, Cl^{-}, and Zn. Among them, Na^{+} + K^{+}, Ca^{2+}, Mg^{2+}, and heavy metal ions, such as Fe, Mn, Zn, and Pb, have a longer projection on the horizontal axis with conventional ions and have a relatively high contribution to PC1. Considering that the upstream of Jinjiang watershed is rich in iron, manganese, zinc, and lead metal mines, and silicate rocks are mainly weathered by the surface water of Jinjiang watershed [59,60], the PC1 is believed to reflect the influence of rock weathering and leaching release from metal minerals mining on the chemical composition of the water. The projections of TN, SO_4^{2-}, and TP on the vertical axis are long and its contribution to PC2 is relatively high. These projections are associated with human activities. Therefore, PC2 mainly explains the influence of agricultural and domestic pollutant discharges on the chemical composition of the water body.

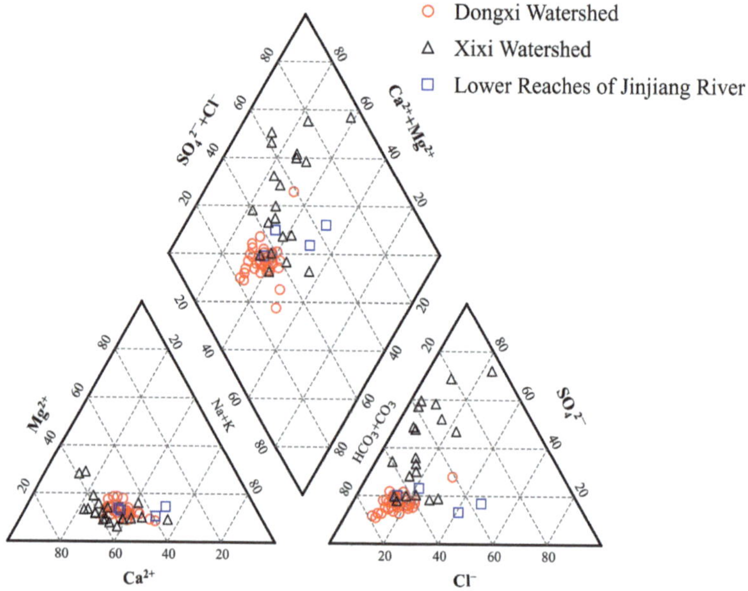

Figure 4. Piper trilinear diagram demonstrating the hydrochemical composition of river water in Jinjiang watershed.

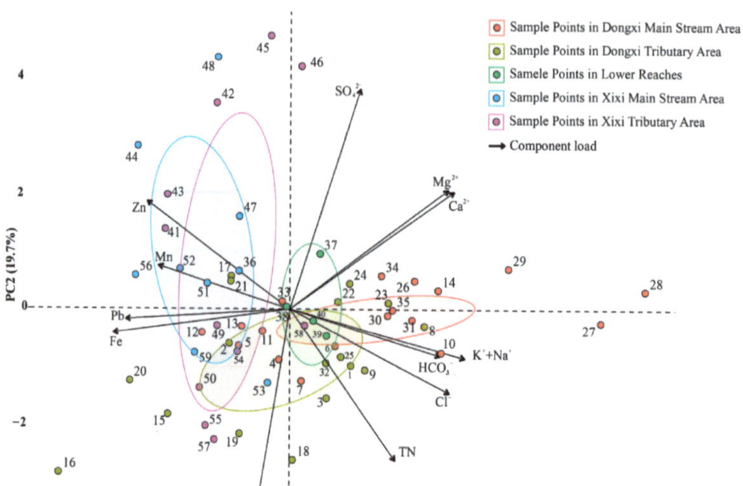

Figure 5. The biplot of all sampled river waters in Jinjiang watershed. The numbers in the figure represent the sample numbers.

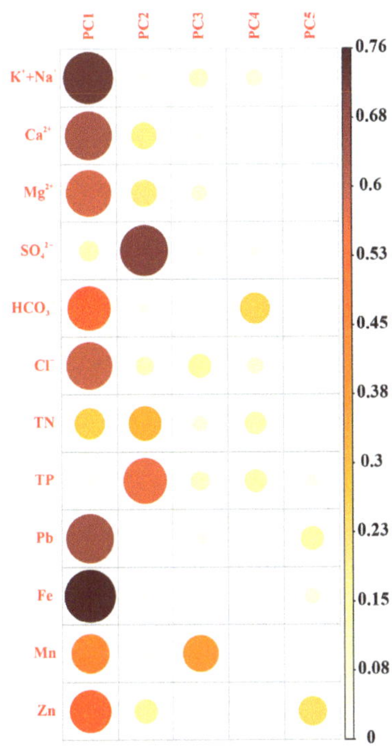

Figure 6. The distribution map of the contribution rate of each hydrochemical index for river water in Jinjiang watershed.

To highlight the trend of data concentration, the samples in different parts are indicated by different colors, and the confidence ellipse with a 99.7% confidence level is constructed by the sample distribution and corresponds to the sampling points by color. The distribution of samples shows that there are obvious differences in the characteristic components in different areas. TP, TN, and Cl^- are characteristic components in the Dongxi watershed, indicating that the chemical composition variation in Dongxi is mainly influenced by the residents and their agricultural production, especially the overdevelopment of the upstream basin of Shanmei reservoir is a direct contributor to nitrogen, phosphorus, and organochlorine pesticides [61,62]. The characteristic components in the Xixi watershed are Fe, Mn, Zn, Pb, etc., which is mainly due to the distribution of minerals containing related elements. A large number of metal mines are distributed from Yidu town in upstream Yongchun County to Jiandou town in the middle reaches of Xixi and the urban area of Anxi County [63], and these mines are the main sources of Fe, Mn, Zn, and Pb in the river. The water samples downstream of Jinjiang watershed are influenced by the complex human activities in the area with the mixed influence of the confluence components of the East and West Creeks. There is a large variability in components, and the main characteristic components are Ca^{2+}, Mg^{2+}, $Na^+ + K^+$, TN, and HCO_3^-.

4.3. Identification of the Water Chemical Composition Change

In Figure 7, the results of the calculation of MD between the basin-wide hydrochemical composition data and the "original composition" are displayed. It reveals that the MD values among the "original composition" samples are quite small, ranging from 2.64 to 3.32. This suggests that the selected original samples exhibit a similar combination of chemical composition. The variation in MD in the Xixi basin ranges from 3.32 to 23.5, indicating relatively minor differences in hydrochemical composition during the rainy season. However, in the Dongxi watershed, except for the samples from the Shanmei reservoir, the MD ranges from 2.64 to 47.4, indicating a significant variation in composition during the rainy season. Prior to the convergence of the East and West Creeks, the MD values between sample 35 of the Dongxi (before its confluence with Xixi) and sample 36 of the Xixi, as well as the "original composition", are all low, which shows little difference in composition. Following the confluence of Dongxi and Xixi, the MD rapidly increased from 3.55 (sample No. 37) to 19.96 (sample No. 39), eventually reaching 86.06 at the estuary. This sharp increase demonstrates a dramatic change in composition, which may be influenced by combined factors of human activities in the downstream city of Quanzhou along Jinjiang and the mixing of seawater where the river meets the sea.

In the Xixi watershed, the Houjing village in Jiandou town, which falls within this sub-basin, is home to a concentration of both coal mines and limestone mines. As the mainstream passes through this area, the mining activities have led to an increase in MD at sampling point 42. Additionally, the concentrations of Ca^{2+}, Mg^{2+}, and $SO_4{}^{2-}$ in the water composition have also increased. When the tributaries of Shuangxi and Qiyangxi flow into this area, MD at sampling point 44 has consistently remained high. This can be attributed primarily to the presence of numerous sulfur iron ore and coal mines distributed within the watersheds of Shuangxi Creek and Qiyangxi Creek. Additionally, the extensive tea planting in the area contributes to elevated levels of TN, Mn, and Zn in the water. Sampling point 43 represents the composition of Shuangxi Creek and Qiyang Creek. At the confluence of the two tributaries, it is observed that the levels of TN, Mn, and Zn in its composition increase, and its MD continuously fluctuates. The elevated MD at sampling point 50 is attributed to the abnormal increase in TN and TP levels. This phenomenon can be attributed to the sampling site's location in downtown Shangqing town, where it is subject to the impact of domestic sewage from the town. Similarly, sampling point 59 in the downtown city of Nan'an also had an increase in MD due to the increase in the level of TP and TN caused by the influence of domestic sewage.

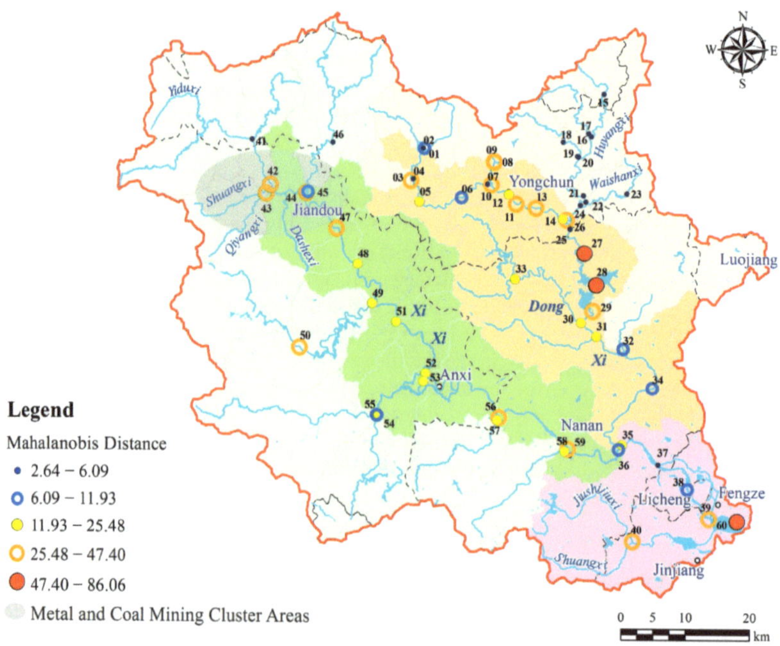

Figure 7. The Mahalanobis distance map of sampled river water in Jinjiang watershed. The numbers in the figure represent the sample numbers.

The same pattern is found in the Dongxi watershed. The elevated MD at sampling point 5 of the mainstream is caused by the confluence of tributaries at sampling point 3, which has different compositions. The elevated MD at sampling point 12 reflects the interference of human activities in Yongchun County with the chemical composition of the river water. The abnormally high values of MD at sampling points 27, 28, and 29 of Shanmei reservoir are mainly due to the large difference between the chemical composition of reservoir water and that of the river water [64]. This proportional difference, an increase when represented in MD, is not due to pollution. Then the decrease in MD at sampling point 31 is because of the recovery of the river's chemical composition. As observed, the MD of the east–west confluence initially exhibits a minor change, but it experiences a significant increase when flowing downstream. This indicates that the human activities and seawater disturbance in the downstream area of Jinjiang watershed results in a substantial alteration in the composition's proportion.

Water samples of the mainstream were added one by one from upstream to downstream, and the centers of components for the mainstreams of Xixi and Dongxi, respectively, were iteratively calculated. The MD from that center at each sampling point was also determined. Using these results, clustered line charts of MD for the mainstem of Dongxi and Xixi were drawn (Figure 7). The mean polyline (black polyline in Figure 8) shows that both Dongxi and Xixi watersheds exhibit a smooth MD for the upstream sample components, indicating that there is less variation among the upstream sample components. The gradual decrease in the middle stream MD shows a trend of gradually decreasing component differences, which indicates that human activities have not yet had a serious impact on the river's chemical composition. In the downstream area of Jinjiang watershed, where human activities are carried out frequently, the MD increases rapidly, reflecting the dramatic changes in the chemical composition caused by the combined disturbance of complex human activities and seawater. The overall covariance structure of the data and the component centers are changing with the continuous addition of samples from the mainstream.

As shown in Figure 8, the clustered line charts of MD in the basin are used to intuitively exhibit the variation in MD along the runoff. In the figure, each polyline is the sample Mahalanobis distance calculated from a single iteration. To highlight the trend produced by the iterations, the color of the polyline in the figure becomes lighter as subsequent sample data are added, where the black polyline is the average of the Mahalanobis distance of the polyline cluster. The horizontal coordinate is the sample number from the upstream to the downstream mainstem.

It can be seen in the clustered line charts of MD in the mainstream of Xixi (Figure 8a) that the variation trend of MD at different places in the mainstream of Xixi shows different or even opposite results. According to the trend of variation of MD, the Xixi is divided from sample point 48 into two areas. The area (I) in the upper reaches mainly represents that the area is little affected by human activities and its chemical composition is close to the "original composition". With the iterative addition of the water chemistry of the mainstream samples, the component centers of the area (I) are gradually moving away from the "original composition". This change results in the different degrees of increase in MD of all samples in the upstream area (I). The area (II) below sample point 48 is greatly affected by human activities. With the iterative addition to the hydrochemical composition of the mainstream samples, the component centers are getting closer to the "non-original composition", resulting in a gradual decrease in MD of the sample composition in the area (II). The rapid increase in component variation in the downstream area of Jinjiang watershed, where the tributaries converge, is reflected in the rapid increase in MD due to the interference of various factors.

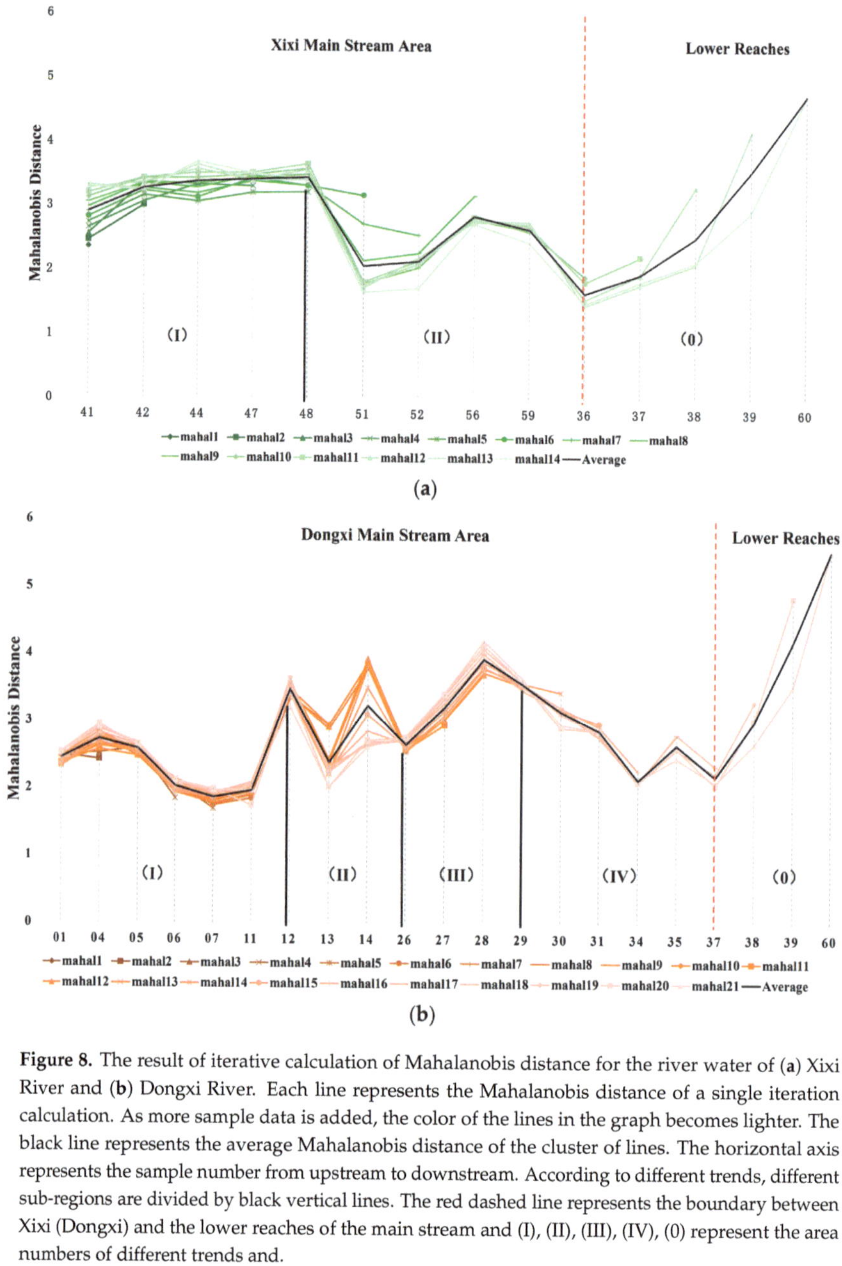

Figure 8. The result of iterative calculation of Mahalanobis distance for the river water of (**a**) Xixi River and (**b**) Dongxi River. Each line represents the Mahalanobis distance of a single iteration calculation. As more sample data is added, the color of the lines in the graph becomes lighter. The black line represents the average Mahalanobis distance of the cluster of lines. The horizontal axis represents the sample number from upstream to downstream. According to different trends, different sub-regions are divided by black vertical lines. The red dashed line represents the boundary between Xixi (Dongxi) and the lower reaches of the main stream and (I), (II), (III), (IV), (0) represent the area numbers of different trends and.

The situation of the mainstream of Dongxi is more complex (Figure 8b), and it can be classified into four distinct areas delineated by sampling points 12, 26, and 28. Area (I) is less disturbed by humans and has a chemical composition that is closer to the "original composition". MD values of samples located in this area will continue to increase with the addition of downstream data, indicating that the river's chemical compositions are constantly being influenced by various integrated factors and therefore move away from the component center. The sampling points 13 and 14 belong to the area (II) and are located within Yongchun County. With the addition of the data, the MD of sample 13 is showing a downward trend, while the MD of sample 14 is increasing. This is mainly because the urban area of Yongchun County is the most densely populated region in the northern mountainous area, and the component changes are generated by the inflow of industrial and domestic pollutants. MD of area (III) is at a high level, which reflects the large difference in chemical composition among the reservoir, lake water, and river water.

The (0) region in Figure 8a,b shows the variational tendency of Mahalanobis distance in lower reaches. It can be seen in both figures that the overall trend of Mahalanobis distance in the region (0) increases sharply, and the single-point Mahalanobis distance decreases with the addition of downstream data. Due to the fact of more interference from human activities, there are sharp differences between the compositions of each water sample, resulting in a sharp increase in Mahalanobis distance. At the same time, since the addition of various downstream data, the composition center deviates from the "original composition" to the downstream composition, so it also shows a decrease in the Mahalanobis distance for a single point in the lower reaches.

5. Conclusions

The compositional data analysis method (CoDA) was proposed from the perspective of compositional data analysis theory, considering the geochemical riverine system as a whole to detect changes in the composition of the entire watershed. A total of 60 water samples were collected and analyzed using traditional analytical methods. Based on this, further analysis of the transformed water chemistry composition data was carried out using RPCA, biplot, and Mahalanobis distance, leading to the following conclusions:

(1) The characteristic components of water bodies in Jinjiang watershed show obvious variability and complex evolutionary characteristics from upstream to downstream. Influenced by the weathering of silicate rocks, the water chemical composition of Jinjiang is mainly composed of Ca^{2+} and Na^{+} cations. The anion of East River is mainly made up of HCO_3^-, and the water chemistry is of HCO_3-Ca type. The anions of the West Stream are dominated by SO_4^{2-}, and the water chemistry of the upper and middle streams is of HCO_3-Ca type with high SO_4^{2-}. The water chemistry type gradually changes to Cl-Ca type in the lower reaches of Bailai Township. The estuary of Jinjiang, due to the influence of seawater, has a relatively higher proportion of Cl-Na type of water chemistry.

(2) The CoDA method served as a valuable tool for identifying and explaining anomalies in the hydrochemistry components within the basin. By comparing the MD value with the central "original component" of Jinjiang, it accurately identifies areas in the basin with outlier water chemistry compositions. The results obtained from RPCA analysis reveal that rock weathering, mining, and agricultural pollution are the primary factors that influence the hydrochemistry components. It is remarkable that the elements such as Fe, Mn, Zn, and Pb originating from mining activities become the distinctive components in the Xixi River basin. The Dongxi River, on the other hand, exhibits characteristic components including TP, TN, and Cl^-, indicating the influence of domestic and agricultural pollution.

This paper reveals the spatial variability in surface water in hydrochemical terms in Jinjiang watershed and discusses its underlying controlling factors. The findings can be utilized as a reference for investigating the hydrochemical spatial differences in other river basins.

Supplementary Materials: The following supporting information can be downloaded at: https://www.mdpi.com/article/10.3390/w16010045/s1, Figure S1: Gibbs Map of all samples in Jinjiang River Basin, Figure S2: End-member diagram of all samples in Jinjiang River Basin.

Author Contributions: Y.Z. (Yuchen Zhu): Field investigation, data curation, formal analysis, and writing—original draft. H.Y.: Formal analysis and writing—original draft and editing. Y.X.: Conceptualization and writing—review and editing. Q.H.: Conceptualization and writing—review and editing. Y.L.: Conceptualization, funding acquisition, and supervision. J.L.: Methodology and formal analysis. L.W.: Field investigation and data curation. Y.Z. (Yuqing Zhang): Field investigation, data curation and methodology. W.H.: Field investigation, data curation, and formal analysis. J.W.: Field investigation and data curation. All authors have read and agreed to the published version of the manuscript.

Funding: This research was funded by [Fujian Provincial Key Laboratory of Water Cycling and Eco-Geological Processes] grant number [SK202305KF01], [the China Geological Survey's project] grant number [DD20190303, DD20230459], [the Natural Science Foundation of China] grant number [42007183], [the Natural Science Foundation of Sichuan Province] grant number [2022NSFSC1084], [the Student Research Training Program of Southwest Jiaotong University] grant number [202310613077], [the Fundamental Research Funds for the Central Universities] grant number [2682022ZTPY002, 2682022ZTPY088].

Data Availability Statement: The data presented in this study are available on request from the first author.

Conflicts of Interest: The authors declare that they have no known competing financial interests or personal relationships that could have appeared to influence the work reported in this paper.

References

1. Yang, H.J.; Xiao, Y.; Hao, Q.C.; Wang, L.W.; Zhang, Y.Q.; Liu, K.; Zhu, Y.C.; Liu, G.X.; Yin, S.Y.; Xie, Z. Geochemical characteristics, mechanisms and suitability for sustainable municipal and agricultural water supply of confined groundwater in central North China Plain. *Urban Clim.* **2023**, *49*, 101459. [CrossRef]
2. Shi, W.; Xia, J.; Gippel, C.J.; Chen, J.X.; Hong, S. Influence of disaster risk, exposure and water quality on vulnerability of surface water resources under a changing climate in the Haihe River basin. *Water Int.* **2017**, *42*, 462–485. [CrossRef]
3. Yuan, Z.; Yan, D.H.; Yang, Z.Y.; Yin, J.; Zhang, C.; Yuan, Y. Projection of surface water resources in the context of climate change in typical regions of China. *Hydrol. Sci. J.* **2017**, *62*, 283–293. [CrossRef]
4. Xiao, Y.; Hao, Q.C.; Zhang, Y.H.; Zhu, Y.C.; Yin, S.Y.; Qin, L.M.; Li, X.H. Investigating sources, driving forces and potential health risks of nitrate and fluoride in groundwater of a typical alluvial fan plain. *Sci. Total Environ.* **2022**, *802*, 149909. [CrossRef]
5. Ewuzie, U.; Nnorom, I.C.; Ugbogu, O.C.; Onwuka, C.V. Hydrogeochemical, microbial and compositional analysis of data from surface and groundwater sources in Southeastern Nigeria. *J. Geochem. Explor.* **2021**, *224*, 106737. [CrossRef]
6. Wang, H.; Sun, L.; Du, P. Surface Water Quality Assessment in Main Stream of Liaohe River. *Adv. Mater. Res.* **2013**, *788*, 354–358. [CrossRef]
7. Gao, F.; Feng, G.; Han, M.; Dash, P.; Jenkins, J.; Liu, C. Assessment of Surface Water Resources in the Big Sunflower River Watershed Using Coupled SWAT–MODFLOW Model. *Water* **2019**, *11*, 528. [CrossRef]
8. Xiao, Y.; Liu, K.; Hao, Q.C.; Li, Y.S.; Xiao, D.; Zhang, Y.J. Occurrence, Controlling Factors and Health Hazards of Fluoride-Enriched Groundwater in the Lower Flood Plain of Yellow River, Northern China. *Expo. Health* **2022**, *14*, 345–358. [CrossRef]
9. Xiao, Y.; Liu, K.; Hao, Q.C.; Xiao, D.; Zhu, Y.C.; Yin, S.Y.; Zhang, Y.H. Hydrogeochemical insights into the signatures, genesis and sustainable perspective of nitrate enriched groundwater in the piedmont of Hutuo watershed, China. *CATENA* **2022**, *212*, 106020. [CrossRef]
10. Buccianti, A. The FOREGS repository: Modelling variability in stream water on a continental scale revising classical diagrams from CoDA (compositional data analysis) perspective. *J. Geochem. Explor.* **2015**, *154*, 94–104. [CrossRef]
11. Guggenmos, M.R.; Daughney, C.J.; Jackson, B.M.; Morgenstern, U. Regional-scale identification of groundwater-surface water interaction using hydrochemistry and multivariate statistical methods, Wairarapa Valley, New Zealand. *Hydrol. Earth Syst.* **2011**, *15*, 3383–3398. [CrossRef]
12. Zhang, B.; Song, X.F.; Zhang, Y.H.; Ma, Y.; Tang, C.Y.; Yang, L.H.; Wang, Z.L. The interaction between surface water and groundwater and its effect on water quality in the Second Songhua River basin, northeast China. *J. Earth Syst. Sci.* **2016**, *125*, 1495–1507. [CrossRef]
13. Zhao, W.; Lin, Y.Z.; Zhou, P.P.; Wang, G.C.; Dang, X.Y.; Gu, X.F. Characteristics of groundwater in Northeast Qinghai-Tibet Plateau and its response to climate change and human activities: A case study of Delingha, Qaidam Basin. *China Geol.* **2021**, *4*, 377–388. [CrossRef]
14. Dall'Agnol, R.; Sahoo, P.K.; Salomão, G.N.; de Araújo, A.D.M.; da Silva, M.S.; Powell, M.A.; Junior, J.F.; Ramos, S.J.; Martins, G.C.; da Costa, M.F.; et al. Soil-sediment linkage and trace element contamination in forested/deforested areas of the Itacaiúnas River Watershed, Brazil: To what extent land-use change plays a role? *Sci. Total Environ.* **2022**, *828*, 154327. [CrossRef] [PubMed]
15. Ma, H.Y.; Zhu, G.F.; Zhang, Y.; Pan, H.X.; Guo, H.W.; Jia, W.X.; Zhou, J.J.; Yong, L.L.; Wan, Q.Z. The effects of runoff on Hydrochemistry in the Qilian Mountains: A case study of Xiying River Basin. *Environ. Earth Sci.* **2019**, *78*, 385. [CrossRef]
16. Goonetilleke, A.; Vithanage, M. Water Resources Management: Innovation and Challenges in a Changing World. *Water* **2017**, *9*, 281. [CrossRef]
17. Zhai, R.; Tao, F.L.; Chen, Y.; Dai, H.C.; Liu, Z.W.; Fu, B.J. Future water security in the major basins of China under the 1.5 °C and 2.0 °C global warming scenarios. *Sci. Total Environ.* **2022**, *849*, 157928. [CrossRef]

18. Yang, L.; Zhao, G.J.; Tian, P.; Mu, X.M.; Tian, X.J.; Feng, J.H.; Bai, Y.P. Runoff changes in the major river basins of China and their responses to potential driving forces. *J. Hydrol.* **2022**, *607*, 127536. [CrossRef]
19. Hao, Q.C.; Li, Y.S.; Xiao, Y.; Yang, H.J.; Zhang, Y.Q.; Wang, L.W.; Liu, K.; Liu, G.X.; Wang, J.; Hu, W.X.; et al. Hydrogeochemical fingerprint, driving forces and spatial availability of groundwater in a coastal plain, Southeast China. *Urban Clim.* **2023**, *51*, 101611. [CrossRef]
20. Vareda, J.; Valente, A.; Durães, L. Assessment of heavy metal pollution from anthropogenic activities and remediation strategies: A review. *J. Environ. Manag.* **2019**, *246*, 101–118. [CrossRef]
21. Yang, Z.; Zhou, Y.X.; Wenninger, J.; Uhlenbrook, S.; Wang, X.S.; Wan, L. Groundwater and surface-water interactions and impacts of human activities in the Hailiutu catchment, northwest China. *Hydrogeol. J.* **2017**, *25*, 1341–1355. [CrossRef]
22. Kumar, V.; Parihar, R.; Sharma, A.; Bakshi, P.; Sidhu, G.P.S.; Bali, A.; Karaouzas, I.; Bhardwaj, R.; Thukral, A.; Gyasi-Agyei, Y.; et al. Global evaluation of heavy metal content in surface water bodies: A meta-analysis using heavy metal pollution indices and multivariate statistical analyses. *Chemosphere* **2019**, *236*, 124364. [CrossRef] [PubMed]
23. Gad, M.; Elsayed, S.; Moghanm, F.S.; Almarshadi, M.H.; Alshammari, A.S.; Khedher, K.M.; Eid, E.M.; Hussein, H. Combining Water Quality Indices and Multivariate Modeling to Assess Surface Water Quality in the Northern Nile Delta, Egypt. *Water* **2020**, *12*, 2142. [CrossRef]
24. Zheng, L.L.; Jiang, C.L.; Chen, X.; Li, Y.H.; Li, C.; Zheng, L.G. Combining hydrochemistry and hydrogen and oxygen stable isotopes to reveal the influence of human activities on surface water quality in Chaohu Lake Basin. *J. Environ. Manag.* **2022**, *312*, 114933. [CrossRef] [PubMed]
25. Benadela, L.; Bekkoussa, B.; Gaidi, L. Multivariate analysis and geochemical investigations of groundwater in a semi-arid region, case of superficial aquifer in Ghriss Basin, Northwest Algeria. *J. Groundw. Sci. Eng.* **2022**, *10*, 233–249.
26. Ismail, A.H.; Shareef, M.A.; Mahmood, W. Hydrochemical characterization of groundwater in Balad district, Salah Al-Din Governorate, Iraq. *J. Groundw. Sci. Eng.* **2018**, *6*, 306–322. [CrossRef]
27. Blake, S.; Henry, T.; Murray, J.; Flood, R.H.; Muller, M.R.; Jones, A.G.; Rath, V.J.A.G. Compositional multivariate statistical analysis of thermal groundwater provenance: A hydrogeochemical case study from Ireland. *Appl. Geochem.* **2016**, *75*, 171–188. [CrossRef]
28. Bahrami, M.; Khaksar, E.; Khaksar, E. Spatial variation assessment of groundwater quality using multivariate statistical analysis(Case Study: Fasa Plain, Iran). *J. Groundw. Sci. Eng.* **2020**, *8*, 230–243. [CrossRef]
29. Flem, B.E.; Reimann, C.; Fabian, K.; Birke, M.; Filzmoser, P.; Banks, D. Graphical statistics to explore the natural and anthropogenic processes influencing the inorganic quality of drinking water, ground water and surface water. *Appl. Geochem.* **2017**, *88*, 133–148. [CrossRef]
30. Gozzi, C.; Filzmoser, P.; Buccianti, A.; Vaselli, O.; Nisi, B. Statistical methods for the geochemical characterisation of surface waters: The case study of the Tiber River basin (Central Italy). *Comput. Geosci.* **2019**, *131*, 80–88. [CrossRef]
31. Ahmed, M.T.; Monir, M.U.; Hasan, M.Y.; Rahman, M.M.; Rifat, M.S.I.; Islam, M.N.; Khan, A.S.; Rahman, M.M.; Islam, M.S. Hydro-geochemical evaluation of groundwater with studies on water quality index and suitability for drinking in Sagardari, Jashore. *J. Groundw. Sci. Eng.* **2020**, *8*, 259–273.
32. Wu, H.F.; Chen, Y.; Chen, X.W.; Liu, M.B.; Gao, L.; Deng, H.J. A new approach for optimizing rain gauge networks: A case study in the Jinjiang Basin. *Water* **2020**, *12*, 2252. [CrossRef]
33. Meng, H.B.; Zhang, J.; Zheng, Z. Retrieving Inland Reservoir Water Quality Parameters Using Landsat 8-9 OLI and Sentinel-2 MSI Sensors with Empirical Multivariate Regression. *Int. J. Environ. Res. Public Health* **2022**, *19*, 7725. [CrossRef] [PubMed]
34. Lin, F.; Chen, X.W.; Yao, H.X.; Lin, F.Y. SWAT model-based quantification of the impact of land-use change on forest-regulated water flow. *CATENA* **2022**, *211*, 105975. [CrossRef]
35. Liu, Y.C.; Wu, L.; Shi, G.W.; Cao, S.W.; Li, Y.S. Characteristics and sources of microplastic pollution in the water and sediments of the Jinjiang River Basin, Fujian Province, China. *China Geol.* **2022**, *5*, 429–438. [CrossRef]
36. Liu, X.Y.; Wei, M.; Zeng, J. Simulating urban growth scenarios based on ecological security pattern: A case study in Quanzhou, China. *Int. J. Environ. Res. Public Health* **2020**, *17*, 7282. [CrossRef]
37. Li, C.J.; Pan, C.Z.; Teng, Y.G. Rainfall-runoff and pollution characteristics of tea garden in Xixi watershed within Jinjiang River basin. *Chin. J. Environ. Eng.* **2013**, *7*, 2909–2914.
38. Lu, Y.M.; Zhang, H.; Shao, W. Prediction of river water quality considering spatiotemporal correlation and meteorological factors. In Proceedings of the 2021 9th International Conference on Agro-Geoinformatics (Agro-Geoinformatics), Shenzhen, China, 26–29 July 2021; pp. 1–5.
39. Wu, J.H.; Wang, Z.C. A hybrid model for water quality prediction based on an artificial neural network, wavelet transform, and long short-term memory. *Water* **2022**, *14*, 610. [CrossRef]
40. Li, Y.S.; Liu, C.L.; Hao, Q.C. *Report on Comprehensive Geological Survey of Xiamen-Zhangzhou-Quanzhou City*; Chinese Academy of Geological Sciences, The Institute of Hydrogeology and Environmental Geology: Shijiazhuang, China, 2021.
41. Nong, X.Z.; Shao, D.G.; Zhong, H.; Liang, J.K. Evaluation of water quality in the South-to-North Water Diversion Project of China using the water quality index (WQI) method. *Water Res.* **2020**, *178*, 115781. [CrossRef]
42. Wu, Y.H.; Zhou, N.Q.; Wu, Z.J.; Lu, S.S.; Cai, Y. Carbon, nitrogen and phosphorus coupling relationships and their influencing factors in the critical zone of Dongting Lake wetlands, China. *J. Groundw. Sci. Eng. Geol.* **2022**, *10*, 250–266.
43. Zhou, Y.H.; Li, P.Y.; Chen, M.J.; Dong, Z.H.; Lu, C.Y. Groundwater quality for potable and irrigation uses and associated health risk in southern part of Gu'an County, North China Plain. *Environ. Geochem. Health* **2021**, *43*, 813–835. [CrossRef] [PubMed]

44. Rollinson, H.R.; Rollinson, H.; Pease, V. *Using Geochemical Data: To Understand Geological Processes*; Cambridge University Press: Cambridge, UK, 2021.
45. Pawlowsky-Glahn, V.; Egozcue, J.J.; Tolosana-Delgado, R. *Modeling and Analysis of Compositional Data*; John Wiley & Sons: Hoboken, NJ, USA, 2015.
46. Aitchison, J. The statistical analysis of compositional data. *J. R. Stat. Soc. Ser. B Methodol.* **1982**, *44*, 139–160. [CrossRef]
47. Buccianti, A.; Mateu-Figueras, G.; Pawlowsky-Glahn, V. *Compositional Data Analysis in the Geosciences: From Theory to Practice*; Geological Society of London: London, UK, 2006.
48. Gabriel, K.R. The biplot graphic display of matrices with application to principal component analysis. *Biometrika* **1971**, *58*, 453–467. [CrossRef]
49. Greenacre, M.; Aitchison, J. Biplots of compositional data. *Appl. Stat.* **2002**, *51*, 375–392.
50. Daunis-i-Estadella, J.; Barceló-Vidal, C.; Buccianti, A. Exploratory compositional data analysis. *Geol. Soc. Lond. Spec. Publ.* **2006**, *264*, 161–174. [CrossRef]
51. De Maesschalck, R.; Jouan-Rimbaud, D.; Massart, D.L. The Mahalanobis Distance. Chemometrics and Intelligent Laboratory Systems. *Am. J. Ind. Bus. Manag.* **2000**, *50*, 1–18.
52. Filzmoser, P.; Ruiz-Gazen, A.; Thomas-Agnan, C. Identification of local multivariate outliers. *Stat. Pap.* **2014**, *55*, 29–47. [CrossRef]
53. Ghorbani, H. Mahalanobis distance and its application for detecting multivariate outliers. *Facta Univ. Ser. Math. Inform.* **2019**, 583–595. [CrossRef]
54. Jouan-Rimbaud, D.; Massart, D.; Saby, C.; Puel, C. Characterisation of the representativity of selected sets of samples in multivariate calibration and pattern recognition. *Anal. Chim. Acta* **1997**, *350*, 149–161. [CrossRef]
55. Templ, M.; Hron, K.; Filzmoser, P. robCompositions: An R-package for robust statistical analysis of compositional data. *Compos. Data Anal. Theory Appl.* **2011**, 341–355.
56. Yang, H.J.; Xiao, Y.; Zhang, Y.Q.; Wang, L.W.; Wang, J.; Hu, W.X.; Liu, G.X.; Liu, F.; Hao, Q.C.; Wang, C. Lithology controls on the mixing behavior and discharge regime of thermal groundwater in the Bogexi geothermal field on Tibetan Plateau. *J. Hydrol.* **2023**, *628*, 130523. [CrossRef]
57. Gibbs, R.J. Mechanisms Controlling World Water Chemistry. *Science* **1970**, *170*, 1088–1090. [CrossRef] [PubMed]
58. Gaillardet, J.; Dupré, B.; Louvat, P.; Allegre, C. Global silicate weathering and CO_2 consumption rates deduced from the chemistry of large rivers. *Chem. Geol.* **1999**, *159*, 3–30. [CrossRef]
59. Yu, R.L.; Hu, G.R.; Lin, C.Q.; Yang, Q.L.; Zhang, C.C.; Wang, X.M. Contamination of heavy metals and isotopic tracing of Pb in intertidal surface sediments of Jinjiang River Estuary, SE China. *Appl. Geochem.* **2017**, *83*, 41–49. [CrossRef]
60. Chen, H.Y.; Teng, Y.G.; Wang, J.S.; Song, L.T. Source apportionment of water pollution in the Jinjiang River (China) using factor analysis with nonnegative constraints and support vector machines. *Environ. Forensics* **2012**, *13*, 175–184. [CrossRef]
61. Yang, D.; Qi, S.H.; Zhang, J.Q.; Wu, C.X.; Xing, X.L. Organochlorine pesticides in soil, water and sediment along the Jinjiang River mainstream to Quanzhou Bay, southeast China. *Ecotoxicol. Environ. Saf.* **2013**, *89*, 59–65. [CrossRef]
62. de Azeredo Morgado, M.G.; Passos, C.J.S.; Garnier, J.; de Lima, L.A.; de Alcântara Mendes, R.; Samson-Brais, É.; Lucotte, M. Large-scale agriculture and environmental pollution of ground and surface water and sediment by pesticides in the Brazilian Amazon: The case of the Santarém region. *Water Air Soil Pollut.* **2023**, *234*, 150. [CrossRef]
63. The People's Government of Quanzhou City. *The Overall Plan of Mineral Resources(2016–2020)*; The People's Government of Quanzhou City: Quanzhou, China, 2017.
64. Yu, N.X.; Qin, Y.; Hao, F.; Lang, Y.C.; Wang, F.S. Using seismic surveys to investigate sediment distribution and to estimate burial fluxes of OC, N, and P in a canyon reservoir. *Acta Geochim.* **2019**, *38*, 785–795. [CrossRef]

sustainability

MDPI

Article

Numerical Simulation of Carbon Tetrachloride Pollution-Traceability in Groundwater System of an Industrial City

Benli Guo [1,2,3,4], Peng Yang [1,2,3,4], Yan Zhou [5,*], Hongjian Ai [1,2,3,4], Xiaodong Li [1,2,3,4], Rifei Kang [1,2,3,4] and Youcheng Lv [1,2,3,4]

1 No.8 Institute of Geology and Mineral Resources Exploration of Shandong Province, Rizhao 276826, China
2 Key Laboratory of Nonferrous Metal Ore Exploration and Resource Evaluation of Shandong Provincial Bureau of Geology and Mineral Resources, Rizhao 276826, China
3 Rizhao Big Data Research Institute of Geology and Geographic Information, Rizhao 276826, China
4 Rizhao Key Laboratory of Land Quality Evaluation and Pollution Remediation, Rizhao 276826, China
5 School of Water Resources and Environment, China University of Geosciences (Beijing), Beijing 100083, China
* Correspondence: yzhou@cugb.edu.cn

Abstract: Carbon Tetrachloride (CCL4) is a colorless, volatile, and toxic liquid. Once it pollutes groundwater, it will not only destroy the ecological environment but also negatively affect the functioning of the human liver. An industrial city in eastern China has been contaminated with carbon tetrachloride (CCL4). Due to the complex hydrogeological conditions, it is difficult to determine the pollution source by a single hydrochemical analysis. In order to solve the traceability problem, in this work we established a traceability system by combining hydrochemical analysis, backward tracing, and forward transport, and analyzed the pollution distribution, pollution-source location and pollution-transport characteristic in groundwater, which provided technical support for CCL4-pollution control of groundwater in this area. (1) Groundwater samples were analyzed using the gas chromatography/mass spectrometry method. Through the pollution concentration-field, we identified the location of the pollution center and the concentration value in the northeast and southwest, screening out the monitoring wells exceeding the criteria: HF#1, DJ#19, DJ#7, SS#4, ZF#2, and DY#3. (2) Backward tracing over the past 30 years and 50 years was conducted through MODPATH. Potential polluting factories were identified by comparing the capture area with the historical distribution of factories. (3) Forward solute-transport was performed at the location of these potentially polluted factories. The distribution characteristics of pollutants in limestone layers and the Quaternary layer were analyzed using a simulated concentration-field for August 2012 and October 2016. (4) Comparing the simulated concentration with the observed concentration, the source of CCL4 pollution was inferred to be the textile mill in the 1990s, the steelworks in the 1970s, and the machine-tool plant in the 1970s and 1990s. Based on the concentration–duration curve at the pollution source, the transport characteristics and the transport speed of the pollutants in the study area were analyzed. This work not only successfully found the location of CCL4-pollution sources, but also helped the local government to analyze the year of pollutant release and recognize the transport pattern of CCL4 in aquifers.

Keywords: traceability; numerical simulation; MODPATH; carbon tetrachloride

1. Introduction

With the development of the social economy, the impact of human activities on groundwater pollution is more and more obvious [1]. Groundwater pollution has the characteristics of invisibility, complexity and delay in discovery. Pollutants can usually be adsorbed in the soil for a long time, and will continue to diffuse and infiltrate into the groundwater to form pollution, threatening human health [2]. Among them, Carbon Tetrachloride pollution is particularly concerning [3].

Carbon Tetrachloride (CCL4) is a toxic liquid, which is colorless, volatile, slightly soluble in water, and easily soluble in most organic solvents [4]. It has been widely used in industry, for decades as an industrial degreasing-agent, pesticide, flame retardant, and for dry cleaning [5]. Studies have shown that prolonged exposure to the CCL4 compound on the human body can cause a number of complications, negatively affecting the functioning of the liver, and the occurrence of malignant tumors, which take several months to develop [6]. Since 1960, there have been studies on carbon tetrachloride pollution in the United States, Finland and China. Research shows that CCL4 pollution is generally widespread, the pollution source unclear, and it is difficult to treat [7–9].

The first step in groundwater CCL4-pollution control is to trace the source of pollutants, which is to find out the location of pollution sources and the history of pollutant migration and transformation, through limited observational-data [10,11]. Groundwater-tracing techniques mainly include the geochemical-footprint method and numerical simulation. Among these, the geochemical-footprint method uses isotopic fingerprints or geostatistical analysis to obtain pollution-source information [12]. However, using the geochemical-footprint method alone cannot completely trace the location of pollution sources and obtain the transport path of pollutants over time, and, especially when the source of pollution is not specified, this method is difficult to apply [13].

With the development of computing technology, numerical simulation has gradually become an important tool for hydrological study. The solute-transport model is an effective method to trace the source of pollution caused by point sources [14]. For example, the MODPATH module can solve the pollution-source identification problems using inverse-solution approaches [15]. In the Lombardy region, one of the most industrialized areas in Italy, MODPATH was used to find multiple-point sources of groundwater pollution [16]. In central-eastern Tuscany, the source of Boron pollution in the shallow groundwater was traced [17]. The accuracy of MODPATH tracing is closely related to the location of contamination. If the polluted monitoring-point is not a typical site along the pollution-transport route, it will lead to the deviation of the source location. In addition, too many pollution monitoring points will also lead to the problem of multiple pollution sources, making it impossible to effectively search for pollution sources.

MT3D are well-known three-dimensional groundwater flow and pollution-transport modules. It has many applications in pollutant forward-transport and pollution plume-prediction. Zhang et al. used the MT3DMS module in GMS numerical-simulation software to establish the migration and transformation model of typical chlorinated hydrocarbons [18]. M Wen et al. established a solute-transport model based on the sampled data to predict CCL4 pollution in the study area over the next 15 years [19]. Therefore, MT3D is suitable for contamination-transport modeling of the watershed under transient conditions. MODPATH can achieve the target of source tracing, but it has uncertainty and cannot analyze the transport of pollutants [20,21]. Therefore, the combination of MODPATH and MT3D is an effective traceability method, and can greatly reduce the uncertainty of the simulation.

In an industrial city in China, CCL4 pollution was found in samples of groundwater. In the study area, single hydrochemical-analysis has been unable to obtain the location of the source of pollution. Using the MODPATH tool to carry out the backward-tracing work produces great uncertainty. Therefore, in this work, we adopted the method of combining hydrochemical analysis, backward tracing, and forward simulation, to trace the source of pollutants over the past 30 and 50 years. Based on the historical factory-distribution, the location of the pollution source was delineated, and the transport characteristic of pollutants was analyzed. This work provides scientific evidence for pollutant treatment in the study area.

2. Material and Methods

2.1. Study Area

The study area is located in the economically developed area of eastern China, a political, economic, and cultural center, with a catchment area of 400 km^2. The elevation of this area is between 24 and 387 m. The climate type in the area is a continental-monsoon climate, with an average annual temperature of 13.7 °C and an average annual rainfall of 651 mm. Rainfall is concentrated from June to August, accounting for 65% of annual rainfall. The average annual evaporation is 1759 mm, and the maximum humidity is in July and August, with relative humidity 75~80%.

The flow direction of groundwater in the study area is generally from south to north, as shown in Figure 1. Affected by the high-intensity pumping of the water-supply source, there is a groundwater-drawdown cone in the north. In the south part, there are low hills with outcropping limestone. Because the overlying Quaternary sediments are very thin, it is a single-karst aquifer. The north part is a piedmont inclined-plain with magmatic intrusions located in the west. The Quaternary sediments gradually become thicker, with a thickness between 20 and 100 m, so there is an exchange between the lower karst-aquifer and upper Quaternary-aquifer. At the northwestern boundary, there develops a main river about 4 m deep and 100 m wide. Tributaries 1 and 2 originate from the mountainous area in the south, and both flow into the main river. Near the northern border, there are spring groups exposed in the northern part of the study area, and the flow rate is 3000–4000 m^3/d.

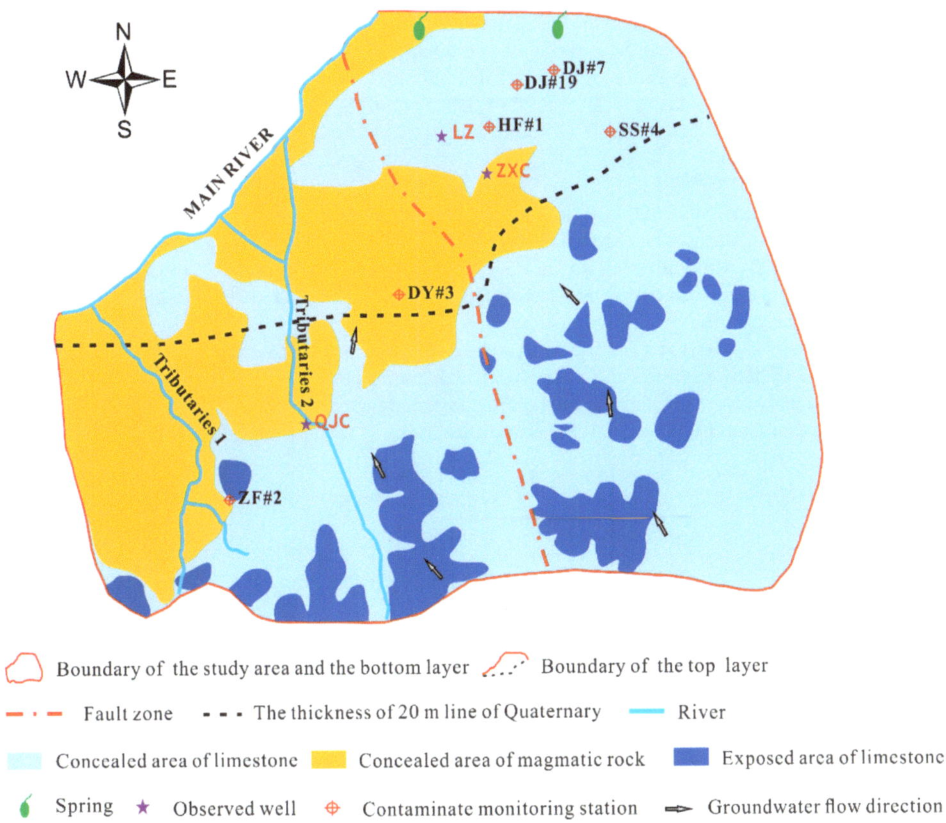

Figure 1. The location of the study area.

2.2. Factory Distribution

The northeastern part of the study area is an important industrial area, with a dense and wide distribution of factories throughout history. Since the 1970s, some fertilizer plants were set up to produce agricultural fertilizers such as ammonium chloride. Some steelworks were used to produce light-solvent oil and benzene-series products. A machine-tool plant was used to produce and clean machine parts. A textile mill needed to use rubber solvents and resin solvents in its production. Some petrochemical works produced gasoline and diesel oil. In addition, there have been cement mills, brickyards, plastic factories, chemical works and other factories in the study area. CCL4 was used or formed in the production and processing of these industrial products. The residual CCL4 on the plant surface infiltrates easily through the loose layer and enters the karst aquifer with the rainfall, resulting in serious pollution problems. Therefore, these factories are a major source of CCL4 pollution.

2.3. Field Investigation

A field investigation was conducted in the study area. A total of 151 water samples were collected, including 68 samples in the dry season and 83 samples in the wet season. Water samples were detected using the gas chromatography/mass spectrometry method. Gas chromatography is based on the fact that different substances in the solid phase and liquid phase have different partition-coefficients, so that different compounds from the column outflow-time are not the same, to achieve the purpose of separating compounds. Mass spectrometry is based on the use of the motion law of charged particles in the magnetic field or electric field, according to the mass-to-charge ratio, to achieve separation analysis and to determine the mass and intensity-distribution of ions. In 2012 and 2016, the positions that exceed the criteria are located at HF#1, DJ#19, DJ#7, SS#4, DY#3, and ZF#2, as shown in Figure 1.

2.4. Groundwater Modeling

In this study, MODFLOW is used for flow simulations and MT3D is used for pollution-transport simulations. The code was run and visualized within the Groundwater Modelling System (GMS). Although there are inhomogeneous dissolution-fractures in the study area, the karst-development degree is relatively low. The groundwater surface is uniform and continuous, and the groundwater-flow movement conforms to Darcy's law. Extensive studies have shown that karst aquifers can be reasonably generalized to equivalent porous media with homogeneous anisotropy, and simulated in MODFLOW [22–24]. The three-dimensional movement of groundwater of a constant density through porous earth-material may be described by the partial differential equation:

$$S_s\frac{\partial h}{\partial t} = \frac{\partial}{\partial x}(K_{xx}\frac{\partial h}{\partial x}) + \frac{\partial}{\partial y}(K_{yy}\frac{\partial h}{\partial y}) + \frac{\partial}{\partial z}(K_{zz}\frac{\partial h}{\partial z}) + \varepsilon \; x, y, z \in \Omega$$

$$\mu\frac{\partial h}{\partial t} = K_{xx}(\frac{\partial h}{\partial x})^2 + K_{yy}(\frac{\partial h}{\partial y})^2 + K_{zz}(\frac{\partial h}{\partial z})^2 - \frac{\partial h}{\partial z}(K_Z + p) + p$$

where S_s is the specific storage of the porous material (L^{-1}); t is Time (T); h is the hydraulic head (L); K_{xx}, K_{yy} and K_{zz} are values of hydraulic conductivity along the x, y, and z coordinate axes, which are assumed to be parallel to the major axes of hydraulic conductivity (L/T); ε is a volumetric-flux-per-unit volume representing the source and sink of water (T^{-1}).

MT3D uses the velocity values calculated by MODFLOW as inputs to solve the transport equation [25]. The equation describing three-dimension solute-transport in groundwater is:

$$\frac{\partial(\theta c)}{\partial t} = \frac{\partial}{\partial x_i}(\theta D_{ij}\frac{\partial C}{\partial x_j}) - \frac{\partial}{\partial x_i}(\theta v_i c) \pm q_s c_s$$

where θ is porosity (dimensionless); c is solute concentration (mg·L^{-1}); D_{ij} is the dispersion tensor (m^2·d^{-1}); V_i is the actual pore-water flow rate (m·d^{-1}); q_s is the value of the recharge and discharge of the unit volume of the aquifer (m^3·d^{-1}) and c_s is the solution concentration at the source and sink (mg·L^{-1}).

To estimate the transport of pollutants in the aquifer, the particle-tracking post-processing program MODPATH was used. The MODPATH program was developed by the United States Geological Survey (USGS), which was designed to work with MODFLOW [26]. Using the MODFLOW cell-by-cell mass-balance files, the MODPATH can track through the simulated flow-field and calculate the time-related capture zones of particles representing the pollution [27]. This program does not take chemical reactions, biodegradation, and concentration changes into account; it only considers advection transport.

MODPATH assumes the validity of Darcy's law and the law of conversation of mass in the same way as MODFLOW. MODFLOW models provide the velocities at the midpoints of the cell boundaries, and MODPATH uses simple linear interpolation to compute the velocity components within the cell [28]. Since the velocity component of a particle is a known function of the coordinates of the particle, the coordinate can be calculated at any time in the future. Particles can be tracked backward in the "upstream" direction by running the particle-tracking algorithm in reverse. Backward tracking is accomplished by reversing the sign of all the velocity components and then applying the algorithm in the same way as for forwarding tracking [29].

3. Results

3.1. Model Setup and Calibration

In line with the hydrogeological data and field survey, the model of the study area is divided into two layers; the top layer is the Quaternary layer, dominated by pore water, and the southern boundary of the top layer is the thickness of the 20 m line of the Quaternary. The bottom layer is the limestone layer, dominated by fractured-karst water. The southern boundary of the bottom layer is consistent with the southern boundary of the study area. These layer boundaries are marked in Figure 1, which is discretized into a 100 m * 100 m grid in the horizontal direction. There are 5210 effective grids in the top layer and 16,401 effective grids in the bottom layer. The simulation period is set from 1 June 2002, to 1 December 2016. Each month is a stress period, with a total of 175 stress periods, and each stress period is divided into two timesteps.

In the horizontal direction, the north is the spring discharge-zone, which is generalized as the general-head boundary. The northwest is the main river, which is generalized as a boundary of the constant head. The east and west boundaries are perpendicular to the water-table contour, and there is almost no exchange, which is generalized as the zero-flow boundary. The southern boundary of the top layer can be recharged by rainfall infiltration and is generalized as the general-head boundary. The southern boundary of the bottom layer can be recharged by piedmont lateral-runoff, and is generalized as the general-head boundary, too. In the vertical direction, the top layer is a convertible aquifer and the upper boundary is a free surface that can receive precipitation recharge, agricultural-irrigation recharge, and evaporation-loss discharge. The bottom layer is a confined aquifer, and its lower boundary is an impermeable floor. There exists leakage between the convertible aquifer and the confined aquifer.

In line with the meteorological-station precipitation data from the local meteorological bureau, we divided the study area into six precipitation zones, and the precipitation-infiltration coefficient ranges from 0.2 to 0.63, gradually decreasing from south to north. The evaporation data is from the evaporation-pan experimental data and the limit of the evaporation depth is 3.5 m. There are 27 groundwater pumping-wells to irrigate wheat and corn during the growing season, with a pumping volume of about 12×10^4 m^3/d. We use the WELL module of MODFLOW to read the monthly pumping-volume data. Agricultural pumping is regarded as surface pumping, and the division is the same as the division of the precipitation-infiltration coefficient.

Groundwater-level observation data from three monitoring wells: ZXC, LZ, and QJC, were collected. The positions of the monitoring wells are shown in Figure 1. Well ZXC had the longest time-series of observed data, from July 2002 to April 2016. The data series from LZ and QJC was relatively short, from September 2015 to December 2016, and from December 2015 to December 2016, respectively. We verified the applicability of the simulation results by comparing the simulated water-level with the observed water-level of these three monitoring wells, as shown in Figure 2. We calculated the error coefficients for the three monitoring wells: the error coefficient for ZXC was 2.8%, for LZ it was 6.5%, and for QJC it was 7.14%. These error coefficients are all within 10%, which means the simulated groundwater-level was consistent with the measured groundwater-level [30].

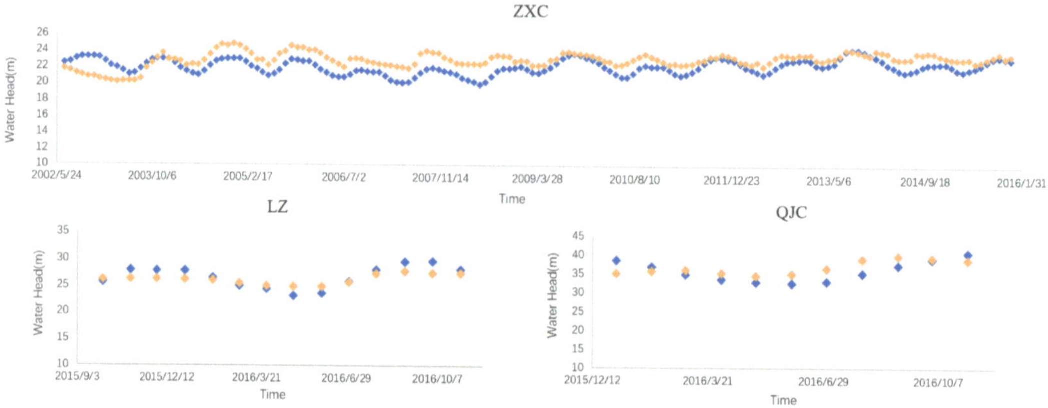

Figure 2. Comparison curve between observed water-level and simulated water-level.

In addition, we also compared the simulated flow-field in the year 2016 with the actual flow-field diagram, as shown in Figure 3. It can be seen that although there are some differences between the simulated water-levels and the observed water-levels, it conforms to the dynamic characteristics of groundwater. The groundwater flow generally flows from south to north. The hydraulic gradient of groundwater flow in mountainous areas is relatively large, and the water level is relatively dense. The hydraulic gradient in the plain area is small, and the water level is sparse. In the simulated flow-field, the groundwater-drawdown cone and fault zone are reflected.

3.2. Analysis of CCL4 Concentration

The concentration-distribution map showed the concentration range and variation trend of CCL4 in 2012 and 2016, as shown in Figure 4. In the northeast and southwest of the study area, there were two centers of pollution concentration. In 2012, the concentration center in the southwest area was near the monitoring well ZF#2, with a value of about 20 ug/L. In 2016, the concentration center in the southwest had barely changed, with a central concentration decrease of about 15 ug/L. In the northeast area, the highest concentration of CCL4 was about 25 ug/L, in the year 2012. By 2016, the concentration center had shifted to the south, near the monitoring well HF#1, and the concentration was around 35 ug/L. There was an area near the monitoring well DY#3 where CCL4 concentrations were around 2 ug/L in both 2012 and 2016.

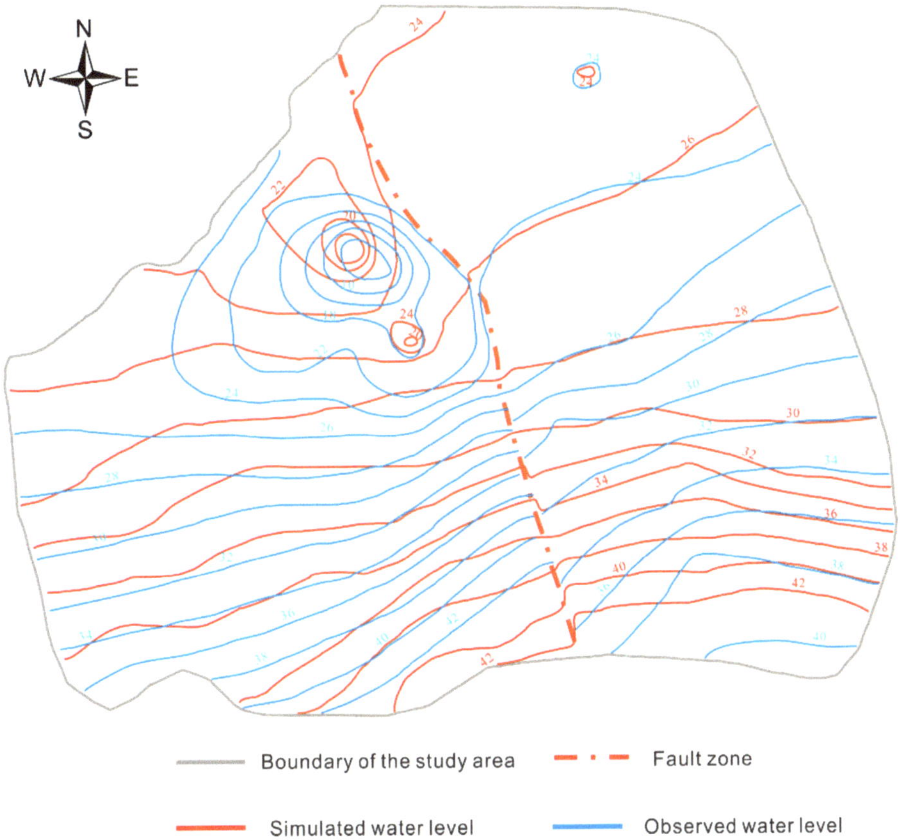

Figure 3. Diagram of the simulated flow-field and observed flow-field in 2016.

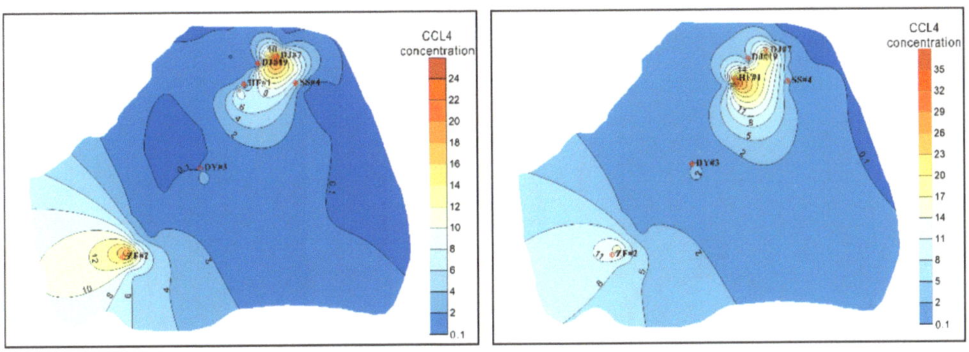

Figure 4. CCL4 concentration distribution map in 2012 (**left**) and 2016 (**right**).

According to the distribution map, parts of the river segment have been affected by CCL4 pollution in the southern area. Compared with the northeast, the pollution scope in the southwest is larger, but the concentration in the center is smaller. In the northwest, a large area of contamination has formed around the monitoring well HF#1-DJ#7-DJ#19-SS#4, which is adjacent to the historic industrial area, and which is largely attributable to industrial pollution. With the flow of groundwater, the scope of pollution continues to expand, threatening the quality of rivers and springs. It is necessary to carry out backward tracing of CCL4 pollution, to delineate the location of the pollution source and control the diffusion of pollutants.

3.3. Backward Tracing of the Polluted Zone

We used MODPATH to carry out the backward particle-trace and capture the pollution zone for the past 30 years and 50 years, respectively, to obtain the possible location of the potentially polluted factories.

According to where the CCL4 pollution was detected in 2012 and 2016, we placed the particles at the location of HF#1, DJ#19, DJ#7, SS#4, ZF#2, and DY#3. Restricted by detection conditions, the location of these polluted points may not be the real center of pollutant concentration. Therefore, with the detected point as the center, the particles were placed in the surrounding area of 250 m. The particle-tracing duration was 30 years, from January 1987 to December 2016, and 50 years, from January 1967 to December 2016, to investigate the potential pollution-factories in the 1990s and 1970s, as shown in Figures 5 and 6. Due to the data required by the model before 2002 being missing, the steady-flow simulation was carried out before the year 2002, and the transient-flow simulation was carried out in the model from the year 2002 to 2016.

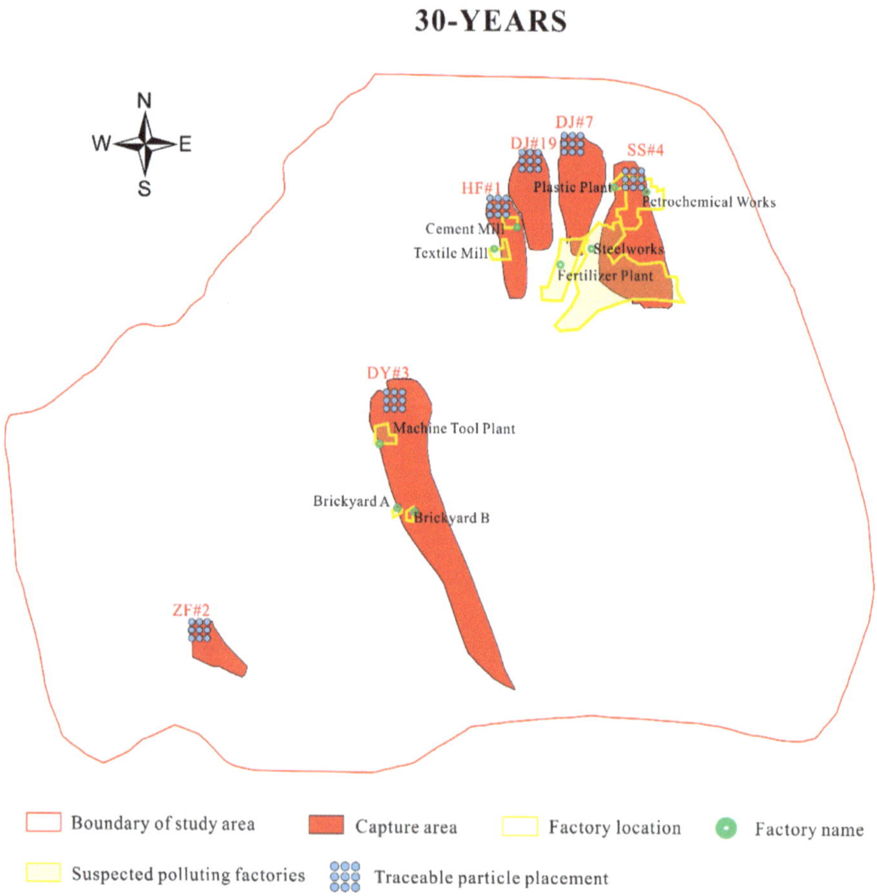

Figure 5. Pollution zone captured in the 1990s (30 years ago).

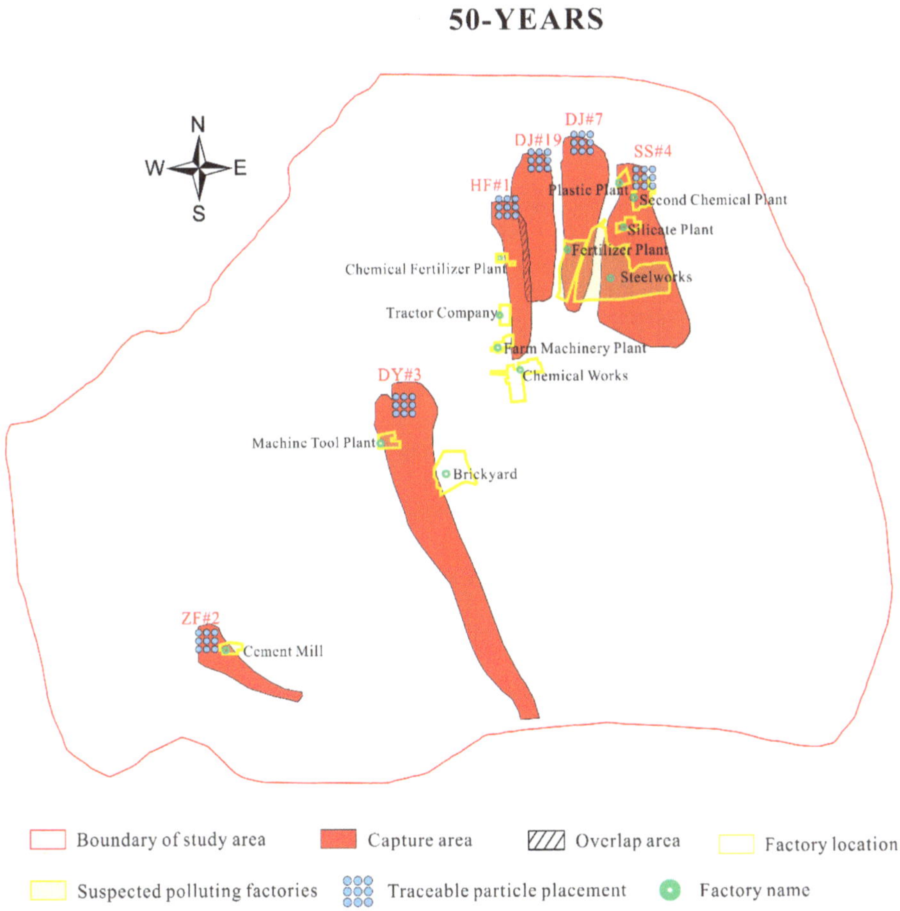

Figure 6. Pollution zone captured in the 1970s (50 years ago).

The results of the 30-year backward tracing are shown in Figure 5. Table 1 shows the areas of pollutant ranges of the six pollution-detected points in the 1990s. The captured zone around HF#1 was about 1900 m long and 500 m wide, and there was a cement mill and textile mill that existed in the 1990s. The captured area around DJ#19 was about 1800 m long and 700 m wide, but there was no historical factory in its capture zone. There was no intersection between the polluted areas captured between HF#1 and DJ#19. The captured zone around DJ#7 was about 2300 m long and 800 m wide, and a fertilizer plant existed. The captured zone around SS#4 was about 2700 m long and 1000 m wide, and the historical factories in the capture zone were the plastic plant and fertilizer plant. The captured area around DY#3 was about 6200 m long and 800 m wide; within the capture zone, there were two brickyards and a machine-tool plant. The area of ZF#2 was the smallest, and there was no historical factory in ZF#2′s capture zone. The capture zone of DY#3 was much larger than the other places, because the hydrogeological conditions of DY#3 are favorable for groundwater flow, and with the existence of a water-supply company, the amount of water pumping in the DY#3 area is considerable. Combined with the products produced by these factories, in the 1990s the potential polluting factories were the textile mill, fertilizer plant, steelworks, and machine-tool plant.

Table 1. Area of pollution-range captured by 6 pollution-detected points in the 1990s.

Well Name	ZF#2	DY#3	HF#1	DJ#19	DJ#7	SS#4
Capture area (km^2)	0.62	4.43	0.88	1.46	1.13	2.60

The results of the 50-year backward tracing are shown in Figure 6. Table 2 shows the areas of pollutant ranges of the six pollution-detected points in the 1970s. The capture zone of each monitoring well has increased, especially the capture zone of DY#3, which has exceeded the southern boundary. Within its capture zone there was a machine-tool plant and a brickyard in the 1970s. An overlapping area occurred between the capture zones of HF#1 and DJ#19, but the overlapping part had no factory in the 1970s. The captured zone around HF#1 was about 3000 m long and 450 m wide. In the 1970s, there was a chemical-fertilizer plant in the capture zone, and there was a tractor company, a farm machinery plant, and a chemical works in the vicinity. There was no historical factory in DJ#19′s capture zone, and a fertilizer plant existed in DJ#7′s capture zone. The factories that existed within the capture zone of SS#4 were a plastic plant, a second chemical plant, a silicate plant, and steelworks. There was a cement mill in the capture zone of ZF#2 in the 1970s. Combined with the products produced by these factories, in the 1970s the potential polluting factories were the chemical-fertilizer plant, the fertilizer plant, the steelworks, the machine-tool plant, and the cement mill.

Table 2. Area of pollution range captured by 6 pollutions detected points in the 1970s.

Well Name	ZF#2	DY#3	HF#1	DJ#19	DJ#7	SS#4
Capture area (km^2)	0.88	>4.77	1.27	2.12	1.72	3.90

3.4. Forward Simulation of CCL4 Transport

By backward tracing, we identified a number of potential polluted factories within the capture zone. To further pinpoint the location of the polluting factories, we performed forward solute-transport at the location of these potential polluted factories, and compared the simulated concentration of forwarding simulation with the observed concentration-data.

Two scenarios were set up, assuming that CCL4 leaks occurred in the 1970s and 1990s. Based on the existing flow-model (transient flow from 2002 to 2016 and steady flow before 2002), the forward solute-transport simulation was carried out. The release of pollutants was instantaneous, without subsequent replenishment. The initial concentration was 785 mg/L, and this value is the solubility of CCl4 in water at 25 °C. Potentially polluting factories are located in the Quaternary layer or the limestone layer, with the parameters related to solute transport in these two layers shown in Table 3.

Table 3. Pollutant-transport parameters of the Quaternary layer and the limestone layer.

	Longitudinal Dispersion	Transverse Dispersion	Molecular Diffusivity
Quaternary	1 m	0.16 m	1.57×10^{-7} m^2/d
Limestone	1.48 m	0.26 m	1.75×10^{-7} m^2/d

Scenario 1: Assuming that the source of pollutant leakage was factories in the 1990s. Potentially polluting factories are the textile mill, the fertilizer plant, the steelworks, and the machine-tool plant. We put pollution particles where these factories were located; among them, the placements of the textile mill and the fertilizer plant were in the Quaternary layer, and the steelworks and machine-tool plant were in the limestone layer. The simulation started in the year 1987. The distribution range of pollutants in August 2012 and October 2016 is shown in Figure 7.

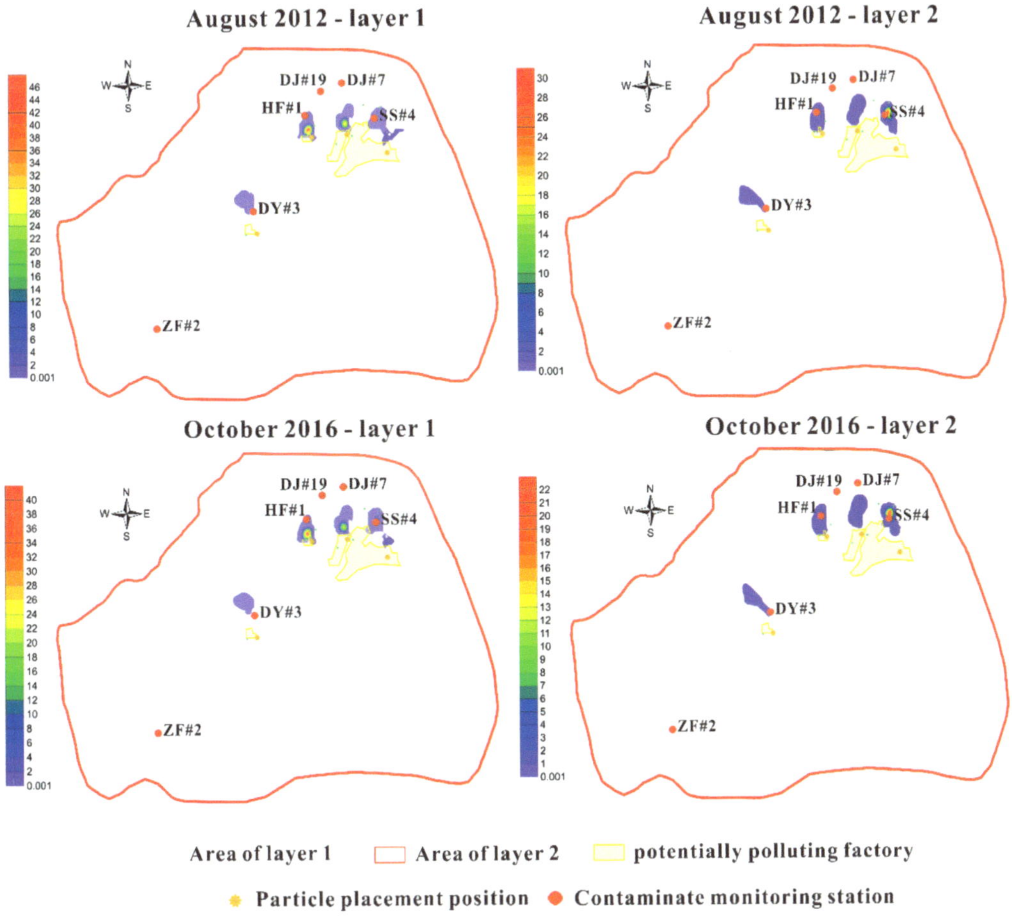

Figure 7. Pollutant-concentration distribution in August 2012 and December 2016 in Scenario 1.

The Quaternary layer is also layer 1 of the model. In August 2012, the concentration of CCL4 at 200 m distances from the textile mill was the highest, at 47 mg/L. As it is 300 m away from the fertilizer plant, the CCL4 concentration was 27 mg/L. The concentration near the steelworks was 2.24 mg/L. CCL4 appeared at a distance of 1500 m from the machine-tool plant, with a concentration of about 0.85 mg/L. The concentration of CCL4 in the textile mill and fertilizer plant was higher, and the distance from the particle placement-point was closer. In October 2016, compared with August 2012, the concentration of CCL4 did not change much, and the distribution range maintained the original form. The highest concentration was still 200 m away from the textile mill, but the concentration was slightly reduced, to 43 mg/L.

The limestone layer is also layer 2 of the model. In August 2012, the highest-concentration point appeared at 2000 m from the steelworks, at 30 mg/L. The CCL4 concentration of other polluted areas was around 0.05 mg/L. In October 2016, compared with August 2012, the highest concentration in the steelworks was reduced to 22 mg/L, and the distribution of CCL4 did not change much. In general, CCL4 in limestone layers transports faster than in the Quaternary layer.

Scenario 2: Assuming that the source of pollutant leakage was factories in the 1970s. Potentially polluting factories are the chemical-fertilizer plant, the fertilizer plant, the steelworks, the machine-tool plant, and the cement mill. We put pollution particles where these factories are located, and among them, the placement of the chemical-fertilizer plant and the fertilizer plant were in the Quaternary layer, and the steelworks, machine-tool plant, and cement mill were in the limestone layer. The simulation started in the year 1967. The distribution range of pollutants in August 2012 and October 2016 is shown in Figure 8.

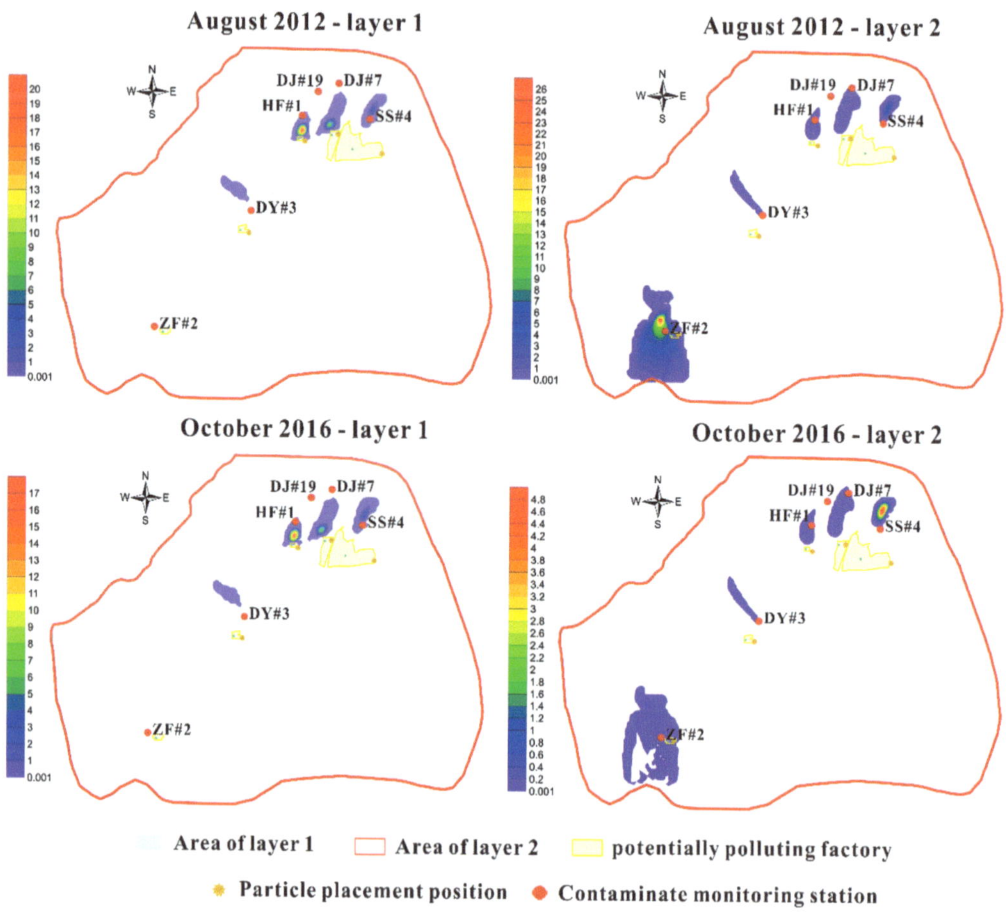

Figure 8. Pollutant-concentration distribution in August 2012 and December 2016, in Scenario 2.

The Quaternary layer is also layer 1 of the model. In August 2012, the concentration of CCL4 in the chemical-plant area was the highest, at 20 mg/L. In the north of the fertilizer plant, the CCL4 concentration was 8 mg/L. At 2500 m from the steelworks, the concentration of CCL4 was 4 mg/L. The concentration near the machine-tool plant was 2.24 mg/L. In October 2016, compared with August 2012, the concentration of CCL4 slightly decreased, and the distribution range maintained its original form.

The limestone layer is also layer 2 of the model. In August 2012, the highest-concentration point appeared at 1000 m from the cement mill, and reached 26 mg/L. The distribution range was large, which is associated with frequent groundwater head-dynamics in this area. In October 2016, compared with August 2012, the maximum CCL4 concentration point changed from the machine-tool-plant area to the steelworks area. The highest concentration near the steelworks was 4.8 mg/L.

3.5. Characteristics of CCL4 Transport

The observed concentrations of CCL4 pollution in HF#1, DJ#19, DJ#7, SS#4, ZF#2, and DY#3 in the years 2012 and 2016 are shown in Table 4 Comparing the simulated- and observed-concentration, in 2012 at point HF#1, regardless of whether it were Scenario 1 or Scenario 2, the simulated concentration of layer 1 was closer to the observed value, and layer 2 was slightly higher. Meanwhile, in 2016 it was the opposite: layer 2 was closer to the observed value, and the concentration value of scenario 1 was 34.36 ug/L in the year 2016, which was almost consistent with the observed value. This indicates that the textile mill is more likely to be a potential source of CCL4 pollution. The steelworks in both Scenario 1 and Scenario 2 are at point SS#4, but the simulated concentration value of Scenario 1 was high, and the concentration of Scenario 2 was significantly decreased, especially the concentration of layer 1 in 2016, which was almost the same as the observed concentration. This difference may be related to the high initial concentration at the particle-delivery point. The difference between the simulated concentration at DY#3 and the observed value was relatively small, especially in Scenario 1, which was the same as the observed value in 2016, which means that the machine-tool plant is likely to be a source of pollution. Since there are no potentially polluting factories at DJ#19 and no factory at ZF#2 for Scenario 1, no pollution was simulated during the simulation period. In Scenario 2, the simulated CCL4 value was very high at ZF#2, far exceeding the observed value. It can be inferred that the pollution detected at DJ#19 and ZF#2 may not be caused by a leak from factories in their pollution-capture zone, but by the superposition of pollutants in other areas. No matter which scenario it is, the pollution at DY#3 transports faster, about 130 m/year. The pollution-transport speed at ZF#2 was the slowest, about 30 m/year. In the HF#1-DJ#19-DJ#7-SS#4 region, the pollution-transport speed was about 50 m/year.

Table 4. The concentrations of CCL4 pollution at detected points in the year 2012 and 2016 (units: ug/L).

Factory Name	August 2012					October 2016				
	Observed	Scenario 1		Scenario 2		Observed	Scenario 1		Scenario 2	
		Layer 1	Layer 2	Layer 1	Layer 2		Layer 1	Layer 2	Layer 1	Layer 2
HF#1	3.07	2.05	8.82	2.74	10.53	35.43	1.69	34.36	5.13	15.03
DJ#19	0. 64	-	-	-	-	11.67	-	-	-	-
DJ#7	-	-	-	-	1.33	23.6	-	-	-	16.8
SS#4	10.39	82	8301	1.22	232.5	1.77	498.6	6651	1.76	13.34
ZF#2	20.22	-	3690	-	-	15.8	-	20.59	-	-
DY#3	1.19	2.0	4.4	-	5.75	3.08	-	3.08	-	1.88

In order to further analyze the pollutant transport, the simulated concentration–duration curve at the monitoring well near the pollution source is drawn.

By analyzing the concentration–duration curve of the HF#1 well near the textile mill since 1986 (Figure 9), it can be found that the pollutants are mainly located in layer 2, which is limestone aquifer. The simulated concentration of HF#1 shows a gradual upward trend. The simulated concentration is close to the observed concentration in 2016, but the concentration in 2012 is much higher than the observed concentration. The observed concentration also shows an upward trend, and it can be speculated that the pollution is still in the rising period, that is, pollution may occur after 1986, or closer to the detection point.

By analyzing the concentration–duration curve of the SS#4 well near the steelworks since 1966 (Figure 10), it can be found that the pollutant concentration in layer 2 is higher, and the migration speed is faster. Compared with the observed concentration, the simulated concentration of the layer 2 is significantly higher, while the concentration of layer 1 is closer to the observed value. It can be inferred that the pollutants may diffuse through the Quaternary aquifer. As the simulated value of the layer 1 is slightly lower than the observed value, the pollutant-leakage time should be slightly later than the year 1966.

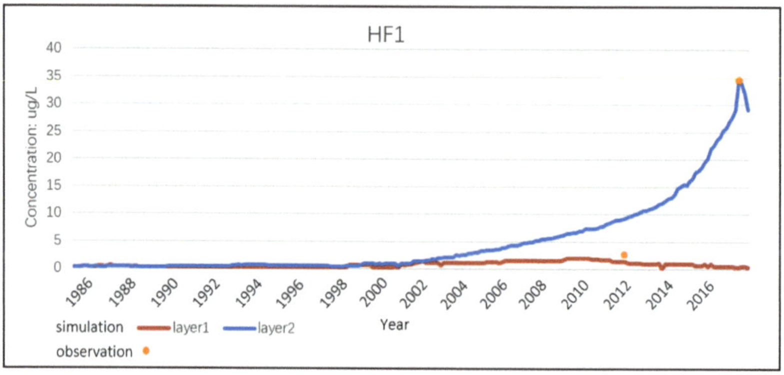

Figure 9. CCL4 concentration– duration curve of the HF#1 well.

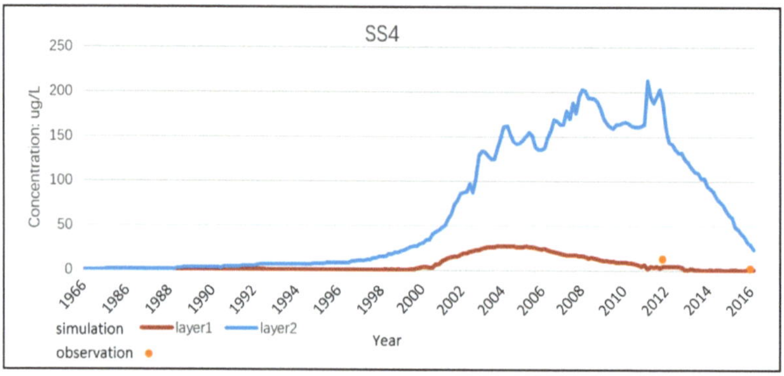

Figure 10. CCL4 concentration–duration curve of the SS#4 well.

The machine-tool plant is located on the limestone aquifer, and its pollutants migrate in layer 2. By analyzing the concentration–duration curve of the DY#3 well near the machine-tool plant (Figure 11), it can be found that whether the pollution leakage occurred in 1966 or 1986, the peak value will appear after a period of time, and the peak shape is similar. Compared with the simulated concentration, the observed concentration in 2012 and 2016 show little difference between Scenario 1 and Scenario 2. Therefore, it is inferred that both scenarios may be pollution-source years.

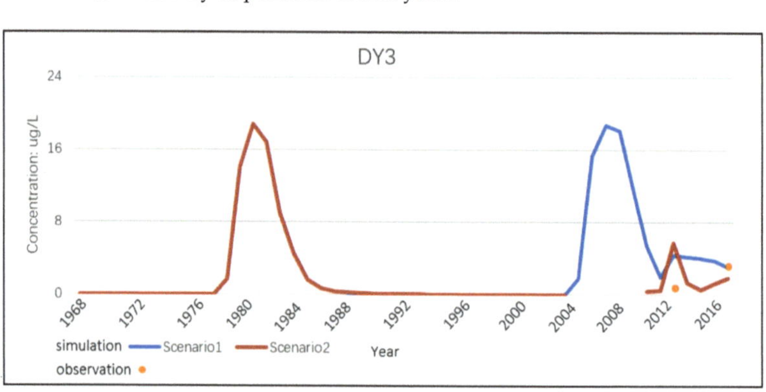

Figure 11. CCL4 concentration duration curve of the DY#3 well.

4. Discussion

Large-scale use of CCL4 will cause significant environmental soil and groundwater contamination, resulting in great health-risks, such as carcinogenic, teratogenic, and mutagenic effects, and can damage the human central nervous system, liver organs, and kidneys [6]. Therefore, it is important and urgent to control carbon tetrachloride pollution in groundwater.

In previous studies in the study area, due to insufficient hydrochemical data, based on solely hydrochemical analysis it was impossible to effectively delineate the zone of pollution sources. The complex hydrogeological conditions in the study area and the existence of fault zones made it impossible to trace the source of pollutants using the isotope method. As a result, tracing the source of CCL4 pollutants has been a major problem for a long time.

The most commonly used numerical-simulation method to trace the source of pollution is the MODPATH tool. There are some research cases; for example, Alberti, L et al. applied the MODPATH code to generate back-traced advective-flow paths to find multiple-point sources of groundwater pollution [15]. Lazzaroni, M et al. used the MODFLOW computer program to identify where the source of contamination is likely located [16]. However, this method only uses MODPATH to trace the source, which is more applicable to cases where the location of pollution sources is relatively clear or the pollution investigation is very detailed. In our study area, the tracing time-span is very large and the long-term observation data of pollutants is scarce. If we only use MODPATH, the delineated pollution range will be very uncertain. In addition, if every contaminated well is traced, the problem of multiple point-sources will be generated, which is not conducive to the development of traceability work.

In this work, we adopted the method of combining hydrochemical analysis, backward tracing, and forward simulation, to trace the source of pollutants. Through hydrochemical analysis, the locations of typical contaminated wells were screened out through the concentration field. Contamination in these typical wells is caused by CCL4 leakage and transport, and these wells are helpful in accurately tracing the source. MT3D has many applications in pollutant transport; for example, Guo Ying et al. analyzed the form of the contaminant plume as well as the changing features under the influence of multiple factors [31]. M Wen et al. established a solute-transport model based on the sampled data to predict carbon tetrachloride pollution in the study area over the next 15 years [18]. Islam, MR et al. simulated CCL4 contaminant plumes in the aquifer with MT3D and observed the dynamic equilibrium of the plume after approximately 12 years, assuming constant sources [32]. These works are of great significance, but they focus more on analyzing and predicting the transport of pollutants, without considering the traceability of pollutants. In our research area, the forward-transport simulation of MT3D is used in the traceability work. On the one hand, it is combined with backward tracing to further determine the specific location of the pollution source, so as to reduce uncertainty. On the other hand, it is helpful to analyze the characteristics of pollutant transport, to determine the pollution layer and pollutant-transport speed.

5. Conclusions

In this work, we established a traceability-method system by combining hydrochemical analysis, backward tracing, and forward transport, and successfully delineated the source of CCL4 pollution in an eastern city of China.

The CCL4 concentration-field in the ear 2012 and 2016 showed that in the northeast and southwest of the study area, there were two centers of pollution concentration. In the middle of the study area, there is a small regional-pollution center, where CCL4 is about 2 ug/L. These pollution centers were caused by CCL4 leakage and transport. Six monitoring wells: HF#1, DJ#19, DJ#7, SS#4, ZF#2, and DY#3 are located in the contaminated center, and exceed the CCL4-concentration criteria.

Using MODPATH for backward tracing, we delineated the pollution zone of six CCL4-pollution-detected points: HF#1, DJ#19, DJ#7, SS#4, ZF#2, and DY#3 30 years ago (the

1990s) and 50 years ago (the 1970s). The results showed that the capture zone of DY#3 is the largest, being 4.43 km^2 in the 1990s and exceeding 4.77 km^2 in the 1970s, and the potential polluting factory in its capture zone was the machine-tool plant. The potential polluting factory was the textile mill and the chemical-fertilizer plant in HF#1′s area, the fertilizer plants in DJ#7′s area, the steelworks in SS#4′s area, and the cement mill in the 1970s in ZF#2′s area.

We performed forward simulations and set Scenario 1: delivery of pollutant particles to potentially polluting factories in the 1990s. Scenario 2: delivery of pollutant particles to potentially polluting factories in the 1970s. By comparing the simulated concentration with the observed concentration, we infer that the textile mill in 1990s, the steelworks in the 1970s, and the machine-tool plant in the 1970s and 1990s are likely to be the source of pollution, and the contaminants at ZF#2 and DJ#19 are likely to be generated by the migration or superposition of pollutants in other areas. The contaminant-transport speed at DY#3 was 130 m/year, while at ZF#2 it was 30 m/year. In the HF#1-DJ#19-DJ#7-SS#4 region, the pollution-transport speed was about 50 m/year.

By analyzing the concentration–duration curve of HF#1, SS#4, and DY#3 wells near the pollution source, it can be found that CCL4 in limestone layers transports faster than in the Quaternary layer. The pollutant in the textile mill was from the limestone layer, and was released slightly later than 1986. The pollutant at the steelworks originated from the Quaternary layer, and was released slightly later than 1966.

This study delineated the contaminated zone of CCL4 and provided a possible source of CCL4, which helps to further identify the contamination source and remediate pollutants. However, the pollution-concentration observed data is lacking, leading to the pollution-transport simulation results showing uncertainty. In addition, the exact us-ime and pollution-storage location in the factories need further investigation.

Author Contributions: Conceptualization, B.G.; data curation, H.A. and Y.L.; investigation, P.Y. and H.A.; methodology, B.G. and X.L.; supervision, P.Y. and R.K.; visualization, Y.Z. and Y.L.; writing—original draft, B.G. and Y.Z. All authors have read and agreed to the published version of the manuscript.

Funding: This study was supported by the "Rizhao City Urban Geological Survey Project", from the "Rizhao City financial fund" (SDGP371100202102000475).

Institutional Review Board Statement: Not applicable.

Informed Consent Statement: Not applicable.

Data Availability Statement: The datasets generated and analyzed during the current study are available from the corresponding author upon reasonable request.

Acknowledgments: The authors would like to thank No.8 Institute of Geology and Mineral Resources Exploration of Shandong Province for their research data.

Conflicts of Interest: The authors declare no conflict of interest.

References

1. Rathi, B.S.; Kumar, P.S.; Show, P.L. A review on effective removal of emerging contaminants from aquatic systems: Current trends and scope for further research. *J. Hazard. Mater.* **2021**, *409*, 124413. [CrossRef] [PubMed]
2. Stuart, M.; Lapworth, D.; Crane, E.; Hart, A. Review of risk from potential emerging contaminants in UK groundwater. *Sci. Total Environ.* **2012**, *416*, 1–21. [CrossRef] [PubMed]
3. Liang, Z.; Han, B.; Jie, M.A.; Liang, X.; Liu, X.; Zhao, H.; Weili, W.U.; Qiu, C.; Xiao, Y. Simulation on Treatment of Carbon Tetrachloride in Groundwater with Bimetal Dielectric Chemical Reaction Wells. *Environ. Sci. Technol.* **2014**, *37*, 141–146.
4. Penny, C.; Gruffaz, C.; Nadalig, T.; Cauchie, H.M.; Vuilleumier, S.; Bringel, F. Tetrachloromethane-Degrading Bacterial Enrichment Cultures and Isolates from a Contaminated Aquifer. *Microorganisms* **2015**, *3*, 327–343. [CrossRef]
5. Doherty, R.E. A History of the Production and Use of Carbon Tetrachloride, Tetrachloroethylene, Trichloroethylene and 1,1,1-Trichloroethane in the United States: Part 1–Historical Background; Carbon Tetrachloride and Tetrachloroethylene. *Environ. Forensics* **2000**, *1*, 69–81. [CrossRef]

6. Liu, D.; Liu, B.; Zhan, H. The Health Risk Assessment of CCl_4 in groundwater of a Certain Chemical Plant. *IOP Conf. Ser. Earth Environ. Sci.* **2018**, *153*, 052011. [CrossRef]
7. Sautter, S. BHI ERC team zeroes in on sources of carbon tetrachloride contamination Hanford Rench. *Hanford Rench March* **2001**, *19*, 1–17.
8. Skrökki, A. Quality of drinking water from one surface water plant and six groundwater plants in the town of Kajaani in Finland. *Vatten* **1992**, *48*, 78–81.
9. Jiang, P.; Ma, Z.M.; Yu, W.W.; Wen, M. Carbon tetrachloride pollution pathway of groundwater a typical contaminated site in the east of the city. *IOP Conf. Ser. Earth Environ. Sci.* **2017**, *82*, 012058. [CrossRef]
10. Datta, B.; Prakash, O.; Cassou, P.; Valetaud, M. Optimal Unknown Pollution Source Characterization in a Contaminated Groundwater Aquifer—Evaluation of a Developed Dedicated Software Tool. *J. Geosci. Environ. Prot.* **2014**, *2*, 11. [CrossRef]
11. Wang, H.; Shi, Z.; Jiang, Y.; Lian, X.; Yang, Y.; Feng, F.; Jia, Y. Research Progress on Indicator of Groundwater Pollution Identification and Traceability. *Res. Environ. Sci.* **2021**, *34*, 1886–1898.
12. Mansuy, L.; Philp, R.P.; Allen, J. Source identification of oil spills based on the isotopic composition of individual components in weathered oil samples. *Environ. Sci. Technol.* **1997**, *31*, 3417–3425. [CrossRef]
13. Atmadja, J.; Bagtzoglou, A.C. State of the art report on mathematical methods for groundwater pollution source identification. *Environ. Forensics* **2001**, *2*, 205–214. [CrossRef]
14. Gu, X.; Shao, J.; Cui, Y.; Hao, Q. Calibration of two-dimensional variably saturated numerical model for groundwater flow in arid inland basin, China. *Curr. Sci. A Fortn. J. Res.* **2017**, *113*, 403–412. [CrossRef]
15. Banejad, H.; Mohebzadeh, H.; Ghobadi, M.H. Numerical Simulation of Groundwater Flow and Contamination Transport in Nahavand Plain Aquifer, West of Iran. *J. Geol. Soc. India* **2014**, *83*, 83–92. [CrossRef]
16. Alberti, L.; Colombo, L.; Formentin, G. Null-space Monte Carlo particle tracking to assess groundwater PCE (Tetrachloroethene) diffuse pollution in north-eastern Milan functional urban area. *Sci. Total Environ.* **2018**, *621*, 326–339. [CrossRef]
17. Lazzaroni, M.; Ceccatelli, M.; Rossato, L.; Nisi, B.; Venturi, S.; Fanti, R.; Tassi, F.; Vaselli, O. Boron pollution in the shallow groundwater system from Isola di Castelluccio (central-eastern Tuscany): Evidences from a geochemical survey and new remediation perspectives from a recently-installed hydraulic barrier and hydrogeological modelling. *Ital. J. Geosci.* **2021**, *140*, 121–140. [CrossRef]
18. Zhang, S.; Ma, Z.; Liu, S. Numerical simulation and prediction of chlorinated hydrocarbon migration in eastern Jinan. *IOP Conf. Ser. Earth Environ. Sci.* **2020**, *531*, 012028. [CrossRef]
19. Wen, M.; Ma, Z.; Jiang, p.; Yu, W. Study on the Law of Carbon Tetrachloride Migration in Groundwater System. In Proceedings of the 2017 2nd International Conference on Advanced Materials Science and Environment Engineering(AMSEE2017), Sanya, China, 25 June 2017; pp. 214–219.
20. Eryigit, M.; Engel, B. Spatiotemporal Modelling of Groundwater Flow and Nitrate Contamination in An Agriculture-Dominated Watershed. *J. Environ. Inform.* **2022**, *39*, 125–135. [CrossRef]
21. Ayvaz, M.T. A hybrid simulation–optimization approach for solving the areal groundwater pollution source identification problems. *J. Hydrol.* **2016**, *538*, 161–176. [CrossRef]
22. Wei, J.; Guo, Y.; Wang, R.; Jiang, Y. Recent advances in simulation approaches of the complex karst medium. *Hydrogeol. Eng. Geol.* **2015**, *42*, 23–34.
23. Abdelaziz, R.; Bambi, C.K.M. Groundwater Assessment of the Bleleone Catchment Karst Aquifer in Southern France. In Proceedings of the World Multidisciplinary Earth Sciences symposium (WMESS), Prague, Czech Republic, 5–9 September 2016.
24. Baalousha, H.M.; Fahs, M.; Ramasomanana, F.; Younes, A. Effect of Pilot-Points Location on Model Calibration: Application to the Northern Karst Aquifer of Qatar. *Water* **2019**, *11*, 679. [CrossRef]
25. Lautz, L.K.; Siegel, D.I. Modeling surface and ground water mixing in the hyporheic zone using MODFLOW and MT3D. *Adv. Water Resour.* **2006**, *29*, 1618–1633. [CrossRef]
26. Garfias, J.; Llanos, H.; Martel, R.; Salas-Garcia, J.; Bibiano-Cruz, L. Assessment of vulnerability and control measures to protect the Salbarua ecosystem from hypothetical spill sites. *Environ. Sci. Pollut. Res.* **2018**, *25*, 26228–26245. [CrossRef]
27. Barry, F.; Ophori, D.; Hoffman, J.; Canace, R. Groundwater flow and capture zone analysis of the Central Passaic River Basin, New Jersey. *Environ. Geol.* **2009**, *56*, 1593–1603. [CrossRef]
28. Shamsuddin, M.K.N.; Suratman, S.; Zakaria, M.P.; Aris, A.Z.; Sulaiman, W.N.A. Particle tracking analysis of river-aquifer interaction via bank infiltration techniques. *Environ. Earth Sci.* **2014**, *72*, 3129–3142. [CrossRef]
29. Pollock, D.W. *User Guide for MODPATH Version 7—A Particle-Tracking Model for MODFLOW: U.S. Geological Survey Open-File Report 2016-1086*; U.S. Geological Survey: Reston, VA, USA, 2016; p. 35. [CrossRef]
30. Yang, W.; Mao, W.; Yang, Y.; Zhu, Y.; Yang, J. Optimizing Conjunctive Use of Groundwater and Cannel Water in Hetao Irrigation District Aided by MODFLOW. *J. Irrig. Drain.* **2021**, *40*, 93–101. [CrossRef]
31. GUO, Y.; TIAN, X.; Zhang, L. Characteristics of Carbon Tetrachloride Pollution in Groundwater in a Water Source. *South North Water Divers. Water Sci. Technol.* **2011**, *9*, 109–113. [CrossRef]
32. Islam, M.R.; Jewett, D.; Williams, J.R.; Hantush, M. Calibration of groundwater flow and contaminant transport modeling for Ogallala, Nebraska. In Proceedings of the International Water Resources Engineering Conference on Water Resources Development and Protection, Memphis, TN, USA, 3–7 August 1998; pp. 105–110.

water

MDPI

Article

Study on Natural Attenuation of Groundwater Organic Pollutants by Integrating Microbial Community Dynamics and Isotope Analysis

Jian Shi [1,2], Yiyin Zhang [2,3], Yuyang Lai [2,4], Ruilu Yang [1,2], Meng Cai [1,2], Susu Fan [1,2] and Xiaomin Gu [1,2,*]

1 Analysis and Testing Center, Nantong University, Nantong 226019, China; shi.j1@ntu.edu.cn (J.S.); yangruilu1987@126.com (R.Y.); caim@ntu.edu.cn (M.C.); fansusu828@ntu.edu.cn (S.F.)
2 Key Laboratory of Water Pollution Control, Nantong 226019, China; z980317yy@sina.com (Y.Z.); 2321110135@stmail.ntu.edu.cn (Y.L.)
3 School of Transportation and Civil Engineering, Nantong University, Nantong 226019, China
4 School of Geographic Science, Nantong University, Nantong 226000, China
* Correspondence: ntugxm@ntu.edu.cn; Tel.: +86-0513-85012965

Abstract: This study focuses on a typical chemically contaminated site in the southeastern coastal region of China, investigating the natural attenuation mechanisms of benzene and o-toluidine in groundwater through high-throughput sequencing, physicochemical analyses, and stable isotope techniques. The results demonstrate significant heterogeneity in the spatial distribution and degradation processes of pollutants within the contaminated zones (W27, W28, W31). Environmental factors such as HCO_3^-, SO_4^{2-}, and ORP predominantly influence the microbial community structure and functional distribution. Stable isotope data reveal that δD and $\delta^{13}C$ enrichment effects are most pronounced in the deep layer (W28_40m), indicating active pollutant degradation, while degradation in the deeper layers of W27 and W31 is constrained by anaerobic conditions and reduced microbial activity. The combined analysis of hydrogen and carbon isotopes elucidates the degradation pathways and dynamic processes of pollutants within the contaminated zones, providing quantitative evidence for natural attenuation mechanisms and scientific support for optimizing site remediation strategies.

Keywords: natural attenuation; microbial community; isotope; barrier technology

1. Introduction

Groundwater contamination is an important environmental problem globally, especially in areas of high industrial activity [1]. Chronic chemical spills and improper waste management lead to the accumulation of organic contaminants in groundwater, posing a significant threat to ecosystems and public health [2]. Organic contaminants, such as benzene and o-toluidine, are particularly hazardous due to their high toxicity, persistence, and mobility [3]. These contaminants can penetrate aquifers and remain in groundwater for long periods, potentially causing serious harm to ecosystems and human health [4,5]. In addition, contaminant transport and degradation exhibit significant spatial and temporal heterogeneity, a process that is influenced by the complex hydrogeochemical conditions of the groundwater system, further increasing the difficulty of management [6,7].

In recent years, many physical, chemical, and biotechnological approaches have been applied to groundwater contamination management, including pump and treat, chemical oxidation, thermal desorption, and bioremediation. However, these technologies are often costly and cause greater disturbance to the ecosystem, especially in low-permeability formations with limited effectiveness [8,9]. Therefore, as a more sustainable and cost-effective alternative, natural attenuation (NA) is gradually becoming a technique of interest, favored for its low intervention, cost effectiveness, and environmental friendliness. Natural attenuation refers to the degradation or transformation of pollutants through physical, chemical, and biological processes without human intervention. Microbiologically mediated biodegradation is a key process in the removal of organic pollutants. Microorganisms convert pollutants into harmless products through complex metabolic pathways in which environmental factors such as redox potential, dissolved oxygen, and the availability of carbon sources play an important role in the degradation process [10,11]. Although the degradation pathways and key microbial species of single contaminants have been investigated, there is limited research on the natural attenuation mechanisms of co-existing contaminants. At contaminated sites, contaminants are often present in mixtures, and interactions between contaminants may affect the structure and function of microbial communities [12,13]. In addition, the spatiotemporal variation in environmental factors and their effects on pollutant degradation and microbial dynamics remain understudied. Several studies have shown that spatially varying physicochemical parameters (such as retardation) can significantly affect the migration of pollutants in groundwater, complicating pollution management [14]. For example, parameters such as permeability, pH, ion concentration, and changes in the redox environment can affect the spatial distribution and migration of contaminants, with more mobile contaminants being more easily transported in certain groundwater zones [15]. Changes in groundwater flow, temperature, and chemical composition can lead to different degradation rates and mechanisms at different locations within a site [16,17]. The hydrogeochemical conditions of a contaminated site can create environments that enhance or inhibit contaminant degradation. For example, in areas with high ion concentrations, microbial activity may be promoted due to nutrient availability, while in low-conductivity environments, microbial metabolism may be limited, reducing degradation efficiency [18]. In this context, understanding how physicochemical factors such as pH, electrical conductivity (EC), redox potential (ORP), and ion concentration affect contaminant distribution and microbial activity is critical to developing effective natural attenuation strategies [19]. While previous research has advanced our understanding of groundwater contamination and natural attenuation, key mechanisms remain underexplored, particularly regarding the interactions of co-existing contaminants and microbial community responses. Most studies have focused on single contaminants or single-method approaches, which fail to capture the full complexity of these processes. Notably, the influence of horizontal and vertical barriers on contaminant transport and distribution has not been adequately addressed. To bridge these gaps, this study integrates microbial community analysis with stable isotope techniques, providing a more holistic approach to understanding the degradation dynamics of co-existing pollutants and their interaction with environmental factors, often overlooked in traditional studies.

This study focuses on a typical chemically contaminated site in the southeastern coastal area of China, which is characterized by horizontal and vertical barriers designed to effectively control the spread of contaminants. These barriers provide a unique environment for studying natural attenuation mechanisms in groundwater. The main goal of this study is to explore the natural attenuation process of co-existing pollutants such as benzene and o-toluidine by combining high-throughput sequencing, physicochemical analysis, and stable isotope technology. Specific objectives include (1) analyzing hydrogeochemical data, studying the spatial heterogeneity of pollutant migration and microbial activity, and revealing the impact of physicochemical factors on natural attenuation; (2) using high-throughput sequencing and stable isotope technology to gain a deep understanding of the dynamics of microbial communities and pollutant degradation pathways in complex multi-pollutant environments; and (3) combining environmental factor analysis, microbial ecology, and isotope technology to provide a scientific basis for optimizing the management of similar contaminated sites with barriers and promote the management and remediation of similar sites.

2. Materials and Methods

2.1. Study Area

The study area is located on the site of a former chemical plant on the southeast coast of China, which used to produce a wide range of chemicals, including acetic acid, tetra sulphonyl pyrrolidone, pyrazolone, diacetyl, diene ketone, and butyrolactone. The area is a typical industrial area and has been subjected to industrial activities for a long time, resulting in serious groundwater contamination. Regarding hydrogeological conditions, the lithology of the aquifer in the study area mainly consists of flat fill, sandy chalk, silt, and powdery clay and the water table is high in the area. The horizontal and vertical permeability coefficients of the aquifer range from 2.0×10^{-5} to 3.53×10^{-6} and 5.0×10^{-5} to 7.53×10^{-6}, respectively. The groundwater flow direction is mainly from southwest to northeast, and the migration of contaminants in the groundwater shows significant spatial heterogeneity due to the influence of topography and regional hydrological conditions. The layout of the monitoring wells is shown in Figure 1. The study area is surrounded by a river on the east side, which is an important artificial waterway connecting the Yangtze River and the Grand Canal; industrial areas on the south and west sides; and residential areas adjacent to the north side. To control the spread of pollution, several preventive and control measures have been taken, including the construction of vertical and horizontal barriers. The vertical barriers are made of materials with a permeability coefficient of less than 1×10^{-7} cm/s to a depth of about 40 m and are embedded in an impermeable layer of at least 2 m, while the horizontal barriers are made of compacted clay or soil mixed with bentonite, not less than 46 cm thick, with a permeability coefficient of no more than 10^{-6} cm/s. These measures have played an important role in effectively limiting the spread of pollutants to the surrounding area.

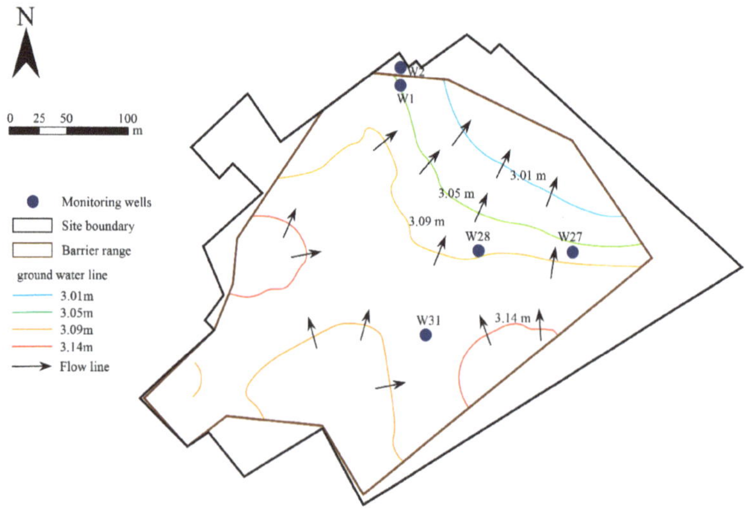

Figure 1. The site and the locations of monitoring wells.

According to the preliminary investigation data, the main pollutants in the study area are chlorinated hydrocarbons and benzene compounds, and benzene and o-toluidine, in particular, are the most dangerous pollutants. The distribution of these pollutants in groundwater shows obvious spatial differences, and the concentration varies greatly in soil layers at different depths. The distribution of pollutants in aquifers at different depths varies considerably due to hydrogeological conditions and the migration characteristics of the pollutants, which further affects the degradation and natural attenuation process of the pollutants.

2.2. Groundwater Sample Collection

Sampling work was conducted in April 2024, using sampling equipment to collect groundwater samples. Before sampling, the groundwater was drawn until the pH value, electrical conductivity (EC), oxidation–reduction potential (ORP), temperature, and dissolved oxygen (DO) measurements reached a stable state. To analyze the spatial differences in the super-risk contaminants and microbial communities in the area, each sampling point was sampled at depths of 6 m, 15 m, and 40 m below the ground surface. Groundwater samples were collected in sealed bottles and transported to the laboratory for further analysis.

2.3. Physicochemical Analysis

This study analyzed two volatile organic pollutants, benzene and o-toluidine, in the groundwater. Both compounds were extracted using liquid–liquid extraction with dichloromethane as the solvent and analyzed by gas chromatography–mass spectrometry (GC-MS). A DB-25 capillary column was used, with a flow rate of 2 mL/min. The oven temperature was initially set to 40 °C for 2 min, followed by a ramp to 100 °C at 5 °C/min and then to 200 °C at 10 °C/min, and held at 200 °C for 8 min. The temperature was then increased to 280 °C at 30 °C/min and held for 3 min. Samples were injected in a 50:1 split mode. Mass spectrometry conditions included a solvent delay of 2 min. The recovery rates, determined through both the blank-spiking and matrix-spiking experiments, were all at least 80% for each compound.

2.4. DNA Extraction, PCR, and High-Throughput Sequencing

In March 2024, 2 L groundwater for microbial community analysis was filtered on the spot using a self-assembled portable filter system (−90 kPa) and microfiltration membranes (0.22 µm) in triplicate. The entire experimental procedure encompassed sample preparation, DNA extraction, PCR amplification, library construction, and high-throughput sequencing, all aimed at ensuring the acquisition of high-quality sequencing data. First, after extracting the genomic DNA, its quality was assessed using 1% agarose gel electrophoresis. Next, PCR amplification was performed using specific primers with barcodes or fusion primers with mismatched bases to target the specified sequencing region. To ensure the accuracy and reliability of subsequent data analysis, amplification was performed under the following two conditions: using the lowest possible cycle number for amplification, and ensuring that all samples were amplified at the same cycle number. Representative samples were randomly selected for pre-testing to ensure that the majority of samples could amplify PCR products of appropriate concentration at the lowest cycle number. PCR amplification was performed using TransGen AP221-02: TransStart Fastpfu DNA Polymerase, and the amplified PCR products were detected by agarose gel electrophoresis with 1% agarose to check the amplified band sizes. The amplified PCR products were detected by 1% agarose gel electrophoresis to determine the size of the amplified bands and purified using an Agencourt AMPure XP Nucleic Acid Purification Kit. Following this, library construction for the second-generation high-throughput sequencing platform was performed, including the ligation of Y-junction adapters, removal of self-ligated adapter fragments, PCR amplification to enrich templates, and the generation of single-stranded DNA fragments by sodium hydroxide denaturation. Finally, sequencing was performed on a second-generation high-throughput sequencing platform, where DNA clusters were generated through bridge PCR amplification, and DNA sequences were synthesized using fluorescently labeled dNTPs. Each step of the experimental process was rigorously controlled to ensure data quality, providing a solid and reliable foundation for subsequent microbial community analysis.

3. Results and Discussion

3.1. Hydrogeochemical Variations and Their Impact on Pollutant Distribution in Groundwater

The hydrogeochemical conditions of groundwater are critical in influencing the distribution of contaminants and microbial activity [20]. In March 2024, hydrogeochemical parameters and contaminant concentrations in the study area showed significant spatial variation (Table 1). The pH of groundwater samples ranged from 6.14 to 6.68, generally approaching neutrality, which favors microbial growth and metabolism [21]. However, within the contaminated zone inside the barrier (W31), the pH at the shallow sampling point W31_6m was 6.14, slightly lower than at other locations, likely due to localized acidification from the anaerobic degradation of contaminants. In contrast, sampling points outside the barrier (W2) exhibited higher pH values, with the shallow point W2_6m reaching 6.67, indicating that the barrier effectively restricted contaminant diffusion.

ORP is a key indicator among hydrogeochemical parameters, reflecting the redox state of groundwater, which in turn affects the metabolic activity of microbial communities. In the study area, ORP values ranged from 39 mV to 125 mV, showing significant spatial variation [22]. In uncontaminated areas (such as W1), the ORP value was high (125 mV), indicating that the oxidative environment promotes the activity of aerobic microorganisms and can degrade organic pollutants more efficiently. Taking W1_6m as an example, the ORP was 125 mV and the DO concentration was 2.28 mg/L, indicating that there is sufficient oxygen, which is conducive to aerobic degradation [16]. In contrast, the ORP value in the deep layer of the contaminated area (such as W31_40m) was low (39 mV), indicating that the area is in a reducing environment, which may be conducive to the growth of anaerobic microorganisms and the occurrence of anaerobic degradation [23]. EC is an indicator of the concentration of dissolved ions in groundwater and has an important impact on the metabolism of microbial communities. Areas with higher EC values generally indicate higher ion concentrations, which help support the growth of certain microbial groups. For example, the EC value of W1_6m was 1.84 mS/cm, which is consistent with the higher DO concentration (2.28 mg/L) and the higher degradation rate of pollutants [15]. Higher EC may indicate that there are more dissolved ions in the water, providing more nutrient sources for microorganisms, which is conducive to promoting the degradation of pollutants. However, areas with lower EC values, such as W31_40m (EC of 4.62 mS/cm), indicate that there are fewer dissolved ions here and microbial activity may be limited to a certain extent, which may also affect the degradation efficiency of pollutants [24]. Low-conductivity environments with limited microbial metabolism often mean fewer available ions and nutrients, thereby reducing the degradation rate [22]. The distribution of total organic carbon (TOC) and total nitrogen (TN) in the study area is closely related to microbial activity and pollutant degradation. The data showed that in polluted areas (such as W31_15m), TOC and TN concentrations were high, 578.5 mg/L and 548.5 mg/L, respectively, indicating that these areas have higher concentrations of organic matter and nitrogen sources, providing sufficient nutrients for microorganisms and promoting the degradation of organic pollutants [23]. In contrast, at W2_6m outside the isolation layer, TOC was 55.00 mg/L and TN was 48.58 mg/L, showing less pollution impact [15].

Table 1. Groundwater hydrogeochemical parameters and characteristic contaminants in March 2024.

		T	pH	DO	ORP	EC	K^+	Ca^{2+}	Na^+	Mg^{2+}	HCO_3^-	Cl^-	SO_4^{2-}	Mn	Fe	TOC	TN	Benzene	o-toluidine
Number	Depth	°C	/	mg/L	mV	mS/cm	mg/L	mg/L	mg/L	mg/L	mg/L	mg/L	mg/L	mg/L	mg/L	mg/L	mg/L	mg/L	mg/L
W1	6 m	18.3	6.58	2.28	125	1.84	26.20	164.8	122.7	33.33	487.5	93.13	350.6	17.46	3.091	36.42	23.87	0.089	0.014
	15 m	18.5	6.46	2.25	125	2.57	12.46	431.9	32.81	17.26	361.3	481.8	63.64	23.03	5.192	77.50	73.00	0.089	0.011
	40 m	18.6	6.44	2.65	125	3.63	16.27	377.8	100.9	165.4	521.9	755.7	32.26	33.05	16.34	35.30	26.41	0.089	0.008
W2	6 m	17.4	6.67	1.63	109	1.86	21.61	381.7	43.68	39.69	430.1	188.7	230.0	60.75	10.29	55.00	48.58	0.089	0.088
	15 m	18.7	6.68	2.45	91	1.76	11.70	294.0	16.25	4.720	177.8	417.5	57.00	5.25	4.771	94.50	90.50	0.089	0.089
	40 m	18.6	6.48	1.58	61	4.30	19.32	933.4	123.8	138.0	470.3	1153	80.74	35.59	30.54	45.46	41.23	0.089	0.088
W27	6 m	16.3	6.35	1.62	57	3.67	15.50	415.5	294.9	59.03	625.2	231.9	963.4	90.26	9.009	70.50	60.00	0.092	2.168
	15 m	15.8	6.47	2.15	76	3.01	4.11	36.55	155.4	16.18	1141	199.9	91.71	8.285	2.672	134.5	122.0	0.113	42.97
	40 m	16.1	6.52	2.32	42	3.22	13.22	55.89	224.2	24.89	1135	251.7	175.6	17.46	2.254	56.50	50.00	0.118	34.45
W28	6 m	16.8	6.38	1.48	47	3.01	12.46	541.3	109.9	90.37	1026	178.9	1017	181.9	30.09	149.5	134.0	0.100	0.804
	15 m	16.4	6.4	1.86	57	3.28	13.22	39.87	116.3	23.23	1130	83.70	372.2	4.050	5.192	82.00	72.00	0.126	21.89
	40 m	16.3	6.44	1.36	86	2.81	19.32	75.17	127.1	43.22	986.5	129.1	234.5	60.76	10.72	65.00	53.00	0.091	0.198
W31	6 m	16.5	6.14	0.86	49	3.51	10.18	957.7	66.84	77.63	957.9	139.5	1180	78.05	10.72	320.0	305.0	0.090	0.094
	15 m	16.6	6.32	0.78	46	6.46	10.18	32.12	325.4	50.42	1594	686.8	228.6	12.55	14.59	578.5	548.5	0.102	1.640
	40 m	16.6	6.40	1.27	39	4.62	34.63	63.96	160.5	47.40	1451	427.9	218.9	33.69	6.037	198.0	174.5	0.118	0.1167

As shown in Figure 2, the spatial distribution of characteristic pollutants shows significant differences between polluted and unpolluted areas. Benzene concentrations ranged from 0.089 to 0.126 mg/L, while o-toluidine concentrations ranged from 0.014 to 42.97 mg/L. Contaminant concentrations were significantly higher in contaminated areas within the barrier (W27, W28, W31) compared to uncontaminated areas (W1, W2). For example, at W27_15m, o-toluidine concentrations reached 42.97 mg/L, several times higher than the maximum concentrations observed outside the barrier (e.g., W2_6m). Benzene concentrations were the highest at W28_15m (0.126 mg/L), which coincided with low dissolved oxygen (DO) levels (1.86 mg/L) and elevated bicarbonate (HCO_3^-) concentrations, indicating that active degradation of the contaminant was occurring [25,26]. Similarly, studies have demonstrated that changes in groundwater chemistry significantly influence the mobility of pollutants and their degradation rates across different environments [27].

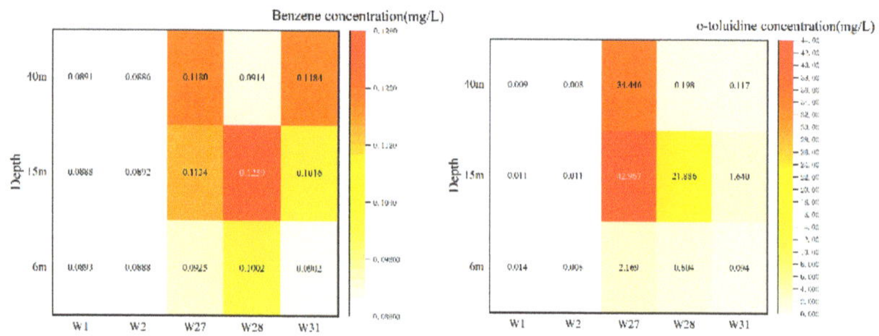

(a) (b)

Figure 2. Characteristic contaminant concentration heat map: (**a**) benzene concentration; (**b**) o-toluidine concentration.

Aquifer heterogeneity also significantly affects the transport and distribution of pollutants. At our study site, the horizontal hydraulic conductivity at the 6 m and 15 m depths was relatively high (7.53×10^{-3} and 6.72×10^{-3} m/s), allowing contaminants to migrate more quickly in the horizontal direction. In contrast, at the 40 m depth, the horizontal conductivity dropped sharply to 7.53×10^{-6} m/s, indicating slower lateral pollutant movement. Similarly, vertical hydraulic conductivity was higher at 6 m (3.53×10^{-3} m/s) and 15 m (2.73×10^{-3} m/s), but significantly lower at 40 m (3.53×10^{-6} m/s). This difference in permeability suggests that contaminant transport is more restricted in the deeper layers. At the 6 m and 15 m depths, pollutants such as benzene and o-toluidine were detected at higher concentrations, reflecting more efficient horizontal and vertical transport. Conversely, at the 40 m depth, where permeability is lower, pollutant concentrations were comparatively lower, indicating reduced mobility in these layers.

3.2. Microbial Community Structure and Functional Diversity in Polluted and Unpolluted Zones

The abundance and diversity of microbial communities are important indicators for assessing the ecosystem health of water bodies [28]. The microbial diversity coverage of all monitoring wells in the study area was above 0.95, indicating that most of the microbial community diversity was captured. The composition of microbial communities at the phylum level and genus level at each sampling site is shown in Figures 3–6. As shown in Table 2, α diversity indices revealed significant differences in species richness and community diversity in different zones and depth samples. The shallow (6 m) layer of the non-polluted area (W1) exhibited the highest species richness (1015 OTUs), Shannon index (7.49), and Simpson index (0.98), which suggests that environments with lower pollution pressures are favorable for the formation of resource-rich and stable ecosystems [29,30]. However, microbial diversity was significantly reduced in the polluted areas (W27, W28, W31). The Shannon index was only 2.37, the Simpson index was 0.58 in W27_15m, and the species richness was also reduced to 178 OTUs [31]. The higher pollutant concentrations (o-toluidine 42.97 mg/L, benzene 0.113 mg/L) significantly suppressed the community diversity of the polluted areas in W27_15m. The higher pollutant concentrations significantly suppressed community diversity while enriching specific degrading bacteria (Acidovorax and Pseudomonas), suggesting that pollution pressure is significantly selective for microbial communities [32,33]. Moreover, studies on bacterial consortia developed for BTEX degradation also show that the enrichment of specific microbial species like Pseudomonas was linked to the presence of toxic pollutants [34]. Specifically, Acidovorax, a genus often associated with the degradation of aromatic compounds, was enriched in the polluted zones [35]. Similarly, Pseudomonas, known for its ability to degrade a wide range of pollutants including hydrocarbons and aromatic compounds, was also dominant in the contaminated areas, reflecting the selective pressure of the pollutants [28,36].

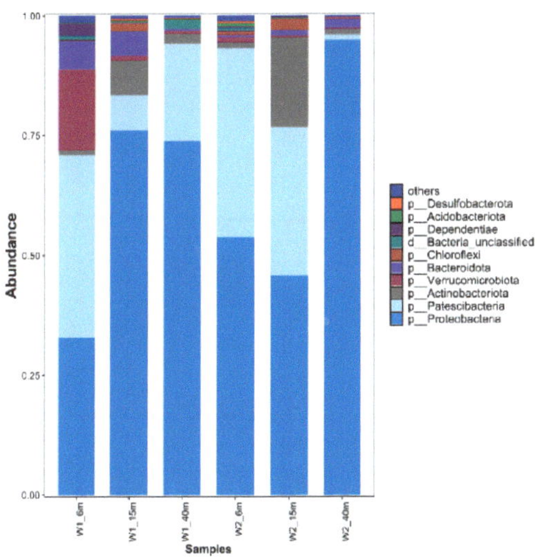

Figure 3. Composition of microbial communities at phylum level in non-contaminated area samples.

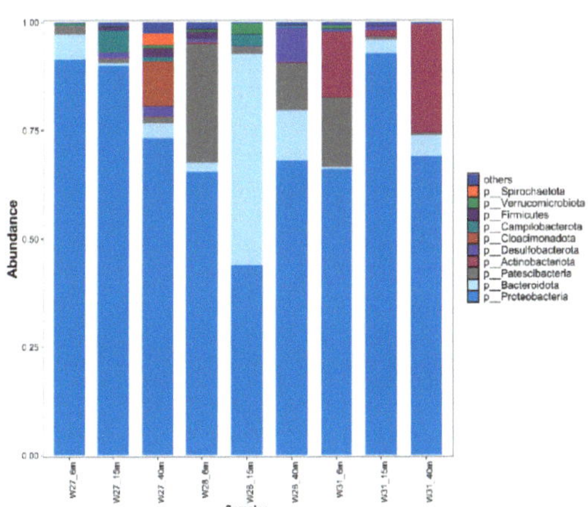

Figure 4. Composition of microbial communities at phylum level in contaminated area samples.

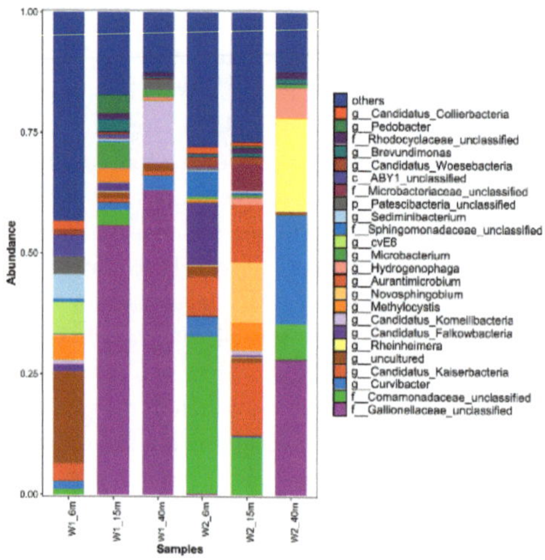

Figure 5. Composition of microbial communities at genus level in non-contaminated area samples.

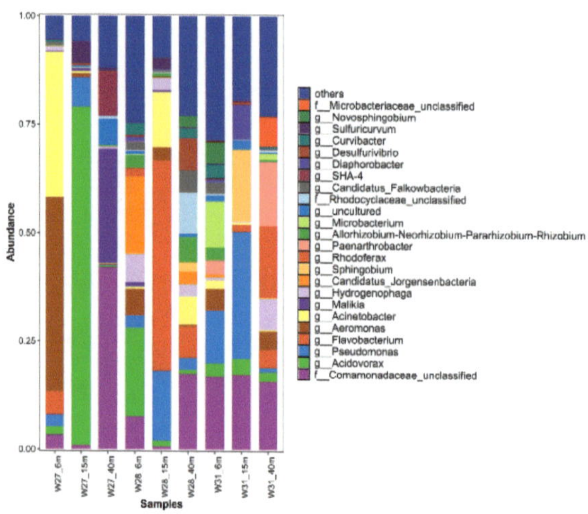

Figure 6. Composition of microbial communities at genus level in contaminated area samples.

Table 2. Index of alpha diversity of groundwater samples.

Sample	OTUs	Shannon	Chao	Simpson	ACE
W1_6m	1015	7.49	1030.9	0.98	1032.6
W1_15m	551	4.19	567.29	0.69	568.98
W1_40m	246	2.65	255.60	0.58	256.21
W2_6m	829	6.09	839.73	0.93	841.86
W2_15m	378	5.67	383.26	0.95	382.86
W2_40m	183	3.71	186.03	0.84	186.64
W27_6m	162	3.20	163.29	0.77	163.65
W27_15m	178	2.37	180.38	0.58	181.03
W27_40m	205	3.25	208.14	0.75	208.51
W28_6m	387	5.72	398.58	0.94	397.64
W28_15m	178	3.23	188.65	0.73	184.85
W28_40m	316	5.80	325.93	0.96	326.66
W31_6m	640	6.84	647.06	0.97	645.95
W31_15m	204	4.11	206.02	0.86	206.49
W31_40m	186	4.89	192.99	0.93	191.08

The depth variation of microbial communities also reflects the effect of environmental conditions on community distribution in different zones. In the non-polluted zone (W2), shallow samples (6 m) exhibited a high species richness (829 OTUs) and Shannon index (6.09); however, in the deeper layer (40 m), the OTUs were reduced to 183, and Shannon's index decreased to 3.71, suggesting that microbial diversity is limited by the nutrient-poor environment at the deeper level [37,38]. In contrast, W28 points within the contaminated area showed different depth distribution characteristics. The shallow (W28_6m) and middle (W28_15m) layers were enriched with aerobic degrading bacteria such as Pseudomonas and Acinetobacter [28,39]. In contrast, the microbial community in the deep layer (W28_40m) tended to be homogenized, suggesting that pollutant degradation in this region may be driven by a combination of metabolic pathways. In particular, the deeper community was enriched with flora capable of participating in anaerobic degradation (Gallionellaceae and Candidatus Jorgensenbacteria), reflecting the regional specificity of pollutant degradation [40,41].

As shown in Figure 7, the redundancy analysis (RDA) revealed significant relationships between environmental factors and microbial community composition across the groundwater samples from different zones (W1, W2, W27, W28, and W31). The first two axes (RDA1 and RDA2) explain 30.54% and 23.32% of the total variation, respectively, with a cumulative explanation of 53.86%, indicating that environmental factors significantly shape the microbial community structure in this study area [42].

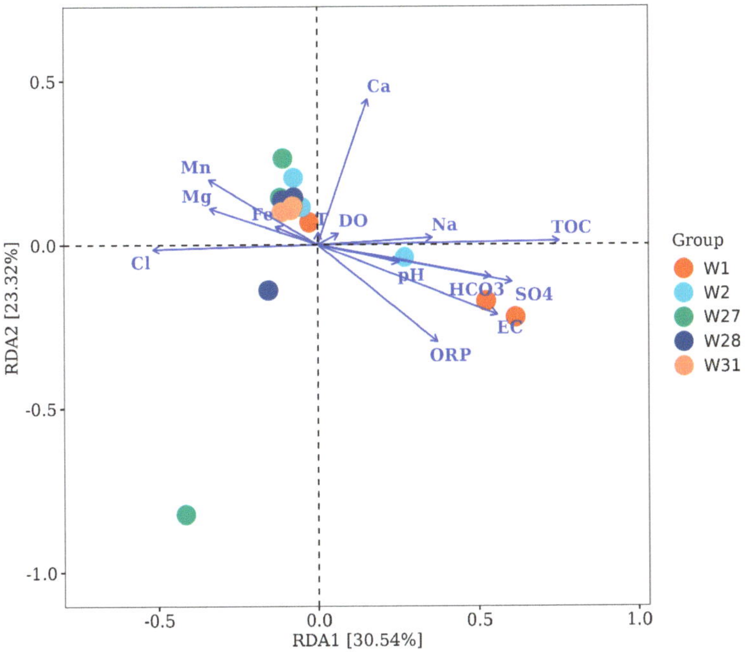

Figure 7. Redundancy analysis (RDA) of environmental factors influencing microbial community structure in groundwater samples.

The clustering patterns in the RDA plot reveal distinct environmental gradients associated with polluted (W27, W28, W31) and unpolluted zones (W1, W2). The polluted zones (W27, W28, and W31) are strongly associated with environmental variables such as HCO_3^-, SO_4^{2-}, EC, and pH, reflecting their critical roles in driving the microbial community structure under contamination stress [43]. Specifically, the high correlation of W27 with sulfate indicates the presence of sulfate-reducing conditions, which are favorable for anaerobic metabolic processes and degradation pathways dominated by sulfate-reducing bacteria [44]. The close alignment of W28 with HCO_3^- and EC suggests that carbonate buffering and ionic strength are key drivers of microbial dynamics in this zone, particularly influencing aerobic and anaerobic degradation processes [45]. In contrast, unpolluted zones (W1 and W2) are closely associated with variables such as ORP, TOC, and DO. These associations highlight the dominance of aerobic microbial communities in these zones, supported by higher oxidizing conditions and nutrient availability [46]. The clustering of W1 around ORP and DO indicates that oxygen availability plays a significant role in maintaining microbial diversity and activity, promoting aerobic degradation mechanisms [47]. The deep layer of W31 is characterized by strong associations with Ca, Mg, and Mn, suggesting that these ions influence the microbial community structure in subsurface environments. The association of these variables with W31 highlights the potential role of metal-reducing bacteria and their contribution to pollutant degradation in anoxic conditions. Interestingly, Cl^- appears to be a distinguishing factor for W2, reflecting its role in shaping microbial communities in the unpolluted zones. The strong correlation with ORP in W2 further suggests that oxidative conditions in this zone facilitate the dominance of aerobic metabolic pathways [48].

The FAPROTAX analysis results further supported these findings, as shown in Figure 8. FAPROTAX is a tool suitable for predicting the functional annotations of biogeochemical cycling processes, particularly for elements such as carbon, nitrogen, phosphorus, sulfur, and hydrogen, in environmental samples like those from marine, freshwater, or terrestrial ecosystems. The analysis provides a robust prediction of microbial community functions, highlighting the importance of biogeochemical cycles in the degradation processes observed in this study [49,50]. The microbial communities in the non-polluted areas (W1, W2) mainly performed carbon and nitrogen cycle-related functions such as aerobic_chemoheterotrophy and nitrogen_fixation [47]. Especially at W1_6m, the high abundance of Methylocystis and Gallionellaceae reflected the diversity of ecological functions under aerobic and hypoxic conditions at this site [51,52]. In contrast, functions within the contaminated zone (W27, W28, W31) were significantly focused on pollutant degradation-related pathways. The shallow (W27_6m) community of W27 was enriched in aromatic compound degradation, driven by Acidovorax and Aeromonas, while the mesocosm (W27_15m) exhibited enhanced sulfate respiration, which corresponded to Desulfobacter enrichment [44]. In addition, the community in the deeper W31 layer (W31_40m) exhibited photoheterotrophy and sulfide metabolism, suggesting adaptive changes in the microbial community under pollution stress [39].

Overall, the microbial communities in the non-polluted and polluted areas showed significant differences in diversity, functional distribution, and environmental response. The non-polluted zones (W1, W2) exhibited high diversity and ecological health, whereas the polluted zones (W27, W28, W31) showed significant selective pressures, resulting in a decreased community diversity and functional concentration, on pollutant degradation. RDA analyses showed that HCO_3^-, SO_4^{2-}, and other environmental factors drove the distribution of degradation-associated flora in the polluted zone, while ORP and Cl^- dominated the distribution of aerobic functional flora in the non-polluted zone. Combined with the functional prediction, these results clearly reveal the complex regulatory mechanisms of environmental factors in pollutant degradation and microbial community function, which provide an important theoretical basis for the study of natural pollutant attenuation.

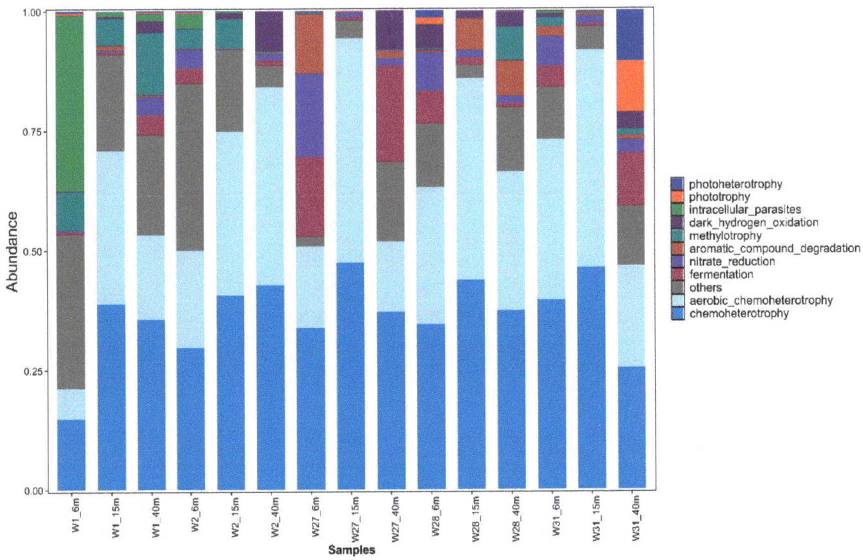

Figure 8. The community functions predicted by FAPROTAX for the groundwater samples.

3.3. Isotope Analysis Reveals Evidence of Natural Attenuation of Pollutants

Isotope analysis is an important tool for assessing the natural attenuation of pollutants, and the dynamic processes and mechanisms of pollutant degradation can be clarified by resolving hydrogen isotope (δD) and carbon isotope ($\delta^{13}C$) data (Table 3) at points in the contaminated area (W27, W28, W31) [53]. During microbial degradation, lighter isotopes are preferentially utilized, leading to the relative enrichment of heavier isotopes in residual contaminants [54]. Therefore, changes in isotope ratios serve as critical indicators of pollutant degradation. The fractionation effect of hydrogen isotopes reflects metabolic pathways such as water molecule cleavage, while carbon isotope variations are indicative of ring-opening and oxidative metabolic processes of aromatic compounds, and the combination of the two provides quantitative evidence of the degradation behaviors in the contaminated area [55]. De Vera et al. assessed the degradation of CBs and benzene in the groundwater of an industrial site located in the USA and found that a significant enrichment of ^{13}C occurred for CBs and benzene, with the greatest degree of enrichment in 1,2-DCB (~10‰). This may be due to the relatively large differences in hydrogeochemical conditions and microorganisms between the two sites [56].

Table 3. Isotopic data of groundwater samples.

Sample	δD	$d(^{13}C_{benzene}/^{12}C_{benzene})$	$d(^{13}C_{o\text{-toluidine}}/^{12}C_{o\text{-toluidine}})$
W27_6m	−45.4	−31.76	−30.00
W27_15m	−44.6	−29.30	−28.39
W27_40m	−46.5	−35.04	−28.40
W28_6m	−46.1	−32.64	−30.27
W28_15m	−46.3	−34.86	−28.45
W28_40m	−41.6	−30.70	−31.99
W31_6m	−47.4	−34.23	−33.13
W31_15m	−44.5	−32.51	−29.50
W31_40m	−47.0	−33.60	−33.32

The study data showed that δD values varied from −47.4‰ to −41.6‰ within the polluted area, while carbon isotopes ($\delta^{13}C$) varied from −35.04‰ to −29.30‰ and from

−33.32‰ to −28.39‰ between benzene and o-toluidine, respectively. These variations showed significant spatial heterogeneity between different points and depths.

At point W27, the δD and $\delta^{13}C$ data reflect significant differences between shallow, intermediate, and deep layers. The $\delta^{13}C$ value for benzene in the shallow layer (W27_6m) was −31.76‰, while o-toluidine had a $\delta^{13}C$ value of −30.00‰, and the relatively low enrichment effect suggests that the degradation process is more active in the shallow layer [54]. However, in the middle layer (W27_15m), the enrichment effect of carbon isotopes was weakened (benzene: −29.30‰; o-toluidine: −28.39‰), while the deeper layer (W27_40m) showed the most significant carbon isotope fractionation (benzene: −35.04‰; o-toluidine: −28.40‰). Combined with the hydrogen isotope data (δD values varying from −45.4‰ to −46.5‰), it can be surmised that the degradation of pollutants in the deeper environment of W27 is somewhat limited, possibly due to the decrease in microbial activity under anaerobic conditions or the inhibitory effect of high pollution concentrations on the degrading bacteria [57]. The data from point W28, on the other hand, show different characteristics, especially the δD value of −41.6‰ in the deep layer (W28_40m), which was the most enriched point in the polluted area, and the carbon isotope variations (benzene: −30.70‰; o-toluidine: −31.99‰) further indicated that the pollutant degradation process was active at this point. This phenomenon may be related to the water chemistry conditions in the deeper layer of W28, e.g., the higher HCO_3^- concentration provided a rich carbon source for microorganisms, and the appropriate redox potential promoted the metabolism of aromatic compounds [58]. In contrast, although the shallow (W28_6m) and middle (W28_15m) layers showed some carbon isotope fractionation effects, the enrichment degree (shallow benzene: −32.64‰; middle benzene: −34.86‰) was slightly lower than that of the deeper layer, suggesting that pollutant degradation in the shallow and middle layers was somewhat limited. The isotope data at point W31 further reveal the complexity of pollutant degradation. The δD value of the shallow layer (W31_6m) was −47.4‰, while the $\delta^{13}C$ value of benzene was −34.23‰ and that of o-toluidine was −33.13‰, which showed a more significant fractionation effect, suggesting that the pollutants underwent an active degradation process in the shallow layer. However, in the deeper layer (W31_40m), although the δD values showed some enrichment (−47.0‰), the magnitude of carbon isotope variations was smaller (benzene: −33.60‰; o-toluidine: −33.32‰), suggesting that the deeper degradation activities were relatively weak. This may be related to the lower redox conditions in the deeper layers and the reduced diversity of microbial communities. Taken together, the δD and $\delta^{13}C$ data within the contaminated area show significant spatial and depth heterogeneity. Isotopic enrichment effects were most pronounced in the deeper layers of W28, suggesting that the contaminant degradation process was most active in this area, which may be dominated by highly efficient functional bacterial communities (Acidovorax and Pseudomonas). In contrast, the degradation activities in the deeper layers of W27 and W31 were limited, mainly by environmental conditions. The combined analysis of hydrogen and carbon isotopes reveals the degradation pathways and dynamic processes at different locations within the contaminated area, which provides a reliable basis for the quantitative assessment of the natural attenuation of pollutants and scientific support for the optimization of the remediation strategy of the contaminated site.

4. Conclusions

This study combined high-throughput sequencing, physicochemical analysis, and stable isotope technology to explore the natural attenuation process and driving mechanism of benzene and o-toluidine in groundwater in typical chemical-contaminated sites in southeastern China. The results showed that the degradation process of pollutants in the contaminated areas (W27, W28, and W31) was significantly affected by environmental factors. HCO_3^- and SO_4^{2-} were the main driving factors affecting the distribution and degradation function of microbial communities, while ORP and Cl^- dominated the microbial community structure in the uncontaminated areas (W1 and W2). Stable isotope analysis further confirmed the heterogeneity of pollutant degradation. The deep layer (W28_40m) showed the strongest δD and $\delta^{13}C$ enrichment effects, indicating that there was the most active degradation process in this layer; meanwhile, the degradation process in the deep layers of W27 and W31 was limited, mainly due to anaerobic conditions and reduced microbial diversity. These findings clarify the degradation dynamics and driving mechanisms of different contaminated areas and provide a theoretical basis for the quantitative evaluation of natural attenuation, as well as practical guidance for the optimization of remediation strategies for contaminated sites.

In the process of contaminated site remediation, natural attenuation should be used as a low-intervention and cost-effective remediation technology, especially in sites with complex geological structures and slow pollutant decay rates. In particular, in sites with horizontal and vertical barriers, the attenuation efficiency should be regularly evaluated in combination with environmental factor monitoring and stable isotope analysis to ensure the scientificity and feasibility of the remediation strategy. In addition, this study emphasizes that in the remediation of contaminated sites, the distribution of microbial communities and environmental factors are closely related, and understanding the changing trends of these factors is crucial for predicting the pollutant decay process. Future research should continue to explore how to further improve the efficiency of natural attenuation through the adjustment of microbial community structure and the optimization of environmental factors, and provide empirical support for the management of similar contaminated sites.

Author Contributions: Conceptualization, J.S. and Y.Z.; methodology, R.Y. and Y.Z.; software, Y.Z. and M.C.; validation, X.G., J.S. and Y.Z; formal analysis, J.S.; investigation, Y.Z. and Y.L.; resources, J.S.; data curation, Y.Z. and S.F.; writing—original draft preparation, Y.Z.; writing—review and editing, X.G.; visualization, J.S.; supervision, X.G.; project administration, J.S.; funding acquisition, Y.Z. All authors have read and agreed to the published version of the manuscript.

Funding: This research was funded by the Special Program for National Science and Technology Major project of China (No. SQ2024AAA060115), Natural Science Foundation of Nantong (No.20241160939), and Key Laboratory of Water Pollution Control of Nantong.

Data Availability Statement: All relevant data are contained within this paper.

Conflicts of Interest: We declare that we do not have any commercial or associative interests that represent conflicts of interest in connection with the work submitted.

References

1. Karunanidhi, D.; Subramani, T.; Roy, P.D.; Li, H. Impact of groundwater contamination on human health. *Environ. Geochem. Health* **2021**, *43*, 643–647. [CrossRef] [PubMed]
2. Li, P.; Karunanidhi, D.; Subramani, T.; Srinivasamoorthy, K. Sources and Consequences of Groundwater Contamination. *Arch. Environ. Contam. Toxicol.* **2021**, *80*, 1–10. [CrossRef] [PubMed]
3. Saeedi, M.; Malekmohammadi, B.; Tajalli, S. Interaction of benzene, toluene, ethylbenzene, and xylene with human's body: Insights into characteristics, sources and health risks. *J. Hazard. Mater. Adv.* **2024**, *16*, 100459. [CrossRef]
4. Cordiano, R.; Papa, V.; Cicero, N.; Spatari, G.; Allegra, A.; Gangemi, S. Effects of benzene: Hematological and hypersensitivity manifestations in resident living in oil refinery areas. *Toxics* **2022**, *10*, 678. [CrossRef] [PubMed]
5. Chandnani, G.; Gandhi, P.; Kanpariya, D.; Parikh, D.; Shah, M. A comprehensive analysis of contaminated groundwater: Special emphasis on nature-ecosystem and socio-economic impacts. *Groundw. Sustain. Dev.* **2022**, *19*, 100813. [CrossRef]
6. Singh, P.K.; Kumar, U.; Kumar, I.; Dwivedi, A.; Singh, P.; Mishra, S.; Seth, C.S.; Sharma, R.K. Critical review on toxic contaminants in surface water ecosystem: Sources, monitoring, and its impact on human health. *Environ. Sci. Pollut. Res.* **2024**, *31*, 56428–56462. [CrossRef] [PubMed]

7. Ravindiran, G.; Rajamanickam, S.; Sivarethinamohan, S.; Karupaiya Sathaiah, B.; Ravindran, G.; Muniasamy, S.K.; Hayder, G. A Review of the Status, Effects, Prevention, and Remediation of Groundwater Contamination for Sustainable Environment. *Water* **2023**, *15*, 3662. [CrossRef]
8. Nguyen, V.-H.; Smith, S.M.; Wantala, K.; Kajitvichyanukul, P. Photocatalytic remediation of persistent organic pollutants (POPs): A review. *Arab. J. Chem.* **2020**, *13*, 8309–8337. [CrossRef]
9. Zhang, B.; Zhang, H.; He, J. Microbial Community Responses and Bioremediation. In *Vanadium in Soils and Plants*; CRC Press: Boca Raton, FL, USA, 2022; pp. 223–267.
10. Feng, H.; Yang, W.; Zhang, Y.; Ding, Y.; Chen, L.; Kang, Y.; Huang, H.; Chen, R. Electroactive microorganism-assisted remediation of groundwater contamination: Advances and challenges. *Bioresour. Technol.* **2023**, *377*, 128916. [CrossRef] [PubMed]
11. Barrientos-Moreno, L.; Espinosa-Urgel, M. Biofilm stress responses associated to aromatic hydrocarbons. Cellular ecophysiology of microbe: Hydrocarbon and lipid interactions. In *Handbook of Hydrocarbon and Lipid Microbiology*; Springer: Cham, Switzerland, 2018; pp. 105–115.
12. Liu, Y.; Ren, Y.; Su, J.; Liu, S.; Xu, L.; Wang, Y. Simultaneous remediation of ciprofloxacin and nitrate threats in groundwater through manganese redox cycling: Performance optimization, mechanisms, and potential for community assembly. *J. Water Process Eng.* **2024**, *64*, 105629. [CrossRef]
13. Chen, D.; Wang, G.; Chen, C.; Feng, Z.; Jiang, Y.; Yu, H.; Li, M.; Chao, Y.; Tang, Y.; Wang, S.; et al. The interplay between microalgae and toxic metal(loid)s: Mechanisms and implications in AMD phycoremediation coupled with Fe/Mn mineralization. *J. Hazard. Mater.* **2023**, *454*, 131498. [CrossRef] [PubMed]
14. Zhang, J.; Chen, L.; Hou, X.; Li, J.; Ren, X.; Lin, M.; Zhang, M.; Wang, Y.; Tian, Y. Effects of multi-factors on the spatiotemporal variations of deep confined groundwater in coal mining regions, North China. *Sci. Total Environ.* **2022**, *823*, 153741. [CrossRef]
15. Yang, X.; Jia, C.; Yang, F.; Yang, H.; Yao, Y. Spatio-temporal variation of groundwater pollution in urban wetlands and management strategies for zoning. *J. Environ. Manag.* **2023**, *342*, 118318. [CrossRef]
16. Zhan, Y.; Guo, Z.; Ruzzante, S.; Gleeson, T.; Andrews, C.B.; Babovic, V.; Zheng, C. Assessment of spatiotemporal risks for nationwide groundwater nitrate contamination. *Sci. Total Environ.* **2024**, *947*, 174508. [CrossRef] [PubMed]
17. Zhang, Q.; Wang, L.; Wang, H.; Zhu, X.; Wang, L. Spatio-temporal Variation of Groundwater Quality and Source Apportionment using Multivariate Statistical Techniques for the Hutuo River Alluvial-Pluvial Fan, China. *Int. J. Environ. Res. Public Health* **2020**, *17*, 1055. [CrossRef]
18. Yang, J.; Ye, M.; Tang, Z.; Jiao, T.; Song, X.; Pei, Y.; Liu, H. Using cluster analysis for understanding spatial and temporal patterns and controlling factors of groundwater geochemistry in a regional aquifer. *J. Hydrol.* **2020**, *583*, 124594. [CrossRef]
19. Guo, X.; Zuo, R.; Meng, L.; Wang, J.; Teng, Y.; Liu, X.; Chen, M. Seasonal and Spatial Variability of Anthropogenic and Natural Factors Influencing Groundwater Quality Based on Source Apportionment. *Int. J. Environ. Res. Public Health* **2018**, *15*, 279. [CrossRef]
20. Mohapatra, B.; Saha, A.; Chowdhury, A.N.; Kar, A.; Kazy, S.K.; Sar, P. Geochemical, metagenomic, and physiological characterization of the multifaceted interaction between microbiome of an arsenic contaminated groundwater and aquifer sediment. *J. Hazard. Mater.* **2021**, *412*, 125099. [CrossRef] [PubMed]
21. Preziosi, E.; Frollini, E.; Zoppini, A.; Ghergo, S.; Melita, M.; Parrone, D.; Rossi, D.; Amalfitano, S. Disentangling natural and anthropogenic impacts on groundwater by hydrogeochemical, isotopic and microbiological data: Hints from a municipal solid waste landfill. *Waste Manag.* **2019**, *84*, 245–255. [CrossRef]
22. Guo, L.; Xie, Q.; Sheng, Y.; Wang, G.; Jiang, W.; Tong, X.; Xu, Q.; Hao, C. Co-variation of hydrochemistry, inorganic nitrogen, and microbial community composition along groundwater flowpath: A case study in Linzhou-Anyang area, Southern North China plain. *Appl. Geochem.* **2022**, *140*, 105296. [CrossRef]
23. Xiao, H.; Mao, C.; Wang, S.; Jia, Z.; Rao, W. Seasonal variation and provenance of organic matter in the surface sediments of the three gorges reservoir: Stable isotope analysis and implications for agricultural management. *Sci. Total Environ.* **2023**, *870*, 161886. [CrossRef]
24. Kim, K.-H.; Kim, H.-R.; Oh, J.; Choi, J.; Park, S.; Yun, S.-T. Predicting leachate impact on groundwater using electrical conductivity and oxidation–reduction potential measurements: An empirical and theoretical approach. *J. Hazard. Mater.* **2024**, *474*, 134733. [CrossRef]
25. Zhou, Q.; Zhang, J.; Xing, K.; Wei, J.; Yao, Y. Groundwater nitrate contamination in China: Spatial distribution, temporal trend, and driver analysis. *Environ. Res.* **2024**, *262*, 119932. [CrossRef] [PubMed]
26. Farzaneh, G.; Khorasani, N.; Ghodousi, J.; Panahi, M. Application of geostatistical models to identify spatial distribution of groundwater quality parameters. *Environ. Sci. Pollut. Res. Int.* **2022**, *29*, 36512–36532. [CrossRef]
27. Jia, Z.; Bian, J.; Wang, Y. Impacts of urban land use on the spatial distribution of groundwater pollution, Harbin City, Northeast China. *J. Contam. Hydrol.* **2018**, *215*, 29–38. [CrossRef] [PubMed]

28. Hemme, C.L.; Deng, Y.; Gentry, T.J.; Fields, M.W.; Wu, L.; Barua, S.; Barry, K.; Tringe, S.G.; Watson, D.B.; He, Z.; et al. Metagenomic insights into evolution of a heavy metal-contaminated groundwater microbial community. *ISME J.* **2010**, *4*, 660–672. [CrossRef] [PubMed]
29. Wang, Y.; Sheng, H.-F.; He, Y.; Wu, J.-Y.; Jiang, Y.-X.; Tam, N.F.-Y.; Zhou, H.-W. Comparison of the levels of bacterial diversity in freshwater, intertidal wetland, and marine sediments by using millions of illumina tags. *Appl. Environ. Microbiol.* **2012**, *78*, 8264–8271. [CrossRef] [PubMed]
30. Tang, X.; Xie, G.; Shao, K.; Hu, Y.; Cai, J.; Bai, C.; Gong, Y.; Gao, G. Contrast diversity patterns and processes of microbial community assembly in a river-lake continuum across a catchment scale in northwestern China. *Environ. Microbiome* **2020**, *15*, 10.
31. Yan, C.; Wang, F.; Geng, H.; Liu, H.; Pu, S.; Tian, Z.; Chen, H.; Zhou, B.; Yuan, R.; Yao, J. Integrating high-throughput sequencing and metagenome analysis to reveal the characteristic and resistance mechanism of microbial community in metal contaminated sediments. *Sci. Total Environ.* **2020**, *707*, 136116. [CrossRef]
32. Li, S.; Chen, J.; Ma, J.; Lu, Y. Functional Microbial Communities Involved in As(III) Oxidation Coupled with Nitrate Reduction in a Paddy Soil. *Water Air Soil Pollut.* **2023**, *234*, 598. [CrossRef]
33. Banerjee, S.; Bedics, A.; Harkai, P.; Kriszt, B.; Alpula, N.; Tancsics, A. Evaluating the aerobic xylene-degrading potential of the intrinsic microbial community of a legacy BTEX-contaminated aquifer by enrichment culturing coupled with multi-omics analysis: Uncovering the role of Hydrogenophaga strains in xylene degradation. *Environ. Sci. Pollut. Res. Int.* **2022**, *29*, 28431–28445. [CrossRef]
34. Szentgyorgyi, F.; Benedek, T.; Fekete, D.; Tancsics, A.; Harkai, P.; Kriszt, B. Development of a bacterial consortium from *Variovorax paradoxus* and *Pseudomonas veronii* isolates applicable in the removal of BTEX. *AMB Express* **2022**, *12*, 4. [CrossRef]
35. Elliott, A. Microbial Communities Associated with Hydraulic Fracturing Fluids from Shale Gas Fields in Western Canada. Master's Thesis, University of Calgary, Calgary, AB, Canada, 2015.
36. Hussain, B.; Chen, J.-S.; Huang, S.-W.; Tsai, I.S.; Rathod, J.; Hsu, B.-M. Underpinning the ecological response of mixed chlorinated volatile organic compounds (CVOCs) associated with contaminated and bioremediated groundwaters: A potential nexus of microbial community structure and function for strategizing efficient bioremediation. *Environ. Pollut.* **2023**, *334*, 122215. [CrossRef] [PubMed]
37. Ni, S.; Teng, Y.; Zhang, G.; Xia, W.; Shu, Y.; Ren, W. Exploring bacterial community assembly in vadose and saturated zone soil for tailored bioremediation of a long-term hydrocarbon-contaminated site. *J. Environ. Manag.* **2024**, *360*, 121114. [CrossRef] [PubMed]
38. Reboleira, A.S.; Bodawatta, K.H.; Ravn, N.M.R.; Lauritzen, S.-E.; Skoglund, R.Ø.; Poulsen, M.; Michelsen, A.; Jønsson, K.A. Nutrient-limited subarctic caves harbour more diverse and complex bacterial communities than their surface soil. *Environ. Microbiome* **2022**, *17*, 41. [CrossRef] [PubMed]
39. Jiao, S.; Liu, Z.; Lin, Y.; Yang, J.; Chen, W.; Wei, G. Bacterial communities in oil contaminated soils: Biogeography and co-occurrence patterns. *Soil Biol. Biochem.* **2016**, *98*, 64–73. [CrossRef]
40. Mejia, M.P.; Rojas, C.A.; Curd, E.; Renshaw, M.A.; Edalati, K.; Shih, B.; Vincent, N.; Lin, M.; Nguyen, P.H.; Wayne, R.; et al. Soil Microbial Community Composition and Tolerance to Contaminants in an Urban Brownfield Site. *Microb. Ecol.* **2023**, *85*, 998–1012. [CrossRef]
41. Costeira, R.; Doherty, R.; Allen, C.C.R.; Larkin, M.J.; Kulakov, L.A. Analysis of viral and bacterial communities in groundwater associated with contaminated land. *Sci. Total Environ.* **2019**, *656*, 1413–1426. [CrossRef]
42. Wang, H.; Li, P.; Liu, X.; Wang, Y.; Stein, L.Y. Groundwater flow regime shapes nitrogen functional traits by affecting microbial community assembly processes in the subsurface. *Sci. Total Environ.* **2024**, *949*, 175083. [CrossRef]
43. Fillinger, L.; Hug, K.; Griebler, C. Selection imposed by local environmental conditions drives differences in microbial community composition across geographically distinct groundwater aquifers. *FEMS Microbiol. Ecol.* **2019**, *95*, fiz160. [CrossRef]
44. Zhi, C.; Hu, X.; Yang, F.; Huang, X.; Chen, H.; Chen, L.; Chen, G.; Wu, Z.; Wang, S. Unraveling microbial community variation along a salinity gradient and indicative significance to groundwater salinization in the coastal aquifer. *J. Hydrol.* **2024**, *642*, 131893. [CrossRef]
45. Sha, C.; Wu, J.; Shen, C.; Wu, J.; Yan, Z.; Wang, M. The ecology of bacterial communities in groundwater of industrial areas: Diversity, composition, network, and assembly. *Environ. Pollut.* **2023**, *322*, 121207. [CrossRef]
46. Liu, S.; Chen, Q.; Li, J.; Li, Y.; Zhong, S.; Hu, J.; Cai, H.; Sun, W.; Ni, J. Different spatiotemporal dynamics, ecological drivers and assembly processes of bacterial, archaeal and fungal communities in brackish-saline groundwater. *Water Res.* **2022**, *214*, 118193. [CrossRef] [PubMed]
47. Ben Maamar, S.; Aquilina, L.; Quaiser, A.; Pauwels, H.; Michon-Coudouel, S.; Vergnaud-Ayraud, V.; Labasque, T.; Roques, C.; Abbott, B.W.; Dufresne, A. Groundwater isolation governs chemistry and microbial community structure along hydrologic flowpaths. *Front. Microbiol.* **2015**, *6*, 1457. [CrossRef]
48. Sheng, Y.; Li, G.; Dong, H.; Liu, Y.; Ma, L.; Yang, M.; Liu, Y.; Liu, J.; Deng, S.; Zhang, D. Distinct assembly processes shape bacterial communities along unsaturated, groundwater fluctuated, and saturated zones. *Sci. Total Environ.* **2021**, *761*, 143303. [CrossRef]

49. Ma, J.; Liu, H.; Chen, H.; Xiong, H.; Tong, L.; Guo, G. Is redox zonation an appropriate method for determining the stage of natural remediation in deep contaminated groundwater? *Sci. Total Environ.* **2024**, *928*, 172224. [CrossRef] [PubMed]
50. Yang, Z.; Peng, C.; Cao, H.; Song, J.; Gong, B.; Li, L.; Wang, L.; He, Y.; Liang, M.; Lin, J.; et al. Microbial functional assemblages predicted by the FAPROTAX analysis are impacted by physicochemical properties, but C, N and S cycling genes are not in mangrove soil in the Beibu Gulf, China. *Ecol. Indic.* **2022**, *139*, 108887. [CrossRef]
51. Gürgan, M.; İrez, E.İ.; Adiloğlu, S. Understanding Bioremediation of Metals and Metalloids by Genomic Approaches. In *Omics Insights in Environmental Bioremediation*; Kumar, V., Thakur, I.S., Eds.; Springer Nature: Singapore, 2022; pp. 375–392. [CrossRef]
52. Morrissy, J.G.; Reichman, S.M.; Currell, M.J.; Surapaneni, A.; Megharaj, M.; Crosbie, N.D.; Hirth, D.; Aquilina, S.; Rajendram, W.; Ball, A.S. Differentiation between Impacted and Unimpacted Microbial Communities of a Nitrogen Contaminated Aquifer. *Environments* **2022**, *9*, 128. [CrossRef]
53. Blázquez-Pallí, N.; Shouakar-Stash, O.; Palau, J.; Trueba-Santiso, A.; Varias, J.; Bosch, M.; Soler, A.; Vicent, T.; Marco-Urrea, E.; Rosell, M. Use of dual element isotope analysis and microcosm studies to determine the origin and potential anaerobic biodegradation of dichloromethane in two multi-contaminated aquifers. *Sci. Total Environ.* **2019**, *696*, 134066. [CrossRef]
54. Feisthauer, S.; Seidel, M.; Bombach, P.; Traube, S.; Knöller, K.; Wange, M.; Fachmann, S.; Richnow, H.H. Characterization of the relationship between microbial degradation processes at a hydrocarbon contaminated site using isotopic methods. *J. Contam. Hydrol.* **2012**, *133*, 17–29. [CrossRef] [PubMed]
55. Thullner, M.; Centler, F.; Richnow, H.-H.; Fischer, A. Quantification of organic pollutant degradation in contaminated aquifers using compound specific stable isotope analysis—Review of recent developments. *Org. Geochem.* **2012**, *42*, 1440–1460. [CrossRef]
56. Wang, M.; Jiang, D.; Yang, L.; Wei, J.; Kong, L.; Xie, W.; Ding, D.; Fan, T.; Deng, S. Natural attenuation of BTEX and chlorobenzenes in a formerly contaminated pesticide site in China: Examining kinetics, mechanisms, and isotopes analysis. *Sci. Total Environ.* **2024**, *918*, 170506. [CrossRef]
57. Meckenstock, R.U.; Morasch, B.; Griebler, C.; Richnow, H.H. Stable isotope fractionation analysis as a tool to monitor biodegradation in contaminated acquifers. *J. Contam. Hydrol.* **2004**, *75*, 215–255. [CrossRef] [PubMed]
58. Richnow, H.H.; Annweiler, E.; Michaelis, W.; Meckenstock, R.U. Microbial in situ degradation of aromatic hydrocarbons in a contaminated aquifer monitored by carbon isotope fractionation. *J. Contam. Hydrol.* **2003**, *65*, 101–120. [CrossRef] [PubMed]

water

MDPI

Article

Spatial Distribution Characteristics and Genetic Mechanism of the Metasilicate-Rich Groundwater in Ji'nan Rock Mass Area, Shandong Province, China

Meng Xu [1], Caiping Hu [2,*], Lixin Zhu [3], Guangzeng Song [1], Wenquan Peng [1], Shijiao Yang [1] and Jinyu Song [1]

1 No. 1 Institute of Geology and Mineral Resources of Shandong Province, Shandong Provincial Bureau of Geology & Mineral Resources, Ji'nan 250100, China
2 Shandong Provincial Geo-Mineral Engineering Exploration Institute, Shandong Provincial Bureau of Geology & Mineral Resources, Ji'nan 250014, China
3 China Geological Survey, Beijing 100037, China
* Correspondence: caipinghu126@126.com

Abstract: Metasilicate-rich groundwater could meet the high demand of the international community for high-quality water. In order to comprehensively analyze the genetic mechanism of metasilicate-rich groundwater, and help human communities effectively exploit and utilize high-quality water resources, taking the Ji'nan rock mass area as an example, this study carried out systematic research on the spatial distribution and genetic mechanism of the metasilicate-rich groundwater Based on the regional hydrogeological conditions, the influencing factors on the spatial distribution characteristics of the metasilicate-rich groundwater in the study area were systematically sorted out by means of petrogeochemistry, hydrochemistry, and chemical weathering index analysis, and the accumulation mechanism of the metasilicate-rich groundwater was discussed from the perspective of water-rock interaction. The results show that: (1) On the northwest side and part of the northeast side of Ji'nan rock mass, the metasilicate content of the groundwater samples exceeded 25 mg/L; the metasilicate content on the south, west, and east sides were relatively low; (2) Ji'nan rock mass is mainly composed of gabbro easily weathered, with high SiO_2 content and high weathering degree showing obvious characteristics of desilication. FeS_2 developing along the contact zones between the rock mass and surrounding rocks was easily oxidized to form H_2SO_4, which enhanced the solubility of silicate minerals in the groundwater. Ji'nan rock mass was located in the low-lying position of the monocline structure, which presented better water conservation and recharges conditions. The above factors resulted in the metasilicate-rich groundwater accumulating in the area of Ji'nan rock mass and showed different spatial distribution characteristics.

Keywords: Ji'nan rock mass; metasilicate-rich groundwater; water-rock interaction; solution of silicate rock; human factor

1. Introduction

The global demand for water has grown approximately six-fold in the last century [1,2]. Continued population growth and climate change in the 21st century will have a severe impact on global water resources [3,4]. Water scarcity occurs in many countries, particularly in the Mediterranean, Middle Eastern, and African countries, and many of these countries are confronted with a crucial combination of a severe lack of an increasing demand for high-quality water [5]. Groundwater has significant advantages of spatial-temporal availability compared to surface water, and it is the most important water resource for more than half of the global population [6,7]. Consequently, high-quality groundwater has played and will continue to play an essential role in social, economic, agricultural and industrial development for the international community.

The metasilicate-rich groundwater is one important kind of high-quality groundwater, which has a significant effect on softening human blood vessels, and relieving arteriosclerosis and heart disease in medicine [8]. Because catering to the rapid demand for high-quality groundwater and being beneficial for human health, metasilicate-rich groundwater has become a hot topic in hydrogeology research.

The hydrochemical formation of metasilicate-rich groundwater is affected by multiple factors, such as geological structure, lithofacies composition, hydrodynamic conditions, and hydrochemical characteristics [9–11]. Due to its important economic and medical value, its genetic mechanism has attracted the attention of many scholars. However, most of the previous studies focused on the influence of extrusive rocks (especially basalt) on the formation of metasilicate, or focused on the discussion about the metasilicate from the perspective of hydrochemical evolution. The research on how the intrusive rocks, geochemical characteristics, and weathering of surrounding rocks affected the metasilicate-rich groundwater from the perspective of water-rock interaction was relatively few, which has limited the effective exploration and utilization of metasilicate-rich groundwater. In the Ji'nan rock mass area, parts of the groundwater are enriched with H_2SiO_3. Moreover, the intrusive rocks are widely distributed, with various mineral types, structural and weathered fractures, and metal sulfide deposits are widely distributed. These have determined the water-rock interaction processes of the H_2SiO_3 enrichment in groundwater.

Taking the Ji'nan rock mass area as an example, this study discusses the distribution characteristics and formation mechanism of the metasilicate-rich groundwater from the perspective of analyzing the coupling relationship between rock weathering and water chemical formation. It not only comprehensively considers the potential influencing factors of the metasilicate-rich groundwater, but also contributes to the effective exploitation and utilization of the metasilicate-rich groundwater, and helps meet the highing demand of the international community for high-quality water.

2. Geological Background

Located in the northern wing of the Taishan Dome, Ji'nan is a north-dipping monoclinic structure dominated by Paleozoic carbonate strata. The southern is mountainous, with exposed limestones distributing continuously, and acts as the recharge area of the groundwater. The central hills become slope, and the main lithology includes Cambrian limestone, shale, and middle and lower Ordovician limestone, and acts as the recharge-runoff area of the groundwater. The northern is the Yellow River alluvial plain, which is the buried limestone distribution area and the groundwater discharge area, where the Ji'nan Rock mass area (study area) is located (Figure 1). Ji'nan is located in the mid-latitude inland zone, and belongs to a temperate monsoon climate, with an average annual precipitation of 669.91 mm.

In the Ji'nan area, the Ordovician limestone forms the main aquifer group, and the groundwater flows from south to north. In the process of northward runoff, after encountering the barrier of Ji 'nan rock mass, part of the groundwater flows out of the surface along the faults and other weak zones to form famous Ji'nan spring groups, and part enters the deep circulation to form the geothermal field on the north of the Yellow River [12].

The study area is based on the Archean Taishan Group (Art), which is overlaid by Cambrian, Ordovician, Carboniferous, Permian, Cretaceous, Neogene and Quaternary strata from bottom to up, and the lithology mainly includes limestone, dolomite, and clastic rock. The faults in the area are mainly composed of three groups of NS, NNE, and nearly NW trending faults, such as the Sangzidian Fault, Qianfoshan Fault, etc. In the early Cretaceous, the Ji'nan rock mass was formed by the eruption of magmatic activities, mainly composed of gabbro and pyroxene monzonite [13]. At present, most of the Ji 'nan rock mass is covered by the Quaternary system, and it only appears sporadically in Yaoshan, Huashan, and Woniushan Mountains, etc. On the plane, it is roughly elliptical in the direction of EW, and the spatial form presents giant rock falx.

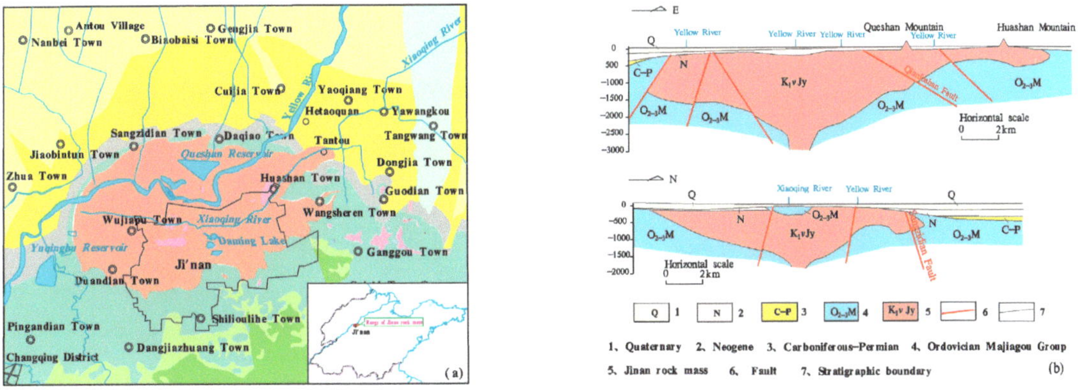

Figure 1. The geological background of the Ji'nan rock mass area. (**a**)—The location and hydrogeological conditions of Ji'nan rock mass area; (**b**)—Plane and longitudinal profile of Ji'nan rock mass.

3. Data Sources and Research Methods

Based on the collection of regional geological and hydrogeological data, 29 groundwater samples in the study area were collected successively in July 2021 from Ordovician limestone, which is a confined aqufier. The sampling points are distributed evenly in the study area (see Figure 2 for the sampling locations). The Technical Regulations on Groundwater Environmental Monitoring (2004) were strictly implemented for sampling and analysis. Each groundwater sample was collected by using two polyethylene bottles, which were cleaned, rinsed, and filtered before sampling. One bottle of the sample was added 65% nitric acid to make the pH < 2, so as to test the cationic concentration, and the other bottle of the sample was used to test anion concentration without adding nitric acid. The samples were sent to the Ji'nan Mineral Resources Supervision and Testing Center of the Ministry of Land and Resources for water quality analysis within 24 h after collection. K^+ and Na^+ were tested by inductively coupled plasma emission spectrometry, and both the detection limits were 0.01 mg/L. Ca^{2+} and Mg^{2+} were tested by ethylenediamine tetraacetate disodium titration, and the detection limits were 2 mg/L and 1 mg/L, respectively. Cl^- was tested with the silver nitrate volumetric method, the detection limit was 1 mg/L. $SO_4{}^{2-}$ was tested with the barium sulfate turbidimetric method, the detection limit was 3 mg/L. $HCO_3{}^-$ was tested by acid-base titration with a detection limit of 5 mg/L. The soluble SiO_2 was tested by a UV spectrophotometer with a detection limit of 0.1 mg/L. The error analysis of the tested results was carried out according to the principle of ion balance, which was calculated to be less than 5%.

Rock samples of different lithologies in the study area were collected and thinly sliced. According to the micropetrographic observation, 11 representative samples without alteration were selected and broken under pollution-free conditions for testing the major elements, including SiO_2, Al_2O_3, Fe_2O_3, MgO, CaO, Na_2O, K_2O, and P_2O_5. Using an X-ray fluorescence (XRF) spectrometer, the test was carried out at the Shandong Testing Center of China Metallurgical Geology Bureau. First, 0.7 g of the sample was weighed, then boric acid was added to melt the sample into a glass sheet at a high temperature, finally, the oxide content was determined on XRF. The error of the total oxide content is 1–3%.

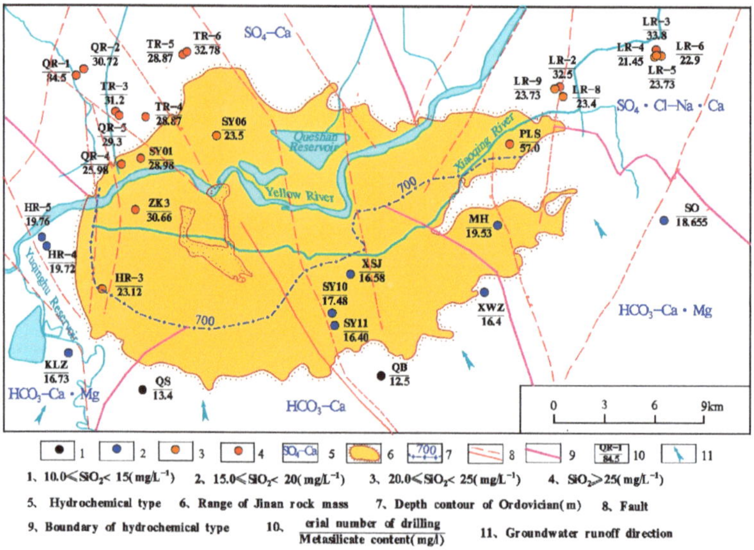

Figure 2. The metasilicate content, hydrochemical types, and distribution of the groundwater in the study area (The content of H_2SiO_3 is the concentration unit mg·L^{-1}).

4. Results and Discussion

4.1. Distribution of the Metasilicate-Rich Groundwater

Full analysis tests were conducted on the groundwater samples from 29 wells in the study area. The results showed that the content of H_2SiO_3 in 12 water samples exceeds 25 mg/L, accounting for 41.38% of the total samples. There were six samples with more than 20~25 mg/L of H_2SiO_3, accounting for 20.69%. There were 11 samples less than 20 mg/L, accounting for 37.93% (Table 1). The samples with more than 25 mg/L of H_2SiO_3 are mainly concentrated in the northwest and northeast sides of the Ji'nan rock mass, among which, nine samples are distributed on the northwest side and three samples are on the northeast side. The content of H_2SiO_3 is low in the south, west, and east sides of the rock mass, indicating that the spatial distribution of the metasilicate-rich groundwater presents regional characteristics. The content of H_2SiO_3 increased gradually from south to north, which was consistent with the groundwater runoff direction (Figure 2).

4.2. Discussion on the Origin of H_2SiO_3 Enrichment

Previous studies have shown that, only when the silicate minerals in surrounding rocks present wide distribution and strong solubility, the surrounding rocks become weathered strongly, the water-rock interaction is intense, the groundwater conservation and recharge conditions are well, then the H_2SiO_3 can be accumulated, and enriched in the groundwater [8,14]. The four above factors about the study area will be discussed in detail in the following paper to clarify the genetic mechanism of the metasilicate-rich groundwater.

4.2.1. Geochemical Conditions of Surrounding Rocks

Due to the difference in the content and the weathering sensitivity of the silicate minerals in the surrounding rocks, there are great differences in the supply degree and rate of soluble SiO_2 to the groundwater from the surrounding rocks with different geochemical characteristics [15]. The surrounding rocks of the groundwater in the study area mainly include gabbro, diorite, and carbonate rocks. In this study, 11 rock samples were collected for testing the major elements (Figure 3). The results shew that the gabbro contained large SiO_2 content, accounting for 50.1~58.5%, with an average of 52.9%. The main minerals were plagioclase and pyroxene which were easily weathered, among which the plagioclase accounted for 42~70%, and the pyroxene accounted for 23~55%. In the diorite samples, the content of SiO_2 was 50.14%, and the main minerals were plagioclase and amphibole, among which the plagioclase, the amphibole and the pyroxene respectively accounted for 63–65%, 33–35%, and 2–3%. The carbonate rocks mainly included dolomite and limestone, and the main mineral component was dolomite or calcite, accounting for about 90%, whereas, the content of SiO_2 accounted for about 0~22.4%, with an average of 15.4%. The content of SiO_2 in carbonate rocks was low, and most of them were free quartz, which was less soluble in water, so they cannot act as the main material sources of the H_2SiO_3 in the groundwater in the study area [15,16]. Therefore, the gabbro and diorite are the main lithologies of Ji'nan rock mass, containing a large number of easily weathered silicate minerals such as plagioclase, pyroxene, and amphibole, which constitute favorable potential sources of H_2SiO_3 in groundwater.

Table 1. The H_2SiO_3 content of the groundwater samples was collected from different sides of the Ji'nan rock mass (See Figure 2 for the locations of the boreholes).

Position	Borehole Number	Content of H_2SiO_3	Borehole Number	Content of H_2SiO_3
West side	HR-5	19.76	HR-3	23.12
	HR-4	19.72		
East side	SO	18.656	MH	19.53
Northeast side	LR-2	32.5	LR-6	22.9
	LR-3	33.8	LR-8	23.4
	LR-4	21.45	PLS	57
	LR-5	23.73		
South side	QS	13.4	SY-11	16.4
	QB	12.5	XSJ	16.58
	XWZ	16.4	KLZ	16.73
	SY-10	17.48		
Northwest side	QR-2	30.72	QR-5	29.3
	TR-5	28.87	SY-1	28.98
	TR-6	32.78	SY-6	23.5
	TR-3	31.2	ZK3	30.66
	TR-4	28.87	QR-4	25.98

Note: The content of H_2SiO_3 is the concentration unit $mg \cdot L^{-1}$.

In the diagram of the water-rock interactions and the distribution of the sample points in the study area (Figure 4), almost all the sample points with more than 25 mg/L content of H_2SiO_3 are located inside or near the end-member of silicate rock, and a few sample points are located in the secondary end-member of amphibole, which proves that Ji'nan rock mass is indeed the main source of the H_2SiO_3 in groundwater. The samples with less than 25 mg/L content of H_2SiO_3 are located in the secondary end-member of amphibole or near the silicate end-member, indicating that the geochemical characteristics of the groundwater in the study area are mainly affected by silicate rocks, but less affected by carbonate rocks or evaporites.

4.2.2. Weathering Degree and Mechanism of Surrounding Rocks

Weathering and desilication of the surrounding rocks act as the basic source of H_2SiO_3 in groundwater, and the weathering degree could affect the reaction equilibrium process of the water-rock interaction [11,17]. As the source of the H_2SiO_3 in groundwater, the Ji'nan rock mass developed abundant structural fractures and weathered fractures (Figure 5a–c), and the main lithologies included easily weathered gabbro and diorite. The above two aspects comprehensively caused the surrounding rocks of the groundwater to be easily weathered, which can be proved by the spherical weathering phenomenon only formed by strong weathering (Figure 5d).

Figure 3. Microscopical features of the rock samples in the study area (**a–e**) are gabbros, (**f**) is diorite, and (**g–i**) are carbonate rocks). (**a**)—Plagioclase: 48–50%; Pyroxene: 43–45%, cleavage developed; Biotite: 3–5%; Opaque minerals: 1–2%; (**b**)—Plagioclase: 43–45%; Pyroxene: 53–55%; Mica: 1–2%; Opaque minerals: 1–2%; (**c**)—Plagioclase: 50–53%; Pyroxene: 43–45%; Mica: 2–3%; Opaque minerals: 1–2%; (**d**)—Plagioclase: 58–60%; Pyroxene: 38–40%; Biotite: 2–3%; Opaque minerals: 1–2%; (**e**)—Plagioclase: 50–53%; Pyroxene: 43–45%; Biotite: 2–3%; Opaque minerals: 1–2%; (**f**)—Plagioclase: 63–65%; Hornblende: 33–35%; Pyroxene: 2–3%; Opaque minerals: 1–2%; (**g**)—Dolomite: 95%±; Calcite: 3–5%; Quartz: 1–2%; Opaque minerals: less; (**h**)—Calcite: 90%±; Dolomite: 3–5%; Quartz: 3–5%; Tremolite: 2–3%; Diopside: little; Opaque minerals: 1%; (**i**)—Calcite: 78–80%; Dolomite: 3–5%; Quartz: 2–3%; Goldmica: 13–15%; Opaque minerals: less. Pl—plagioclase; Px—pyroxene; Hbl—amphibole; Cal—calcite; Dol—Dolomite.

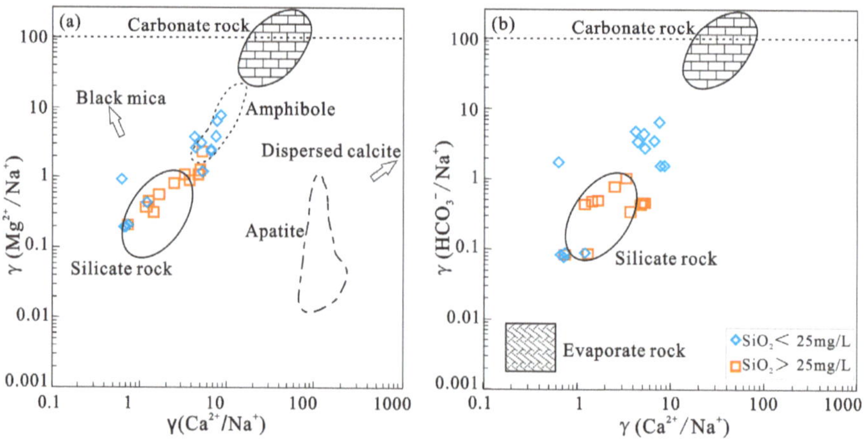

Figure 4. The relationship of Ca^{2+}/Na^{+} and Mg^{2+}/Na^{+} (**a**), Ca^{2+}/Na^{+} and HCO_3^{-}/Na^{+} (**b**) in the study area.

Figure 5. Fracture development and spherical weathering of the Ji'nan rock mass. (**a**)—X type joint and weathered fractures; (**b**)—Structural fracture zone; (**c**)—X type joint and weathered fissures; (**d**)—Spherical weathering.

The weathering degree of rocks is often characterized by the chemical weathering index, that is, the ratio of different specific major element oxides is used to indicate the weathering degree of the surrounding rocks [18–20]. Based on the characteristics of main lithology and major elements of the rock mass, and referring to the methods adopted by predecessors in studying the weathering degree of magmatic rocks [21–23], this study selected index of lateralization (IOL) and chemical alteration index (CIA) to quantitatively characterize the weathering mechanism and degree of Ji'nan rock mass. The two weathering indexes present high sensitivity to rock weathering, and the calculation formula is as follows respectively:

$$\text{IOL}\ [(Al_2O_3 + Fe_2O_3)/(SiO_2 + Al_2O_3 + Fe_2O_3)] \times 100 \quad (1)$$

$$\text{CIA}\ [Al_2O_3/(Al_2O_3 + CaO^* + Na_2O + K_2O)] \times 100 \quad (2)$$

note: The IOL index was calculated using the mass fraction of oxide; while the CIA index was calculated using the molecular mole number of oxide; CaO^* refers to the molar content of the silicate minerals, excluding the CaO content of carbonate minerals. Since the molar ratio of CaO and Na_2O in silicate is usually 1:1, when the molar number of CaO is greater than the molar number of Na_2O, the molecular mole of CaO^* is equal to that of Na_2O to carry out the calculation, and when the molar number of CaO^* is less than that of Na_2O, the molar number of CaO^* is equal to that of CaO when used to calculate.

The IOL index indicates the process that Al_2O_3 and Fe_2O_3 replace SiO_2 in silicate minerals, and the larger the value is, the lower the weathering degree will be. CIA index is the mole percentage ratio between the oxides of alkali metal elements K, Ca, and Na, which are easy to migrate, and the oxides of Al, which are difficult to migrate, during the process of rock weathering. CIA is used to characterize the chemical weathering intensity of the rocks, and the larger the index value is, the stronger the weathering effect will be [24].

By using the above two calculation methods, combined with the tested data of the major elements of Ji'nan rock mass (Table 2), 8 groups of gabbro and diorite samples were randomly selected to calculate the chemical weathering index of Ji'nan rock mass (Table 3).

The calculated IOL values range from 26.46 to 34.56, with an average value of 31.81. As can be seen from the S-A-F diagram in Figure 6, the weathering is at the stage of kaolinization, which shows obvious characteristics of desilication, iron-rich and aluminum-rich, indicating a high degree of weathering. CIA values range from 48.68 to 57.26, with an average value of 53.18. The rock samples are all located near A-CN in the A-CN-K plot, and the weathering process is at the stage of kaolinization, which is consistent with the conclusion presented in the S-A-F diagram.

Table 2. Major elements of surrounding rocks.

No.	Lithology	K2O	Al2O3	CaO	TFe2O3	MgO	SiO2	TiO2	L.O.I	FeO
1	Gabbro	1.13	16.34	8.51	4.71	2.66	58.50	0.99	0.69	2.80
2	Gabbro	1.45	17.21	7.54	8.30	4.74	55.86	0.75	0.31	4.57
3	Gabbro	1.27	16.21	9.07	10.06	6.33	51.81	0.84	0.65	6.43
4	Gabbro	0.76	14.70	10.22	10.26	9.75	51.40	0.49	−0.07	6.47
5	Gabbro	1.16	14.44	9.37	10.03	8.98	52.28	0.68	0.03	6.55
6	Gabbro	0.47	14.65	9.32	10.31	8.89	52.46	0.74	−0.23	7.67
7	Gabbro	0.80	13.42	11.00	10.06	11.20	50.73	0.48	−0.01	7.31
8	Diorite	1.12	17.39	7.91	9.09	6.59	50.14	0.84	2.52	4.86
9	Marble	0.24	3.96	32.40	1.33	10.10	23.86	0.17	28.09	0.08
10	Limestone	0.48	3.70	32.93	2.25	7.92	22.35	0.16	30.52	1.55
11	Domolite	0.05	0.60	31.49	0.16	19.19	4.13	0.03	44.74	0.15

Note: The unit is $\omega(B)/10^{-2}$.

Table 3. The calculation result of chemical weathering index in the study area.

No.	1	2	3	4	5	6	7	8
IOL	34.56	26.46	31.35	33.65	32.69	31.89	32.24	31.64
CIA	56.83	50.75	57.26	54.16	52.10	52.13	53.48	48.72

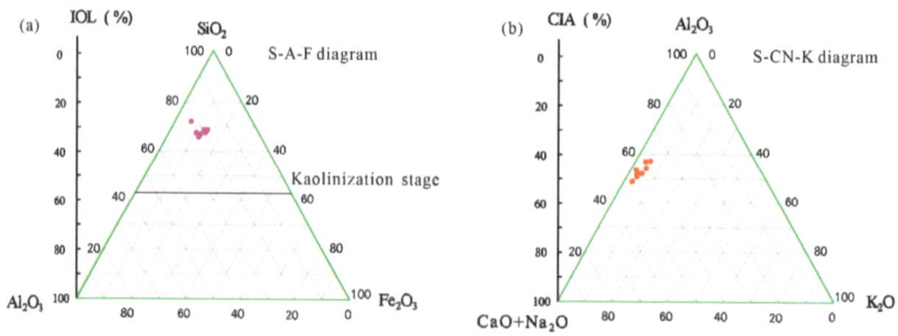

Figure 6. Ternary diagram of IOL and CIA indexes of the rock samples in the study area. (**a**): S-A-F diagram; (**b**): A-CN-K diagram.

4.2.3. Water-Rock Interaction

The weathering and fracture development degree of Ji'nan rock mass present intense, so it could obtain a wide contact area with groundwater and lead to strong water-rock interaction to occur, together with the desilication weathering mechanism, which jointly led the rocks could become intensively leached and supply the source of H_2SiO_3 in the groundwater.

In the study area, the silicate minerals related to the groundwater runoff and occurrence are mainly pyroxene, feldspar, amphibole, etc., whose dissolution can provide not only H_2SiO_3 to the groundwater, but also Ca^{2+}, Mg^{2+} and Na^+, etc. It is commonly H_2CO_3 in groundwater that provides solubility, and may also be other acids formed by water-rock interaction or groundwater pollution, such as H_2SO_4 and HNO_3 [25,26].

Correlation analysis between hydrochemical components is often used to reveal the origin relationship between different components [27,28]. As can be seen from the correlation matrix of each chemical component in the groundwater of the study area (Table 4), H_2SiO_3 has a significant positive correlation with SO_4^- ($p < 0.01$), and the correlation coefficient is 0.705, while the correlation coefficient with $H_2CO_3^-$ is only 0.081, showing no correlation, indicating that in the formation process of H_2SiO_3, it is mainly H_2SO_4 involved in the dissolution of silicate minerals, but not H_2CO_3. During the intrusion of the Ji'nan rock mass, the contact zones between Ji'nan rock mass and the previously deposited Cambrian-Ordovician carbonate rocks developed skarn iron ores rich in FeS_2 [29]. In the field, large amounts of pyrite nodules can be seen clearly at the outcrops of this kind of contact zones (Figure 7). With the participation of groundwater, the FeS_2 can be easily oxidized to generate H_2SO_4, which has enhanced the dissolution ability of the groundwater and increased the solubility of the main silicate minerals in Ji'nan rock mass. H_2SiO_3 also presents a significant positive correlation with Ca^{2+} and Mg^{2+} ($p < 0.01$), and the correlation coefficients are respectively 0.635 and 0.696, but poor correlation with Na^+ or K^+ ($p > 0.05$), indicating that the pyroxenes (silicate minerals containing calcium and magnesium) in the study area are the main mineral components providing H_2SiO_3, and albite or potassium feldspars are the secondary mineral components. TDS present a significant positive correlation with H_2SiO_3 ($p < 0.05$), but a negative correlation with HCO_3^-, indicating that the increase of TDS in groundwater is attributed to the dissolution of silicate rocks rather than carbonate rocks. Cations and TDS are negatively correlated with HCO_3^-, indicating that the production of cations inhibited the dissolution of carbonate rocks. As the main carbonate minerals in the study are calcite ($CaCO_3$) and dolomite ($CaMg(CO_3)_2$), it implies that Ca^{2+} and Mg^{2+} produced during the formation of H_2SiO_3 inhibited the dissolution of calcite and dolomite due to the common ion effect. According to the Debye-Hückel Formula, borehole SY12 in the north of the study area was selected randomly to calculate the saturation index of $CaCO_3$ based on the tested hydrochemistry data, the result was $SI_{cal} = 0.36 > 0$, which indicated that groundwater in the north of the study area was saturated with carbonate minerals, which proved that the carbonate deposits were generated due to the common ion effect, and further the content of HCO_3^- became reduced. The main reaction processes involved in these series of water-rock interactions are as follows:

Pyrite oxidizes to produce acid: $4FeS_2 + 15O_2 + 8H_2O \rightarrow 2Fe_2O_3 + 8SO_4^{2-} + 16H^+$ (Sulfuric acid)

Pyroxene dissolution: $H_2SO_4 + Ca(Mg)Al_2Si_2O_8 + 2H_2O \rightarrow Ca^{2+}(Mg^{2+}) + SO_4^{2-} + 2H_2SiO_3 + 2AlOOH$

Table 4. Correlation coefficient matrix of hydrochemical parameters of the groundwater in the study area.

	H_2SiO_3	TDS	Ca^{2+}	Mg^{2+}	Na^+	K^+	HCO_3^-	Cl^-	SO_4^{2-}
H_2SiO_3	1								
TDS	0.463 *	1							
Ca^{2+}	0.635 **	0.951 **	1						
Mg^{2+}	0.696 **	0.905 **	0.986 **	1					
Na^+	0.228	0.948 **	0.806 **	0.726 **	1				
K^+	0.368	0.981 **	0.881 **	0.812 **	0.986 **	1			
HCO_3^-	0.081	−0.396	−0.31	−0.276	−0.431	−0.404	1		
Cl^-	0.2	0.939 **	0.789 **	0.708 **	0.998 **	0.979 **	−0.432	1	
SO_4^{2-}	0.705 **	0.907 **	0.991 **	0.992 **	0.730 **	0.821 **	−0.259	0.708 **	1

Note: ** indicates a significant precorrelation at 0.01 level, and * indicates a significant correlation at 0.05 level.

It is generally believed that with the participation of H_2CO_3, if the H_2SiO_3 in groundwater is mainly sourced from the dissolution of silicate minerals such as feldspar, pyroxene, and amphibole, then $\gamma[(Ca^{2+} + Mg^{2+} + Na^{2+} + K^{2+})/HCO_3^-] = 1.0$, which is called the weathering line of silicate rock [30]. If there are other acids in the water to participate in the dissolution of silicate minerals, the SO_4^{2-}, Cl^- can be provided to participate in the anion balance, and the samples will deviate from the line of $\gamma[(Ca^{2+} + Mg^{2+} + Na^{2+} + K^{2+})/HCO_3^-] = 1.0$. According to the relationship diagram of $\gamma[(Ca^{2+} + Mg^{2+})/HCO_3^-]$ and $\gamma[(Na^+ + K^+)/(HCO_3^-)]$ (Figure 8), it can be seen that the groundwater samples in the study area are located near or above the weathering line of the silicate rock, indicating that the chemical characteristics of the groundwater in the study area are mainly controlled by the weathering of the silicate rocks and affected by other acids, which is consistent with the conclusion presented in Figure 4. The samples with H_2SiO_3 content greater than 25 mg/L are located above the weathering line of the silicate rocks, and the deviation from the weathering line is large, indicating that the metalicate-rich groundwater is mainly formed by the dissolution of silicate minerals with the participation of other acids. Considering the hydrochemical analysis data of the study area, it indicates that the ionic balance of the groundwater is maintained by SO_4^{2-}. That is, H_2SO_4 is mainly involved in the leaching of silicate minerals. Amounts of samples are located in the region of $\gamma[(Ca^{2+} + Mg^{2+})/HCO_3^-] > 1.0$, indicating that the water environment of the samples has been affected by human activities, which is presumed to be related to the participation of SO_4^{2-} and Cl^- in water ion balance provided by the mining of Zhangmatun Iron Mine on the northeast side of Ji'nan rock mass [31]. The value of $\gamma[(Ca^{2+} + Mg^{2+})/HCO_3^-]$ in all samples was much higher than that of $\gamma[(Na^+ + K^+)/HCO_3^-]$, reflecting that the minerals involved in the leaching process were mostly calcium and magnesia-containing minerals, namely pyroxene in the study area, which was consistent with the above analysis results.

Figure 7. The pyrite nodules developed along the contact zones between the igneous rocks and the surrounding carbonate rocks. (**a**): The contact zone between the igneous rock and the surrounding carbonate rock; (**b**): The pyrite nodules developed along the contact zone.

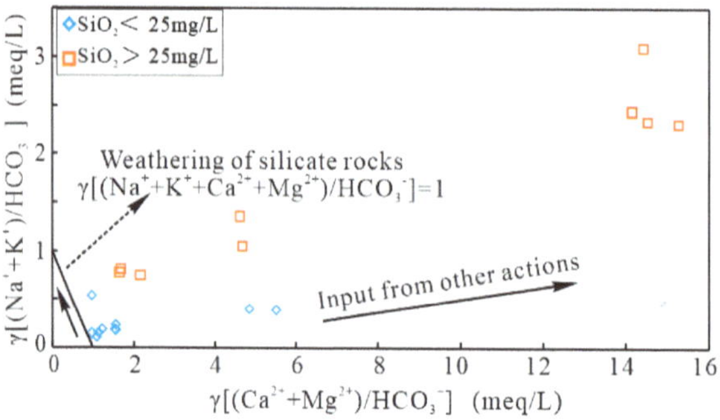

Figure 8. Relative contributions from silicate and carbonate weathering by H_2CO_3.

4.2.4. Conservation and Recharge Conditions of the Groundwater

Influenced by the Yanshan Movement, the terrain of Ji'nan City is high in the south and low in the north, presenting a monoclinic structure. As located in the north of Ji'nan City, the groundwater in the study area can be rapidly recharged by the Wohushan Reservoir, Jinxiuchuan Reservoir and atmospheric precipitation in the southern mountainous area [32]. In addition, the atmospheric precipitation can also rapidly recharge the groundwater along the weak zones, such as the Sangzidian fault, especially the contact zones between

Ji'nan rock mass and the surrounding rocks, along which the FeS_2 could be oxidized to form H_2SO_4, leading the solubility of silicate minerals in groundwater is enhanced. So in the study area, the interaction between atmospheric precipitation, surface water and groundwater is intense, which provided better water source conservation and recharge conditions for the groundwater. As a result, the groundwater runoff is relatively fast, the regeneration rate is quick, and the surrounding rocks or minerals are pretty soluble, thus guaranteeing the enrichment of H_2SiO_3 in groundwater.

The main factors controlling the equilibrium between dissolution and precipitation of the H_2SiO_3 in groundwater are the supply source of metasilicate minerals (solid amorphous SiO_2) and the solubility of secondary minerals formed during weathering. Through the equilibrium system of dissolution-precipitation of metasilicate minerals, not only can we identify the source of H_2SiO_3, but also can indicate the relative saturation of metasilicate minerals in the groundwater, and then help judge the solubility of metasilicate minerals in the groundwater [30]. As shown in Figure 9, in the $[Na^+]/[H^+]$-H_2SiO_3 system (Figure 9a), the groundwater sample points in the study area are generally located in the stable area of kaolinite, indicating that the water-rock interaction is in the process of forming kaolinite by the inhomogeneous dissolution of plagioclase, which is consistent with the above analysis that the rock weathering is mostly at the stage of kaolinization (CIA, IOL) (Figure 6). In the system of $[Mg^{2+}]/[H^+]$-H_2SiO_3 (Figure 9b), the samples are generally located in the stable end-member of chlorite, and some samples with $SiO_2 < 25$ mg/L are located in the stable end-member of kaolinite. In the $[Ca^{2+}]/[H^+]$-H_2SiO_3 system (Figure 9c), all the samples are located in the stable end-member of kaolinite. All the groundwater samples did not reach the saturation state of metasilicate minerals, indicating that all the samples have the ability to continue to dissolve metasilicate minerals, which reflected that the groundwater in the study area was rapidly recharged so that it could maintain the persistent ability to dissolve the metasilicate minerals.

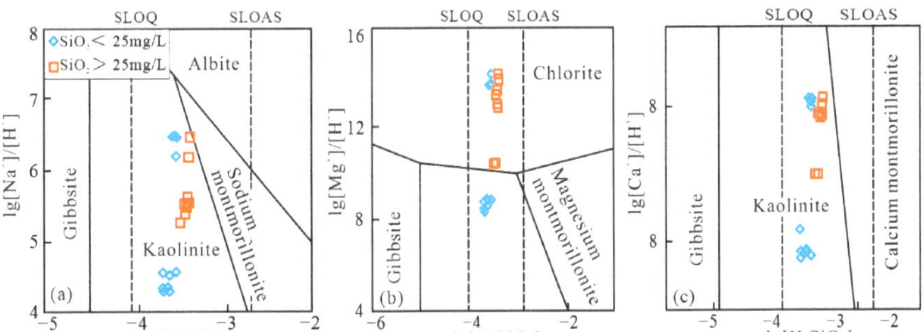

Figure 9. Mineral equilibrium phase diagram for the groundwater in the study area. SLOQ = saturation line of quartz; SLOAS = saturation line of amorphous SiO_2. (**a**): the $[Na^+]/[H^+]$-H_2SiO_3 system; (**b**): the $[Mg^{2+}]/[H^+]$-H_2SiO_3 system; (**c**): the $[Ca^{2+}]/[H^+]$-H_2SiO_3 system.

4.3. Genetic Model of the Metasilicate Enrichment

Based on the spatial distribution characteristics and hydrogeological conditions of metasilicate-rich groundwater in the Ji'nan rock mass area obtained by the above research, and combined with the genetic mechanism of the Ji'nan spring area analyzed by previous scholars [33,34], the causes of the enrichment of H_2SiO_3 in groundwater in the study area are summarized as the following: After the infiltration of atmospheric precipitation, the water flew from south to north through the Ordovician limestone. During the runoff process, parts of the water got blocked by the Ji'nan rock mass and then migrated upward to form Baotu Spring, Heihu Spring, etc., and parts of the water entered the deep circulation, and participated in the strong water-rock interaction along the weathering zones of the rock mass, the weak zones of fractures, and especially the contact zones between Ji'nan rock mass and surrounding rocks, where the FeS_2 were oxidized to produce amounts of H_2SO_4 in groundwater, making the groundwater in the study area present strong dissolution ability. The metasilicate minerals such as pyroxene, plagioclase, and amphibole in the silicate rocks were dissolved and desilicated, and the silicon elements entered the groundwater and moved along the runoff direction of the groundwater, and finally accumulated in the downstream of Ji'nan rock mass (i.e., the north side of Ji'nan rock mass), which led to the enrichment of H_2SiO_3 in groundwater (Figure 10). However, on the northeast side of Ji'nan rock mass, there are human production activities such as the mining of Zhangmatun Iron Mine [31], which caused sulfur pollution to the groundwater and further inhibited the dissolution of silicate minerals, so the water-rock interaction became weakened, resulting only part of the sample sites developed the metasilicate-rich groundwater in this area. More samples would be more beneficial to conclude the reliable genetic model.

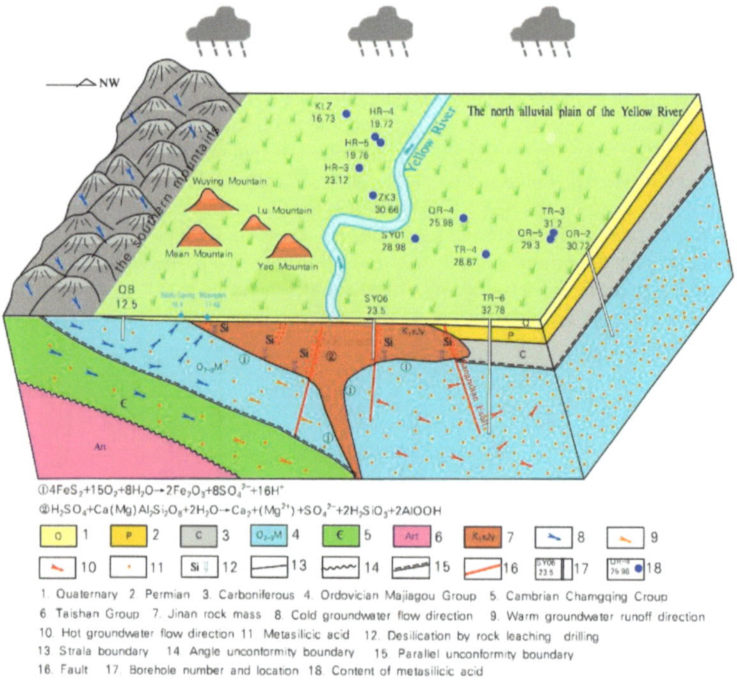

Figure 10. Genetic model of the metasilicate-rich groundwater around Ji'nan rock mass.

5. Conclusions

(1) The distribution of the metasilicate-rich groundwater in the Ji'nan rock mass area was spatially different. On the northwest side and part of the northeast side, the content of H_2SiO_3 in groundwater was large, forming metasilicate-rich groundwater. However, on the east, west, and south sides of the rock mass, little metasilicate-rich groundwater developed.

(2) Ji'nan rock mass is mainly composed of gabbro and contains amounts of silicate minerals that easily weathered. The weathering was at the stage of kaolinization showing a high degree, and presented the desilication mechanism, which acted as the main source of H_2SiO_3 in the groundwater.

(3) FeS_2 developed along the contact zones between the Ji'nan rock mass and surrounding rocks, which were easily oxidized to produce H_2SO_4 and enhanced the ability of groundwater to dissolve silicate minerals. The study area is located at the low position of the monocline structure, where the groundwater could be recharged and renewed quickly.

(4) Lithology, structure and hydrochemical characteristics constitute the favorable factors for the formation of metasilica-rich groundwater, whereas, sulfur pollution caused by human activities acted as the negative factor. In the future, more samples are suggested to be collected and tested, in order to verify the controlling factors and make up the genetic model.

Author Contributions: Conceptualization, L.Z.; methodology, C.H.; software, S.Y. and J.S.; validation, G.S.; formal analysis, M.X.; investigation, M.X.; resources, W.P.; data curation, W.P.; writing—original draft preparation, M.X.; writing—review and editing, M.X. and C.H.; visualization, G.S.; supervision, L.Z.; project administration, C.H.; funding acquisition, G.S. All authors have read and agreed to the published version of the manuscript.

Funding: This research was funded by National Natural Science Foundation of China, grant number 42002127.

Data Availability Statement: Not applicable.

Conflicts of Interest: The authors declare no conflict of interest.

References

1. Wada, Y.; Flörke, M.; Hanasaki, N.; Eisner, S.; Fischer, G.; Tramberend, S.; Satoh, Y.; Van Vliet, M.T.H.; Yillia, P.; Ringler, C.; et al. Modeling Global Water Use for the 21st Century: The Water Futures and Solutions (WFaS) Initiative and Its Approaches. *Geosci. Model Dev.* **2016**, *9*, 175–222. [CrossRef]
2. Liu, Y.Q.; Wen, D.G.; Lü, L.; Li, W.; Zhang, F.C.; Wang, X.F.; Meng, S.X. Characteristics of karst groundwater flow systems of typical faulted basins in Yimeng Mountain area: A case study of Laiwu Basin. *Bull. Geol. Sci. Technol.* **2022**, *41*, 157–167.
3. Xiao, Y.; Hao, Q.C.; Zhang, Y.H.; Zhu, Y.C.; Yin, S.Y.; Qin, L.M.; Li, X.H. Investigating sources, driving forces and potential health risks of nitrate and fluoride in groundwater of a typical alluvial fan plain. *Sci. Total Environ.* **2022**, *802*, 149909. [CrossRef]
4. McDonald, R.I.; Green, P.; Balk, D.; Fekete, B.M.; Revenga, C.; Todd, M.; Montgomery, M. Urban Growth, Climate Change, and Freshwater Availability. *Proc. Natl. Acad. Sci. USA* **2011**, *108*, 6312–6317. [CrossRef]
5. Polemio, M.; Voudouris, K. Groundwater Resources Management: Reconciling Demand, High Quality Resources and Sustainability. *Water* **2022**, *14*, 2107. [CrossRef]
6. Xiao, Y.; Liu, K.; Hao, Q.C.; Li, Y.S.; Xiao, D.; Zhang, Y.J. Occurrence, Controlling Factors and Health Hazards of Fluoride Enriched Groundwater in the Lower Flood Plain of Yellow River, Northern China. *Expo. Health* **2022**, *14*, 345–358. [CrossRef]
7. Xiao, Y.; Liu, K.; Hao, Q.C.; Xiao, D.; Zhu, Y.C.; Yin, S.Y.; Zhang, Y.H. Hydrogeochemical insights into the signatures, genesis and sustainable perspective of nitrate enriched groundwater in the piedmont of Hutuo watershed, China. *Catena* **2022**, *212*, 106020. [CrossRef]
8. Zhou, C.S.; Zou, S.Z.; Zhu, D.N.; Lin, Y.S.; Wang, J.; Fan, L.J.; Li, J.; Lan, F.N.; Li, Y.Q.; Deng, R.X.; et al. Characteristics, causes and development suggestions of high quality groundwater containing metasilicate in Zhaojue area, Sichuan Province. *Geol. China* **2022**, *49*, 849–859.
9. Kim, S.H.; Choi, B.Y.; Lee, G.Y.; Yun, S.T.; Kim, S.O. Compositional data analysis and geochemical modeling of CO2- water- rock interactions in three provinces of Korea. *Environ. Geochem. Health* **2019**, *41*, 357–380. [CrossRef]
10. Sun, Q.F.; Tian, H.; Guo, X.D.; Yu, H.M.; Li, L.J. The discovery of silicic acid and strontium enrichment areas in groundwater of Changchun area, Jilin Province. *Geol. China* **2017**, *44*, 1031–1032.
11. Sun, H.Y.; Wei, X.F.; Sun, X.M.; Jia, F.C.; Li, D.J.; He, Z.X.; Li, J. Formation mechanism and geological construction constraints of metasilicate mineral water in Yudaokou, Hannuoba basalt area. *Earth Sci.* **2020**, *45*, 4236–4253.
12. Li, C.S.; Wu, X.C.; Sun, B.; Sui, H.B.; Geng, F.Q.; Qi, H.; Ma, X.Y. Hydrochemical characteristics and formation mechanism of geothermal water in northern Ji'nan. *Earth Sci.* **2018**, *43*, 313–325.
13. Wang, S.J.; Zhang, C.J.; Yang, E.X.; Song, Z.Y.; Wang, L.F.; Xu, K.M. Division of Mesozoic intrusive stages in Luxi area. *Shandong Land Resour.* **2009**, *25*, 18–23.
14. Shan, T.T.; Xu, S.G.; Fan, Z.G.; Ruan, W. Characteristics and formation mechanism of metasilicate mineral water in Xishan mountain of Kunming. *J. Kunming Univ. Sci. Technol.* **2019**, *44*, 39–47.
15. Sun, H.Y.; Sun, X.M.; Wei, X.F.; Chen, Z.R.; Liu, W.; Huang, X.K.; Li, X.; Yin, Z.Q.; Liu, W.B. Formation mechanisms of metasilicate mineral water in Chengde, Hebei Province: Evidence from rock weathering and water-rock interaction. *Geol. China* **2022**, *49*, 1088–1113.
16. Ye, N.; Li, Y.T.; Huang, B.W.; Xi, B.B.; Jiang, H.; Lu, Z.Y.; Chen, Q.L.; You, D.H.; Xu, J. Hydrothermal silicification and its impact on Lower–Middle Ordovician carbonates in Shunnan area, Tarim Basin, NW China. *Geol. J.* **2022**, *57*, 3538–3557. [CrossRef]
17. Liang, C.C.; Wang, W.; Ke, X.M.; Ou, A.F.; Wang, D.H. Hydrochemical Characteristics and Formation Mechanism of Strontium-Rich Groundwater in Tianjiazhai, Fugu, China. *Water* **2022**, *14*, 1874. [CrossRef]
18. Moses, C.; Robinson, D.; Barlow, J. Methods for measuring rock surface weathering and erosion: A critical review. *Earth-Sci. Rev.* **2014**, *135*, 141–161. [CrossRef]

19. Wu, B.J.; Peng, B.; Zhang, K.; Kuang, X.L.; Fang, X.H.; Zeng, D.Z. A new chemical index of identifying the weathering degree of black shales. *Acta Geol. Sin.* **2016**, *90*, 818–832.
20. Ofili, S.; Soesoo, A.; Panova, E.G.; Hints, R.; Hade, S. Geochemical Reconstruction of the Provenance, Tectonic Setting and Paleoweathering of Lower Paleozoic Black Shales from Northern Europe. *Minerals* **2022**, *12*, 602. [CrossRef]
21. Nesbitt, H.W.; Young, G.M. Early Proterozoic climates and plate motions inferred from major element chemistry of lutites. *Nature* **1982**, *299*, 715–717. [CrossRef]
22. Babechuk, M.G.; Widdowson, M.; Kamber, B.S. Quantifying chemical weathering intensity and trace element release from two contrasting basalt profiles, Deccan Traps, India. *Chem. Geol.* **2014**, *363*, 56–75. [CrossRef]
23. Romero-Mujalli, G.; Hartmann, J.; Hosono, T.; Louvat, P.; Okamura, K.; Delmelle, P.; Amann, T.; Böttcher, M.E. Hydrothermal and magmatic contributions to surface waters in the Aso caldera, southern Japan: Implications for weathering processes in volcanic area. *Chem. Geol.* **2022**, *588*, 120612. [CrossRef]
24. Rouzaut, S.; Campodonico, V.A.; Pasquini, A.I. Weathering and paleoprecipitation indices in a Late Pleistocene–Holocene loess–paleosol sequence in central Argentina. *Environ. Earth Sci.* **2021**, *80*, 28. [CrossRef]
25. Gao, Z.J.; Wang, Z.S.; He, K.Q.; Victor, K.; Liu, J.T. Hydrochemical characteristics and controlling factors of karst groundwater in middle and upper reaches of Dawen River basin. *Bull. Geol. Sci. Technol.* **2022**, *41*, 264–272.
26. Hu, X.B.; Fang, J.C.; Zhai, H.W.; Zhang, K.; Ma, Y.M.; Jin, J.H.; Gao, X.B. Application of sulfur and oxygen isotopes in identifying the source of sulfate in karst water from Xin'an spring area. *Bull. Geol. Sci. Technol.* **2022**, *41*, 333–340.
27. Kale, A.; Bandela, N.; Kulkarni, J.; Sahoo, S.K.; Kumar, A. Hydrogeochemistry and multivariate statistical analysis of groundwater quality of hard rock aquifers from Deccan trap basalt in Western India. *Environ. Earth Sci.* **2021**, *80*, 288. [CrossRef]
28. Liu, P.; Yang, M.; Sun, Y.J. Hydro-geochemical processes of the deep Ordovician groundwater in a coal mining area, Xuzhou, China. *Hydrogeol. J.* **2019**, *27*, 2231–2244. [CrossRef]
29. Deng, J.; Wang, C.M.; Bagas, L.; Santosh, M.; Yao, E.Y. Crustal architecture and metallogenesis in the south-eastern North China Craton. *Earth-Sci. Rev.* **2018**, *182*, 251–272. [CrossRef]
30. Guo, Q.H.; Wang, Y.X. Simulation of geochemical processes affecting groundwater in Quaternary porous aquifers of Taiyuan basin: A typical Cenozoic rift basin. *Earth Sci. Front.* **2014**, *21*, 83–90.
31. Wang, H.H.; Li, X.Z.; Shen, L.J.; Zhu, Y.Z.; Zhou, M.L. Geological characteristics and metallogenic model of "Yucheng Type" rich ion deposit in Qihe Yucheng area in Shandong Province. *Shandong Land Resour.* **2021**, *37*, 26–35.
32. Jia, C.; Wang, C.; Liu, S.; Yang, X.; Liu, W.; Gao, S.; Zhu, H.H. Study on hydrogeochemical characteristics of groundwater in the alluvial plain of western Ji'nan. *Water Resour. Hydropower Eng.* **2022**, *53*, 49–60.
33. Wang, J.L.; Jin, M.G.; Lu, G.P.; Zhang, D.L.; Kang, F.X.; Jia, B.J. Investigation of discharge-area groundwaters for recharge source characterization on different scales: The case of Ji'nan in northern China. *Hydrogeol. J.* **2016**, *24*, 1723–1737. [CrossRef]
34. Xing, L.T.; Yu, M.; Su, Q.W.; Zhao, Z.H.; Gao, Y.; Zhang, Y.F. Influence and repair of underground engineering construction on karst flow field. *Bull. Geol. Sci. Technol.* **2022**, *41*, 242–254.

water

MDPI

Article

Geochemical Characteristics and Genesis of Brine Chemical Composition in Cambrian Carbonate-Dominated Succession in the Northeastern Region of Chongqing, Southwestern China

Zhi-lin Zheng [1], Bin Xie [1], Chun-mei Wu [1], Lei Zhou [1], Ke Zhang [2], Bin-chen Zhang [3] and Ping-heng Yang [4,*]

[1] No. 208 Hydrogeological and Engineering Geological Team, Chongqing Bureau of Geological and Mineral Resource Exploration and Development, Chongqing 400711, China; 15923347226@163.com (Z.-l.Z.); xiebin106256@163.com (B.X.); 18602310430@163.com (C.-m.W.); 13657677378@163.com (L.Z.)

[2] College of Humanities and Development Studies, Beijing 100091, China; kecorrine94chaie@163.com

[3] Chongqing Institute of Geological Survey, Chongqing 401121, China; 13629792756@163.com

[4] Chongqing Jinfo Mountain Karst Ecosystem National Observation and Research Station, School of Geographical Sciences, Southwest University, Chongqing 400715, China

* Correspondence: balance@swu.edu.cn

Abstract: Deeply situated brine is abundant in rare metal minerals, possessing significant economic worth. To the authors' knowledge, brine present within the Cambrian carbonate-dominated succession in the northeastern region of Chongqing, Southwestern China, has not been previously reported. In this investigation, brine samples were collected from an abandoned brine well, designated as Tianyi Well, for the purpose of analyzing the hydrochemical characteristics and geochemical evolution of the brine. Halide concentrations, associated ions, and their ionic ratios within the sampled brine were analyzed. The brine originating from the deep Cambrian aquifer was characterized by high salinity levels, with an average TDS value of 242 ± 11 g/L, and was dominated by a Na-Cl facies. The studied brine underwent a moderate degree of seawater evaporation, occurring between the saturation levels of gypsum and halite, accompanied by some halite dissolution. Compared to modern seawater evaporation, the depletion of Mg^{2+}, HCO_3^-, and SO_4^{2-} concentrations, along with the enrichment of Ca^{2+}, Li^+, K^+, and Sr^{2+}, is likely primarily attributed to water–rock interactions. These interactions include dolomitization, combination of halite dissolution, upwelling of lithium- and potassium-bearing groundwater, calcium sulfate precipitation, biological sulfate reduction (BSR), and the common ion effect within the brine system. This research offers valuable insights into the genesis of the brine within the Cambrian carbonate succession and provides theoretical backing for the development of brine resources in the future.

Keywords: brine; genesis; chemical composition; water–rock interactions; Cambrian carbonate

1. Introduction

The deeply buried regions within most sedimentary basins contain substantial volumes of brines [1–4], which exert pivotal influences on diverse geological processes—including natural resource development—while holding significant implications for both scientific research and economic utilization [5]. The basinal brines (also referred to as formation waters or oil-field saline solutions) typically exhibit salinity levels ranging from 10 to 300 g/L [6]. The majority of brines belong to either Na-Cl or Ca-Cl facies, distinguished chemically by their elevated concentrations of Ca^{2+}, surpassing combined levels of SO_4^{2-}, HCO_3^-, and CO_3^{2-} ions compared to seawater and other common surface/near-surface water types (e.g., Na-Cl-SO_4 or Ca-HCO_3 facies). Evaporite deposits found within the basins commonly originated from the evaporation of isolated seawater [7]. The brines are either connate ones that have persisted since the early Paleozoic era—which have been trapped since the early Paleozoic era in these closed basins with restricted groundwater circulation [8,9]—or brines that have emerged from the dissolution of the evaporite deposits themselves [10,11]. The chemical compositions of brines offer valuable insights into geochemical, hydrological, and tectonic developments within the Earth's crust as well as providing an understanding of several pertinent issues relating specifically to metal ore deposits, dolomitized reservoirs, and hydrocarbon exploration/production [12,13]. Notably, dissolved metals tend to be prevalent within saline solutions exceeding 200 g/L [6].

The sedimentary brines present themselves as exceptional natural phenomena shrouded in unresolved enigmas concerning their origin and evolution of chemical compositions. Genesis behind the brines poses an essential and captivating challenge within contemporary hydrogeological/hydrogeochemical communities [14]. Potential hypotheses regarding the genesis of brine chemical compositions include evaporite dissolution, subaerial seawater evaporation, interactions between rocks and water, osmotic/membrane filtration, and ion exchanges [9,15–20]. The divergent viewpoints on the genesis of brine arise not only from their unique chemical compositions and extreme salinity but also from their extensive distribution across various continental crustal blocks on Earth.

The chemistry of groundwater is significantly influenced by the mineralogical composition of the aquifer it traverses. The chemistry of major ions and their interrelationships in groundwater are effective in identifying the origins of solutes and elucidating the evolution of groundwater [21]. The chemical composition reflects the evolution of groundwater and serves as a crucial method for investigating the formation of brines. This includes the application of ion ratio methods such as Cl^-/Br^- [22–24], Ca excess and Na deficit relative to modern seawater reference ratios [25], and comparison of cation and anion abundances in brines with modern seawater [26]. Significant advancements have been achieved in understanding the genesis of brine composition through hydrochemical studies [5,12,13,27,28].

To the authors' knowledge, Cambrian-era sedimentary brine occurrences remain unreported in the region of the northeastern Chongqing municipality, Southwestern China. Such occurrences are part of this enigma since questions about their genesis remain unresolved. This study aims to investigate the hydrochemical characterization of brine in northeastern Chongqing to understand its geochemical evolution by utilizing the halide concentrations and associated ions from an abandoned brine well, Tianyi Well. This reconnaissance hyrdogeochemical study may provide valuable insights into economic explorations within deep geological environments in the study area.

2. Description of the Area Adjacent to the Studied Brine

2.1. Location and Geographical Settings

The study area is located in Wuxi County, positioned in northeastern Chongqing, SW China (Figure 1a–c). It is characterized by mountainous terrain with significant relief, ranging from 320 to 1790 m above sea level.

It is within a subtropical monsoon climate, with an average annual temperature of 18 °C. In January, the average temperature is 8 °C, while in July it rises to 35 °C. The region receives an average annual precipitation of 1450 mm, concentrated mainly between May and October each year, peaking in September.

2.2. Geological Settings

Apart from the absence of Devonian and Carboniferous strata due to geological denudation, the exposed formations adjacent to the studied Tianyi Well range from the Cambrian to Triassic (Figure 1d). The Triassic strata consist mainly of limestone, with minor occurrences of dolomitic limestone and gray-to-gray-purple calcareous shale, reaching an overall thickness exceeding 350 m. The Permian strata are predominantly composed of limestone, accompanied by iron clay and coal seams, with a total thickness ranging from 510 to 931 m. The Silurian strata are characterized by quartz sandstone, mudstone, and shale, measuring between 852 and 1053 m in thickness. The Ordovician strata primarily comprises limestone, dolomite, argillaceous dolomite, and silty shale, with an aggregate thickness of 304–391 m. The Cambrian strata are chiefly composed of dolomite interbedded with evaporites, with a total thickness surpassing 775 m. The region is mainly underlain by Archean deeply metamorphosed basic–ultrabasic metamorphic rocks [29].

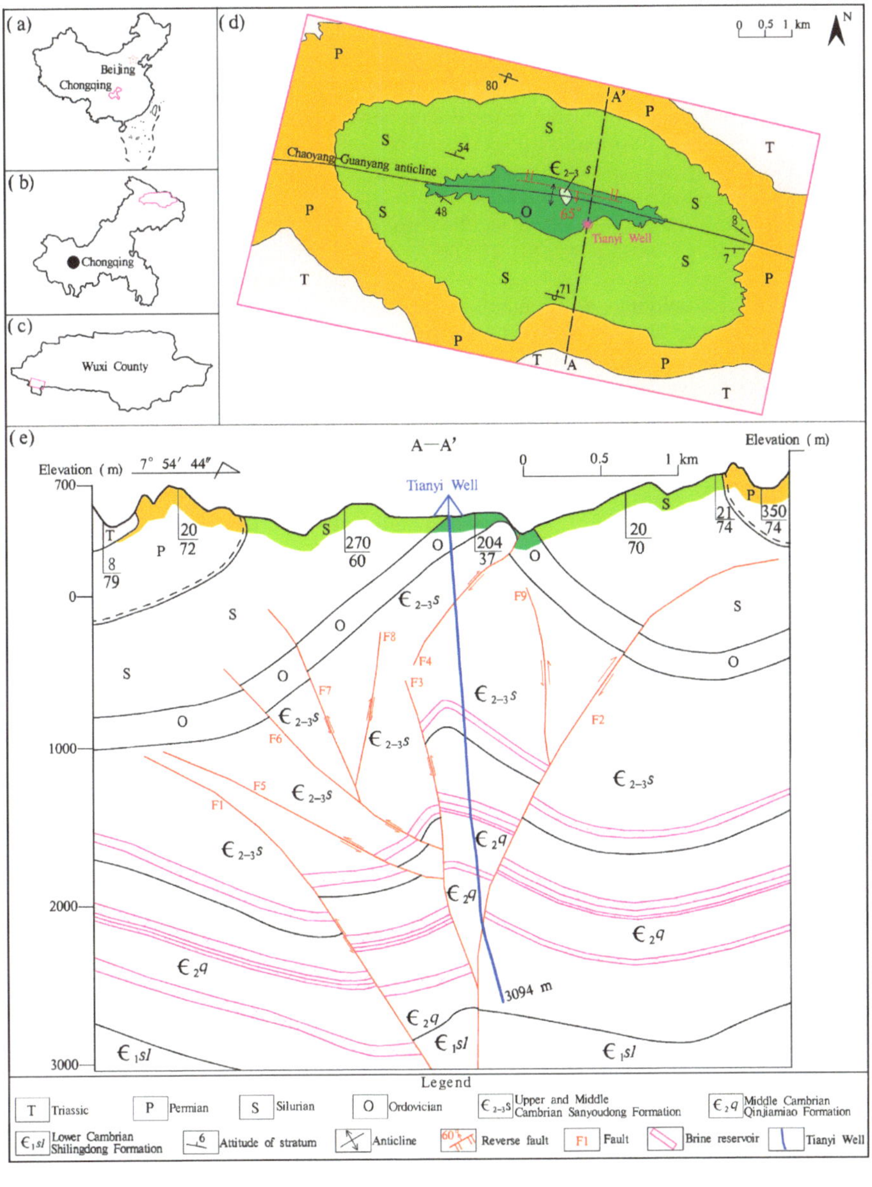

Figure 1. (**a**–**c**) Location of the studied brine of Tianyi Well, (**d**) its adjacent outcrops, and (**e**) cross-section A-A′ showing an overview of the general lithology of the well and surrounding successions, in which the faults were identified through seismic wave analysis.

The primary tectonics in this area is a steeply dipping reverse fault, measuring 1.7 km in length, which has developed within the paraxial part of the Chaoyang–Guanyang anticline extending northwestward (Figure 1d). This fault exhibits a dip angle of 50–70°

and a dip direction at approximately 190°. The upper hanging wall consists of Middle and Upper Cambrian Sanyoudong Formation ($\in_{2\text{–}3}$s) carbonate, displaying distinct features of thrusting and extrusion, whereas the footwall comprises shale of the Ordovician Formation. The fracture bandwidth measures between 2 and 5 m with landforms aligning along its trend for about the maximum distance. Additionally, seismic wave analysis indicates the significant presence of regional underground reverse faults (Figure 1e), which may restrict fluid movement in the vicinity. The Tianyi Well—the building of which commenced in 1964—was drilled to a depth of 3094 m along the anticlinal axis, traversing through the Upper and Middle Cambrian Sanyoudong Formation ($\in_{2\text{–}3}$s) as well as the Middle Cambrian Qinjiamiao Formation ($\in_2$q) (Figure 1e).

The strata closely associated with brine primarily belong to the Cambrian succession. The following provides a lithological overview of the Upper and Middle Cambrian Sanyoudong Formation ($\in_{2\text{–}3}$s) and Middle Cambrian Qinjiamiao Formation ($\in_2$q).

Sanyoudong Formation ($\in_{2\text{–}3}$s): The upper portion of this formation consists of gray-to-dark-gray medium-thick bedded dolomite interbedded with gray-yellow medium-thick bedded breccia dolomite, and some bioclastic limestone. Horizontal bedding prevails occasionally accompanied by sutures and mold structures. The thickness of the formation ranges from 235.1 to 297 m in the area; however, drilling at Tianyi Well encountered a fault resulting in repeated formation with a drilling thickness of 1173.5 m and a true vertical thickness of 4277 m (Figure 1e). At the lower end lies a thick-bedded dolomitic medium-grained quartz sandstone layer with an approximate thickness of 2 m. Based on Tianyi Well's drilling data, the underground brine reservoir was discovered between depths of 1239 and 1278 m below the wellhead (Figure 1e).

Qinjiamiao Formation ($\in_2$q): This formation remains unexposed on the regional surface but the drilling core indicates predominantly thin-layer argillaceous dolomite alongside light gray and yellow-gray thin-to-medium-layer sandy dolomite embedded with dolomite, gypsum, and salt soluble breccia. Drilling at Tianyi Well reached depths of 1752 m with a true vertical thickness of 385.2 m. This formation yielded three brine reservoirs, located at depths of 1790–1839, 1895.4–1912.0, and 2250.4–2288.7 m, respectively, below the wellhead (Figure 1e). The porosity of the brine reservoirs ranges from 1.0 to 24.0%, with a weighted average porosity of 2.0%.

From a stratigraphic perspective, during the epoch spanning the Middle- and Upper-Cambrian periods, a considerable regression in seawater occurred in a hot climate, which facilitated the development of lagoon-like sedimentary facies on confined platforms. Following this, the intensified evaporation process led to the emergence of Sabkha environments, conducive to the substantial deposition of salt minerals. Specifically, the carbonates were initially deposited in shallow waters, likely in tidal flat or lagoonal environments. As evaporation continued to intensify, gypsum and halite successively precipitated and deposited. However, at Tianyi Well, a clean water drilling technique was utilized below a depth of 340 m under the wellhead, thereby impeding the acquisition of rock cores essential for confirming the existence of halite layers within the Sanyoudong and Qinjiamiao Formations. Approximately 150 km from this location, a drilled well struck a layer of halite deposit measuring 120 m in thickness within the Cambrian carbonate strata beneath the surface [30]. This confirmed the presence of salt minerals in the area.

From a hydrogeological perspective, the deep-seated structural faults in this region are predominantly reverse faults (Figure 1e), formed under compressive stress conditions. As these faults evolved, the rocks along the fault plane and within the fault zone underwent significant compression, resulting in the formation of dense rock strata that serve as a barrier to shield the underlying brine. However, it is possible for rainwater to infiltrate into the subterranean realm via the exposed Ordovician carbonates, as depicted in Figure 1e. This infiltration process may exert a discernible dissolving influence upon the brine, potentially altering its composition.

3. Methodology and Materials

3.1. Field Sampling

Brine samples were collected at the surface from the Tianyi Well by pumping. The pumping operation was conducted within the timeframe of 23–28 November 2023, wherein the brine was extracted from a depth of 1008 m beneath the surface of the wellhead, and the entire process lasted for a duration of 66 h. The pumping discharge ranged from 0 to 13 m^3/h and it stabilized at approximately 11 m^3/h. Upon cessation of pumping, the recovery period of the water level took 47.9 h (Figure 2). The extracted brine was the result of the mixture of the entire well, representing the overall condition of the deep brine system. The pumped brine, characterized by its gray-black coloration, was translucent yet cloudy, bearing a saline character with suspended particles. It also released a rotten egg smell, which is from hydrogen sulfide (H_2S) gas. The brine samples were collected in strict adherence to protocol after being pumped to the surface. During the pumping activity, one brine sample was collected every 2 h within the first 6 h of pumping, followed by samples collected every 4 h (Figure 2). A total of 19 brine samples were obtained. High-density poly ethylene bottles, 5 L in size, were used for brine sample collection. To prevent contamination, sampling bottles were rinsed with the sampled brine five times prior to collection. All brine samples were immediately stored in a cool and sealed environment after sampling, and then transported back to the laboratory within 24 h for measurement.

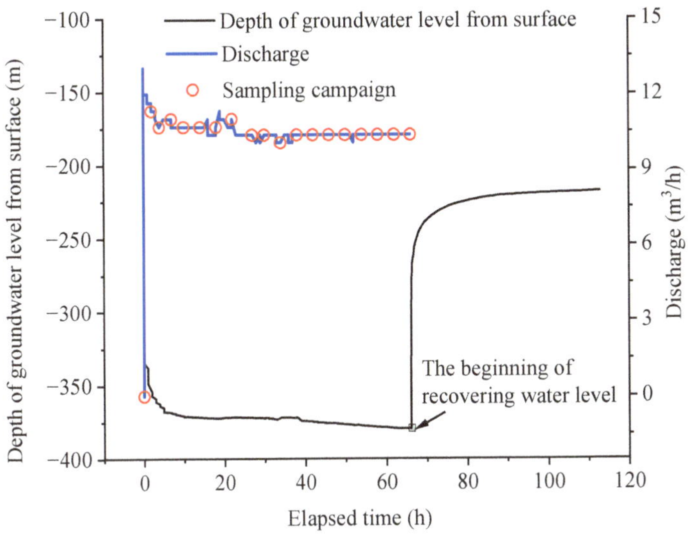

Figure 2. Pumping, recovering water level, and sampling campaign of the studied brine.

3.2. Measurement

Na^+, K^+, Ca^{2+}, Mg^{2+}, Li^+, SO_4^{2-} concentrations were analyzed using an ICP-OES Avio 200 (PerkinElmer Company, Waltham, MA, USA). Cl^- and HCO_3^- concentrations were determined using brown and white acid burettes, respectively. Br^- concentration was measured using an Ultraviolet spectrophotometer UV2700 (Shimadzu Corporation, Kyoto, Japan). The accuracy of ion concentration testing is better than 0.01 mg/L. The measurements were carried out at the Salt Lake Chemical Analysis Test Center of Qinghai Institute of Salt Lake Research, Chinese Academy of Sciences, under laboratory conditions with a temperature of 20 °C and relative humidity of 29%.

4. Result and Discussion

4.1. Hydrochemistry

The main concentrations of cations and anions in the studied brine from the Tianyi Well are listed in Table 1. From the dynamic variations in TDS and ion concentration during the pumping process, the TDS value of the studied brine increased from 211 to 247 g/L within 10 h of the initial pumping, and then stabilized at ~246 g/L during the subsequent pumping process (Figure 3). This stabilization of the TDS values suggests a relatively closed brine aquifer. However, the average TDS value of 242 ± 11 g/L significantly exceeds that of modern seawater with a value of 3.3 g/L, and the geothermal water at a circulation depth of 1.3 km around the study area within 100 km exhibiting concentrations ranging from 57.6 to 88.7 g/L [31]. The pattern of total milliequivalents of cations and anions, and concentrations for cations and anions in brine were similar to TDS, which increased within 10 h at the beginning of pumping and then stabilized in the region (Table 1; Figure S1).

Table 1. Physical and chemical parameters of the studied brine.

Elapsed Time	K^+	Na^+	Ca^{2+}	Mg^{2+}	Li^+	Sr^{2+}	Br^-	$CO_3{}^{2-}$	HCO_3	Cl^-	$SO_4{}^{2-}$	pH	TDS	Na^+: Cl^-	TMMC	TMMA	Balance Test
0	2992	72,159	6562	1028	89	226	357	78.09	154	128,196	1223	7.88	211	0.87	3650	3631	0.25
2	2645	73,010	5657	972	79	183	279	20.13	196	124,732	4780	7.70	210	0.90	3625	3571	0.75
4	3410	82,816	7243	1205	101	249	384	<0.43	148	146,290	1905	7.33	241	0.87	4175	4149	0.32
6	3366	81,077	7027	1195	99	244	354	<0.43	66	142,966	2147	7.33	236	0.87	4086	4056	0.37
10	3500	85,051	7186	1244	103	249	402	<0.43	150	149,126	2875	7.08	247	0.88	4275	4238	0.44
14	3455	84,655	7380	1243	103	247	375	<0.43	150	148,864	2808	7.10	247	0.88	4266	4230	0.43
18	3533	84,726	7252	1264	101	246	364	<0.43	151	148,963	2675	7.02	247	0.88	4266	4232	0.41
22	3481	83,501	7113	1241	101	243	363	<0.43	152	146,152	3440	7.03	245	0.88	4203	4160	0.51
26	3523	84,664	7404	1298	104	253	382	<0.43	153	148,574	3584	7.05	249	0.88	4275	4230	0.51
30	3571	83,368	7228	1255	103	250	387	<0.43	156	146,176	3587	7.28	247	0.88	4207	4163	0.53
34	3573	83,425	7215	1285	102	249	371	<0.43	152	146,631	3167	7.33	247	0.88	4211	4171	0.47
38	3475	82,861	7047	1266	102	249	363	<0.43	156	145,470	2965	7.34	243	0.88	4174	4136	0.45
42	3532	84,076	7153	1257	102	246	358	<0.43	164	146,211	4789	7.31	249	0.89	4233	4176	0.67
46	3507	83,765	7252	1272	101	249	364	<0.43	153	146,754	3663	7.28	247	0.88	4225	4180	0.53
50	3470	83,340	7092	1278	102	249	391	<0.43	152	145,950	3465	7.33	244	0.88	4198	4155	0.52
54	3531	84,822	7166	1251	103	244	383	<0.43	149	148,133	3752	7.30	249	0.88	4265	4219	0.54
58	3567	83,197	7157	1279	103	247	370	<0.43	152	146,173	3154	7.27	246	0.88	4198	4158	0.47
62	3508	84,283	7054	1245	103	246	397	<0.43	156	147,365	3351	7.27	247	0.88	4236	4194	0.50
66	3535	84,153	7270	1260	103	246	395	<0.43	156	147,021	4152	7.34	248	0.88	4243	4192	0.60
	3430 *	825,76 *	7077 *	1228 *	100 *	243 *	370 *		151 *	144,724 *	3236 *	7.29 *	242 *	0.88 *	4158 *	4118 *	

Note: elapsed time is in h; the concentrations of major ions are in mg/L; TDS is in g/L; Na^+: Cl^- is in molar ratio; TMMC and TMMA stand for the total milliequivalent of major cations, total milliequivalent of major anions, respectively; balance test is in %; * is denoted as average value.

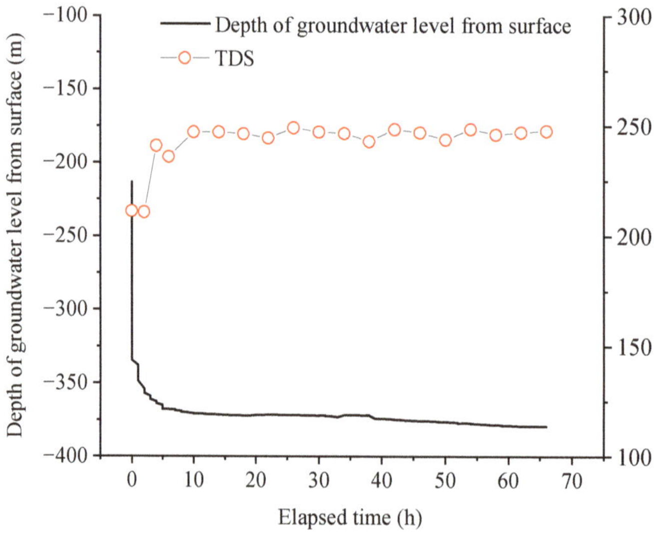

Figure 3. Variation in TDS values during pumping.

From the perspective of mean concentration, the milliequivalent of major cations ($Na^+ + K + Ca^{2+} + Mg^{2+} + Sr^{2+} + Li^+$) in the studied brine was 4154 ± 184 mEq/L, while it was 4118 ± 182 mEq/L for anions (Cl^- + SO_4^{2-} + HCO_3^-). The milliequivalent of major ions significantly exceeded that of geothermal water in both the northeastern region and the main urban area, ranging from 24.3 to 45.6 mEq/L for cations and from 24.3 to 47.6 mEq/L for anions [31,32]. The trends in anion and cation concentrations corresponded with the TDS levels, maintaining a relatively stable state (Table 1; Figure S1). It is noteworthy to mention that the presence of potassium (averaging at 3430 ± 223 mg/L), lithium (averaging at 100 ± 6 mg/L), and bromine (averaging at 370 ± 26 mg/L) in brine possesses significant industrial mining potential. However, the economic exploitation of these ions does not constitute the primary objective of this study, and therefore, will not be further discussed.

The hydrochemical facies serves as a fundamental method for conducting hydrogeochemical and environmental geochemical investigation, providing crucial insights into the characteristics of groundwater's chemical components. Currently, the Piper trilinear diagram introduced by Piper in 1944 as a renowned graphical tool for water chemistry analysis [33] has been predominantly employed to investigate hydrochemical facies. The diagram has been extensively utilized within the field of hydrochemistry [34–37]. By projecting the chemical component data of the studied brine samples onto the Piper trilinear diagram as shown in Figure 4, it can be observed that the dominant hydrochemical facies of the studied brine is Na-Cl. The studied brine, comprising of Na-Cl facies and having an average TDS value of 242 ± 11 g/L, falls within the range of chloride-dominated, halite-undersaturated waters as outlined by Hanor (1994) [6]. This range varies between 10 g/L and the upper threshold of 250–300 g/L [6].

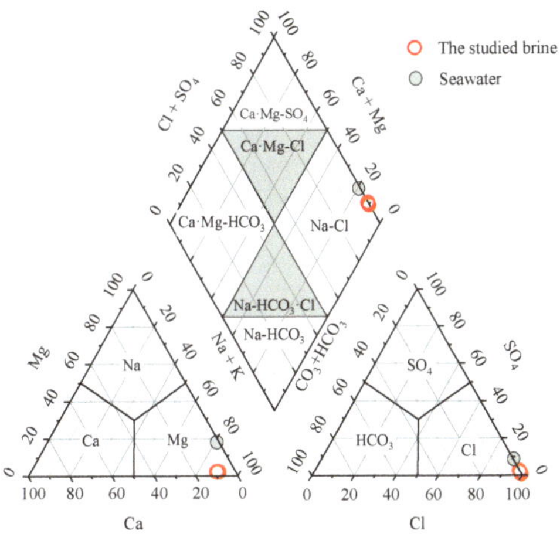

Figure 4. Piper trilinear diagram shows that the studied brine was a Na-Cl hydrochemical facies.

In 1970, Gibbs [38] introduced a graphical method to identify the sources of the main chemical elements in water bodies. This diagram categorizes the main formation mechanisms of water chemical components into three categories: atmospheric precipitation, rock weathering dominance, and evaporation–crystallization process. The method primarily utilizes the relationship between the TDS value and (K^+ + Na^+)/(K^+ + Na^+ + Ca^{2+}) as well as Cl^-/(Cl^+ + HCO_3^-) in the mEq/L ratio. Although initially proposed for surface water chemistry research [38,39], Gibbs's diagram has been widely adopted in groundwater research [40–43]. All the studied brine samples were positioned within the realm of the evaporation–crystallization process (Figure 5). This observation together with the extremely high TDS values indicates that the brine originated from evaporated ancient seawater.

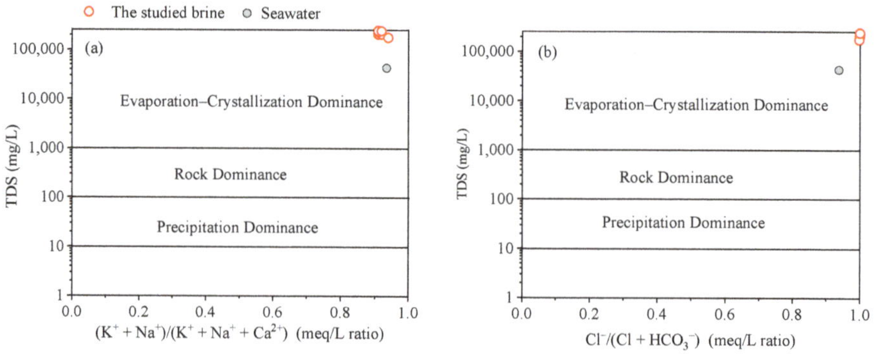

Figure 5. Gibbs's diagram depicts the range of brine studied, falling within the evaporation–crystallization process. The plot was adapted from Gibbs (1970) [38].

4.2. Stages of Brine Evolution

Chloride and bromine are halogen elements found in most groundwater. Cl^- and Br^- emerge as ideal indicators for studying the chemical evolution of brines due to their relatively inert and conservative characteristics [22,23]. Cl^- and Br^- concentrations were widely used as indicators of evaporation and halogen dissolution in formation waters (Carpenter, 1978) [19]. The Cl^-/Br^- ratio maintains its stability during the evaporation of seawater, reaching a halt at the point of halite saturation (Figure 6). After this saturation, Cl^- is preferentially eliminated via halite precipitation, resulting in a decrease in the Cl^-/Br^- ratio, as Carpenter (1978) [19] observed. Conversely, during halite dissolution, the Cl^-/Br^- ratio rises, shifting the data points to the right of the seawater evaporation trajectory. When a brine, evaporated beyond the halite saturation point, is diluted by meteoric water, the mixture shifts to the left of the seawater evaporation trajectory (Figure 6). Based on this understanding, the data of the study area plot to the right of the seawater evaporation trajectory, situated between the points of gypsum and halite saturation (Figure 6). This suggests that the studied brine underwent a moderate level of seawater evaporation, specifically taking place at a phase when the brine's saturation point lay somewhere between the saturation levels of gypsum and halite, together with some extent of halite dissolution.

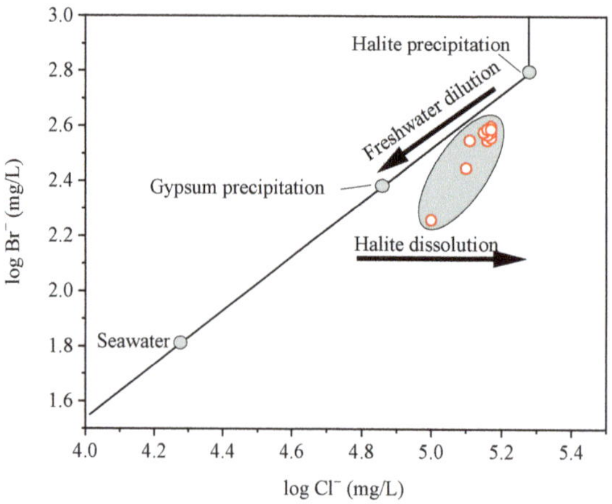

Figure 6. Cl^- vs. Br^- concentrations suggests a moderate level of evaporation between gypsum and halite saturation and some degree of halite dissolution.

The molar ratio of Na^+ to Cl^- in halite upon dissolution is 1:1. Conversely, the average molar ratio recorded in our studied brine samples exhibited a slight deviation, measuring 0.88, marginally surpassing the 0.86 ratio observed in seawater and indicating a subtle trend toward the dissolution ratio. This observation suggests that the composition of the studied brine primarily results from the evaporation of ancient seawater, with only a minor contribution from the dissolution process. Furthermore, these findings corroborate the validity of insights derived from the aforementioned Cl^-/Br^- ratio analysis. Additionally, the Giggenbach triangle plot (Figure S2) clearly indicates that the studied brine was saturated, providing further evidence that the primary source of the studied brine was from evaporated seawater, with halite dissolution as a secondary factor.

4.3. Potential Mechanisms Underlying the Ionic Enrichment or Depletion

To evaluate the extent of enrichment or depletion relative to evaporated modern seawater, Mirnejad et al. (2011) [26] utilized the abundances of cations and anions present in the brines as a means of evaluation. In this method, the average Cl^- concentration of all studied brine samples (144,724 mg/L) was initially divided by the Cl^- concentration in seawater from the Huanghai Sea, China (17,530 mg/L [44]), yielding a concentration factor. This factor was then multiplied with the concentration of each ion in seawater to estimate the concentration of these ions in evaporated seawater. Finally, the mean concentrations of the ions of the studied brine were normalized to the evaporated seawater concentrations by dividing the concentration of each ion by the corresponding concentration in the previously calculated evaporated seawater. The abundances of these ions in the studied brine are presented in Figure 7. The concentrations of Ca^{2+}, Li^+, Sr^{2+}, and Na^+ in the studied brine—standardized with respect to the chlorinity of modern seawater—were observed to be higher than those in evaporated modern seawater characterized by a value of 1, whereas the concentrations of K^+, Mg^{2+}, Br^-, HCO_3^-, and SO_4^{2-} were significantly lower. The anomalous patterns observed in ion concentrations suggest with certainty that the genesis of the studied brine cannot be solely attributed to seawater evaporation and halite dissolution but also involves distinct patterns of geochemical process that modify the brine's chemical composition in the study area.

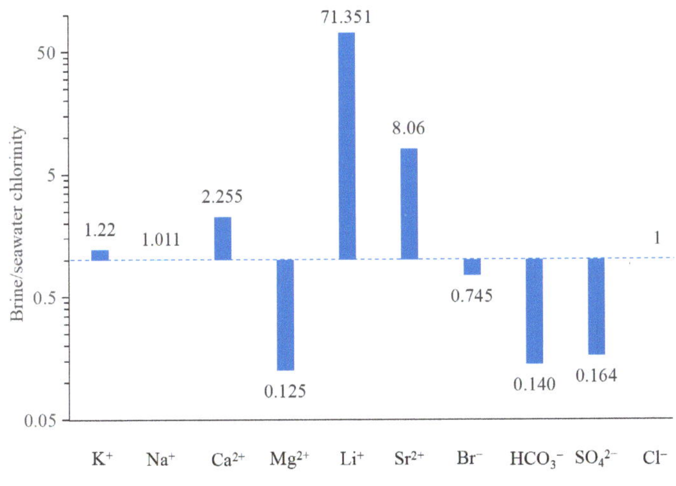

Figure 7. The ionic patterns of the studied brine standardized with respect to the chlorinity of modern seawater (concentrations from Chen, 1983), showing the brine cannot be attributed solely to the evaporation of seawater but also water–rock interactions. The value of dash line is 1 and indicates concentrated seawater. A value of 1 signifies equivalence to seawater, a value greater than 1 denotes an enrichment relative to seawater, whereas a value less than 1 indicates depletion compared to seawater.

The observed enrichment in Ca^{2+}, Li^{+}, K^{+}, Sr^{2+}, and Na^{+} concentrations and depletion in Mg^{2+}, Br^{-}, HCO_3^{-}, and SO_4^{2-} (Figure 7) is likely primarily attributable to the dolomitization process, combination of halite dissolution, upwelling of lithium- and potassium-bearing groundwater, calcium sulfate precipitation, biological sulfate reduction (BSR), and the common ion effect in the brine environments. This comprehension will be thoroughly examined and deliberated in the subsequent sections.

The Cambrian carbonate formations of the studied brine provide abundant $CaCO_3$ for dolomitization. The dolomitization involves the replacement of calcium carbonate (limestone) with calcium magnesium carbonate (dolomite), as expressed by Equation (1), thus resulting in a decrease in Mg^{2+} concentration and a dramatic increase in Ca^{2+} concentration that far exceeds the combined concentrations of SO_4^{2-} and HCO_3^{-}. The dolomitization ultimately made the studied brine calcium-rich. This phenomenon has been observed and documented in brines throughout the world, as evidenced by the studies conducted by Carpenter (1978) [22] and Gavrieli and Stein (2006) [45].

$$2CaCO_3 + Mg^{2+} = CaMg(CO_3)_2 + Ca^{2+} \quad (1)$$

The enrichment of Li^{+} concentration in the studied brine is likely due to a combination of halite dissolution and the influence of the crystalline basement. Although pegmatites and granites are the primary sources of Li^{+} [46], host rocks containing high-dissolved lithium brines are predominantly composed of carbonates, likely due to their inherently limited sorption capacities [27]. The lithium isotopic compositions (δ^7Li) of two brine samples were determined to be 12.58 and 12.8‰, respectively, significantly lower than the values observed in seawater (31‰ [47]). These measurements also fell below the range found in deep saline groundwater from the arid Atacama Desert region in Northern Chile, where values ranged between 23.5 and 31.5‰, suggesting a contribution from marine-derived salts and mineral alteration processes [48]. The δ^7Li values of the studied brine were much higher than those of deep-seated magmatic rocks in the region of western Sichuan, China (1.1‰ [49]). The lithium isotope values observed in the analyzed brine samples were situated between those of seawater and magmatic sources, suggesting that the Li^{+} likely originated from a combination of halite dissolution and the upwelling of lithium-bearing groundwater derived from alteration of minerals in the magmatic rocks within the region's crystalline basement. In a study conducted by Dugamin et al. [18], it was proposed that substantial enrichment of lithium occurs in brines that are primary in nature and initially in equilibrium with halite, and optimal reservoirs for lithium-rich brines are characterized by the accumulation of brine derived from evaporites during tectonic events, such as compressional events or salt tectonics, located in close proximity to the current reservoir. To maintain the integrity of these reservoirs and prevent dilution or depletion of lithium, it is crucial to shield them from convection and infiltration of surface water via recharge zones. Further investigation is warranted to elucidate the precise sources of these ions in the future utilizing lithium isotope technology.

The observed enrichment of K^{+} in the studied brine may align with the mechanism underlying Li^{+} enrichment. In addition to contributions from halite dissolution, it could also originated from the upwelling of potassium-rich fluids from the region's crystalline basement.

As indicated by Carpenter (1978) [22], the presence of Sr^{2+} in the brines exhibits a strong correlation with the recrystallization process of aragonite transforming into calcite. However, the Sr^{2+} ion possesses a radius comparable to that of Ca^{2+} and holds an identical valence, thereby facilitating its substitution for Ca^{2+} in the crystal lattices of diverse minerals, including kainite, plagioclase, gypsum, calcite, and dolomite [50]. Hence, the presence of Sr^{2+} in the studied brine is probably linked to the dolomitization process, which ultimately leads to the enrichment of Sr^{2+}.

The depletion of SO_4^{2-} in studied brine can be attributed to two primary mechanisms. Firstly, there is direct precipitation, where SO_4^{2-} combines with enriched Ca^{2+} to form calcium sulfate, which can manifest as either gypsum or anhydrite. This enrichment of Ca^{2+} is a direct consequence of the dolomitization process within the brine as previously mentioned (Equation (1)), enhancing the availability of Ca^{2+} for precipitation reactions. This precipitation process is described by Equation (2):

$$Ca^{2+} + SO_4{}^{2-} = CaSO_4 \quad (2)$$

Secondly, sulfate reduction is another mechanism that contributes to the loss of $SO_4{}^{2-}$ from the brine. There are two types of sulfate reduction processes in the lithosphere [51]. The first type is biological sulfate reduction (BSR), which is mediated by microorganisms such as sulfate-reducing bacteria and occurs at relatively low temperatures, generally below 100 °C, and at depth of the strata less than 2000–2500 m. The second type is thermochemical sulfate reduction (TSR) involving hydrocarbons and sulfates, which typically takes place at temperatures exceeding 120 °C [51]. In this study of Tianyi Well, the water temperature at a depth of 1008 m was 45 °C, which suggests that BSR was the main process responsible for the $SO_4{}^{2-}$ depletion occurring there. A pronounced H_2S odor was detected from the brine during sampling, verifying the occurrence of BSR. The occurrence of sulfate reduction is convincingly demonstrated by the exceedingly low desulfurization coefficient (approaching a value of 0, Table S1) observed in the studied brine, thereby providing empirical support for the process. This BSR process, often accompanied by the precipitation of calcite, involves a complex set of reactions that reduce sulfate while producing a range of byproducts, including H_2S, water, CO_2, and ultimately, $CaCO_3$, as expressed by Equation (3). While the precise stoichiometry and intermediate steps of sulfate reduction may vary depending on specific conditions, a generalized representation of this process could be encapsulated by the following reaction scheme, as expressed in Equation (3) [22], with the understanding that the coefficients are illustrative and may not precisely reflect all scenarios.

$$Ca^{2+} + SO_4{}^{2-} + 1.33CH_2 = H_2S + CaCO_3 + 0.33H_2O + 0.33CO_2 \quad (3)$$

The depletion of $HCO_3{}^-$ concentration observed in the studied brine can be rationalized through the mechanism of the common ion effect. In carbonate aquifers, a high concentration of $HCO_3{}^-$ is expected; however, the brine in the study area exhibited a very low $HCO_3{}^-$ concentration averaging at 151 ± 22 mg/L (Table 1), even lower than that found in the karst shallow groundwater (~300 mg/L) located 400 km away from the Tianyi Well [52]. This substantial disparity may imply that $HCO_3{}^-$ in the studied brine precipitated with Ca^{2+} ions from the dissolution of gypsum/anhydrite within the geological formation to trigger calcite precipitation. This phenomenon, commonly referred to as the common ion effect, is frequently encountered in deep aquifers [32,53–55]. The chemical reaction governing this process can be expressed as follows:

$$2HCO_3{}^- + CaSO_4{\cdot}nH_2O \rightarrow CaCO_3 + CO_2 + SO_4{}^{2-} + (1 + n)H_2O \quad (4)$$

where n equals 2 for gypsum and 0 for anhydrite, highlighting the dissolution reactions specific to these minerals.

The mechanism of bromine depletion in brine in the studied brine is unclear. Theoretically, bromine is a highly stable element, and its concentration tends to remain nearly constant. It precipitates only when the evaporation rate reaches 90 times that of seawater [23]. However, the brine in this study was situated between gypsum saturation (3.8 times seawater [23]) and halite saturation (10.6 times seawater [23]) as aforementioned, indicating that the depletion of Br^- was likely influenced by additional factors. Further investigation is warranted.

The slight elevation in Na^+ concentration, as illustrated in Figure 7, suggests that the salinity of the brine has remained largely stable, thereby indicating an absence of significant cation exchange within the carbonate brine system.

5. Conclusions

The brine of Na-Cl type in Cambrian carbonate-dominated rocks in the northeastern Chongqing region of Southwestern China is potentially sourced from interstitial fluids present within formations abundant in halite. The brine exhibited an average TDS value of 242 ± 11 g/L. It is noteworthy that the presence of potassium, lithium, and bromine in this brine holds significant potential for industrial mining.

The Cl^-/Br^- ratio indicates that the studied brine underwent moderate seawater evaporation, occurring between the saturation levels of gypsum and halite, accompanied by some degree of halite dissolution. Upon standardization with respect to the chlorinity of modern seawater, the concentrations of Ca^{2+}, Li^+, K^+, Sr^{2+}, and Na^+ in the studied brine were observed to be elevated compared to those in evaporated modern seawater; conversely, the concentrations of Mg^{2+}, Br^-, HCO_3^-, and SO_4^{2-} were notably lower. These anomalous patterns observed in ion concentrations are attributed to various water–rock interactions, including the dolomitization process, combination of halite dissolution and upwelling of lithium- and potassium-bearing groundwater, calcium sulfate precipitation, biological sulfate reduction (BSR), and the common ion effect within the brine. This brine underwent compositional evolution through its interactions with minerals in the surrounding rocks, ultimately resulting in its current chemical composition.

While this study has largely elucidated the sources of the main chemical components in the studied brine, uncertainties remain that necessitate additional isotopic data for validation. Furthermore, both the origin of the water and the chemical components in brine are critically important. Further investigation is warranted to ascertain the specific origins of brine water.

Supplementary Materials: The following supporting information can be downloaded at: https://www.mdpi.com/article/10.3390/w16192859/s1, Figure S1: Trend in anion and cation concentrations, maintaining a relatively stable state over time; Figure S2: Giggenbach triangle plot indicates that the brine was saturated; Table S1: Desulfurization coefficient.

Author Contributions: Conceptualization, Z.-l.Z., K.Z., B.-c.Z. and P.-h.Y.; Data curation, L.Z. and P.-h.Y.; Funding acquisition, Z.-l.Z.; Investigation, Z.-l.Z., B.X., C.-m.W., L.Z. and P.-h.Y.; Methodology, Z.-l.Z. and P.-h.Y.; Project administration, Z.-l.Z. and P.-h.Y.; Resources, Z.-l.Z. and P.-h.Y.; Software, L.Z. and P.-h.Y.; Supervision, P.-h.Y.; Validation and visualization, L.Z. and P.-h.Y.; Writing—original draft, Z.-l.Z., L.Z. and P.-h.Y.; Writing—review and editing, Z.-l.Z., L.Z. and P.-h.Y. All authors have read and agreed to the published version of the manuscript.

Funding: This research was funded by the Chongqing Technology Innovation and Application Development Project, Science and Technology Program from Chongqing Municipal Bureau of Planning and Natural Resources (under grant number: [2023]175-4), and Wuxi County.

Data Availability Statement: Data is contained within the article or Supplementary Material.

Acknowledgments: We express our gratitude to Tao Wang and Jie Feng for their sampling work in the field. We are also deeply appreciative of the valuable and constructive suggestions offered by the four anonymous reviewers, as well as Assistant Editors Keely Zhang and Elma Shi.

Conflicts of Interest: The authors declare no conflicts of interest.

References

1. Person, M.; McIntosh, J.; Kim, J.-H.; Noyes, C.; Bailey, L.; Lingrey, S.; Krantz, R.; Lucero, D.; Reiners, P.; Ferguson, G. Hydrologic windows into the crystalline basement and their controls on groundwater flow patterns across the Paradox Basin, western USA. *Geol. Soc. Am. Bull.* **2024**, *136*, 3156–3168. [CrossRef]
2. Ngombe, O.G.; Walter, J.; Chesnaux, R.; Molson, J. Application of hierarchical cluster analysis and principal component analysis to identify compositional trends of brine in crystalline basements and sedimentary basins. *Appl. Geochem.* **2024**, *169*, 106030. [CrossRef]
3. Marza, M.; Ferguson, G.; Thorson, J.; Barton, I.; Kim, J.-H.; Ma, L.; McIntosh, J. Geological controls on lithium production from basinal brines across North America. *J. Geochem. Explor.* **2024**, *257*, 107383. [CrossRef]
4. Yu, X.C.; Wang, C.L.; Huang, H.; Yan, K. Origin of lithium in oilfield brines in continental petroliferous basin: Insights from Li and Sr isotopes in the Jianghan Basin, central China. *Mar. Pet. Geol.* **2024**, *160*, 106576. [CrossRef]

5. Yoshimura, S. Controls on the Salinity of Sedimentary Basinal Fluids Under Constant Chemogravitational Potential Conditions. *Geochem. Geophys. Geosyst.* **2023**, *24*, e2022GC010628. [CrossRef]
6. Hanor, J.S. *Origin of Saline Fluids in Sedimentary Basins*; Special Publication; The Geological Society of London: London, UK, 1994.
7. Johnson, K.S. Evaporite-karst problems and studies in the USA. *Environ. Geol.* **2008**, *53*, 937–943. [CrossRef]
8. Michael, K.; Machel, H.G.; Bachu, S. New insights into the origin and migration of brines in deep Devonian aquifers, Alberta, Canada. *J. Geochem. Explor.* **2003**, *80*, 193–219. [CrossRef]
9. Ferguson, G.; McIntosh, J.C.; Grasby, S.E.; Hendry, M.J.; Jasechko, S.; Lindsay, M.B.J.; Luijendijk, E. The Persistence of Brines in Sedimentary Basins. *Geophys. Res. Lett.* **2018**, *45*, 4851–4858. [CrossRef]
10. Carreira, P.M.; Marques, J.M.; Nunes, D. Source of groundwater salinity in coastline aquifers based on environmental isotopes (Portugal): Natural vs. human interference. A review and reinterpretation. *Appl. Geochem.* **2014**, *41*, 163–175. [CrossRef]
11. Labotka, D.M.; Panno, S.V.; Locke, R.A.; Freiburg, J.T. Isotopic and geochemical characterization of fossil brines of the Cambrian Mt. Simon Sandstone and Ironton-Galesville Formation from the Illinois Basin, USA. *Geochim. Cosmochim. Acta* **2015**, *165*, 342–360. [CrossRef]
12. Kazemihokmabad, P.; Khamehchi, E.; Kalatehno, J.M.; Ebadi, R. A comparative study of brine solutions as completion fluids for oil and gas fields. *Sci. Rep.* **2024**, *14*, 12628. [CrossRef] [PubMed]
13. Bute, S.I.; Zhou, J.-X.; Luo, K.; Girei, M.B.; Peter, R.T. Pb-Zn-Ba deposits in the Nigerian Benue Trough: A synthesis on deposits classification and genetic model. *Ore Geol. Rev.* **2024**, *166*, 105947. [CrossRef]
14. Stueber, A.M.; Walter, L.M. Origin and chemical evolution of formation waters from Silurian-Devonian strata in the Illinois Basin, USA. *Geochim. Cosmochim. Acta* **1991**, *55*, 309–325. [CrossRef]
15. Billings, G.K.; Hitchon, B.; Shaw, D. Geochemistry and origin of formation waters in the Western Canada Sedimentary. *Chem. Geol.* **1969**, *4*, 211–223. [CrossRef]
16. Graf, D.L. Chemical osmosis, reverse chemical osmosis, and the origin of subsurface brines. *Geochim. Cosmochim. Acta* **1982**, *46*, 1431–1448. [CrossRef]
17. Gianfriddo, C.; Bull, S. Brine reflux as a possible first order control on the alkali geochemistry of baseline sedimentary rocks in the Proterozoic North Australian Zinc Belt. *Precambrian Res.* **2023**, *390*, 107044. [CrossRef]
18. Dugamin, E.J.; Cathelineau, M.; Boiron, M.-C.; Richard, A.; Despinois, F. Lithium enrichment processes in sedimentary formation waters. *Chem. Geol.* **2023**, *635*, 121626. [CrossRef]
19. Warr, O.; Giunta, T.; Onstott, T.C.; Kieft, T.L.; Harris, R.L.; Nisson, D.M.; Lollar, B.S. The role of low-temperature ^{18}O exchange in the isotopic evolution of deep subsurface fluids. *Chem. Geol.* **2021**, *561*, 120027. [CrossRef]
20. Engle, M.A.; Reyes, F.R.; Varonka, M.S.; Orem, W.H.; Ma, L.; Ianno, A.J.; Schell, T.M.; Xu, P.; Carroll, K.C. Geochemistry of formation waters from the Wolfcamp and "Cline" shales: Insights into brine origin, reservoir connectivity, and fluid flow in the Permian Basin, USA. *Chem. Geol.* **2016**, *425*, 76–92. [CrossRef]
21. Saleh, A.; Gad, A.; Ahmed, A.; Arman, H.; Farhat, H.I. Groundwater hydrochemical characteristics and water quality in Egypt's central eastern desert. *Water* **2023**, *15*, 971. [CrossRef]
22. Carpenter, A.B. Origin And Chemical Evolution Of Brines In Sedimentary Basins. In Proceedings of the SPE Annual Fall Technical Conference and Exhibition, Houston, TX, USA, 1–3 October 1978.
23. Mccaffrey, M.A.; Lazar, B.; Holland, H.D. The Evaporation Path of Seawater and the Coprecipitation of Br^- and K^+ with Halite. *J. Sediment. Petrol.* **1987**, *57*, 928–937. [PubMed]
24. Freeman, J.T. The use of bromide and chloride mass ratios to differentiate salt-dissolution and formation brines in shallow groundwaters of the Western Canadian Sedimentary Basin. *Hydrogeol. J.* **2007**, *15*, 1377–1385. [CrossRef]
25. Davisson, M.L.; Criss, R.E. Na-Ca-CI relations in basinal fluids. *Geochim. Cosmochim. Acta* **1996**, *60*, 2743–2752. [CrossRef]
26. Mirnejad, H.; Sisakht, V.; Mohammadzadeh, H.; Amini, A.H.; Rostron, B.J.; Haghparast, G. Major, minor element chemistry and oxygen and hydrogen isotopic compositions of Marun oil-field brines, SW Iran: Source history and economic potential. *Geol. J.* **2011**, *46*, 1–9. [CrossRef]
27. Darvari, R.; Nicot, J.P.; Scanlon, B.R.; Kyle, J.R.; Elliott, B.A.; Uhlman, K. Controls on lithium content of oilfield waters in Texas and neighboring states (USA). *J. Geochem. Explor.* **2024**, *257*, 107363. [CrossRef]
28. Yan, K.; Wang, C.L.; Chen, R.Y.; Liu, C.L.; Wang, J.Y.; Yu, X.C.; Shen, L.J.; Li, R.Q.; Zhou, Y.; Zhou, Q. Origin and evolution of deep lithium-rich brines in the southwest Jianghan Basin, central China: Evidence from hydrochemistry and stable isotopes. *J. Hydrol.* **2023**, *626*, 130163. [CrossRef]
29. Hu, G.; Teng, J.; Ruan, X.; Wang, Q.; Yan, Y.; Wang, P.; Xiong, S. Magnetic anomaly characteristics and crystalline basement variation of the Qinling orogenic belt and its adjacent areas. *Chin. J. Geophys.* **2014**, *57*, 556–571. (In Chinese with English Abstract)
30. Wang, S.; Zheng, M. Cambrian salt-forming environment in northeastern Sichuan basin and Its significance for finding potash. *Sci. Technol. Rev.* **2014**. *32*, 41–49, (In Chinese with English Abstract)
31. Zhang, J.; Yang, P.; Groves, C.; Luo, X.; Wang, Y. Influence of geological structure on the physicochemical properties and occurrence of middle-deep groundwater in Chongqing, Southwest China. *J. Hydrol.* **2022**, *610*, 127782. [CrossRef]
32. Yang, P.; Cheng, Q.; Xie, S.; Wang, J.; Chang, L.; Yu, Q.; Zhan, Z.; Chen, F. Hydrogeochemistry and geothermometry of deep thermal water in the carbonate formation in the main urban area of Chongqing, China. *J. Hydrol.* **2017**, *549*, 50–61. [CrossRef]
33. Piper, M. A graphic procedure in the geochemical interpretation of water-analyses. *Trans. Am. Geophys. Union* **1944**, *25*, 914–928.

34. Kazakis, N.; Matiatos, L.; Ntona, M.M.; Bannenberg, M.; Kalaitzidou, K.; Kaprara, E.; Mitrakas, M.; Loannidou, A.; Vargemezis, G.; Voudouris, K. Origin, implications and management strategies for nitrate pollution in surface and ground waters of Anthemountas basin based on a $\delta^{15}N\text{-}NO_3^-$ and $\delta^{18}O\text{-}NO_3^-$ isotope approach. *Sci. Total Environ.* **2020**, *724*, 138211. [CrossRef] [PubMed]
35. Wu, J.H.; Zhou, H.; He, S.; Zhang, Y.X. Comprehensive understanding of groundwater quality for domestic and agricultural purposes in terms of health risks in a coal mine area of the Ordos basin, north of the Chinese Loess Plateau. *Environ. Earth Sci.* **2019**, *78*, 446. [CrossRef]
36. Chenaker, H.; Houha, B.; Vincent, V. Hydrogeochemistry and geothermometry of thermal water from northeastern Algeria. *Geothermics* **2018**, *75*, 137–145. [CrossRef]
37. Liu, F.; Song, X.; Yang, L.; Zhang, Y.; Han, D.; Ma, Y.; Bu, H. Identifying the origin and geochemical evolution of groundwater using hydrochemistry and stable isotopes in the Subei Lake basin, Ordos energy base, Northwestern China. *Hydrol. Earth Syst. Sci.* **2015**, *19*, 551–565. [CrossRef]
38. Gibbs, R.J. Mechanisms Controlling World Water Chemistry. *Science* **1970**, *170*, 1088–1090. [CrossRef]
39. Marandi, A.; Shand, P. Groundwater chemistry and the Gibbs Diagram. *Appl. Geochem.* **2018**, *97*, 209–212. [CrossRef]
40. Smida, H.; Tarki, M.; Dassi, L. Groundwater quality and mineralization process in the Braga shallow aquifer, Central Tunisia: An overview. *Carbonates Evaporites* **2022**, *37*, 28. [CrossRef]
41. Muhammad, S.; Ullah, I. Spatial and seasonal variation of water quality indices in Gomal Zam Dam and its tributaries of south Waziristan District, Pakistan. *Environ. Sci. Pollut. Res.* **2022**, *29*, 29141–29151. [CrossRef]
42. Ez-zaouy, Y.; Bouchaou, L.; Saad, A.; Hssaisoune, M.; Brouziyne, Y.; Dhiba, D.; Chehbouni, A. Morocco's coastal aquifers: Recent observations, evolution and perspectives towards sustainability. *Environ. Pollut.* **2022**, *293*, 118498. [CrossRef]
43. Shen, B.B.; Wu, J.L.; Zhan, S.; Jin, M.; Saparov, A.S.; Abuduwaili, J. Spatial variations and controls on the hydrochemistry of surface waters across the Ili-Balkhash Basin, arid Central Asia. *J. Hydrol.* **2021**, *600*, 126565. [CrossRef]
44. Chen, Y. Sequence of salt separation and regularity of some trace elements distribution during isothermal evaporation (25 °C) of the Huanghai Sea water. *Acta Geol. Sin.* **1983**, 379–390. (In Chinese with English Abstract)
45. Gavrieli, I.; Stein, M. On the origin and fate of the brines in the Dead Sea basin. In *New Frontiers in Dead Sea Pale-Oenvironmental Research*; Enzel, Y., Agnon, A., Stein, M., Eds.; Geological Society of America Special Paper 401; Geological Society of America: Boulder, CO, USA, 2006; pp. 183–194.
46. Parnell, J.; Armstrong, J.G.T. Surface expression of Late Caledonian magmatic lithium concentration, in the Rhynie Chert, UK. *Geochem. Explor. Environ. Anal.* **2023**, *23*, geochem2023-028. [CrossRef]
47. You, C.F.; Chan, L.H. Precise determination of lithium isotopic composition in low concentration natural samples. *Geochim. Cosmochim. Acta* **1996**, *60*, 909–915. [CrossRef]
48. Herrera, C.; Godfrey, L.; Urrutia, J.; Custodio, E.; Gamboa, C.; Jódar, J.; Lam, E.; Fuentes, J. Origin of old saline groundwater in the deep coastal formations of the Atacama Desert region: Consideration of lithium, boron, strontium and uranium isotopes contents. *J. Hydrol.* **2023**, *624*, 129919. [CrossRef]
49. Yu, F.; Yu, Y.; Wang, D.; Gao, J.; Wang, C.; Guo, W. Application of Li isotope in geothermal fluid-rock interaction: A case study of modern Li-rich geothermal water in western Sichuan. *Acta Petrol. Sin.* **2022**, *38*, 472–482. (In Chinese with English Abstract)
50. Yang, N.; Su, C.L.; Liu, W.B.; Zhao, L. Occurrences and mechanisms of strontium-rich groundwater in Xinglong County, northern China: Insight from hydrogeological and hydrogeochemical evidence. *Hydrogeol. J.* **2022**, *30*, 2043–2057. [CrossRef]
51. Trudinger, P.A.; Chambers, L.A.; Smith, J.W. Low-temperature sulphate reduction: Biological versus abiological. *Can. J. Earth Sci.* **1985**, *22*, 1910–1918. [CrossRef]
52. Yang, P.H.; Yuan, D.X.; Ye, X.C.; Xie, S.Y.; Chen, X.B.; Liu, Z.Q. Sources and migration path of chemical compositions in a karst groundwater system during rainfall events. *Chin. Sci. Bull.* **2013**, *58*, 2488–2496. [CrossRef]
53. Yang, P.H.; Luo, D.; Hong, A.H.; Ham, B.; Xie, S.Y.; Ming, X.X.; Wang, Z.X.; Pang, Z.H. Hydrogeochemistry and geothermometry of the carbonate-evaporite aquifers controlled by deep-seated faults using major ions and environmental isotopes. *J. Hydrol.* **2019**, *579*, 124116. [CrossRef]
54. Blasco, M.; Auque, L.F.; Gimeno, M.J. Geochemical evolution of thermal waters in carbonate—Evaporitic systems: The triggering effect of halite dissolution in the dedolomitisation and albitisation processes. *J. Hydrol.* **2019**, *570*, 623–636. [CrossRef]
55. Gil-Marquez, J.M.; Barbera, J.A.; Andreo, B.; Mudarra, M. Hydrological and geochemical processes constraining groundwater salinity in wetland areas related to evaporitic (karst) systems. A case study from Southern Spain. *J. Hydrol.* **2017**, *544*, 538–554. [CrossRef]

water

MDPI

Article

Analysis of Runoff Changes in the Wei River Basin, China: Confronting Climate Change and Human Activities

Ruirui Xu [1,2,3], Chaojun Gu [4], Dexun Qiu [1,2,3], Changxue Wu [2], Xingmin Mu [1,2,3,*] and Peng Gao [1,2,3,*]

1 State Key Laboratory of Soil Erosion and Dryland Farming on the Loess Plateau, Institute of Soil and Water Conservation, Chinese Academy of Sciences and Ministry of Water Resources, Xianyang 712100, China
2 Institute of Soil and Water Conservation, Northwest A & F University, Xianyang 712100, China
3 University of Chinese Academy of Sciences, Beijing 100049, China
4 Yangtze River Basin Monitoring Center Station for Soil and Water Conservation, Changjiang Water Resources Commission, Wuhan 430012, China
* Correspondence: xmmu@ms.iswc.ac.cn (X.M.); gaopeng@ms.iswc.ac.cn (P.G.)

Abstract: Abrupt runoff reduction in the Wei River Basin (WRB) has attracted extensive attention owing to climate change and human activities. Nevertheless, previous studies have inadequately assessed the respective contributions of climate variability and human activities to runoff change on different spatial scales. Using Mann–Kendall and Pettitt's methods, this study identified long-term (1970–2018) changes in hydro-meteorological variables. Furthermore, the Budyko-based method was used to quantify the influence of climate change and human activities on runoff change at different spatial scales of the WRB, including the whole WRB, three sub-basins, and sixteen catchments. The results show that a significant decrease trend was identified in runoff at different spatial scales within the WRB. Runoff in almost all catchments showed a significant downward trend. Temperature, potential evapotranspiration, and the parameter n showed significant increases, whereas no significant trend in precipitation was observed. The change in runoff was mainly concentrated in the mid-1990s and early 2000s. Anthropogenic activities produced a larger impact on runoff decrease in the WRB (62.8%), three sub-basins (53.9% to 65.8%), and most catchments (–47.0% to 147.3%) than climate change. Dramatic catchment characteristic changes caused by large-scale human activities were the predominant reason of runoff reduction in the WRB. Our findings provide a comprehensive understanding of the dominate factors causing runoff change and contribute to water resource management and ecosystem health conservation in the WRB.

Keywords: runoff; climatic variation; land cover change; Budyko framework; Wei River Basin

1. Introduction

Runoff, a key component of the hydrological cycle, is strongly influenced by climate variability and anthropogenic activities, and therefore, changes dramatically at various spatial and temporal scales [1,2]. Climate change includes the changes in precipitation, temperature, and potential evaporation, which largely affect the future runoff of a basin [3]. In addition to climate change, anthropogenic factors such as ecological restoration programs, urbanization, industrialization, reservoir operation, large-scale irrigation, and drainage have strongly shaped the runoff regime [4]. Previous studies have shown that global runoff has increased over the 20th century [5]. However, river flow has decreased significantly in many basins worldwide, such as the Spanish basins [6], the Lake Issyk-Kul Basin [7], East African basins [8], as well as some basins in China, such as the Yellow River Basin [9] and the Shiyang River Basin [10], especially in arid and semi-arid regions [11,12]. The continuing decline in runoff has caused a severe water crisis and ecosystem imbalance, threatening the high state and sustainable development of the economy and society [13].

According to the IPCC report [14], the global climate has been undergoing extensive and complicated change in the past century, which aggravates the uneven distribution of precipitation (P) [15,16] and potential evapotranspiration (PET) [17], and consequently influences water balance. For rivers in tropical areas at low latitudes, the long-term tendency of runoff change is generally consistent with that of P [18,19]; however, for rivers in middle latitudes and some alpine regions, the correlation between runoff and P is weak [20,21]. The direct or indirect interference of runoff by human activities, including deforestation, afforestation, reservoir operation, large-scale irrigation, and urbanization, is an important reason for this regional difference [9,22,23]. Quantitative assessment of the impacts of climate variability and anthropogenic activities on runoff regimes at the regional scale is therefore vital to water resource management.

Several methods have been applied to analyze the influence of climate variability and human intervention on alterations in runoff [24–26]. Empirical statistical methods can achieve results quickly and easily; however, it is difficult to interpret complex and non-linear hydrological behaviors. Hydrological models, although powerful, can be constrained by the inherent variability and uncertainty in the impact of human activities on hydrological systems [27], and the difficulty in obtaining numerous input data related to human activities, such as water abstraction and irrigation data [28,29]. With the Budyko [30] hypothesis, various Budyko-based approaches have been widely used to explore the non-linear interactions among climate, landscape, and hydrology due to its straightforward calculations and physical mechanisms [31]. It suggests that the ratio of precipitation to potential evapotranspiration (P/PET) is the primary factor determining the hydrological balance at basin level and thereby dictates the dryness or wetness of a basin. Early empirical formulations [30,32,33] represented the general hydrological reactions of catchments to global climate forcing, as they were based on continental or global runoff information. In 1981, Fu [34] incorporated the catchment characteristic parameter (parameter n) into catchment water balance. Parameter n comprehensively reflects catchment characteristics, including the characteristics of soil, topography, geologic properties, and vegetation cover [31,35]. In 2008, Yang et al. [36] derived the Choudhury–Yang formula based on Choudhury's [37] proposed empirical theoretical equation. The early consistency of the rich theory and the later improvements to the methodology increased the credibility of the method. Therefore, coupled with the elasticity coefficient method, the Choudhury–Yang method has been successfully employed to analyze and separate the impacts of climate change and human activities on changes in runoff [3,27,38,39].

As the largest tributary of the Yellow River, the Wei River discharges approximately 19.7% of water into the Yellow River every year, which plays a significant role in the management and sustainable development of water resources [40]. It provides ~85% of the water supply for irrigation, domestic, and industrial use, and nourishes more than 22 million people in this basin, making it an essential factor for regional economic development [41]. However, significant runoff reductions have been reported in many studies [40,42]. With the growth in population and development of the economy, the demand for water resources tends to increase, and water scarcity becomes more severe [43,44]. Moreover, the northern part of the Wei River Basin (WRB) is located in the Loess Plateau, which suffers from serious soil erosion. Catchment characteristics and runoff patterns have been significantly affected by the execution of ecological restoration projects [44–46]. Despite the fact that many studies have separated the influence of climate change and human intervention on WRB runoff [24,40,46], different studies have arrived at disparate conclusions due to their varying methods and periods. Moreover, most studies only focused on alterations in runoff of the entire basin or a single hydrological station but ignored the runoff change in smaller catchments. Therefore, we conducted a comprehensive attribution analysis of runoff change at different catchment scales of the WRB from 1970 to 2018 by using the Budyko-based method. The main objectives of this study were to (1) depict the trend of runoff, P, PET, temperature, and parameter n in different catchments and the change for abrupt year; (2) assess the spatial-temporal variability of the runoff elasticity coefficients to P, PET, and parameter n in different catchments; and (3) quantitatively estimate the changes in runoff attributable to climate variables and land cover and clarify the possible influencing factors at different spatial scales.

2. Study Area and Data

2.1. Study Area

The WRB (34–38° N; 104–111° E) is located in the lower midstream of the Yellow River, covering an area of approximately 1.35×10^5 km^2 (Figure 1). The Wei River rises in the Niaoshu Mountain of Gansu Province, through three provinces (Gansu, Shaanxi, and Ningxia), and across the Loess Plateau before entering the Yellow River. Its two primary tributaries are the Jing River and Beiluo River, which divide the Wei River Basin into three sub-basins: the Jing River (JR) basin, Beiluo River (BLR) basin, and the upstream of the Wei River (UWR) basin. Traditionally, the region is named the Jing–Luo–Wei Region, and the catchment characteristics of the three tributaries are very different in space. The WRB is situated in the continental monsoon climate zone, and experiences high temperatures and abundant P during summer, whereas winter brings low temperatures and scarce P. The average annual P is approximately 536 mm. The distribution of P is uneven across space, with the majority of it (over 60%) being concentrated within the months from June to September (Figure 2).

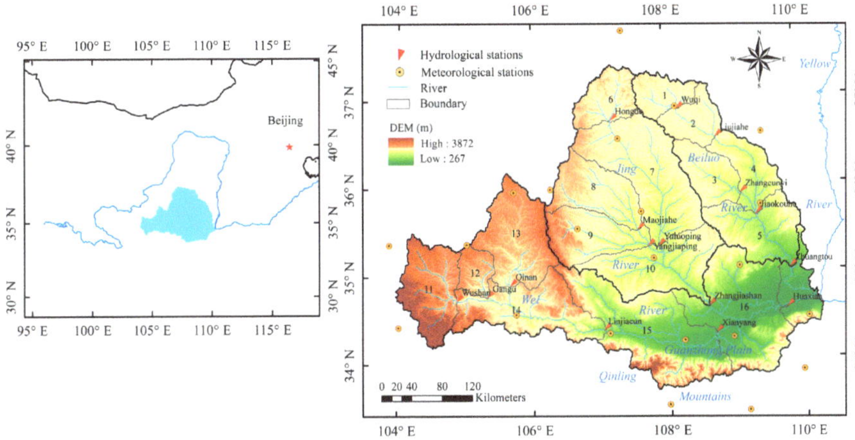

Figure 1. Location of the study region and hydrological and meteorological stations in the Wei River Basin.

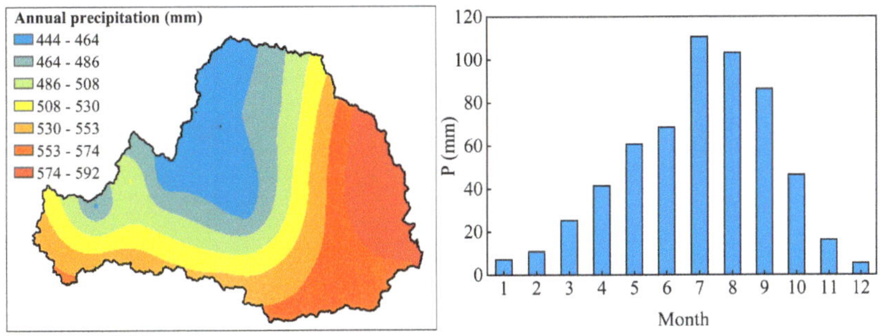

Figure 2. Spatial distribution of multi-year average precipitation and intra-annual distribution of precipitation in the Wei River Basin.

2.2. Data

Annual runoff data for 16 hydrological stations from 1970 to 2018 were made available by the Yellow River Water Resources Commission. Daily meteorological data (1970–2018) from 22 meteorological stations were collected from the National Meteorological Information Centre of China (NMIC) (https://www.nmic.cn/ (accessed on 1 January 2020)), and consisted of precipitation, wind speed, relative humidity, sunshine duration, and average, maximum, and minimum temperature. The information of the hydrological stations is shown in Table S1 in the Supplementary Materials. Land cover datasets from 1980, 1995, and 2018 were available at the Chinese Academy of Sciences Resource and Environmental Science Data Center (http://www.resdc.cn/ (accessed on 1 January 2020)). The land cover was divided into six classifications: cropland, forest, settlement, grassland, water, and others, with a spatial resolution of 1 km. The 15 day Normalized Difference Vegetation Index (NDVI) dataset was acquired from the Global Inventory Modeling and Mapping Studies (GIMMS) NDVI3g vegetation index dataset designed by NASA (https://ladsweb.modaps.eosdis.nasa.gov/ (accessed on 1 January 2020)), with a spatial resolution of 500 m from 1982 to 2015. The annual mean and maximum NDVI values were used in this study.

The entire basin was divided into 16 catchments, considering the distribution of hydrological stations, to examine the spatial features of runoff change. The catchment controlled by a single hydrological station was divided into one catchment, and the runoff in that catchment was equal to the observed runoff of this hydrological station. For instance, the runoff value of catchment 1 was equal to the runoff observed at the Wuqi station. The catchment controlled by adjacent hydrological stations was divided into one catchment, and the runoff value of this catchment was calculated using the runoff difference between the upstream and downstream hydrological stations. For example, the runoff of catchment 2 was the observed runoff of Liujiahe minus that of Wuqi. Table S2 in the Supplementary Materials presented comprehensive details about the catchments.

3. Methodology

3.1. Trend Detection

The non-parametric Mann–Kendall (M–K) method was implemented to detect the trends for annual runoff, P, PET, and parameter n in the WRB [47,48]. This method was selected for trend detection in hydro-meteorological records because its robustness demands non-normality and censoring data.

For a given dataset X (i = 1, 2, 3, ... , n), the test statistic (S) is given by

$$S = \sum_{i=1}^{n-1} \sum_{i+1}^{i} \mathrm{sgn}(X_i - X_j) \quad (1)$$

where

$$\mathrm{sgn}(x_j - x_i) = \begin{cases} 1 & x_j - x_i > 0 \\ 0 & x_j - x_i = 0 \\ -1 & x_j - x_i < 0 \end{cases} \quad (2)$$

A standardized statistic Z is obtained by

$$Z = \begin{cases} (S-1)/\sqrt{n(n-1)(2n+5)/18} & S > 0 \\ 0 & S = 0 \\ (S+1)/\sqrt{n(n-1)(2n+5)/18} & S < 0 \end{cases} \quad (3)$$

To reduce the influence of the serial correlation, the trend-free pre-whitening test was applied to identify any temporal trends in hydro-meteorological variables [49], which is defined as

$$Y_t = X_t - rX_{t-1} \quad (4)$$

where Y_t is the de-trended and pre-whitened series, referred to as the residual series, and X_t represents the original time series. If the lag-one serial coefficient r of the de-trended series is statistically significant at the 5% level, it means that the serial correlation has been eliminated. Additionally, a Mann–Kendall test was conducted on the whitened series to obtain a Z-statistic. A positive Z-value indicates an upward trend, while a negative value indicates a downward trend [40]. $|Z| < 1.96$ means no significant trend was detected ($p > 5\%$). $|Z| \geq 1.96$ and $|Z| \geq 2.58$ indicate that the significance test has passed with a confidence level of 95% and 99%.

3.2. Breakpoint Analysis

In this study [49], Pettitt's test was applied to identify the abrupt year in the hydrological data by utilizing a significance level of 5% through a non-parametric approach [50]. This test detects the occurrence of breakpoints using a non-parametric method at a significance level of 5%. This test applies a function of the Mann–Whitney statistic ($U_{t,n}$), which has been mentioned in the Mann–Kendall test. When a significant breakpoint is detected, it indicates that the two subseries ($x_1, \ldots, x_t$ and $x_{t+1}, x_{t+2}, \ldots, x_n$) are from different populations. $U_{t,n}$ is expressed as follows:

$$U_{t,n} = \sum_{i=1}^{t} \sum_{j=1}^{n} \mathrm{sgn}(X_t - X_j) \quad \text{If } t = 2, \ldots, n \tag{5}$$

The test statistic K_n is defined as maximum value of $|U_{t,n}|$

$$K_n = \max|U_{t,n}| \tag{6}$$

The associated probability (p) with K_n is given by:

$$p = \exp\left(\frac{-6(K_n)^2}{n^3 + n^2}\right) \tag{7}$$

The null hypothesis of Pettitt's test is that there is no change point.

3.3. Assessing the Impacts of Climatic and Anthropogenic Factors on Runoff Change

3.3.1. Budyko Framework

For a given catchment, the long-term water balance equation can be written as

$$R = P - AET - \Delta S \tag{8}$$

where R, P, AET, and ΔS are denoted as the runoff depth (mm), precipitation (mm), actual evapotranspiration (mm), and water storage change, respectively. Here, the variation in water storage can be ignored for the given hydrological year and is assigned a value of zero.

Budyko [30] proposed that the ratio of PET to P over a long-term average can be used to describe the hydrological drought conditions of a region. This basis for the hypothesis is that the water and energy balance processes are regulated by climate and topography, and thus any changes in these factors will lead to corresponding variations in runoff. Over the years, this hypothesis has been widely utilized in hydrology and hydrological modelling to understand the impacts of climate and land surface characteristics on runoff processes [30,36,37]. To separate the influence on runoff independently, this study adopted the elasticity coefficient method coupled with the Budyko frameworks. AET can be derived from the Choudhury–Yang equation:

$$AET = \frac{P \times PET}{(P^n + PET^n)^{1/n}} \tag{9}$$

where PET represents the potential evapotranspiration, and parameter n comprehensively reflects the catchment characteristics, including the characteristics of soil, topography, geologic properties, and vegetation cover [35].

3.3.2. Elasticity Coefficient

Combining Equations (8) and (9), we obtain an equation R = f (P, PET, n). P is the observed annual mean precipitation. PET can be estimated using the FAO Penman–Monteith [51], which is currently recognized as an effective method for estimating PET. The method takes into account variations in meteorological factors such as temperature, humidity, wind speed, and solar radiation, as well as the complexity of vegetation cover, and is used to estimate PET in different regions/basins to help design irrigation systems and improve crop management. The parameter n can be derived from this equation using MATLAB R2022b software.

The total differential equation of R with different factors can be written as

$$dR = \frac{\partial f}{\partial P}dP + \frac{\partial f}{\partial PET}dPET + \frac{\partial f}{\partial n}dn \tag{10}$$

The elasticity of runoff to a specific independent variable change is the degree of the response of runoff to the changes in variables, e.g., P, PET, or n, which can be estimated by the proportional change in simulated runoff. The elasticity of runoff is given as

$$\varepsilon_{x_i} = \frac{\partial R}{\partial_{x_i}} \times \frac{x_i}{R} \tag{11}$$

where ε_{x_i} is the elasticity of runoff to P, PET, or n. Therefore, Equation (10) is written as

$$dR = \varepsilon_{x_P}\frac{R}{x_P}dx_P + \varepsilon_{x_{PET}}\frac{R}{x_{PET}}dx_{PET} + \varepsilon_{x_n}\frac{R}{x_n}dx_n \tag{12}$$

The elasticity of runoff is expressed as [3]:

$$\varepsilon_P = \frac{1 - \left[\frac{(PET/P)^n}{1+(PET/P)^n}\right]^{1/n+1}}{1 - \left[\frac{(PET/P)^n}{1+(PET/P)^n}\right]^{1/n}} \tag{13}$$

$$\varepsilon_{PET} = \frac{1}{1+(PET/P)^n}\frac{1}{1 - \left[\frac{1+(PET/P)^n}{(PET/P)^n}\right]^{1/n}} \tag{14}$$

$$\varepsilon_n = \frac{A - B}{\left[1 + (PET/P)^n\right]^{1/n} - 1} \tag{15}$$

$$A = \frac{P^n \ln(P) + PET \ln(PET)}{P^n + PET} \tag{16}$$

$$B = \frac{\ln(P^n + PET^n)}{n} \tag{17}$$

3.3.3. Attribution Analysis of Runoff Changes

According to the breakpoint, the whole time series can be divided into two periods: "period 1" which represents the base period before the abrupt year, and "period 2" which represents the human impact period after the abrupt year. Thus, the change in annual runoff from period 1 to period 2 is given by

$$\Delta R = R_2 - R_1 \tag{18}$$

where ΔR is the difference between the mean annual observed runoff in the two different periods, R_1 is the average runoff in the base period, and R_2 is the average runoff in the human impact period.

The total runoff change over a long period can be estimated as follows:

$$\Delta R_{total} = \Delta R_{clima} + \Delta R_{human} \quad (19)$$

where ΔR_{clima} is the change in runoff depth due to climatic variables, including the change in runoff caused by P (ΔR_P) and PET (ΔR_{PET}). ΔR_{human} is the runoff change caused by human activities, equal to the change in runoff caused by the catchment characteristics of parameter n. The simulated runoff change can be calculated as follows:

$$\begin{aligned} \Delta R_P &= \varepsilon_P \tfrac{R}{P} \Delta P \\ \Delta R_{PET} &= \varepsilon_{PET} \tfrac{R}{PET} \Delta PET \\ \Delta R_n &= \varepsilon_n \tfrac{R}{n} \Delta n \end{aligned} \quad (20)$$

ΔP, ΔPET, and Δn represent the difference in P, PET, and parameter n from period 1 to period 2.

The relative contribution is the percentage of P, PET, and parameter n to the runoff change, which is expressed as

$$\begin{aligned} C_P &= \tfrac{\Delta R_P}{\Delta R} \times 100\% \\ C_{PET} &= \tfrac{\Delta R_{PET}}{\Delta R} \times 100\% \\ C_n &= \tfrac{\Delta R_n}{\Delta R} \times 100\% \end{aligned} \quad (21)$$

where C_P, C_{PET}, and C_n are the contribution proportions of climate variation (P and PET) and anthropogenic activities to runoff change, respectively.

The performance of the Budyko-based method can be evaluated by Nash–Sutcliffe efficiency values (NSE) and the absolute values of the relative error (R_e).

$$NSE = 1 - \frac{\sum_{i=1}^{n} (\Delta R_1 - \Delta R_2)^2}{\sum_{i=1}^{n} (\Delta R_1 - \Delta R_{1a})^2} \quad (22)$$

$$R_e = \frac{\Delta R_1 - \Delta R_2}{\Delta R_1} \times 100\% \quad (23)$$

ΔR_1 is the observed runoff change, ΔR_2 is the runoff change estimated by Budyko-based method. ΔR_{1a} is the average observed runoff change.

4. Results

4.1. Trends in Hydro-Meteorological Factors

To better understand the variation in runoff and its influencing factors at different scales, we analyzed the trends in the annual runoff, P, PET, temperature, and parameter n within the three sub-basins (UWR, JR, and BLR) of the WRB from 1970 to 2018 (Figure 3). Significantly decreasing trends were observed for runoff ($p < 0.01$), with average rates of 0.87 mm/y, 0.60 mm/y, and 0.19 mm/y in the UWR, JR, and BLR, respectively. P presented downward trends in the UWR and BLR and an upward trend in the BLR, though all trends were found to be non-significant. PET in the UWR and JR showed upward trends (no significance at $p > 0.05$), and both increased by 1.15 mm/y, while the PET of the BLR presented downward trends. Temperature showed a significant increase trend ($p < 0.01$) in all three regions with average rates of 0.03 °C/y, 0.04 °C/y, and 0.02 °C/y in the UWR, JR, and BLR, respectively. In terms of parameter n, the UWR and JR displayed a significant increasing trend ($p < 0.01$), with an average rate of 0.01 mm/y and 0.02 mm/y, respectively, while no significant trend was found for parameter n in the BLR.

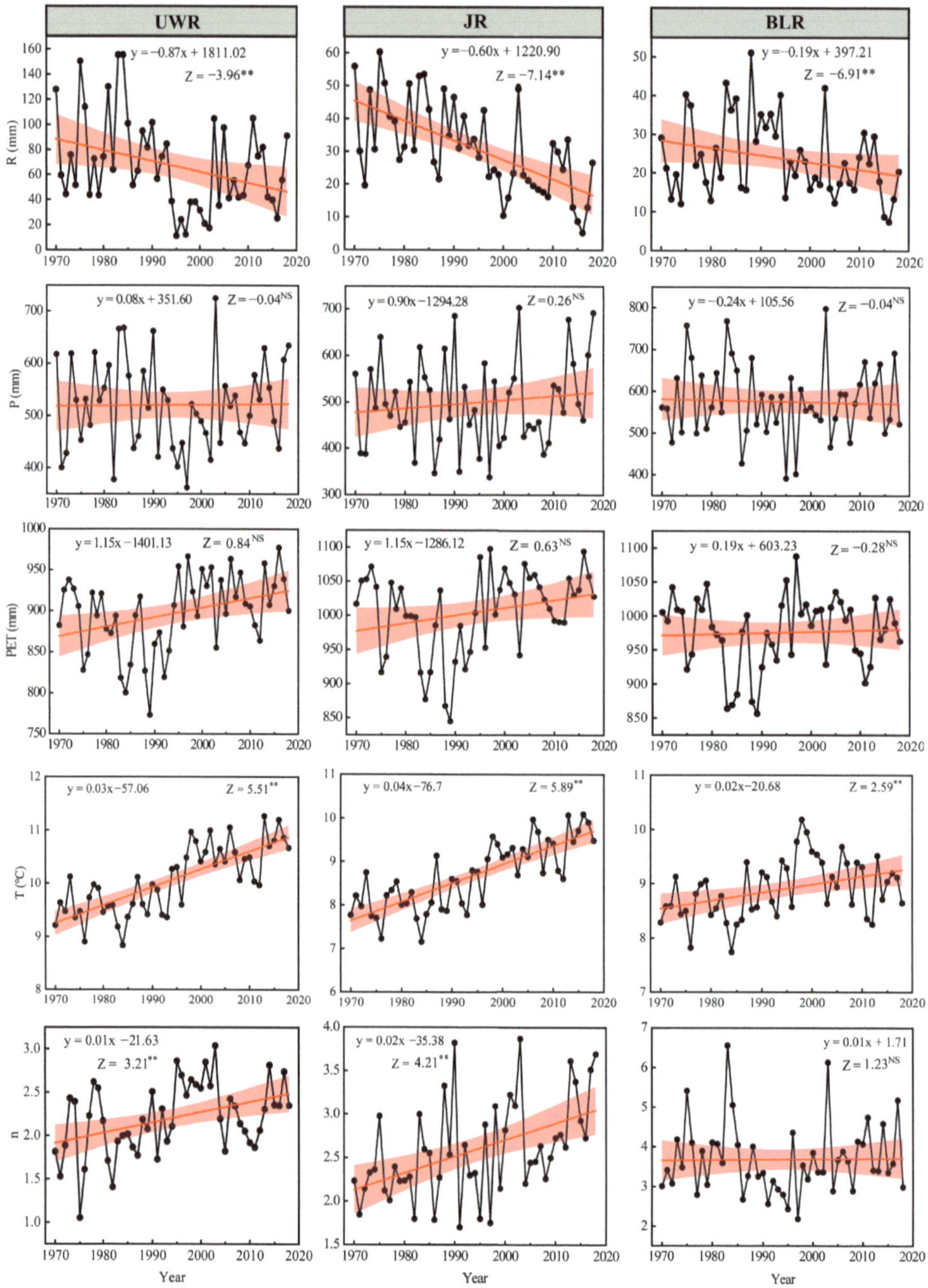

Figure 3. Observed annual runoff depth (R); precipitation (P); potential evapotranspiration (PET); temperature (T) and parameter n in the upstream of the Wei River (UWR), Jing River (JR), and Beiluo River (BLR). The solid red lines indicate the trend; the Z value was detected using the Mann–Kendall test (**: significant at $p < 0.01$; NS: no significance at $p > 0.05$). The 95% confidence interval is denoted by the pink strips.

The trends in annual runoff, P, PET, and parameter n from 1970 to 2018 are shown in Figure 4. Out of all the catchments, 12 catchments (which make up 75% of the catchments) exhibited a notable decrease trend ($p < 0.05$) in annual runoff. P showed an increase trend in most catchments, but all of them were statistically insignificant. PET in half of the catchments showed significant increasing trends. A significant increase trend ($p < 0.05$) in temperature was observed in 15 catchments (accounting for 94% of catchments). In addition, parameter n showed a significant increase trend ($p < 0.05$) in 75% of the catchments. The significant increase trend of parameter n indicates that the disturbance of catchment characteristics is gradually intensified.

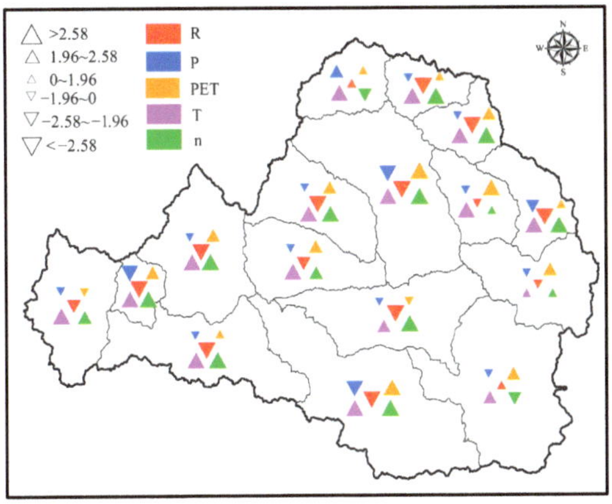

Figure 4. Trend analysis of hydro-meteorological variables in 16 catchments of the Wei River Basin. Trends in annual runoff depth (R); precipitation (P); potential evapotranspiration (PET); temperature (T); and parameter n. △ and ▽ denote increasing and decreasing trends, respectively.

4.2. Abrupt Years of Runoff Change

The abrupt years of annual runoff were determined using Pettitt's method in the WRB and three sub-basins (UWR, JR, and BLR) from 1970 to 2018 (Table 1). In the WRB, the runoff underwent an abrupt change in 1993 ($p < 0.05$). In the three sub-basins, significant abrupt years ($p < 0.01$) were detected in the UWR (1993), JR (1996), and BLR (1994). It was found that the change in runoff and P decreased after the abrupt year, while that in PET and parameter n increased in the WRB and the three sub-basins. 62.5% of the catchments exhibited significant abrupt changes ($p < 0.05$) around the early 1990s and around 2000. The change in hydro-meteorological characteristics and the contribution of climatic and anthropogenic factors to runoff changes in 16 catchments are shown in Table S3 in the Supplementary Materials.

Table 1. The change in the hydro-meteorological characteristics and the elasticity coefficients of runoff to precipitation (P), potential evapotranspiration (PET), and parameter.

Basin	Abrupt Year	Period	Change before and after the Abrupt Year				Elasticity Coefficient		
			R (mm)	P (mm)	PET (mm)	n	ε_P	ε_{PET}	ε_n
WRB	1993 *	1970–1993 1994–2018	−12.3	−3.2	28.7	0.4	3.15 3.28	−2.15 −2.28	−2.12 −2.38
UWR	1993 **	1970–1993 1994–2018	−15.3	11.2	55.7	0.5	2.57 3.07	−1.57 −2.07	−1.69 −2.19
JR	1996 **	1970–1996 1997–2018	−18.3	−1.1	52.5	0.5	3.05 3.64	−2.05 −2.64	−2.37 −2.94
BLR	1994 **	1970–1994 1995–2018	−24.0	−19.0	44.8	0.3	3.31 3.34	−2.31 −2.34	−2.41 −2.46

Notes: abrupt year: **: significant at $p < 0.01$; *: significant at $p < 0.05$.

4.3. The Elasticity Coefficient of Runoff to Climate Change and Anthropogenic Factors

Table 1 presents the elasticity coefficients of runoff with respect to P, PET, and parameter n for the WRB and the three sub-basins. Compared to the base period, there was a noticeable increase in the absolute values of the runoff elasticity coefficient compared to the base period, implying an increasing sensitivity of runoff change to climate variables and parameter n. For the WRB, the elasticity coefficients after the abrupt year were 3.28 for P, –2.28 for PET, and –2.38 for parameter n. These values suggest that a 10% reduction in P would cause a 32.8% decline in runoff, whereas a corresponding decrease in PET or parameter n would result in an increase of 22.8% and 23.8%, respectively. The results indicated that the highest and positive response of runoff was to P, while it exhibited a negative response to PET and parameter n.

The elasticity coefficients of runoff in 16 catchments concerning P, PET, and parameter n are shown in Figure 5. The spatial distribution patterns of the runoff elasticity coefficient in response to P exhibit similarities with those of PET, but are different in polarity. The elasticity coefficient of runoff to P ranged from 2.6 to 4.2 (Figure 5a). Contrary to P, the elasticity coefficients (–3.2 to –1.6) of runoff were negative for PET (Figure 5b). In the middle and downstream regions of the BLR, the absolute elasticity coefficients of runoff with respect to precipitation (P) and potential evapotranspiration (PET) were found to be higher compared to other regions across the WRB. This suggests that changes in P and PET have a greater impact on the amount of runoff in these regions. The elasticity coefficients of runoff to parameter n ranged from –3.4 to –1.6, and higher absolute values were found in the JR and BLR (Figure 5c). This finding suggested that alterations in catchment characteristics have a greater impact on the runoff variations in these specific regions as compared to others.

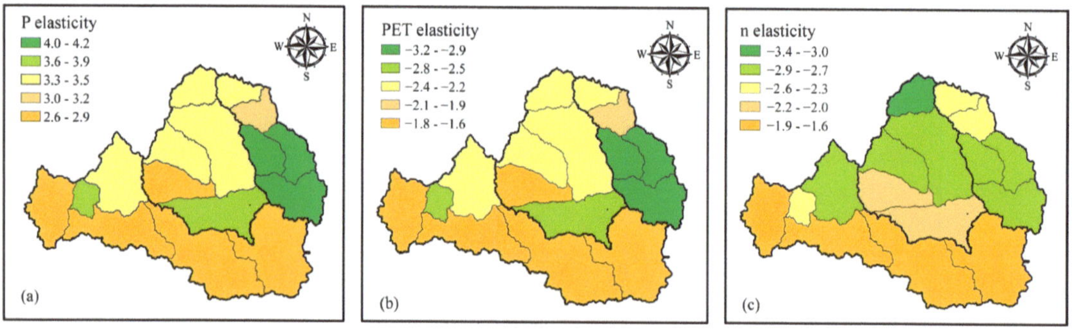

Figure 5. The elasticity of runoff to precipitation (**a**), elasticity of runoff to potential evaporation (**b**), elasticity of runoff to parameter catchment characteristics of parameters (**c**) in 16 catchments of the Wei River Basin. The elasticity of runoff means the percent change in runoff coming from the change in precipitation/potential evaporation/catchment characteristics of parameters by 1%.

4.4. Evaluating the Impacts of Climatic Anthropogenic Factors on Runoff

The performance of the Budyko-based method is shown in Figure 6. Upon comparing the observed and simulated changes in runoff, it was found that the regression slopes of both close to 1. The NSE values were greater than 0.99, and the absolute values of the R_e were less than 5%. This indicates that the method has a strong modeling capability for runoff changes in most catchments.

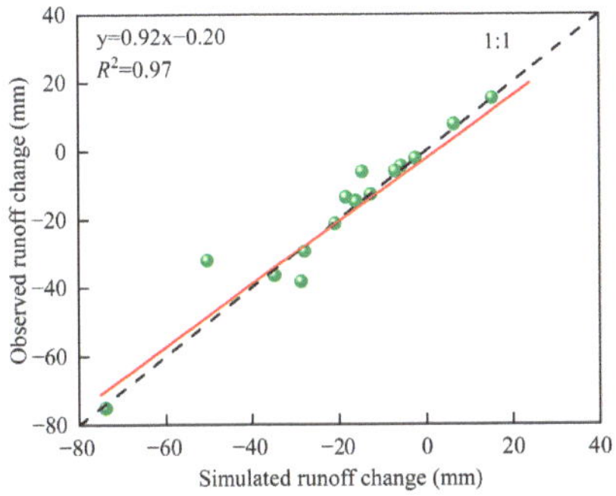

Figure 6. Comparison of the observed and simulated runoff change. The solid red lines mean the regression line of the observed and simulated runoff, and the black dotted line represents the regression line with a slope of 1.

Table 2 demonstrates the runoff alterations caused by hydro-meteorological variables (P and PET) and the catchment characteristics parameter n in the WRB and the three sub-basins. The findings indicate that the decreased runoff, induced by P, PET, and parameter n in the WRB and the three sub-basins, may be affected by the decrease in P and the increase in PET and parameter n. The values obtained for parameter n are related to integrated catchment characteristics, including vegetation cover, soil properties, and topography [52,53]. Regarding soil properties and topography as static variables, land cover change is an important factor affecting catchment hydrological behavior [3]. Thus, the impact of various human activities such as afforestation, deforestation, soil and water conservation projects, and surface water consumption will ultimately be indicated by parameter n. The study found that climate variability, which includes P and PET, contributed to 37.2% of the changes in runoff, while human activities accounted for 62.8%. Furthermore, the impact of human activities on the decrease in runoff was particularly significant in each sub-basin, with a contribution rate exceeding 50%.

Table 2. The runoff change in hydro-meteorological characteristics and the contribution of climate change and human activities to runoff changes in the WRB and three sub-basins.

Region	Runoff Change Induced by P/PET/n (mm)			Contribution Rate to Runoff Change (%)			Contribution Rate of Climate Change (%)	Contribution Rate of Human Activities (%)
	ΔR_P	ΔR_{PET}	ΔR_n	C_P	C_{PET}	C_n		
WRB	−6.0	−5.0	−18.6	20.3	16.9	62.8	37.2	62.8
UWR	−11.2	−9.3	−21.2	26.8	22.3	50.9	49.1	50.9
JR	−0.2	−3.4	−15.3	1.1	17.9	80.9	20.8	79.2
BLR	−3.4	−3.2	−9.7	20.7	19.5	59.8	40.2	59.8

ΔR_P, ΔR_{PET}, and ΔR_n are the runoff changes induced by the change in P, PET, parameter n after the abrupt year, respectively. C_P, C_{PET}, and C_n are the relative contributions of P, PET, parameter n to runoff change.

It can be observed from Figure 7 that the runoff changes before and after the abrupt year induced by hydro-meteorological variables and parameter n is spatially inconsistent. The change in P positively impacted runoff (runoff increase) in most catchments, while changes in PET and parameter n had a negative influence on runoff (runoff decrease) in all catchments. The runoff change induced by P, PET, and parameter n between two periods ranged from −8.5 to 15.6 mm, −7.5 to −1.2 mm, and −38.1 to 12.5 mm, respectively. Figure 8 shows the contribution proportions of climate variables (P and PET) and catchment characteristics (parameter n) to runoff change. Except for catchments 3 and 4, all catchments showed that catchment characteristic changes induced by human activities contributed more than climate variables to runoff change. Therefore, catchment characteristic changes were the dominant factors shaping runoff in most catchments, while climate change contributed the most to runoff in two catchments (3 and 4) in the middle of the BLR. For sub-region 3, the increased PET (38.5 mm) from the base period to the human impact period had the most significant impact on runoff reduction, accounting for 113%. For catchment 4, P (52.2 mm) increased by 52 mm after the abrupt year, and the runoff change induced by P was 8.3 mm, which accounted for 107%.

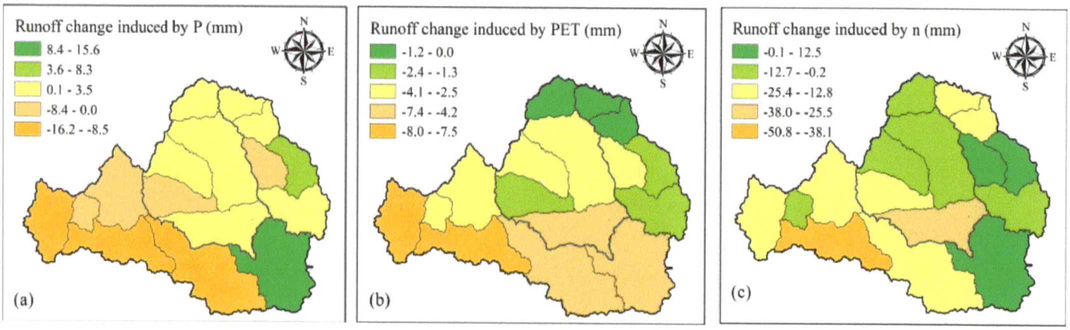

Figure 7. Runoff change induced by P (**a**), PET (**b**) and parameter n (**c**) in 16 catchments of the Wei River Basin. P: precipitation; PET: potential evapotranspiration; n: catchment characteristic parameter.

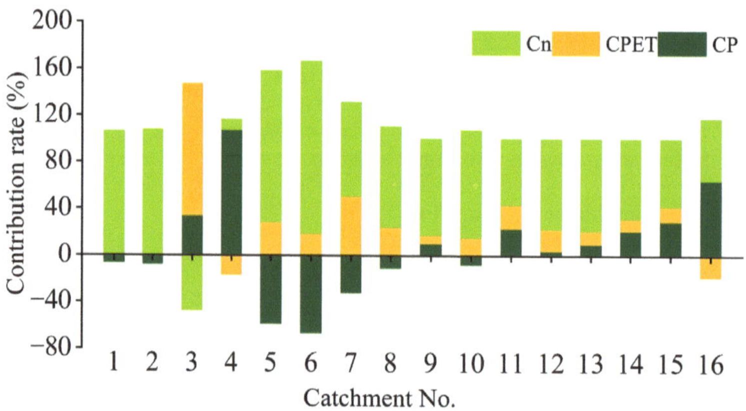

Figure 8. Attribution of the annual runoff change in the 16 catchments in the Wei River Basin.

5. Discussion

5.1. Contribution of the Climatic Factors to Runoff Change

Water resource systems are dramatically influenced by climatic variation, especially in arid and semi-arid areas [7,54]. Previous studies have reported that variations in temperature, P, and PET are manifestations of climate change, which has a significant impact on runoff in the WRB [55,56]. It has been shown that P and temperature are positively associated with terrestrial water storage, and that decreased terrestrial water storage is driven by decreased P and increased PET [57]. Our study indicates that the temperature of the WRB has shown a notable rise. Pang et al. [58] observed that the temperature in the WRB has increased by 0.42 °C/10a. The high temperature accelerates evapotranspiration, reducing the magnitude of water entering the river channel and peak runoff [38]. The drastic runoff reduction caused by increasing PET in the WRB was also observed by Zhao et al. [55]. However, as the main source of runoff, our study showed that the annual P declined 12.3 mm after 1993, but there was no noticeable decreasing trend in the WRB from 1970 to 2018. Similar decreasing trends in P were also detected in other studies in the

WRB [46,59]. Our results indicated that P showed the highest and most positive correlation with runoff change, but the change in P (20.3%) contributing to the decrease in runoff was slightly greater than that for PET (16.9%) in this region. This is in agreement with the previous results, which identified that changes in PET have a weaker impact on runoff change than those in P [52]. In addition, climate change has led to the reduction in runoff in the WRB from 1970 to 2018, although its impact is far less than that of human activities. In contrast, Luo et al. [60] demonstrated that climate change has increased the annual runoff in the source area of the Yangtze River, which has offset a large part of the annual runoff reduced by the ecological protection plan. This may be attributed to the decrease in P and the increase in PET in the WRB.

5.2. Contribution of the Catchment Characteristics' Change to Runoff Change and Potential Anthropogenic Activities

To investigate runoff changes caused by human activities in the WRB, we evaluated three satellite images that recorded land cover. Croplands, forests, and grasslands constituted the major types of land cover in the WRB. Our results indicate the change point in runoff regime in the WRB around 1995, leading to this study focusing on land cover change after the transition (from 1995 to 2018). During this period, the main land cover conversion in the WRB was concentrated from cropland to grassland, settlement, and forest; from grassland to cropland and forest; and from forest to grassland, with a total conversion area of 54,851 km^2 (Figure 9a). Consequently, we selected the six main conversion types and computed the proportion of each land conversion type in the overall conversion area based on the total conversion area (Figure 9b). Conversion between cropland and grassland had the highest percentage (around 30%) followed by conversion between forest and cropland (about 11%), while transformations of some cropland to settlement and forest were also presented, which account for 6.4% and 6.5%. Meanwhile, the net change in four main land cover types underwent greater changes from 1995 to 2018 were analyzed: cropland, settlement, forest, and grassland. During this period, cropland decreased by 4203 km^2, while forest and grassland increased by 1162 km^2 and 1114 km^2, respectively. These changes can be partly attributed to large-scale ecological conservation programs, such as the Grain for Green (GFG) projects implemented in 1999, which led to the reconversion of substantial areas of cropland to forest and grassland, substantially increasing the vegetation cover in the WRB [61]. Chen et al. [62] also observed a dramatic change in land cover owing to an increase in vegetation restoration. Moreover, we observed a significant increase in settlement area, with an increase of 2012 km^2, which may be related to the urbanization of the local area.

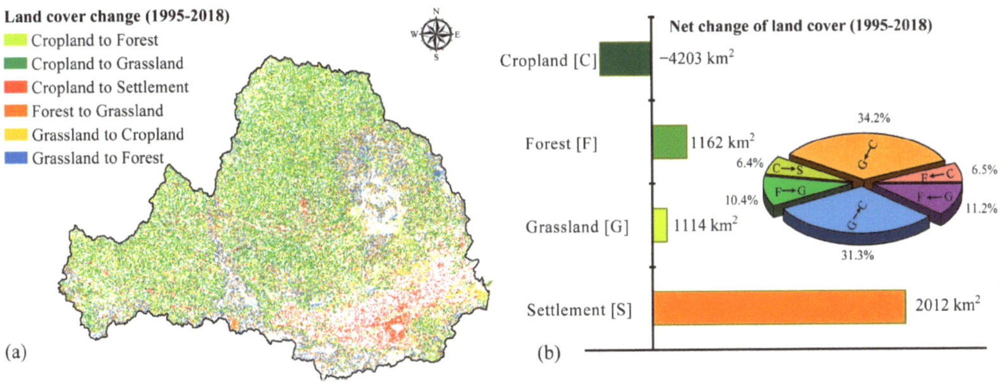

Figure 9. Spatial and temporal variability and net change of land cover in the Wei River Basin from 1995 to 2018. (**a**) Land cover change (1995–2018). (**b**) Net change of land cover.

Given the execution of large-scale ecological conservation programs, we assessed vegetation coverage variations using the annual mean NDVI and annual maximum NDVI of the WRB from 1982 to 2015 (Figure 10). The results show that vegetation cover increased significantly ($p < 0.05$) after 2000, which may be attributed to the GFG project executed after 1999. Research has suggested that large-scale vegetation restoration will locally decrease the annual mean water availability and runoff [63,64]. Increased vegetation cover allows more precipitation to be intercepted by the vegetation canopy and root system [65], thereby diminishing terrestrial runoff [66]. Yang et al. [67] found that increased vegetation coverage was the main reason for runoff reduction in the middle reaches of the Yellow River Basin, which decreased runoff by 35% from 2000 to 2005. Similar results were obtained by Liu et al. [68] and Zuo et al. [42], who concluded that vegetation coverage was the dominant factor responsible for runoff reduction. In addition, large-scale soil and water conservation measures (e.g., afforestation, terraces, fish-scale pits, and grazing ban) have played a vital role in vegetation restoration in the WRB. Since the implementation of these measures in the 1980s, runoff has declined by approximately 9.9×10^8 m^3/year (15.5% of runoff at Huaxian station) in the WRB [55]. Moreover, interesting results were observed showing that 62.5% of the catchments have significant abrupt changes ($p < 0.05$), which may attribute to the soil and water conservations measures implemented after the 1990s and Grain for Green project since 2000 [9]. However, for the runoff regime in the WRB and three sub-basins, there was no significant abrupt change detected around 2000. It may be that vegetation restoration is a slow and long-term process, and the hydrological effect of vegetation on runoff in large catchment has a hysteresis effect [69]. Zhao et al. [70] also found that the response of runoff change to vegetation cover change in large catchment (>1000 km^2) weakened with the increase in catchment scale.

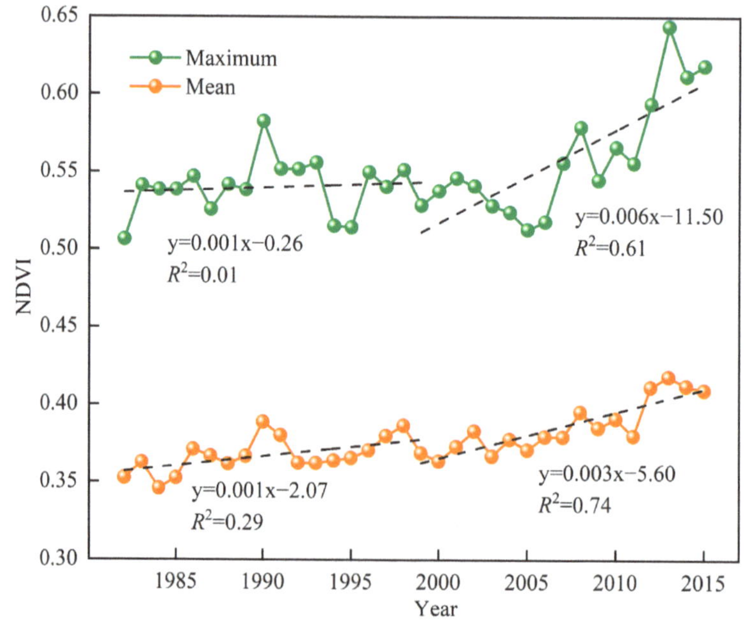

Figure 10. The variations in the mean annual NDVI and maximum NDVI in the Wei River Basin from 1982 to 2015.

The impacts of human activities on catchment characteristics are complex, especially in the context of the extensive implementation of ecological conservation programs. In addition to these ecological conservation programs, the WRB has also witnessed many other intensive human activities in the past 30 years, such as urbanization, and large-scale infrastructure construction (water infrastructures). A large amount of cropland has been converted into settlement (3503 km^2) from 1995 to 2018; this change mainly occurred in the Guanzhong Plain owing to urbanization. Intensified urbanization would have also resulted in changes in catchment characteristics by increasing impervious areas, consequently leading to increases in runoff and decreases in infiltration [22]. Additionally, the reservoir/dams construction has a direct impact on the catchment characteristics of parameter n, mainly by affecting the vegetation coverage in the irrigation area to affect the runoff [38]. Our results showed that human activities play a dominant role in runoff decreasing, which is consistent with the findings in other catchments such as Loess Plateau [22], Yellow River [71], and Yangtze River [60].

5.3. Limitations of the Study

This study has some notable limitations. Equation (10) is a first-order approximation, which may lead to errors in the calculated impacts of climatic variations and anthropogenic activities. In addition, this framework of contribution estimation supposes that anthropogenic activities and climatic variation are completely independent, but it is presently difficult to easily separate and decompose them [71]. Both climate change and human activities will affect the change in catchment characteristic parameter n. Attributing land cover change to human activities will overestimate the impact of local human activities on runoff change [3].

6. Conclusions

This study quantified the impact of climate change and human-induced land cover variations on runoff change in the WRB using the elasticity coefficient method coupled with decomposition methods based on the Budyko framework. The results show the following:

(1) The annual runoff and P presented a statistically downward trend and a significant decrease in runoff was detected in all three sub-basins ($p < 0.01$) and 75% of the catchments ($p < 0.05$). However, PET and parameter n showed an increasing trend in the WRB, three sub-basins, and most catchments.

(2) The significant abrupt years of runoff change were observed in the mid-1990s at the WRB and the three sub-basins ($p < 0.01$); 62.5% of the catchments exhibited significant abrupt changes ($p < 0.05$) in the early 1990s and around 2000.

(3) The results of the elasticity coefficient revealed that P had the highest positive correlation with runoff change, but PET and parameter n had a negative correlation. During the human impact period, a 10% decrease in P would result in a 32.8% drop in runoff, whereas a 10% decrease in PET and parameter n would induce a 22.8% and 23.8% increase in runoff, respectively.

(4) Land cover change induced by human activities had a greater impact on runoff decline than climatic factors in the entire basin, three sub-basins, and most catchments. The impacts of P, PET, and parameter n resulted in average runoff reductions of 20.3, 16.9, and 62.8%, respectively, in the WRB.

These findings highlight the importance of systematically investigating the impacts of climate change and land cover on runoff reduction using the Budyko-based method at different catchment scales. This study could contribute to better quantitative clarification of the driving factors of runoff variation and provide a reference foundation for sustainable water management and ecosystem health conservation across the WRB and midstream of the Yellow River. However, considering the complex interactions among diverse human activities, accurately separating the contributions of other human activities remains challenging. More human activity metrics (e.g., GDP and population) should be considered when interpreting runoff changes, and more methods should be applied to verify the accuracy of the results in future studies.

Supplementary Materials: The following supporting information can be downloaded at: https://www.mdpi.com/article/10.3390/w15112081/s1, Table S1: Information on the hydrological stations;

Table S2: Information on the hydro-meteorological characteristics of the 16 sub-regions in the Wei River Basin; Table S3: The change in hydro-meteorological characteristics and contribution of climatic and human activity factors to runoff changes in catchments.

Author Contributions: Conceptualization, R.X., X.M. and P.G.; Methodology, R.X., C.G., D.Q. and C.W.; Resources, C.G., D.Q., C.W., X.M. and P.G.; Writing—original draft, R.X.; Writing—review & editing, D.Q. and P.G.; Supervision, X.M.; Funding acquisition, P.G. All authors have read and agreed to the published version of the manuscript.

Funding: This research was funded by National Key Research and Development Program of China, Grant/Award Number: 2016YFC0501707.

Data Availability Statement: The daily series of meteorological data used in this study are available on National Meteorological Information Centre of China (NMIC) (https://www.nmic.cn/), while the data sharing of runoff are not applicable.

Conflicts of Interest: The authors declare no conflict of interest.

References

1. PAGES2k Consortium. A global multiproxy database for temperature reconstructions of the Common Era. *Sci. Data* **2017**, *4*, 170088. [CrossRef] [PubMed]
2. Martin, J.T.; Pederson, G.T.; Woodhouse, C.A.; Cook, E.R.; McCabe, G.J.; Anchukaitis, K.J.; Wise, E.K.; Erger, P.J.; Dolan, L.; McGuire, M.; et al. Increased drought severity tracks warming in the United States' largest river basin. *Sci. Data* **2020**, *117*, 11328–11336. [CrossRef]
3. Xu, X.Y.; Yang, D.W.; Yang, H.B.; Lei, H.M. Attribution analysis based on the Budyko hypothesis for detecting the dominant cause of runoff decline in Haihe basin. *J. Hydrol.* **2014**, *510*, 530–540. [CrossRef]
4. Wagener, T.; Sivapalan, M.; Troch, P.A.; McGlynn, B.L.; Harman, C.J.; Gupta, H.V.; Kumar, P.; Rao, P.S.C.; Basu, N.B.; Wilson, J.S. The future of hydrology: An evolving science for a changing world. *Water Resour. Res.* **2010**, *46*, W05301. [CrossRef]
5. Labat, D.; Goddéris, Y.; Probst, J.L.; Guyot, J.L. Evidence for global runoff increase related to climate warming. *Adv. Water Resour.* **2004**, *27*, 631–642. [CrossRef]
6. Pulido-Velazquez, D.; Collados-Lara, A.J.; Pérez-Sánchez, J.; Segura-Méndez, F.J.; Senent-Aparicio, J. Climate change impacts on the streamflow in Spanish basins monitored under near-natural conditions. *J. Hydrol. Reg. Stud.* **2021**, *38*, 100937. [CrossRef]
7. Alifujiang, Y.; Abuduwaili, J.; Groll, M.; Issanova, G.; Maihemuti, B. Changes in intra-annual runoff and its response to climate variability and anthropogenic activity in the Lake Issyk-Kul Basin, Kyrgyzstan. *Catena* **2021**, *198*, 104974. [CrossRef]
8. Guzha, A.C.; Rufino, M.C.; Okoth, S.; Jacobs, S.; Nóbrega, R.L.B. Impacts of land use and land cover change on surface runoff, discharge and low flows: Evidence from East Africa. *J. Hydrol. Reg. Stud.* **2018**, *15*, 49–67. [CrossRef]
9. Zhao, G.; Mu, X.; Wen, Z.; Wang, F.; Gao, P. Soil Erosion, Conservation, and Eco-Environment Changes in the Loess Plateau of China. *Land Degrad. Dev.* **2013**, *24*, 499–510. [CrossRef]
10. Xue, D.X.; Zhou, J.J.; Zhao, X.; Liu, C.F.; Wei, W.; Yang, X.M.; Li, Q.Q.; Zhao, Y.R. Impacts of climate change and human activities on runoff change in a typical arid watershed, NW China. *Ecol. Indic.* **2021**, *121*, 107013. [CrossRef]
11. Wang, W.; Shao, Q.; Yang, T.; Peng, S.; Xing, W.; Sun, F.; Luo, Y. Quantitative assessment of the impact of climate variability and human activities on runoff changes: A case study in four catchments of the Haihe River basin, China. *Hydrol. Process.* **2013**, *27*, 1158–1174. [CrossRef]
12. Giuntoli, I.; Renard, B.; Vidal, J.P.; Bard, A. Low flows in France and their relationship to large-scale climate indices. *J. Hydrol.* **2013**, *482*, 105–118. [CrossRef]
13. Han, H.; Hou, J.M.; Huang, M.S.; Li, Z.B.; Xu, K.Y.; Zhang, D.W.; Bai, G.G.; Wang, C. Impact of soil and water conservation measures and precipitation on streamflow in the middle and lower reaches of the Hulu River Basin, China. *Catena* **2020**, *195*, 104792. [CrossRef]
14. Dong, X.; Yao, Z.; Chen, C. Runoff Variation and Responses to Precipitation in the Source Regions of the Yellow River. *Resour. Sci.* **2007**, *29*, 67–73.
15. Berg, P.; Moseley, C.; Haerter, J.O. Strong increase in convective precipitation in response to higher temperatures. *Nat. Geosci.* **2013**, *6*, 181–185. [CrossRef]
16. Abbott, B.W.; Bishop, K.; Zarnetske, J.P.; Minaudo, C.; Chapin, F.S.; Krause, S.; Hannah, D.M.; Conner, L.; Ellison, D.; Godsey, S.E.; et al. Human domination of the global water cycle absent from depictions and perceptions. *Nat. Geosci.* **2019**, *12*, 533–540. [CrossRef]
17. Yin, Y.H.; Wu, S.H.; Chen, G.; Dai, E.F. Attribution analyses of potential evapotranspiration changes in China since the 1960s. *Theor. Appl. Climatol.* **2009**, *101*, 19–28. [CrossRef]
18. Mengistu, D.; Bewket, W.; Dosio, A.; Panitz, H.-J. Climate change impacts on water resources in the Upper Blue Nile (Abay) River Basin, Ethiopia. *J. Hydrol.* **2021**, *592*, 125614. [CrossRef]
19. Jahfer, S.; Vinayachandran, P.N.; Nanjundiah, R.S. Long-term impact of Amazon river runoff on northern hemispheric climate. *Sci. Rep.* **2017**, *7*, 10989. [CrossRef]

20. Zhang, D.; Hong, H.Y.; Zhang, Q.; Li, X.H. Attribution of the changes in annual streamflow in the Yangtze River Basin over the past 146 years. *Theor. Appl. Climatol.* **2014**, *119*, 323–332. [CrossRef]
21. Liu, X.W.; Xu, Z.X.; Liu, W.F.; Liu, L. Responses of hydrological processes to climate change in the Yarlung Zangbo River basin. *Hydrol. Sci. J.* **2019**, *64*, 2057–2067. [CrossRef]
22. Liang, W.; Bai, D.; Wang, F.; Fu, B.; Yan, J.; Wang, S.; Yang, Y.; Long, D.; Feng, M. Quantifying the impacts of climate change and ecological restoration on streamflow changes based on a Budyko hydrological model in China's Loess Plateau. *Water Resour. Res.* **2015**, *51*, 6500–6519. [CrossRef]
23. Potter, N.J.; Chiew, F.H.S. An investigation into changes in climate characteristics causing the recent very low runoff in the southern Murray-Darling Basin using rainfall-runoff models. *Water Resour. Res.* **2011**, *47*, W00G10. [CrossRef]
24. Huang, S.Z.; Liu, D.F.; Huang, Q.; Chen, Y.T. Contributions of climate variability and human activities to the variation of runoff in the Wei River Basin, China. *Hydrol. Sci. J.* **2016**, *61*, 1026–1039. [CrossRef]
25. Li, C.Y.; Fang, H.Y. Assessment of climate change impacts on the streamflow for the Mun River in the Mekong Basin, Southeast Asia: Using SWAT model. *Catena* **2021**, *201*, 105199. [CrossRef]
26. Wang, G.S.; Xia, J.; Chen, J. Quantification of effects of climate variations and human activities on runoff by a monthly water balance model: A case study of the Chaobai River basin in northern China. *Water Resour. Res.* **2009**, *45*, W00A11. [CrossRef]
27. Song, X.Y.; Sun, W.Y.; Zhang, Y.Q.; Song, S.B.; Li, J.Y.; Gao, Y.J. Using hydrological modelling and data-driven approaches to quantify mining activities impacts on centennial streamflow. *J. Hydrol.* **2020**, *585*, 124764. [CrossRef]
28. Cong, Z.; Yang, D.; Gao, B.; Yang, H.; Hu, H. Hydrological trend analysis in the Yellow River basin using a distributed hydrological model. *Water Resour. Res.* **2009**, *45*, W00A13. [CrossRef]
29. Liu, Q.; Yang, Z.; Cui, B.; Sun, T. The temporal trends of reference evapotranspiration and its sensitivity to key meteorological variables in the Yellow River Basin, China. *Hydrol. Process.* **2010**, *24*, 2171–2181. [CrossRef]
30. Budyko, M.I. *Climate and Life*; Academic Press: Cambridge, MA, USA, 1974.
31. Gan, G.J.; Liu, Y.B.; Sun, G. Understanding interactions among climate, water, and vegetation with the Budyko framework. *Earth-Sci. Rev.* **2021**, *212*, 103451. [CrossRef]
32. Pike, J.G. The estimation of annual run-off from meteorological data in a tropical climate. *J. Hydrol.* **1964**, *2*, 116–123. [CrossRef]
33. Turc, L. The water balance of soils. Relation between precipitation, evaporation and flow. *Ann. Agron* **1954**, *5*, 491–569.
34. Baw-puh, F. ON the calculation of the evaporation from land surface. *Chin. J. Atmos. Sci.* **1981**, *5*, 23–31.
35. Roderick, M.L.; Farquhar, G.D. A simple framework for relating variations in runoff to variations in climatic conditions and catchment properties. *Water Resour. Res.* **2011**, *47*, W00G07. [CrossRef]
36. Yang, H.B.; Yang, D.W.; Lei, Z.D.; Sun, F.B. New analytical derivation of the mean annual water-energy balance equation. *Water Resour. Res.* **2008**, *44*, W03410. [CrossRef]
37. Choudhury, B.J. Evaluation of an empirical equation for annual evaporation using field observations and results from a biophysical model. *J. Hydrol.* **1999**, *216*, 99–110. [CrossRef]
38. Hu, Y.N.; Duan, W.L.; Chen, Y.N.; Zou, S.; Kayumba, P.M.; Sahu, N. An integrated assessment of runoff dynamics in the Amu Darya River Basin: Confronting climate change and multiple human activities, 1960–2017. *J. Hydrol.* **2021**, *593*, 126905. [CrossRef]
39. He, Y.; Qiu, H.; Song, J.; Zhao, Y.; Zhang, L.; Hu, S.; Hu, Y. Quantitative contribution of climate change and human activities to runoff changes in the Bahe River watershed of the Qinling Mountains, China. *Sustain. Cities Soc.* **2019**, *51*, 101729. [CrossRef]
40. Gao, P.; Geissen, V.; Ritsema, C.J.; Mu, X.M.; Wang, F. Impact of climate change and anthropogenic activities on stream flow and sediment discharge in the Wei River basin, China. *Hydrol. Earth Syst. Sci.* **2013**, *17*, 961–972. [CrossRef]
41. Song, J.X.; Xu, Z.X.; Liu, C.M.; Li, H.E. Ecological and environmental instream flow requirements for the Wei River—The largest tributary of the Yellow River. *Hydrol. Process.* **2007**, *21*, 1066–1073. [CrossRef]
42. Zuo, D.; Xu, Z.; Yao, W.; Jin, S.; Xiao, P.; Ran, D. Assessing the effects of changes in land use and climate on runoff and sediment yields from a watershed in the Loess Plateau of China. *Sci Total Environ.* **2016**, *544*, 238–250. [CrossRef] [PubMed]
43. Huang, S.Z.; Huang, Q.; Chang, J.X.; Leng, G.Y.; Xing, L. The response of agricultural drought to meteorological drought and the influencing factors: A case study in the Wei River Basin, China. *Agric. Water Manag.* **2015**, *159*, 45–54. [CrossRef]
44. Li, C.; Zhang, H.B.; Gong, X.H.; Wei, X.W.; Yang, J.T. Precipitation Trends and Alteration in Wei River Basin: Implication for Water Resources Management in the Transitional Zone between Plain and Loess Plateau, China. *Water* **2019**, *11*, 2407. [CrossRef]
45. Zhao, Q.; Wang, L.; Liu, H.; Zhang, Q. Runoff and sediment variation and attribution over 60 years in typical Loess Plateau basins. *J. Soils Sediments* **2019**, *19*, 3631–3647. [CrossRef]
46. Gao, H.D.; Wu, Z.; Jia, L.L.; Pang, G.W. Vegetation change and its influence on runoff and sediment in different landform units, Wei River, China. *Ecol. Eng.* **2019**, *141*, 105609. [CrossRef]
47. Kendall, M.G. *Rank Correlation Measures*; Charles Griffin: Oxford, UK, 1975.
48. Mann, H.B. Nonparametric test against trend. *Econom. Soc.* **1945**, *13*, 245–259. [CrossRef]
49. Yue, S.; Wang, C.Y. The Mann-Kendall Test Modified by Effective Sample Size to Detect Trend in Serially Correlated Hydrological Series. *Water Resour. Manag.* **2004**, *18*, 201–218. [CrossRef]
50. Pettitt, A.N. A Non-parametric Approach to the Change-point Problem. *J. R. Stat. Soc.* **1979**, *28*, 126–135. [CrossRef]
51. Allen, D.G.; Pereira, L.O.; Raes, D.; Smith, M.; Allen, R.; Pereira, L.; Smith, M.A. Crop evapotranspiration: Guidelines for computing crop water requirements. *Irrig. Drain* **1998**, *56*, 300.

52. Zhao, G.J.; Tian, P.; Mu, X.M.; Jiao, J.Y.; Wang, F.; Gao, P. Quantifying the impact of climate variability and human activities on streamflow in the middle reaches of the Yellow River basin, China. *J. Hydrol.* **2014**, *519*, 387–398. [CrossRef]
53. Yang, H.; Yang, D. Derivation of climate elasticity of runoff to assess the effects of climate change on annual runoff. *Water Resour. Res.* **2011**, *47*, W07526. [CrossRef]
54. Hu, W.; Si, B. Technical Note: Improved partial wavelet coherency for understanding scale-specific and localized bivariate relationships in geosciences. *Hydrol. Earth Syst. Sci.* **2021**, *25*, 321–331. [CrossRef]
55. Zhao, G.J.; Mu, X.M.; Tian, P.; Wang, F.; Gao, P. Climate changes and their impacts on water resources in semiarid regions: A case study of the Wei River basin, China. *Hydrol. Process.* **2013**, *27*, 3852–3863. [CrossRef]
56. Zuo, D.; Xu, Z.; Wu, W.; Zhao, J.; Zhao, F. Identification of Streamflow Response to Climate Change and Human Activities in the Wei River Basin, China. *Water Resour. Manag.* **2014**, *28*, 833–851. [CrossRef]
57. Deng, H.J.; Chen, Y.N. Influences of recent climate change and human activities on water storage variations in Central Asia. *J. Hydrol.* **2017**, *544*, 46–57. [CrossRef]
58. Pang, J.T.; Duan, J.L.; Zhang, R.; Lu, J.C. Characteristics of Spatiotemporal evolution and climate response of vegetation cover in the Wei River Basin from 2000 to 2019. *Res. Soil Water Conserv.* **2021**, *28*, 230–237. [CrossRef]
59. Gao, P.; Mu, X.; Wang, F.; Li, R. Changes in streamflow and sediment discharge and the response to human activities in the middle reaches of the Yellow River. *Hydrol. Earth Syst. Sci.* **2011**, *15*, 1–10. [CrossRef]
60. Luo, K. Contribution of ecological conservation programs and climate change to hydrological regime change in the source region of the Yangtze River in China. *Reg. Environ. Chang.* **2022**, *22*, 1–24. [CrossRef]
61. Sun, Q.; Miao, C.; Duan, Q.; Wang, Y. Temperature and precipitation changes over the Loess Plateau between 1961 and 2011, based on high-density gauge observations. *Glob. Planet. Chang.* **2015**, *132*, 1–10. [CrossRef]
62. Chen, Y.; Wang, K.; Lin, Y.; Shi, W.; Song, Y.; He, X. Balancing green and grain trade. *Nat. Geosci.* **2015**, *8*, 739–741. [CrossRef]
63. Wang, D.B.; Hejazi, M. Quantifying the relative contribution of the climate and direct human impacts on mean annual streamflow in the contiguous United States. *Water Resour. Res.* **2011**, *47*, W00J12. [CrossRef]
64. Xu, G.C.; Zhang, J.X.; Li, P.; Li, Z.B.; Lu, K.X.; Wang, X.K.; Wang, F.C.; Cheng, Y.T.; Wang, B. Vegetation restoration projects and their influence on runoff and sediment in China. *Ecol. Indic.* **2018**, *95*, 233–241. [CrossRef]
65. Qiu, D.; Xu, R.; Wu, C.; Mu, X.; Zhao, G.; Gao, P. Vegetation restoration improves soil hydrological properties by regulating soil physicochemical properties in the Loess Plateau, China. *J. Hydrol.* **2022**, *609*, 127730. [CrossRef]
66. Zhou, J.; Fu, B.J.; Gao, G.Y.; Lu, Y.H.; Liu, Y.; Lu, N.; Wang, S. Effects of precipitation and restoration vegetation on soil erosion in a semi-arid environment in the Loess Plateau, China. *Catena* **2016**, *137*, 1–11. [CrossRef]
67. Yang, X.; Sun, W.; Mu, X.; Gao, P.; Zhao, G. Run-off affected by climate and anthropogenic changes in a large semi-arid river basin. *Hydrol. Process.* **2020**, *34*, 1906–1919. [CrossRef]
68. Liu, X.; Jin, S.; Liu, C.; Gao, Y.; Yang, S.; Gao, Y. Influences of shrubs-herbs-arbor vegetation coverage on the runoff based on the remote sensing data in Loess Plateau. *Acta Geogr. Sin.* **2014**, *69*, 1595–1603. [CrossRef]
69. Mo, X.; Liu, S.; Hu, S. Co-evolution of climate-vegetation-hydrology and its mechanisms in the source region of Yellow River. *Acta Geogr. Sin.* **2022**, *77*, 1730–1744. [CrossRef]
70. Zhang, M.F.; Liu, N.; Harper, R.; Li, Q.; Liu, K.; Wei, X.H.; Ning, D.Y.; Hou, Y.P.; Liu, S.R. A global review on hydrological responses to forest change across multiple spatial scales: Importance of scale, climate, forest type and hydrological regime. *J. Hydrol.* **2017**, *546*, 44–59. [CrossRef]
71. Zheng, H.X.; Zhang, L.; Zhu, R.R.; Liu, C.M.; Sato, Y.; Fukushima, Y. Responses of streamflow to climate and land surface change in the headwaters of the Yellow River Basin. *Water Resour. Res.* **2009**, *45*, W00A19. [CrossRef]

hydrology

MDPI

Article

Geoenvironmental Effects of the Hydric Relationship Between the Del Sauce Wetland and the Laguna Verde Detritic Coastal Aquifer, Central Chile

Blanca Gana [1,2], José Miguel Andreu Rodes [3], Paula Díaz [2], Agustín Balboa [2,4], Sebastián Frías [2,4], Andrea Ávila [2,4], Cecilia Rivera [2,5], Claudio A. Sáez [2,5,6] and Céline Lavergne [2,5,7,*]

1 Programa de Doctorado Interdisciplinario en Ciencias Ambientales, Universidad de Playa Ancha, Avenida Leopoldo Carvallo 270, Playa Ancha, Valparaiso 2340000, Chile; blanca.gana@upla.cl
2 HUB Ambiental UPLA, Universidad de Playa Ancha, Valparaiso 2340000, Chile
3 Departamento Ciencias de la Tierra y del Medio Ambiente, Facultad de Ciencias, Universidad de Alicante, 03080 Alicante, Spain
4 Departamento de Geología, Facultad de Ingeniería, Universidad Nacional Andrés Bello, Viña del Mar 2520000, Chile
5 Departamento Ciencias y Geografía, Facultad de Ciencias Naturales y Exactas, Valparaiso 2340000, Chile
6 Departamento Ciencias del Mar, Facultad de Ciencias, Universidad de Alicante, 03080 Alicante, Spain
7 Departament de Biologia Marina i Oceanografia, Institut de Ciències del Mar (ICM), Consejo Superior de Investigaciones Científicas (CSIC), 08003 Barcelona, Spain
* Correspondence: lavergne@icm.csic.es

Abstract: In the central region of Chile, the Mega-Drought together with the demographic increase near the coast threatens groundwater availability and the hydrogeological functioning of coastal wetlands. To understand the hydric relationship between an aquifer and a wetland in a semi-arid coastal region of Central Chile (Valparaíso, Chile), as well as its geoenvironmental effects, four data collection campaigns were conducted in the wetland–estuary hydric system and surroundings, between 2021 and 2022, including physical, hydrochemical, and isotopic analyses in groundwater (n = 16 sites) and surface water (n = 8 sites). The results generated a conceptual model that indicates a hydraulic connection between the wetland and the aquifer, where the water use in one affects the availability in the other. With an average precipitation of 400 mm per year, the main recharge for both systems is rainwater. Three specific sources of pollution were identified from anthropic discharges that affect the water quality of the wetland and the estuary (flow from sanitary landfill, agricultural and livestock industry, and septic tank discharges in populated areas), exacerbated by the infiltration of seawater laterally and superficially through sandy sediments and the estuary, increasing salinity and electrical conductivity in the coastal zone (i.e., 3694 μS/cm). The Del Sauce subbasin faces strong hydric stress triggered by the poor conservation state of the riparian–coastal wetland and groundwater in the same area. This study provides a detailed understanding of hydrological interactions and serves as a model for understanding the possible effects on similar ecosystems, highlighting the need for integrated and appropriate environmental management.

Keywords: wetland; coastal aquifer; groundwater; hydrogeochemistry

1. Introduction

The groundwater connection to land surface processes and surface water bodies is multi-scale. These connections maintain flows along river corridors and into wetlands during droughts, ultimately resulting in continental river flow and its discharge to the oceans. These interactions across spatial scales close the hydrological cycle from the continent to the oceans in a dynamic balance [1]. Wetlands are surfaces covered by saline, brackish or fresh water, permanently or temporarily, artificially or naturally [2]. They have a high economic and social value due to the functions they perform, the goods that can be obtained from them, and their attributes as a part of humanity's cultural heritage, although this has not always been the case. In the past, wetlands were considered unhealthy, dangerous, and economically worthless places, associated with diseases such as malaria due to the number of insects and other hydric diseases. This led to their misuse and abuse to the point of eliminating them. However, from a few decades ago, they began to be seen as areas of great diversity in plants and animals and given an important ecological and economic value. Thus, due to the dual effects of human activities and natural factors, the area of wetlands in the world has been decreasing and their quality has been deteriorating [3–9].

Many wetlands develop in places where groundwater is near the surface or are generally fed jointly by surface water and groundwater, with their water being a mixture of these two components [10]. Often, the aquifer status and the good condition of the wetland ecosystem are linked [11,12]. Therefore, the rational use of wetlands [13] requires that surface water, groundwater, and wetlands are managed in an integrated manner to ensure ecosystem and water sustainability. Hence, this requires an understanding of the interactions between groundwater and surface water, and therefore a robust hydrogeo-ecological framework [14].

Despite the strong dependence between groundwater and humid areas, the relationship between the wetland and the aquifer is often not known and this causes the use of groundwater to affect and deteriorate these damp environments. The use of groundwater for crop irrigation is one of the main problems suffered by aquifers. Overexploitation to extract water produces a drop in the water table and a decrease in the wetland contributions, especially during periods of low water levels when precipitation decreases. Moreover, water pollution in aquifers, for various reasons, especially by irrigation returns, can generate severe impacts in humid areas.

All this means that wetlands are among the most threatened ecosystems worldwide, in such a way that since 1900 more than 64% of the planet's wetlands have been lost, affecting the diversity of species present in these ecosystems [15]. From the study of wetlands of international importance, which integrates 2303 wetlands worldwide, Xu et al. [16] acknowledge that the most significant impact factors are pollution (54%), the use of biological resources (53%), modification of natural systems (53%), and agriculture and aquaculture (42%), with the wetlands' surfaces and environments being the most affected objects. Specifically, riverine wetlands are usually mainly affected by pollutants from domestic sewage, urban wastewater, garbage and solid waste, agricultural and forestry effluents, industrial effluents, etc., and are the most affected in terms of biodiversity and water resource regulations, while marine/coastal wetlands are the most vulnerable to climate change [16].

In Chile, knowledge about groundwater is quite limited and less than that available for other components of the hydrological cycle. This situation further complicates the understanding of the functioning and status of its humid areas. The lack of information on the distribution, quantity, and status of groundwater currently becomes more relevant due to the risk of hydric deficit [17]. According to the Intergovernmental Panel on Climate Change [18], the mean annual precipitation will decrease in mid and low latitudes by the end of the century [18–20], reducing the annual average river runoff and water availability [19], and will increase the sea level, which will extend the salinization of groundwater and estuaries, affecting water quality and further reducing the availability of fresh water for populations and ecosystems in coastal areas [19]. In the coastal basins of Central Chile (30°–38° S) in Southern Latin America, both surface and groundwater resources are highly dependent on rainy periods [21] and are therefore predicted to be under high water pressure. In this central area of Chile, there are diverse types of wetlands, such as high Andean meadows and temporary wetlands (stream ravines and streams) [22]. A large part of these wetlands has been affected since 2010 by a notorious decrease in precipitation, with annual deficits between 25% and 45%, generating a continuous dry period of several years that has been called The Central Chile Mega-Drought [23]. This deficit generated a decrease in river flows, reservoirs, groundwater, and vegetation productivity in the northern sector of Central Chile [24], in addition to a decrease of up to 90% in average river discharge [25].

The Valparaíso Region, located in this area of Chile, and on which this study is focused, is a clear example of the climatic and demographic affectation, in which humid areas have begun to suffer. In 2021, accumulated precipitation as of June 30 reached about 100 mm at the Peñuelas Lake station, while a normal year (calculated for the 1991–2020 period) reaches 300 mm [26]. At the same time, a deficit in the flow levels of more than 45% was recorded [26]. The Laguna Verde sector (33°05′ S–71°35′ W) has faced significant anthropogenic pressure, particularly since 2010, due to rapid growth in the accommodation and tourism industries [27]. Additionally, the expansion of small-scale agriculture and potential industrial surface pollution also affected the Del Sauce stream [28], the main watercourse in the area. Consequently, the quality of this vital water resource has progressively declined [28–31]. However, the water from the estuary continues to be used for irrigation and recreational activities in the coastal zone, as recorded by DGA [32] and Tobar and Torres [33], and has been recognized in the summer period within the framework of this study.

This research responds to the socio-environmental problems of the Laguna Verde town (Valparaíso, Chile), and seeks to understand the relationship between groundwater and surface water of the Del Sauce stream, to identify its conservation status, the causes of this, and its possible geoenvironmental effects. Therefore, it is intended to advance, using knowledge of the groundwater in this region, an accurate definition of the aquifer or aquifers related to the Del Sauce stream, its geometry, typology, flow directions, its waters classification, and its interaction with the wetland. To achieve this goal, a holistic analysis was carried out covering hydrochemical, isotopic, piezometric, and physicochemical measurements in both groundwater and surface water, inside and outside the Del Sauce basin, during a full year of monitoring.

2. Methodology

2.1. Geographical Setting

The study area corresponds to the immediate surroundings of the Del Sauce stream, where the town of Laguna Verde is placed, located on the western margin of South America in Central Chile, and has been delimited approximately between UTM coordinates 6,337,500 N–6,327,500 N and 250,000 E–258,000 E. The town of Laguna Verde is considered a tourist center and is located approximately 9 km from the regional capital Valparaíso, which in the last census of 2017 had 3686 inhabitants.

The region is part of the Peñuelas Lake subbasin, defined within the so-called Coastal Basins between Aconcagua and Maipo [34]. According to the Chilean Ministry for the Environment, it is a continental, riparian, and permanent wetland type, which integrates the Las Cenizas and Del Sauce streams, called Del Sauce Estuary Wetland (https://humedaleschile.mma.gob.cl/ (accessed on 10 July 2024); Figure 1). The Del Sauce stream, (also called El Sauce), which flows into the sea in the Laguna Verde Bay, emerges in the confluence area between the Las Cenizas and the La Luz streams [21] (sporadic contribution from the outflow of the La Luz lagoon that acts as a reservoir, which in return receives water from the Peñuelas Lake; Figure 1). Along the valley of the Del Sauce stream, there is a small detritic aquifer developing, of which waters are used to supply and irrigate the Laguna Verde locality.

The Laguna Verde region is affected by a Mediterranean climate with winter rainfall and coastal influence (Csbn), according to the Köppen–Geiger climate classification, with average annual precipitation of 600 mm and average annual temperatures of 14 °C [35].

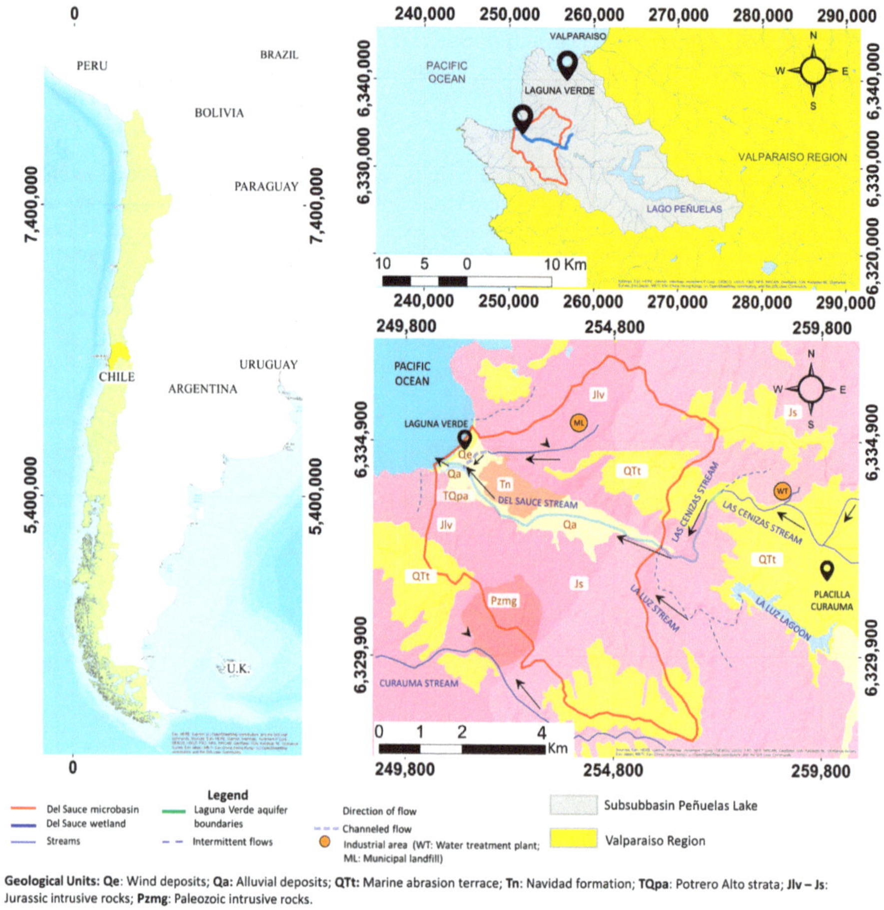

Figure 1. Map of the studied area: Left: Chile in South America and location of Valparaíso region (in yellow). Right–above: location of the Peñuelas Lake basin and the Del Sauce microbasin. Right–below: geologic map of the Del Sauce microbasin, hydrographic network with flow direction, highlighting the location of the Del Sauce wetland and industrial areas.

Upstream of the so-called Del Sauce wetland, in the locality of Placilla, the wastewater treatment plant, owned by the ESVAL company (UTM coordinates: 6,333,341 N and 259,360 E, Zone S19; Figure 1) treats wastewater generated in the localities of Placilla and Curauma (14.906 dwellings as of 2019) to be disposed in the Las Cenizas affluent, complying with the Chilean emission standard for the regulation of pollutants associated with liquid waste discharge to marine and inland surface waters, approved by the D.S. MINSEGPRES No. 90/00 [36]. The treatment consists of a roughing stage for the removal of the garbage that accompanies wastewater, then a stage of biological removal of the organic load in an aerated lagoon, to then move to a settling lagoon, in order to separate the biological sludge by assisted decantation, and finally a stage of effluent disinfection, treated with chlorine to remove the microbiological load, for subsequent disposal in the Las Cenizas affluent [36]. Another relevant industrial point in the basin under study corresponds to the El Molle sanitary landfill, which is called a treatment and final disposal center for household solid waste, solid waste assimilable to household, and hospital waste, designed to use a surface of 56.7 ha, and is placed in the same area of the former El Molle landfill, which covers a surface of approximately 30 ha, covering between both projects a total of 86.1 ha of the El Molle property, located in the Quebrada Verde sector, Colorado Hill, in the coastal area of the Valparaíso Region (UTM coordinates: 254,703.5 N and 6,335,741.6 N, Zone S19).

2.2. Hydrological and Geological Setting

The region has an abrupt relief, with low altitudes where the Del Sauce stream flows, reaching between 1 and 20 m above sea level, surrounded by mountain ranges that rise up to 550 m above sea level (Figure 1). The Del Sauce stream basin has an area of 34.2 km^2, a perimeter of 29.87 km, a width of approximately 6 km, and is defined as a microbasin (according to the classification of Reyes et al. [37]). The compactness coefficient or Gravelius index provides a coefficient Kc = 1.44, indicating an average tendency to flooding (according to the classification of Villela and Matos [38]). The hypsometric curve of this microbasin compared with the curves proposed by Strahler [39] indicates that this basin represents a mature river, being in an intermediate stage, characterized by being in a state of equilibrium. The average slope of the basin reaches 19%, consequently, according to Alcántara's classification [40], it corresponds to a "Moderately steep" basin. The structure degree of the drainage network for the microbasin is of order "3", which with respect to the classification of the stream type determines a "medium" bifurcation degree [41].

From the geological point of view, most part of the Del Sauce basin is characterized by the presence of an intrusive basement, with rocks such as diorites, gabbros, and tonalites of Jurassic age, belonging to the Laguna Verde and Sauce units (Jlv and Js, respectively) [42,43]. In the middle and lower part of the Del Sauce stream, sedimentary rocks of Tertiary age are recognized, such as a relic of the Navidad Formation (Tn; Figure 1), which is made up of sandstones, shales, and siltstones, which have a maximum thickness of 120 m [44], and the Potrero Alto Strata (TQpa). As quaternary deposits, the so-called Abrasion Terrace (QTt) of marine origin are recorded in high sectors. The lower areas of the basin, where part of the Laguna Verde town is placed, are filled with unconsolidated sediments comprising fluvial deposits with interdigitated gravitational deposits, composed of medium sands, fine to medium gravels, and small portions of clays, forming the unit Alluvial Deposits of Holocene age (Qa) [43]; while bordering the coastline, there are unconsolidated beach sediments, composed of sands and gravels of well-rounded boulders that may form active dunes, and are defined as Current Littoral and Wind Deposits, of Holocene age (Qe; Figure 1) [43]. Structurally, the Laguna Verde fault stands out, corresponding to a tectonic fault of normal character, in an NW direction, with a sunken block toward the south (on which the channel of the Del Sauce stream is placed), which extends along the northern limit of the alluvial fill of the Del Sauce stream [43,45].

2.3. Hydrogeological Characterization

To define the hydrogeological units in the study area, after defining the drainage basin, a geological survey was generated. Subsequently, the hydraulic and stratigraphic background and pumping tests present in the records of the wells registered in the Chilean Ministry of Public Works, requested through the Chilean Law on Transparency at http://www.portaltransparencia.cl (accessed on 16 January 2024), were reviewed. To complete the information and obtain current data, two Porchet Tests were performed in the wetland (UTM coordinates 251,347 E; 6,334,158 N and 253,003 E; 6,333,071 N). These data were processed with the open-access AQTESOLV software version 4.0. Based on the geological characteristics of the study area and the permeability or hydraulic conductivity ranges obtained, the different geological formations were hydrogeologically characterized.

2.4. Flow Measurement

To determine the runoff flow in the Del Sauce stream, a gauging point was selected in the upper course of the river, specifically in the so-called Las Cenizas stream (sector S20). All this flow runs over the Del Sauce downstream until it reaches its final section before reaching the sea, where it remains dammed for most of the year by a sandbank, in such a way that it does not discharge directly into the sea. During the campaigns carried out in the area (4 campaigns during the years 2021 and 2022), it was detected as the only estuary providing surface hydric resources. These measurements were performed with a Portable Hydraulic Winch (FP111, Global Water), according to the velocity–area method, or Bührer velocity–area method [46].

2.5. Piezometric Measurement

Piezometric levels were measured in reachable wells, corresponding to 9 wells constructed in the Alluvial Deposit unit (P1, P3, P5, P6, P7, P8, P10, P14, P15), one in the unit Current Littoral and Wind Deposits (P4), one in the Navidad Formation (P2), one in the unit Potrero Alto strata (P26), one in the unit Abrasion Terrace (P27), and two in the Jurassic Intrusives unit (P12 and P13; Figure 2). These measurements were carried out with a 100 m water level meter (Holykell brand). From these measurements, isopiestic maps were generated to establish the flow direction, thus defining groundwater dynamics. Based on the geographic density, an equidistance of 1 m was defined between each piezometric curve.

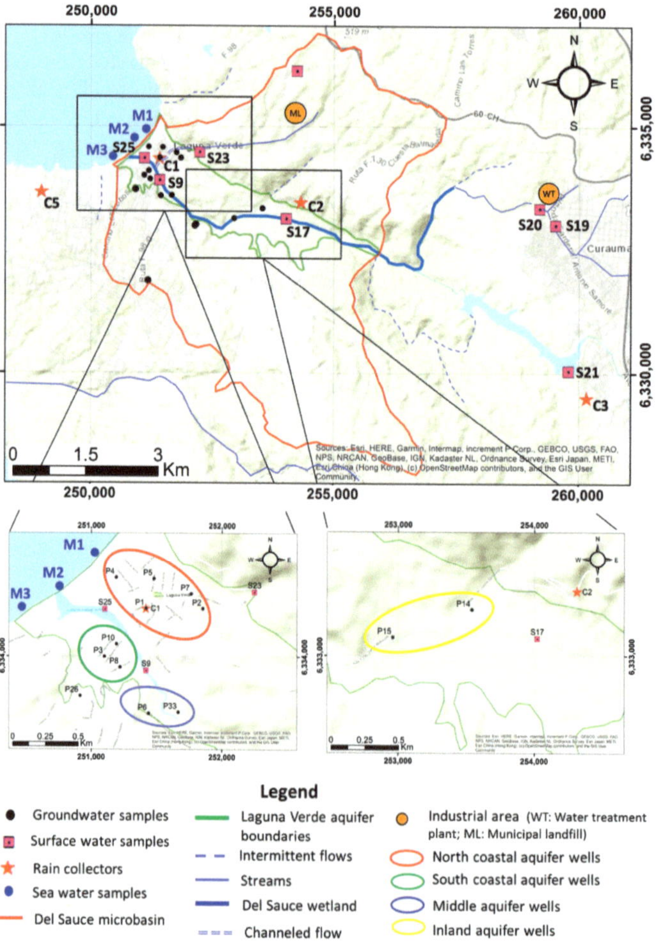

Figure 2. Location of sampling sites and physicochemical data collection: Above: location map of the study area and sampling points for surface water, groundwater, and rainwater collectors. Below–left: details of the coastal and middle zone; below–right: details of the inland zone.

2.6. Water Sampling

To characterize the surface water and groundwater chemistry in the studied area and establish the hydric relationship of groundwater with the Del Sauce stream, four field campaigns were conducted. Water was sampled for the physicochemical and isotopic analysis. Sampling was carried out during the months of September–October 2021 (spring), January–February 2022 (summer, dry season), April–May 2022 (autumn), and July–August 2022 (winter, rainy season).

The hydrogeochemical sampling points selected corresponded to groundwater (n = 16) collected from wells located in the different geological units of the study area, and surface water (n = 8), collected from La Luz Lagoon, Del Sauce stream, Las Cenizas affluent, spring in the upper part adjacent to the sanitary landfill site, and surface runoff from this same industrial area. In addition, during the winter and spring 2022 campaigns, three seawater samples were collected in the northern, middle, and southern zones of the bay (M1, M2, and M3, respectively; Figure 2). At each selected point, pH, total dissolved solids (TDS), temperature, dissolved oxygen (DO), water electrical conductivity (EC), and chemical oxygen demand (COD) were measured in situ with a portable multiparameter meter (HI98194, Hanna) as well as free chlorine with a chlorine meter (Pocket Colorimeter II, HACH). Turbidity was measured with a turbidimeter (HI83414, Hanna) and color was evaluated by visual comparison (St. Method 2120B; Table S1). Water samples were taken in triplicate to be analyzed by major ions (calcium, magnesium, sodium, potassium, chloride, sulfate, bicarbonate) and other secondary ions (such as nitrites, nitrates, phosphates, carbonates, and fluorides), microelements (such as aluminum, bromide, manganese, lead, among others) and microbiological elements (fecal and total coliforms). The elements boron and strontium were measured in one sample from the sea, one from the coastal aquifer, one from the inland aquifer, and one from the estuary at its mouth, for the last two sampling campaigns. All analyses were performed following the methodologies proposed in the 24th Edition of *Standard Methods for the Examination of Water and Wastewater* (details available in Supporting Materials, Table S1).

For the first season sampling (spring: September–October 2021), 25 samples were selected for the ^2H and ^{18}O water isotopic analyses in triplicates and 5 rainwater collectors were arranged in a transect from the lower to the upper part of the basin (samples C1 to C5; Figure 2). During the following field campaigns (summer, autumn, and winter of 2022), all hydrogeochemical sampling points were considered for isotopic analysis. These samples were sent for ^2H and ^{18}O water isotope analysis to the Alfred Wegener Institute for Polar and Marine Research in Potsdam, Germany.

2.7. Data Analysis

For the morphometric, hydrogeological, and hydrodynamic analysis (flow model), as well as for the preparation of thematic maps, ArcMap 10.4 software was used. Regarding the hydrogeochemical information, the data were examined and analytically treated individually and in groups, in order to detect possible pollutant sources and the chemical relationship between groundwater and wetland in their different sections. The data groups correspond to the following (Figure 2): north coastal aquifer (wells P1, P2, P4, P5, P7); south coastal aquifer (wells P3, P8, and P10), middle aquifer (wells P6 and P33), and inland aquifer (wells P14 and P15). A multivariate analysis of variance (MANOVA) was performed to identify the factors that most influenced the physicochemical parameters and major ions. The free-access software Diagrammes (v 6.77) was used. To determine the irrigation water quality, the Riverside diagram [47] and Chilean Standard 1333 [48], which sets out the water quality requirements for different uses, were used. In turn, water–rock interactions and the study of hydrogeochemical processes or pollution problems were established using hydrogeochemical diagrams, such as modified Stiff Diagram [49], or Piper Diagram [50], and the ionic relations $rCl^-/rHCO_3^-$, rSO_4^{2-}/rCl^-, rBr^-/rCl^-, rMg_2^+/rCa_2^+, rK^+/rNa^+, and rNa_2^+/rCl^- [51–53].

For the isotopic analysis of both surface water and groundwater, the $\delta^{18}O$ ‰ v/s $\delta^{2}H$ ‰ V SMOW diagrams were used, performed with the software Diagrammes (v 6.77).

3. Results

3.1. Precipitation and Flows

The average annual precipitation in the Del Sauce region had great variability during the years 2010 and 2022, years that are considered within the so-called Mega-Drought in Central Chile (CR2, 2015). According to data from the meteorological directorate of Chile, the average precipitation measured in the study area surroundings (Peñuelas Lake weather station: UTM coordinates 6,329,722 N and 261,586 E, Zone S 19), for a range of 30 years (1992–2022), reached an annual average of 585 mm, while the precipitation recorded for the range of the years between 2010 and 2022 (range within Mega-Drought) reached an annual average of 406 mm, which represents a 30% decrease in precipitation of the area in relation to a normal year. Within the period called Mega-Drought, in the study area, a maximum annual precipitation of 677 mm was recorded in 2015, and a minimum of 129 mm in 2019. Regarding precipitation distribution over an average year, the maxima are recognized mainly during the austral winter in the months of June, July, and August, with average values of 127 mm, followed by the month of May of the autumn mid-season, with an average of about 100 mm, while the minimum precipitation records are recognized in the months of January, February, and March, corresponding to the dry summer season, with seasonal averages of 1.9 mm, determining that the study area corresponds to a Unimodal precipitation regime (Figure 3).

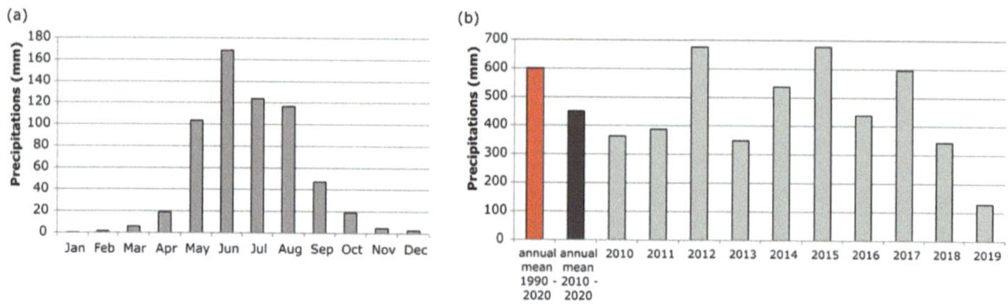

Figure 3. Average precipitation in the studied area: (**a**) monthly average precipitation for the period 1992–2022. (**b**) annual average precipitation for the periods 1992–2022 (red), 2010–2022 (dark gray), and detail of annual precipitation for each year between 2010 and 2022 (light gray).

Regarding Del Sauce stream hydrometry, the flow measured in the summer dry season (March 2022) reached 0.2 m^3/s, and 0.32 m^3/s in the winter rainy season (August 2022), estimating an average annual flow for the Del Sauce wetland of 0.26 m^3/s.

3.2. Hydrogeology

Until now, the only hydrogeological information of the study area resided in the reports conducted by the DGA (2017) [54] in which this zone was included within the so-called Laguna Verde Estuary aquifer that comprises a large part of the Peñuelas Lake basin that extends inland toward the continent from the Pacific Ocean to approximately 27 km in a SE direction. Under this definition, the Peñuelas Lake basin comprises a basin surface of 204 km^2, and an aquifer area of 43.8 km^2 (Figure 1).

Particularly, in the Del Sauce microbasin, based on the geological and hydrogeological review carried out, two sets of varied lithologies have been defined that can be considered as independent units with clearly different hydrogeological behavior (Figure 4).

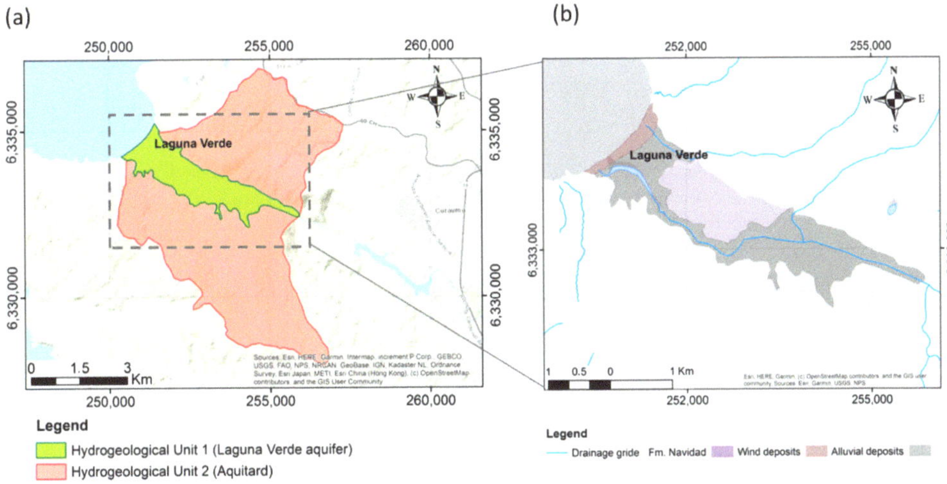

Figure 4. (**a**) Hydrogeological units in the Del Sauce microbasin. (**b**) Details of geological units of HU1.

Hydrogeological Unit 1 (HU1): Denominated in this study as Laguna Verde Detritic Coastal Aquifer or Laguna Verde Aquifer. This is a detritic aquifer formed mainly by alluvial sediments such as medium sands and fine to medium gravels with small portions of clays, belonging to the Alluvial Deposits formation [43] that fill the valley of the Del Sauce; by poorly consolidated detritic rocks that constitute the Navidad Formation located on the northern edge of the valley, with a maximum depth of 120 m [44]; and by quaternary wind sediments (coarse-to-medium sands), which form a small strip of approximately 200 m from the coastline inland, and are called Wind Deposits Unit [43]. Thus, the Laguna Verde Detritic Coastal Aquifer reaches a total area of 5 km^2, and a perimeter of 16,668 m. The aquifer thickness is not known exactly, since most of the points do not reach the impermeable wall. According to the information obtained in this study based on the constructive characteristics of the existing wells, everything indicates that it would present a minimum of 10 m in the coastal zone and 25 m in the inland zone. According to information from some aquifers in adjacent coastal basins with similar characteristics, these small coastal aquifers have a maximum thickness of 30 m [34]. This unit acts as a free detritic aquifer.

The hydraulic characterization of this unit, transmissivity (T) and permeability (k), has been made from the reinterpretation of pumping tests for the geological units. In the Alluvial Deposits, it has been determined that T reaches an average of 200 m^2/day. Considering an approximate saturated thickness of 20 m, the permeability (k) obtained was 10 m/day. Locally, the two Porchet tests performed for this study in this unit revealed an infiltration rate of 124 and 194 mm/h, corresponding to a permeability k = 3 and 4.7 m/day, respectively. In the Navidad Formation, a transmissivity of 195 m^2/day and a permeability of 7.8 m/day were measured, considering an approximate saturated thickness of 25 m. These results indicate that both lithological sections have good hydrogeological characteristics and constitute a good aquifer.

Hydrogeological Unit 2 (HU2): Defined by the Jurassic intrusive rock units (grouping the Laguna Verde and Sauce geological units), which in the study area are recognized as highly weathered and with a high clay content, with estimated permeabilities based on their granulometry close to 10^{-7} m/s. In addition, this unit includes the unit Abrasion Terrace, corresponding to high marine terraces located on the crystalline rocks, the unit Potrero Alto Strata, and Paleozoic intrusives. Although in this unit there are home extraction wells, the difficulty of these wells to contain and provide significant or enough water volumes throughout the year is also recognized. The catchments in this unit presented very depressed piezometric levels, and in many of them, when water was extracted for sampling, these dried up, which shows the low productivity of the catchments. This hydrogeological unit corresponds to a very poor aquifer or aquitard.

3.3. Heads and Relation between Stream and Aquifer

Based on the piezometric levels, the dynamics of groundwater flow in the Laguna Verde aquifer were analyzed. Broadly speaking, the isopiestic maps elaborated for the dry and humid seasons suggest a certain similarity in the behavior throughout the year (Figure 5a,b). The flow of the aquifer waters has a general direction from the southeast, which corresponds to its highest part, toward the northwest, where the discharge into the sea occurs.

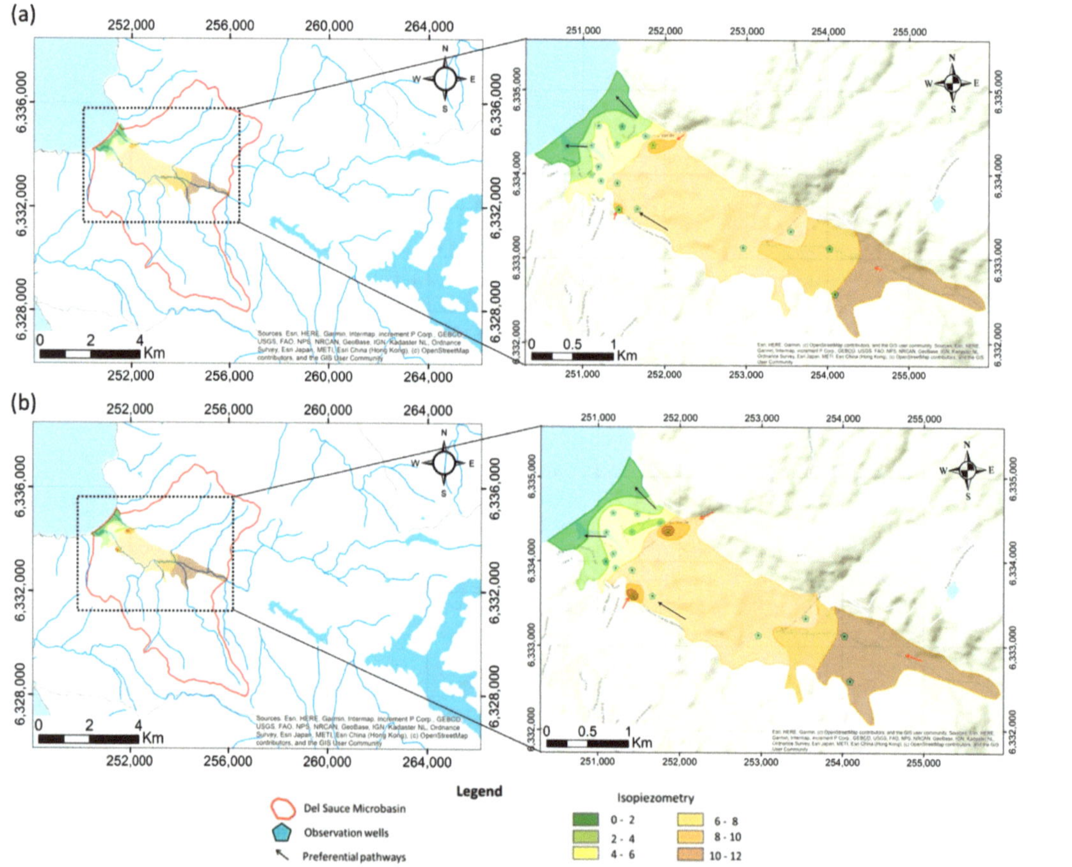

Figure 5. (**a**) Location of the Laguna Verde aquifer, with details of the Piezometric map for summer 2022 (December to March, austral dry season). (**b**) Location of the Laguna Verde aquifer, with details of the Piezometric map winter 2022 (June to August, austral rainy season) isopiestics every 1 m. Red arrows indicate recharge zones to the Laguna Verde aquifer.

A piezometric threshold in the coastal zone stands out, where higher levels are reached in the sector where the Del Sauce stream flows, which is maintained over time, resulting in a groundwater divide that would separate the northern and the southern sectors.

Three sectors were recognized with high piezometric altitudes throughout the year. The most relevant in the interior zone of the aquifer, in the headwaters of the basin, with altitudes above 10 m a.s.l., a second sector of smaller extension, recorded in well P2, and the third in the middle sector recorded in well P6, both with piezometric altitudes above 8 m a.s.l.

3.4. *Hydrogeochemistry of the Hydrological System*

3.4.1. Physicochemical Parameters

Details of the physicochemical parameters of the hydrological system under study are presented in Table 1. The annual average of total dissolved solids (TDS) in the waters of the Del Sauce and Las Cenizas streams reaches 920 mg/L, and increases downstream, recording 652 mg/L in the upper zone S20, until reaching 1263 mg/L in the lower Sauce or mouth. The TDS in the Laguna Verde aquifer has a spatially varying distribution, with 457 mg/L in the inland zone, passing through the middle zone with 633 mg/L, to the coastal or discharge zone, where it reaches 2173 mg/L, a behavior similar to the one presented by electrical conductivity (EC) in the zone. According to the Freeze and Cherry [55] classification, the waters from the north and southern coastal aquifer are classified as brackish (>1000 mg/L TDS). On the other hand, the artificial lagoon S30 and the dump flow S23 present annual averages of 10,567 and 6144 mg/L, respectively. The waters of HU2 reach 657 mg/L.

The pH in the waters of the Del Sauce and Las Cenizas streams ranged between 6 and 8.4, with an annual average of 7.6, while groundwater has a pH between 6.8 and 7.2, with an average of 7. The waters of the lagoon adjacent to the sanitary landfill in the upper sector of the basin (S30) and of the dump flow from the same sector (S23) have an annual average pH of 6.5 and 8.1, respectively. On the other hand, the groundwater that characterizes Hydrogeological Unit 2 (HU2) has an annual average pH of 7.4.

3.4.2. Major and Relevant Ions

Table 1 shows concentrations of the major ions and recommended ones in the waters of the study. During the studied period, Las Cenizas (S20) was the unique affluent feeding the Del Sauce stream. In this sampling site, the major ion is bicarbonate (271 mg/L), with values similar to those found in the upper section of the Del Sauce stream (224 mg/L). In its middle and lower sections, chloride becomes predominant in the wetland waters (329 and 440, respectively). The predominant ion in the inland zone of the aquifer is bicarbonate (159 mg/L), a composition similar to that of the middle zone (159 mg/L), with the difference that there is an increase in chloride ions, from 69 mg/L in the inland zone to 124 mg/L in the middle zone, with the latter ion becoming predominant in the coastal zone of the aquifer (807 mg/L). The samples from the flow coming from the sanitary landfill (S23) have annual averages of chloride and sulfate of 1987 y 1640 mg/L, respectively, while in the lagoon S30, sulfate predominated with 2179 mg/L. On the other hand, HU2 waters present similar concentrations between bicarbonate (166 mg/L), chloride (168 mg/L), and sulfate (106 mg/L) ions.

Nitrates show an annual average of 6.5 mg/L in the Las Cenizas affluent, 13.3 mg/L in the upper section of the Del Sauce stream, 9.4 mg/L in its middle section, and 9.7 mg/L in its lower section. Low nitrate concentration was detected in groundwaters from the inland sector. In its middle section, nitrates in the aquifer reach an annual average of 2.3 mg/L, with the highest concentration in the southern coastal sector with 64 mg/L, followed by the north coastal sector with 51 mg/L. The artificial lagoon S30 presents an annual average concentration of this ion of 21.2 mg/L, while the dump flow S23 reaches the highest concentration of nitrates in the hydric system with 155 mg/L. The nitrates present in HU2 reach 34.5 mg/L.

Table 1. Concentrations of the major ions and recommended ones in the waters of the study area (mean ± SD mg/L) and annual average and standard deviation of physicochemical parameters measured in situ in the study area. TDS: total dissolved solids; T: temperature; EC: electrical conductivity; OD: oxygen demand; COD: Chemical Oxygen Demand; TUR: turbidity; COL: color.

		Major Ions and Recommended										Physicochemical Parameters							
	Parameters	HCO_3^- mg/L	CO_3^{2-} mg/L	Cl^- mg/L	SO_4^{2-} mg/L	Na^+ mg/L	K^+ mg/L	Mg^{2+} mg/L	Ca^{2+} mg/L	NO_3^- mg/L	NO_2^- mg/L	TDS (mg/L)	T (°C)	pH	EC (µS/cm)	DO (mg/L)	COD (mg/L)	TUR NTU	COL Pt-Co
Groundwater	North coastal aquifer	219 ± 44	0.5 ± 0.9	854 ± 469	424 ± 327	149 ± 146	10.5 ± 6	94 ± 81	114 ± 98	51 ± 55	0.6 ± 0.1	2493 ± 1094	15.5 ± 1.3	6.8 ± 0.5	4155 ± 2044	3.6 ± 2.1	31.2 ± 26.2	1.0 ± 1.1	12.8 ± 8
	Southern coastal aquifer	162 ± 39	0.1 ± 0.1	761 ± 666	185 ± 108	219 ± 279	14.8 ± 4.5	55 ± 26	80 ± 28	64 ± 86	0.3 ± 0.6	1853 ± 1186	16.3 ± 1.2	6.9 ± 0.4	2925 ± 2211	5.2 ± 1.4	8.0 ± 5.8	1.4 ± 2.3	5.4 ± 1.5
	Middle aquifer	159 ± 30	0.7 ± 1.4	124 ± 46	99 ± 68	54 ± 24	2.1 ± 0.4	27 ± 10	30 ± 14	24 ± 36	0.2 ± 0.3	633 ± 142	16.5 ± 0.9	7.2 ± 0.6	1309 ± 462	6.6 ± 2	3.6 ± 5.1	17.2 ± 23.6	4.8 ± 1.5
	Inland aquifer	159 ± 30	0.3 ± 0.3	69 ± 9	60 ± 29	59 ± 59	1.7 ± 0.9	36 ± 30	67 ± 74	1.3 ± 1.5	0.01 ± 0	4577 ± 200	16.3 ± 03	7.2 ± 0.6	643 ± 172	4.2 ± 2.5	2.2 ± 2.8	7.9 ± 12.4	10.6 ± 15.9
	HU2	166 ± 48	0.4 ± 0.3	168 ± 75	106 ± 67	71 ± 35	4.6 ± 3	32 ± 1	40 ± 20	35 ± 57	0.2 ± 0.2	657 ± 230	16.9 ± 1.9	7.4 ± 0.7	1134 ± 291	6.2 ± 2.2	2.7 ± 3.5	5.4 ± 11.3	6.7 ± 9.2
Surface Water	La Luz Lagoon	108 ± 6	5.1 ± 8.6	34 ± 3	28 ± 1	41 ± 35	2.2 ± 1.2	37 ± 46	131 ± 199	3 ± 2.9	0.02 ± 0	229 ± 44	16.4 ± 4.9	8.3 ± 0.6	405 ± 26	7.8 ± 2.8	44.6 ± 59.6	4.5 ± 2.8	6.3 ± 2.5
	Las Cenizas (S19)	287 ± 50	1.1 ± 0.7	101 ± 31	183 ± 24	56 ± 24	6.25 ± 5.5	31 ± 9	71 ± 17	24 ± 33	0.1 ± 0	637 ± 75	14.2 ± 4.3	7.4 ± 0.5	1213 ± 288	5.6 ± 3.9	91.7 ± 62.7	22 ± 16.8	29 ± 24.6
	Las Cenizas (S20)	271 ± 23	0.5 ± 0.3	137 ± 58	162 ± 23	59 ± 28	16.7 ± 1.3	20 ± 5	72 ± 23	6.5 ± 6	0.1 ± 0.02	652 ± 139	16.6 ± 4.8	7.2 ± 0.4	1124 ± 288	5.9 ± 1.7	61.5 ± 31.7	16.6 ± 10.1	17.5 ± 5
	High El Sauce	224 ± 44	2.9 ± 2.8	155 ± 17	155 ± 25	62 ± 28	17.4 ± 2.3	22 ± 6	75 ± 23	13 ± 11	3.4 ± 0.5	702 ± 227	12.8 ± 2.6	7.9 ± 0.4	1429 ± 624	5 ± 2.8	58.2 ± 26.2	14.0 ± 7.4	15.0 ± 5.8
	Middle El Sauce	187 ± 62	1.6 ± 1.2	329 ± 125	185 ± 29	134 ± 98	18 ± 3.4	33 ± 11	66 ± 13	9.4 ± 8.5	2.6 ± 2.6	1064 ± 327	16.14 ± 2.4	5.9 ± 3.8	1831 ± 562	8.1 ± 2	57.6 ± 28	13.2 ± 9.1	18.8 ± 2.5
	Lower El Sauce	171 ± 48	3 ± 2.2	440 ± 222	189 ± 36	119 ± 126	20 ± 3.4	37 ± 12	63 ± 15	9.7 ± 6	8.02 ± 7.1	1263 ± 539	15.2 ± 5.2	7.8 ± 0.3	2279 ± 1108	8.2 ± 1.7	62.2 ± 37.9	16.6 ± 9.8	15.0 ± 5.8
	Artificial Lagoon S30	11 ± 5	0 ± 0	6388 ± 376	2179 ± 2000	186 ± 192	69 ± 56	74 ± 32	90 ± 35	21 ± 24	0.3 ± 0.4	10,567 ± 2263	15.3 ± 3.5	6.5 ± 3.5	15,532 ± 2503	7.2 ± 1	223.9 ± 112.7	39 ± 1.7	30 ± 5.4
	Dump Flow S23	356 ± 116	3.7 ± 3	1987 ± 344	1640 ± 421	88 ± 12	34 ± 15.5	90 ± 44	111 ± 25	155 ± 144	10.7 ± 13.6	6144 ± 1567	11.9 ± 1.7	8.1 ± 0.4	8552 ± 1172	7.8 ± 0.7	248.3 ± 56.1	10.3 ± 0.3	113 ± 25.5

Regarding the hydrochemical facies identified by the Piper diagrams for surface water, it can be established that, in general, the waters in the Las Cenizas affluent have clear mixed facies with respect to cations and anions, while in the Del Sauce stream, they are of chloride and mixed types with respect to cations. On the other hand, the waters of the La Luz lagoon, located in the eastern part of the study area, are of the bicarbonate type, while they vary between calcic and mixed with respect to cations (Figure 6).

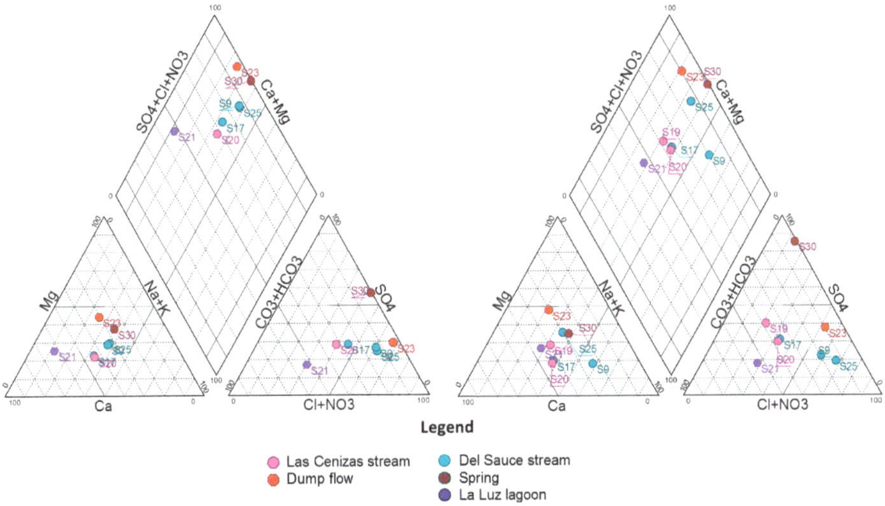

Figure 6. Piper diagrams for surface water of the study area in summer 2022 (left) and winter 2022 (right).

Regarding the other surface sources sampled, the artificial lagoon S30 and the dump flow S23 correspond to sulfate- and chloride-type waters, respectively, a condition that is maintained during the entire year (Figure 6).

Regarding the aquifer waters (Figure 7), mixed chloride and sodium facies are generally observed. However, during spring, the waters in the north coastal (P1) and southern coastal (P3 and P10) sectors are characterized by a sodium chloride water type. In winter, this tendency is observed more locally at points P10 and P33.

As shown in the modified Stiff diagram, the maximum mineralization was reached in spring 2021, while the lowest mineralization was recorded in summer 2022. As a tendency, mineralization gradually increased from summer to winter 2022 (Figure 7).

From the microelements analysis carried out in the waters of the entire hydric system studied (Table 2), the high value of Br stands out for surface water, where this element was highest downstream, reaching the lower Sauce area with concentrations of 7 mg/L. In groundwater, the presence of the minor constituents Al, Ba, Br, Fe, and Mn is observed with higher annual average concentrations than expected (>0.1 mg/L), highlighting the presence of Br, mainly in the north coastal zone, reaching concentrations of 18 mg/L. The inland zone presents concentrations of 1.8 mg/L of Br and 1.3 mg/L of Fe, while in the middle sector, both elements increase to 3.1 and 3.2 mg/L, respectively. In the runoff from the sanitary landfill (S23), Br reaches the highest concentration of the entire hydric system with 59.4 mg/L exhibiting high seasonal variability. On the other hand, the artificial lagoon S30 presents average Br concentrations of 12 mg/L, in addition to 4.5 mg/L of Al and 14 mg/L of Mn. In HU2 waters, the presence of Br (i.e., 3 mg/L) and Fe (i.e., 2 mg/L) stands out. The annual averages of the microelements omitted in the table (B, Cd, Li, Mo, Ni, Pb, Ag) had concentrations below the detection limit in all samples (<0.01 mg/L).

Table 2. Analysis of microelements in waters of the studied hydric system.

	Microelements (mg/L)	F	Al	Ba	Br	Co	Cu	Fe	Mn	V	Zn
Groundwater	North Coastal aquifer	0.2 ± 0.03	0.7 ± 2.4	0.1 ± 0.03	17.9 ± 12.7	0.01 ± 0.001	0.03 ± 0.03	0.5 ± 1.4	1.1 ± 1.6	0.1 ± 0.1	0.03 ± 0.03
	Southern Coastal aquifer	0.2 ± 0.02	0.6 ± 1.5	0.1 ± 0.01	11.5 ± 9.5	0.008 ± 0.004	0.01 ± 0.005	0.2 ± 0.2	0.01 ± 0.007	0.06 ± 0.03	0.03 ± 0.02
	Middle aquifer	0.2 ± 0.05	0.1 ± 0.1	0.04 ± 0.02	3.1 ± 0.8	0.01 ± 0.00	0.04 ± 0.06	3.2 ± 4.7	0.3 ± 0.5	0.1 ± 0.1	0.03 ± 0.03
	Inland aquifer	0.2 ± 0.03	0.2 ± 0.3	0.1 ± 0.1	1.8 ± 1	0.1 ± 0.1	0.02 ± 0.02	1.3 ± 1.3	0.6 ± 1.3	0.01 ± 0.004	0.04 ± 0.05
	UH2	0.3 ± 0.1	0.8 ± 2.4	0.02 ± 0.02	3.0 ± 1.2	0.009 ± 0.003	0.02 ± 0.02	1.9 ± 5.8	0.04 ± 0.05	0.02 ± 0.01	0.04 ± 0.04
Surface Water	La Luz Lagoon	0.19 ± 0.0	0.77 ± 0.5	0.27 ± 0.4	0.83 ± 0.1	0.48 ± 0.8	0.04 ± 0.1	2.47 ± 3.5	0.1 ± 0.1	0.01 ± 0.0	0.021 ± 0.0
	Las Cenizas S19	0.46 ± 0	2.2 ± 2.8	0.05 ± 0.01	1.1 ± 0	0.01 ± 0.01	0.01 ± 0.01	2.3 ± 2.6	0.62 ± 0.3	0.02 ± 0.01	0.03 ± 0.01
	Las Cenizas S20	0.45 ± 0.06	0.64 ± 0.16	0.02 ± 0	1.94 ± 0.7	0.01 ± 0.01	0.24 ± 0.4	2.13 ± 0.2	0.2 ± 0.06	0.01 ± 0	0.07 ± 0.09
	High El Sauce	0.4 ± 0.03	0.5 ± 0.3	0.03 ± 0.005	3.0 ± 0.9	0.007 ± 0.00	0.1 ± 0.1	1.2 ± 0.3	0.3 ± 0.1	0.01 ± 0.0	0.04 ± 0.006
	Middle El Sauce	0.4 ± 0.06	1.6 ± 3.04	0.04 ± 0.006	5 ± 2.5	0.008 ± 0.004	0.03 ± 0.03	1.1 ± 1	0.6 ± 0.3	0.01 ± 0.0	0.01 ± 0.06
	Lower El Sauce	0.4 ± 0.1	0.9 ± 0.3	0.04 ± 0.0	7 ± 3.6	0.01 ± 0.001	0.03 ± 0.00	1.2 ± 0.2	0.6 ± 0.2	0.01 ± 0.0	0.01 ± 0.0
	Artificial Lagoon S30	0.2 ± 0.03	4.5 ± 0.01	0.1 ± 0.005	12.2 ± 0.7	0.01 ± 0.001	0.03 ± 0.00	2.5 ± 0.3	13.9 ± 3.6	0.2 ± 0.1	0.04 ± 0.009
	Dump Flow S23	0.1 ± 0.02	0.2 ± 0.01	0.1 ± 0.02	59.4 ± 13.5	0.01 ± 0.001	0.01 ± 0.00	0.9 ± 0.00	4.1 ± 0.3	0.4 ± 0.2	0.01 ± 0.0

3.4.3. Biological Contamination and Water Quality

Fecal coliforms were detected in all the groundwater samples from the coastal zone, with a particularly high abundance in the P4 well with >2400 MPN/100 mL. Fecal coliforms were also retrieved in the surface water of the artificial lagoon S30 (>2300 MPN/100 mL), in the flow from the sanitary landfill (S23) and at the mouth of the Del Sauce estuary, suggesting that these waters are limited for certain uses and crops.

In comparative terms and based on multiple indicators that include salinity, EC, TDS, Cu, V, and the presence of fecal coliforms, the best water quality for irrigation is recognized in La Luz Lagoon and in waters of the inland part of the aquifer, while the least recommended would be the waters of the north coastal part of the aquifer, those of well P10 (southern coastal aquifer), the waters of the artificial lagoon (S30), and those of the flow from sanitary landfill (S23).

3.4.4. Ionic Ratios

To establish the hydrogeochemical processes occurring in the Del Sauce stream and the interrelationship between the different surface waters and groundwaters, some ionic ratios, such as $rCl/rHCO_3$, rMg/rCa, rK/rNa, rBr/rCl, and rNa/rCl, were calculated for each of the established water subgroups (Figure 8).

Since this is a coastal system, the influence of seawater may have some significance; therefore, ionic relations were determined for the seawater of Laguna Verde Bay. The annual mean values obtained were $rCl/rHCO_3 = 297$; rMg/rCa = 5.9; rK/rNa = 0.02; rBr/rCl = 0.03; and rNa/rCl = 0.84. The spills and HU2 were also characterized. In the flow S23, the ratios calculated for summer and wintr were $rCl/rHCO_3 = 19.3$ and 5.6; rMg/rCa = 1.5 and 1.6; rK/rNa = 0.29 and 0.22; rBr/rCl = 0.01 and 0.02; and rNa/rCl = 0.07 and 0.08, respectively, all indicating some variability of the effluent salinity over time. Regarding HU2, these relations are much lower and show some stability. In the Del Sauce stream, it can be noted that the $rCl/rHCO_3$ and rMg/rCa ratios increase from the upstream until the downstream section. This situation is similar between summer and winter.

Regarding the inland aquifer, it shows low values representative of freshwater in which mineralization is predominantly acquired by rock dissolution. However, groundwaters from the coastal part exhibit a greater variability. Interestingly, the highest rMg/rCa ratio (i.e., 1.84) was calculated in the P4 well from the north coastal aquifer during the dry season (summer). In this same dry season, $rCl/rHCO_3$ and rMg/rCa ratios in groundwaters from the middle part of the aquifer were 1.54 and 1.55, respectively.

During the winter rainy season, the $rCl/rHCO_3$ ratios show a seasonal maximum of 10.3 in well P2 in the north coastal aquifer. The rMg/rCa ratio shows a value of 1.86 in well P4 of the north coastal aquifer, and a ratio of 1.46 in well P10 of the southern coastal aquifer. Finally, the rNa/rCl ratio shows a value of 0.54 in well P2 from the north coastal aquifer, whereas wells P10 and P3 from the southern coastal aquifer exhibit higher ratios of 0.96 and 1.03, respectively. The Del Sauce stream in its lower section shows a rNa/rCl ratio of 0.18.

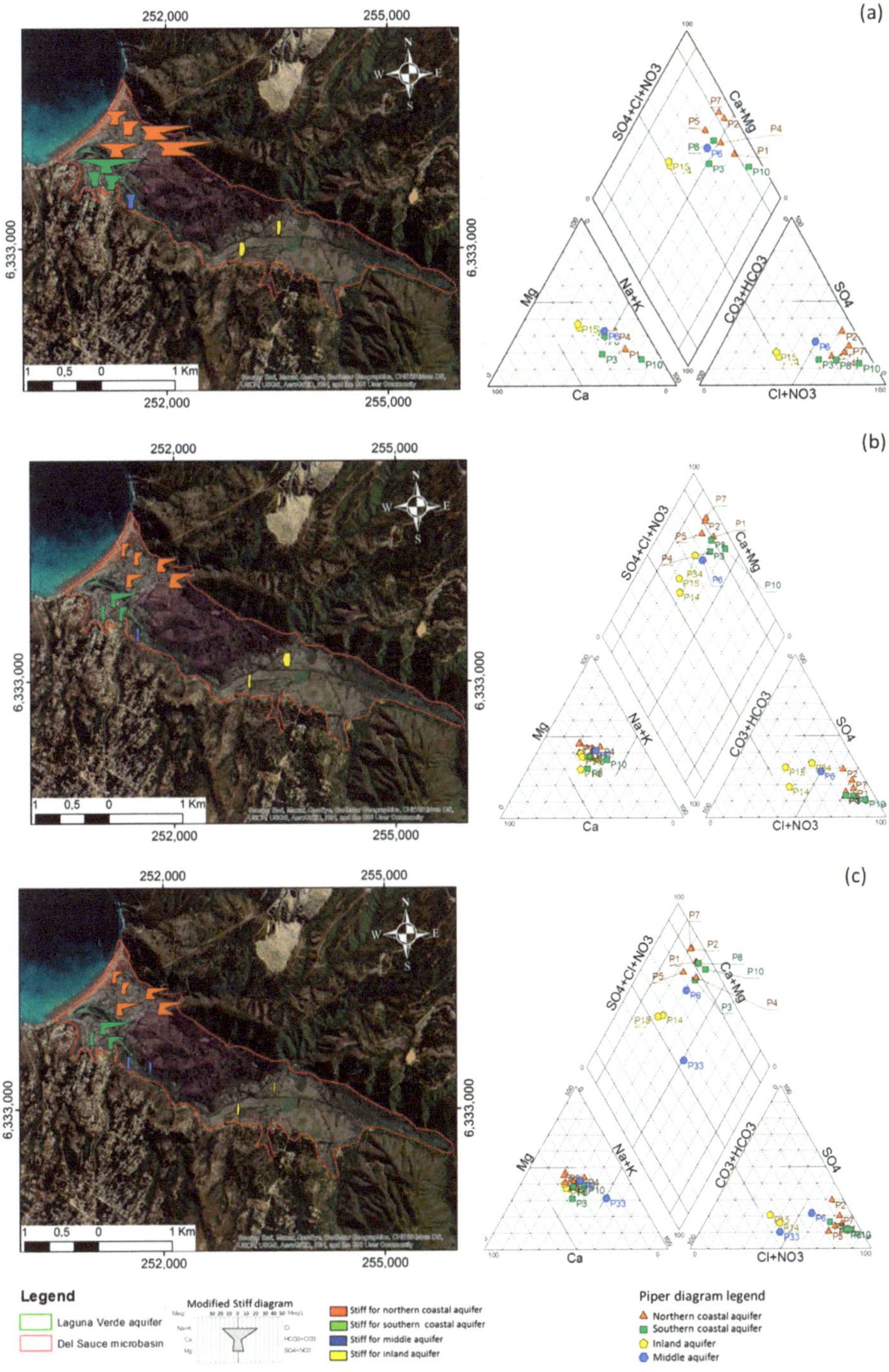

Figure 7. *Cont.*

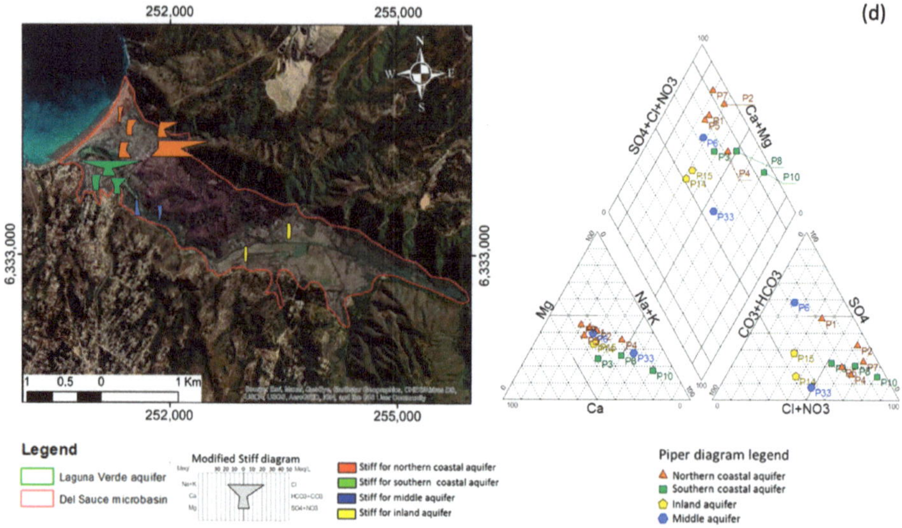

Figure 7. Laguna Verde aquifer hydrochemical maps, modified Stiff (left) and Piper (right) diagrams, for seasons: (**a**) spring 2021; (**b**) summer 2022; (**c**) autumn 2022; (**d**) winter 2022.

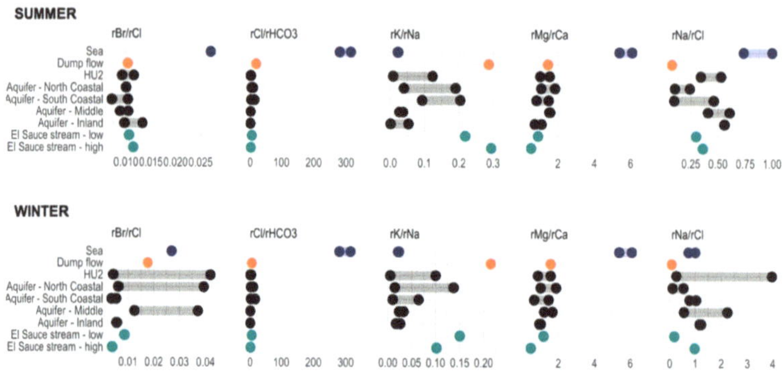

Figure 8. Comparative plot between ionic ratios of the studied hydric system. Superficial waters are highlighted in color and groundwaters are represented in black.

3.4.5. Isotopic Analysis

We observed a correlation between the waters of the estuary–aquifer system and the global meteoric water line (Figure 9). The superficial waters of the mainstream are lighter than the groundwaters of the aquifer. These waters were found to be more depleted in $\delta^{18}O$ ‰ and $\delta^{2}H$ ‰ in the highest sector (Las Cenizas; P20), with −10.4 $\delta^{18}O$ ‰ and −79.7 $\delta^{2}H$ ‰, and it is gradually enriched until the lower zone (lower El Sauce; S25), already close to its mouth, where averages of −7.3 $\delta^{18}O$ ‰ and −58.3 $\delta^{2}H$ ‰ were recorded. During the 2022 sampling, a gradual decrease of $\delta^{18}O$ ‰ and of $\delta^{2}H$ ‰ was observed since the summer season, where a maximum average of −7.3 $\delta^{18}O$ ‰ and −61.5 $\delta^{2}H$ ‰ V SMOW is recorded in these waters, until winter with an average of −9.8 $\delta^{18}O$ ‰ and −70.9 $\delta^{2}H$ ‰ V SMOW. In spring 2021, mean values were −9.2 $\delta^{18}O$ ‰ and −71.6 $\delta^{2}H$ ‰ V SMOW.

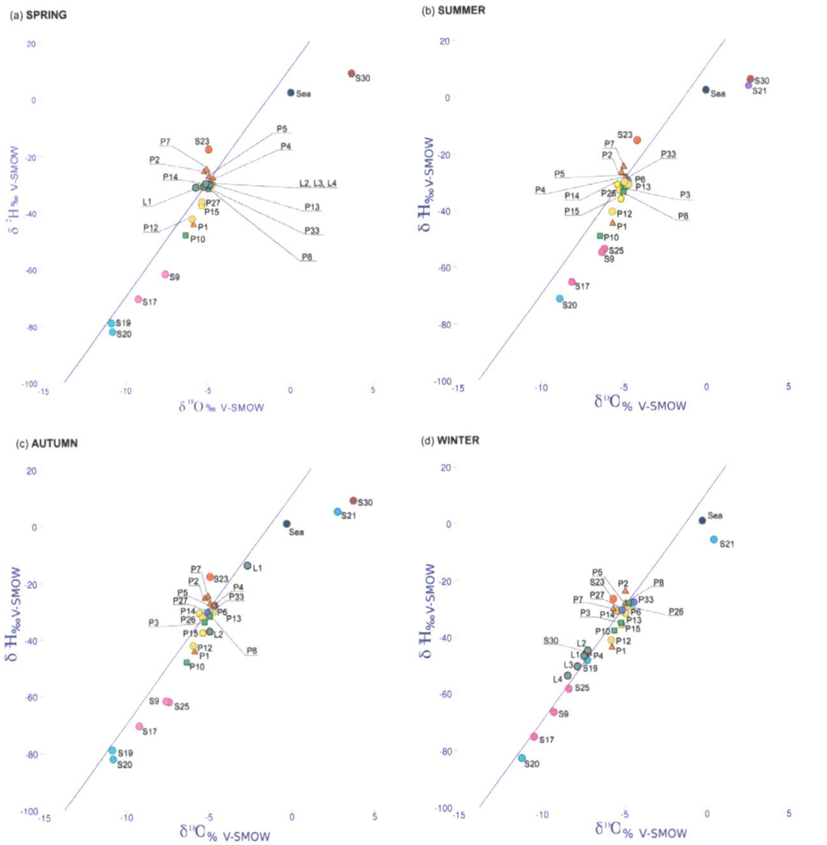

Figure 9. Diagrams of water isotopes $\delta^{18}O$ ‰ and $\delta^{2}H$ ‰ V SMOW for the total waters sampled in the four sampling campaigns: (**a**) spring, (**b**) summer, (**c**) autumn, (**d**) winter. In blue, global meteoric water line (GMWL).

Regarding the groundwaters of the Laguna Verde aquifer, the values are more similar to those of precipitation than to those of the main stream. The rainwater collected and analyzed showed values between −4.81 and −8.41 of $\delta^{18}O$ ‰ V SMOW and between −30.3 and −53.89 of $\delta^{2}H$ ‰ V SMOW, matching those of the GMWL. In groundwater, $\delta^{18}O$ data showed a low dispersion of data during the year (i.e., values between −4.65 and −7.11 $\delta^{18}O$ ‰ V SMOW, averaging −5.64 $\delta^{18}O$ ‰ V SMOW). However, $\delta^{2}H$ showed higher variability with values between −21.9 and −49.33 (average = −35.64 $\delta^{2}H$ ‰ V SMOW; Figure 9). On one hand, the sites the most depleted in both indicators correspond to P1, P10, and P15, showing characteristics of possible evaporation, because they are located slightly below the Global Meteoric Water Line GMWL [56–58]. On the other hand, the water samples of the aquifer corresponding to P7 and P2 are recognized for being located above the Global Meteoric Water Line and with the highest $\delta^{2}H$ ranges of the aquifer (annual averages of −26.5 and −25.7 $\delta^{2}H$ ‰ V SMOW, respectively).

While HU2 has a stable behavior throughout the year, well P12 robustly showed an isotopic signature similar to that of well P1 from the north coastal aquifer, with evidence of evaporation (i.e., slight enrichment in deuterium). Regarding the points associated with spills, the dump flow (S23) presents annual mean values of −4.9 $\delta^{18}O$ ‰ and −19.6 $\delta^{2}H$ ‰, while S30 records 0.7 $\delta^{18}O$ ‰ and −5.8 $\delta^{2}H$ ‰ V SMOW.

4. Discussion

Water scarcity in coastal areas is rising due to the increase in land occupation in these attractive zones. The increasing need for water generates significant pressures on the physical environment, especially on surface and subterranean water, which causes alterations to the natural environment. To carry out suitable hydric management, it is necessary to have a complete knowledge of the environment and of the relationships between water bodies. However, sufficient information is not always available, and in the specific case of the Laguna Verde locality (Valparaíso, Chile), there is hardly any data on the groundwater behavior and its relationship with surface water.

Based on the analysis and interpretation of the hydrological background and the collection of hydrodynamic, hydrochemical, and isotopic data, a conceptual model of the functioning of the Del Sauce wetland–Laguna Verde aquifer hydric system is proposed (Figure 10). The valley through which the Del Sauce stream flows constitute a small detritic aquifer whose main recharge is rainwater [21], while discharge into the sea can occur when the sandbar is opened. Precipitation during 2010–2022 shows the effects of the Mega-Drought that affected this area of the country, whereby precipitation was reduced by a mean of 25%, with occasional losses of 75%, as occurred in 2019 when only 130 mm was received [35]. During the last few years, the Del Sauce stream presented average flows of 0.26 m^3/s, coming from the Las Cenizas affluent. The isotopic signature of the stream waters confirmed that the water from the La Luz reservoir, which is much heavier due to evaporation processes, does not reach the Del Sauce stream, nor does the water from the artificial lagoon S30, which is strongly affected by pollution from the sanitary landfill. The flow values recorded during this study cannot be compared with data prior to the Mega-Drought as there are no previous records; however, hydrometric backgrounds in discharge zones of other natural channels in Central Chile indicate that these values would be reduced by up to 90% of the expected normal value [23].

From the piezometric information, significant anomalies are those shown in the sectors where wells P2 and P6 are located, as they present higher levels than the rest of the catchments. These could be related to possible water transfer areas from the lateral edges to the Laguna Verde aquifer.

Regarding the relationship of the Del Sauce stream with the Laguna Verde detritic coastal aquifer, everything indicates that in the inland zone (high Sauce, upstream), it behaves as a gaining river throughout the year (aquifer provides water to the stream) until it nears points P15 and P14, where the extraction of large water volumes alter this dynamic and cause a descent cone. It is more complex to establish the hydraulic relationship in the middle zone. As mentioned above, the existence of a possible recharge edge in its southern part makes it easier for the stream (middle Sauce) to behave as a gaining river, although pumping at point P33 could affect this relationship. Finally, in the coastal zone, the estuary behaves mainly as a losing stream, and it only occasionally receives some groundwater contributions. This behavior is not very usual, since the wetlands connected to the aquifers tend to feed them in rainy seasons, while in dry seasons, the opposite process occurs. However, the estuary that forms this wetland is dammed by a sandbar for most of the year, breaking up naturally in rainy seasons, which generates an increase in surface runoff water in dry seasons, and a decrease at certain times when the passage to the sea is opened in winter.

The isotopic signatures of the water wells P1 and P10, both located on both sides of the estuary, are somewhat more depleted than the other points of the aquifer in the coastal zone. This could be due to the mixing of the aquifer waters with the infiltrated waters coming from the humid zone.

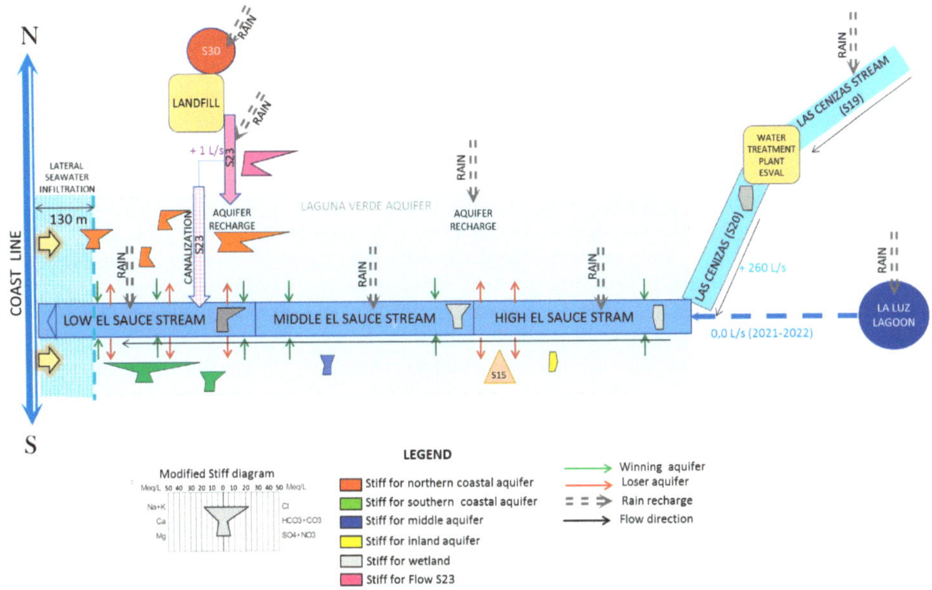

Figure 10. Conceptual hydrogeological–hydrogeochemical model of the Del Sauce wetland–Laguna Verde aquifer hydric system. Schematic plan view.

However, the different uncertainties mentioned above should be verified with longer piezometric monitoring and a higher number of sites. Bianchi et al. [59] and Schiavo et al. [60] suggested the use of stochastic simulations using the Monte Carlo method to capture the inherent variability of aquifer systems by generating multiple scenarios. Given that the available data come from a few wells and the column information shows variable and heterogeneous lithological characteristics, the best approach to establish a good conceptual model is through a stochastic model. Geostatistical methods, which are less demanding as they require little information, based on an estimated structural model with its uncertainty, have been applied with good results in heterogeneous detrital aquifers similar to those of Laguna Verde. We hereby provide the first step of a longer monitoring that will allow, with the use of these powerful models, the efficient assessment of the uncertainty in the geological and piezometric data.

In general, the stream waters vary from bicarbonate to (weakly) chloride in its upper course, with low mineralization, to chloride and sulfate waters with greater mineralization in the mouth zone, and with a progressive increase in nitrates concentration. All this causes the quality to deteriorate downstream. It is noted that the estuary waters have a certain salinity throughout their length and are not recommended for irrigation. This coincides with previous studies in which they were classified as poor-quality waters [28,31]. In its final section, the presence of fecal coliforms was also detected.

Groundwater presents a hydrodynamic and hydrochemical behavior similar to that of the stream waters. Broadly speaking, the water flow is from SW to NE, which is locally altered by the pumping effects. The chemical composition of its waters shows that the dominant facies in the inland part of the aquifer is bicarbonate, while the proportion of salts increases toward the coastal zone, with chloride becoming dominant. Another noteworthy characteristic is the presence of fecal coliforms in the coastal zone. This analysis shows the gradual deterioration of the water system from its source to the area closest to the coast, where high values of chlorides (440 mg/L) and nitrates (9.7 mg/L) stand out in the estuary, and with high values of chlorides (807 mg/L) and nitrites (57 mg/L), as well as Br (18 mg/L), in the aquifer. The highest values in the entire water system were retrieved from runoff S23, which comes from the landfill and delivers its water to the coastal zone with annual averages of chloride and sulfates of 1987 and 1640 mg/L, respectively, 155 mg/L of nitrates and 59.4 mg/L of bromine.

All this indicates that in the coastal zone, both groundwater and surface water are affected by anthropic pollutant sources. First, the most important pollutant source corresponds to the flow coming from the vicinity of the sanitary landfill in the coastal zone (S23), which has notoriously high annual averages in relevant parameters (e.g., TDS, COD, sulfates, chlorides, nitrates, potassium), including the presence of coliforms; all these parameters are in ranges that some authors associate with leachates from sanitary landfills [61]. Their influence on groundwater seems clear because an elevation of the piezometric surface is generated in the environment of the discharge and there is a significant influence on the water quality near the discharge (wells P2 and P7). Also, the values of the ionic ratios $rCl/rHCO_3$ or rMg/Ca showed a similarity with those of the waters of this discharge, and the waters near the discharge present similar isotopic signatures. Finally, this water coming from the upper zone flows through a ravine of an E–W direction that has previously been interpreted as an inferred fault in the intrusive rock unit [43]. Therefore, it is not ruled out that this fluid may also infiltrate in some percentage along its route and flow into the aquifer subsurface through HU2 if the fault is considered to behave hydraulically as conductive, a behavior generally expected in 70% of faults, and in 60% of faults in crystalline rocks [62].

Secondly, the smaller-scale agricultural and livestock industries present in the inland and middle zone of the watershed (avocado plantations, cattle ranches, and alfalfa plantations in the inland zone of the basin; and small-scale crops in the middle zone) are considered another source of pollution. This pollution is detected by the progressive increase of up to 50 times more nitrates in the coastal zone of the aquifer compared to the inland zone. Such behavior is not observed in the surface water of the wetland, which has nitrates in similar concentrations throughout its length, which is interpreted as a direct infiltration of nitrogenous elements from the unsaturated zone into the aquifer, mainly as a result of agricultural and livestock activities in the upper and middle zone of the basin, where these industries are located. This is consistent with the EEA (2023) [63], which recognizes that mineral fertilizers and manure are the main sources of nitrate concentrations in groundwater in the EU, and it is estimated that 80% of the nitrogen spill to the aquatic environment in the EU comes from agriculture.

Finally, another source of pollution may correspond to raw wastewater and septic tank liquid discharged mainly in the populated area, affecting the saturated zone of the aquifer and the Del Sauce stream in its coastal zone. Based on the Metcalf and Eddy (1995) classification [64], this coastal sector presents high levels of wastewater indicators such as the presence of fecal coliforms throughout the year (up to 350 MPN/100 mL), maximum chloride and sulfate concentrations of 660 mg/L and 225 mg/L, respectively, and high TDS contents (annual mean of 1263 mg/L). This type of pollution is also detected in the Las Cenizas affluent, in the locality of Placilla, upstream of the water treatment plant (S19), recognized by the significant presence of fecal coliforms (>1000 MPN/100 mL), TDS annual mean of 637 mg/L, and high sulfate and chloride concentrations. The lack of infrastructure for the management of wastewater in the aquifer zone has led to the accumulation of pollutants in water bodies, which negatively affects public health and the environment, and shows the need to implement an adequate sanitation system.

In general, ionic relations, such as $rCl/rHCO_3$, rBr/Cl, or rMg/Ca, do not seem to show values that make us suspect a generalized marine intrusion in the aquifer. However, a salinity anomaly was observed near the estuary and other specific sites (i.e., P3, P4, and P10) whose ionic ratios are elevated at certain times and could point to some influence of seawater. In winter, storms and rough waters increase in the study area, a sector that has been classified as having a medium-to-high sensitivity or susceptibility to suffer floodings on the coastal edge due to storm surges. Other authors have assigned these processes as responsible for the mixing of fresh and saltwater [65–73].

When the sandbar that during a large part of the year dams the waters of the Del Sauce estuary opens in the sector of its mouth, the surface seawater intake occurs through the estuary, affecting its waters, which in turn infiltrates the nearest areas of the aquifer (i.e., P10). During the dry season (summer), a time of low aquifer recharge, it was detected that the rMg/rCa ionic ratio shows in P4 a ratio of 1.84, a value that is only attributable to a mixing with seawater. This seawater inflow possibly takes place from the coastline to at least 140 m inland (P4), laterally through the sandy sediments present in the coastal strip, an expected event in this type of aquifer [74], and/or horizontally because of flooding due to storm surge events.

The shallow depth of the piezometric level, the increase in rough waters and storms in recent years, and the increase in the population that uses the hydric resource in the Laguna Verde locality, together with the impact of pollutant sources, have reduced the freshwater availability for the populations and ecosystems in this coastal area. And this coincides with the findings reported by the IPCC [19], as this sector is very sensitive to the possible effects of climate change. The above exacerbates the relevance of employing suitable hydric resource management, where the basin as a territorial geographic unit is the most appropriate for the planning and management of water quality [17]. One of the most important management factors is the allocation of sufficient water of a quality suitable for the population and for maintaining the functions of the wetland ecosystems to be conserved. This requires consideration and management of land-based activities that affect coastal wetlands, such as commercial agriculture (which reduces available runoff in the headwaters of the basin), the damages caused to riparian areas (which alter the flow configuration and erosion/deposition patterns), and the extraction of an excessive groundwater volume (which can lower the phreatic zone and thus reduce the baseflow for wetlands and terrestrial ecosystems) [13]. This raises the need to obtain resources from other sources, such as desalination. The intake of seawater and brackish water through the ground, if feasible, presents favorable aspects, since they are susceptible to producing freshwater by membrane processes (desalting) at a lower production cost than seawater desalination [75]; therefore, it is of interest to promptly assess the possible extraction of these waters from the coastal aquifer, associating it with its possible environmental effects.

5. Conclusions

In the coastal region of Central Chile, the physical, hydrochemical, and isotopic analyses have been complementary in achieving a general and specific view of the behavior and quality of a wetland–aquifer hydric system. Under a climatic context of Mega-Drought in a Mediterranean climate, the quality of the hydric resource in the study area is recognized to be very deteriorated. Given the hydraulic connection, the specific pollutant sources have affected both the wetland and the estuary. These sources correspond mainly to the flow from the sanitary landfill and the one generated by the agricultural and livestock industries settled in the valley, in addition to slight seawater intrusion in some specific coastal sectors, which has increased the salinity in this area. This study provides unprecedented information about the physical, hydrometric, hydraulic, and hydrochemical characteristics of a coastal wetland–aquifer hydric system in central Chile, and the effects of its hydraulic connection. It contributes to a better understanding of the hydric dynamics and salinization processes in a coastal sector affected by anthropic pollution, raising concerns about the poor state of conservation of both surface water and groundwater resources, mainly in the coastal sector. We raised the need for integrated management of the sector's hydric resources that internalizes climatic variability, not allowing the direct use of these waters by the population, working on the reduction and elimination of the detected anthropic pollutant sources, and developing strategies for the recovery of the waters of the Del Sauce wetland and the Laguna Verde aquifer.

Supplementary Materials: The following supporting information can be downloaded at: https://www.mdpi.com/article/10.3390/hydrology11100174/s1. Table S1. Detailed list of methods used for physiochemical parameter measurements. Detection limit and accuracy are given. Figure S1. Riverside diagrams (USDA, 1975) for all samples of the hydric system in summer 2022 (left) and winter 2022 (right).

Author Contributions: Conceptualization, B.G., J.M.A.R., C.A.S. and C.L.; methodology, B.G., P.D., C.R., C.A.S. and C.L.; formal analysis, B.G., P.D. and C.L.; validation, J.M.A.R., C.A.S. and C.L.; investigation, B.G., A.B., S.F., A.Á. and C.L.; resources, C.A.S.; data curation, B.G., S.F. and C.L.; writing—original draft preparation, B.G.; writing—review and editing, J.M.A.R., C.A.S. and C.L.; visualization, B.G. and C.L.; supervision, C.A.S. and C.L.; project administration, C.A.S. and C.L.; funding acquisition, B.G., C.A.S. and C.L. All authors have read and agreed to the published version of the manuscript.

Funding: This work has been funded by the Sacyr (Chile and Spain), the FIC-R BIP project grant number 40046077 of the Regional Government of Valparaíso, and by the DGI Project of the Playa Ancha University, grant number CNE 16–20, awarded by D.E. No. 0345/2020, directed by Jean Pierre Francois. Céline L. was funded by the projects ANID InES I + D 2021 grant number INID210013 and Marie Curie Postdoctoral Fellowship HORIZON-MSCA-2022-PF-01 project grant number 101106387.

Data Availability Statement: The original contributions presented in the study are included in the article Supplementary Materials; further inquiries can be directed to the corresponding author.

Acknowledgments: We would like to thank the ideas and logistic support of Domingo Zarzo and Iván Sola. Also, we acknowledge the Laguna Verde community, who made it possible to gather the information presented in this study, mostly represented by Julia Poblete. We would also like to highlight the support we received from the president of the Laguna Verde APR, Colville Smith, to carry out this research. We acknowledge in this instance the technical and human support of the personnel that integrates the HUB Ambiental laboratory of the Playa Ancha University, mainly Fabiola Moenne as well as Sebastián Martorell, who also supported us as a specialist in environmental water sampling. We are grateful for the cooperation of the Stable Isotope Laboratory of the Alfred Wegener Institute (AWI) for Polar and Marine Research, Research Unit Potsdam for the measurement of water isotopes. Finally, we thank the 'Support Network for the Review and Editing of Manuscripts' of the Playa Ancha University. This work acknowledges the Severo Ochoa Centre of Excellence accreditation (CEX2019-000928-S) funded by AEI 10.13039/501100011033.

Conflicts of Interest: The authors declare no conflicts of interest.

References

1. Wörman, A.; Packman, A.I.; Marklund, L.; Harvey, J.W.; Stone, S.H. Fractal topography and subsurface water flows from fluvial bedforms to the continental shield. *Geophys. Res. Lett.* **2007**, *34*, L07402. [CrossRef]
2. Ramsar. The Ramsar Convention: What's It All about? Secretaría de la Convensión Ramsar 2015. Available online: https://www.ramsar.org/sites/default/files/fs_6_ramsar_convention.pdf (accessed on 15 January 2024).
3. Mitsch, W.J.; Gosselink, J.G. *Wetlands*, 2nd ed.; Van Nostrand Reinhold: New York, NY, USA, 1994.
4. Brinson, M.M.; Malvárez, A.I. Temperate freshwater wetlands: Types, status, and threats. *Environ. Conserv.* **2002**, *29*, 115–133. [CrossRef]
5. Finlayson, C.M. Forty years of wetland conservation and wise use. *Aquat. Conserv. Mar. Freshw. Ecosyst.* **2012**, *22*, 139–143. [CrossRef]
6. Davidson, N.C. How much wetland has the world lost? Long-term and recent trends in global wetland area. *Mar. Freshw. Res.* **2014**, *65*, 934–941. [CrossRef]
7. Gardner, R.C.; Barchiesi, S.; Beltrame, C.; Finlayson, C.; Galewski, T.; Harrison, I.; Paganini, M.; Perennou, C.; Pritchard, D.; Rosenqvist, A.; et al. *State of the World's Wetlands and Their Services to People: A Compilation of Recent Analyses (March 31, 2015)*; Ramsar Briefing Note No. 7; Ramsar Convention Secretariat: Gland, Switzerland, 2015. [CrossRef]
8. Gardner, R.C.; Finlayson, C.M. *Global Wetland Outlook: State of the World's Wetlands and Their Services to People*; Ramsar Convention Secretariat: Gland, Switzerland, 2018.
9. Xu, X.; Xiong, G.; Chen, G.; Fu, T.; Yu, H.; Wub, J.; Liu, W.; Su, Q.; Wang, Y.; Liu, S.; et al. Characteristics of coastal aquifer contamination by seawater intrusion and anthropogenic activities in the coastal areas of the Bohai Sea, eastern China. *J. Asian Earth Sci.* **2021**, *217*, 104830. [CrossRef]
10. Kazezyılmaz-Alhan, C.M. Analytical solutions for contaminant transport in streams. *J. Hydrol.* **2008**, *348*, 524–534. [CrossRef]
11. U.S. EPA. *Connectivity of Streams and Wetlands to Downstream Waters: A Review and Synthesis of the Scientific Evidence (Final Report)*; EPA/600/R-14/475F; U.S. Environmental Protection Agency: Washington, DC, USA, 2015.

12. Drexler, J.Z.; Knifong, D.; Tuil, J.L.; Flint, L.E.; Flint, A.L. Fens as whole-ecosystem gauges of groundwater recharge under climate change. *J. Hydrol.* **2013**, *481*, 22–34. [CrossRef]
13. Ramsar Convention Secretariat. *The Ramsar Convention Manual: A Guide to the Convention on Wetlands (Ramsar, Iran, 1971)*, 6th ed.; Ramsar Convention Secretariat: Gland, Switzerland, 2013.
14. Sophocleus, M. Interactions between groundwater and surface water: The state of the science. *Hydrogeol. J.* **2002**, *10*, 52–67. [CrossRef]
15. Ramsar. Humedales: En Peligro de Desaparecer en Todo el Mundo. Convención Sobre Humedales. Ficha Informativa 3. 2011. Available online: www.ramsar.org/es/acerca-de/uso-racional-de-los-humedales (accessed on 15 January 2024).
16. Xu, T.; Weng, B.; Yan, D.; Wang, K.; Li, X.; Bi, W.; Li, M.; Cheng, X.; Liu, Y. Wetlands of international importance: Status, threats, and future protection. *Int. J. Environ. Res. Public Health* **2019**, *16*, 1818. [CrossRef]
17. Stehr, A.; Álvarez, C.; Álvarez, P.; Arumí, J.L.; Baeza, C.; Barra, R.; Berroeta, C.A.; Castillo, Y.; Chiang, G.; Cotoras, D.; et al. Recursos hídricos en Chile: Impactos y adaptación al cambio climático. Informe de la mesa Agua. Santiago: Comité Científico COP25; Ministerio de Ciencia, Tecnología, Conocimiento e Innovación. 2019. Available online: https://cdn.digital.gob.cl/filer_public/e6/ff/e6ff260a-d926-4210-83e6-ad7b840b320c/19agua-recursos-hidricos-stehr.pdf (accessed on 15 January 2024).
18. IPCC. *Climate Change 2013: The Physical Science Basis. Contribution of Working Group I to the Fifth Assessment Report of the Intergovernmental Panel on Climate Change*; Cambridge University Press: Cambridge, UK, 2013.
19. IPCC. *Climate Change 2008: Impacts, Adaptation and Vulnerability. Contribution of Working Group II to the Fourth Assessment Report of the Intergovernmental Panel on Climate Change*; Cambridge University Press: Cambridge, UK, 2008.
20. Horton, B.P.; Rahmstorf, S.; Engelhart, S.E.; Kemp, A.C. Expert assessment of sealevel rise by AD 2100 and AD 2300. *Q. Sci. Rev.* **2014**, *84*, 1–6. [CrossRef]
21. Dirección General de Aguas (DGA). Inventario de Cuencas, Subcuencas y Subsubcuencas de Chile. DGA. División de Estudios y Planificación. 2014. Available online: https://bibliotecadigital.ciren.cl/handle/20.500.13082/32709 (accessed on 25 October 2023).
22. Ministerio del Medio Ambiente de Chile (MMA). *Estrategia Nacional de Humedales 2024*; Ministerio del Medio Ambiente de Chile: Santiago, Chile, 2024.
23. CR2. Centro de Ciencia del Clima y la Resiliencia (CR2). *La Mega Sequía 2010–2015: Una Lección para el Futuro*; Informe a la Nación; CR2. Centro de Ciencia del Clima y la Resiliencia (CR2): Santiago, Chile, 2015.
24. Garreaud, R.D.; Boisier, J.P.; Rondanelli, R.; Montecinos, A.; Sepúlveda, H.H.; Veloso-Aguila, D. The Central Chile Mega Drought (2010– 2018): A climate dynamics perspective. *Int. J. Climatol.* **2019**, *40*, 421–439. [CrossRef]
25. Masotti, I.; Aparicio-Rizzo, P.; Yevenes, M.A.; Garreaud, R.; Belmar, L.; Farías, L. The Influence of River Discharge on Nutrient Export and Phytoplankton Biomass Off the Central Chile Coast (33°–37°S): Seasonal Cycle and Interannual Variability. *Front. Mar. Sci.* **2018**, *5*, 423. [CrossRef]
26. Dirección General de Aguas (DGA). Información Pluviométrica, Fluviométrica, Estado de Embalses y Aguas Subterráneas. Boletín N° 524 Mes Diciembre Año 2021. División de Hidrología (DGA). 2021. SSD: 15614003. Available online: https://dga.mop.gob.cl/productosyservicios/informacionhidrologica/Informacin%20Mensual/Boletin_12_Diciembre_2021.pdf (accessed on 26 July 2023).
27. Zúñiga, M. Caracterización de la Vulnerabilidad Socio Territorial en el área Rural, Forestal y de Expansión Urbana de Laguna Verde, Valparaíso. Bachelor's Thesis, Departamento de Geografía, Facultad de Ciencias Naturales y Exactas, Universidad de Playa Ancha, Valparaíso, Chile, 2015.
28. Rivera Castro, C.A.; Letelier Pino, J.A.; Acevedo Pizarro, B.; Tobar Correa, T.D.P.; Loreto Torres Lepe, C.; Cataldo Figueroa, A.M.; Rivera Castro, M.Á. Water quality in the El Sauce Estuary, Valparaíso, Central Chile. *Rev. Int. De Contam. Ambient.* **2020**, *36*, 261–273. [CrossRef]
29. Millanir, N. Proyecto de Factibilidad de Instalación de Redes de Alcantarillado y Agua Potable Sector Curaumilla-Laguna Verde. Bachelor's Thesis, Departamento de Diseño y Manufactura, Universidad Técnica Federico Santa María, Viña del Mar, Chile, 2003.
30. *AES/02/2012*; Caracterización de aguas, Sector Laguna Verde, Quinta Región. Silob Chile. Informe técnico preparado por el Departamento de Ingeniería Ambiental: Valparaíso, Chile, 2012; 18p.
31. Pozo-Solar, F.; Cornejo-D'Ottone, M.; Orellana, R.; Acuña, C.; Rivera, C.; Aguilar-Muñoz, P.; Lavergne, C.; Molina, V. Microbial and Biogeochemical Shifts in a Highly Anthropogenically Impacted Estuary ("El Sauce" Valparaíso). *Water* **2023**, *15*, 1251. [CrossRef]
32. DGA. *Dirección General de Aguas, Gobierno de Chile. Evaluación de los Recursos Subterráneos de las Cuencas Costeras de la Quinta Región*; Informe Técnico. Serie de Documentos Técnicos S.D.T. N° 201; Dirección General de Aguas, Ministerio de Obras Públicas: Santiago, Chile, 2005; 93p.
33. Tobar, T.; and Torres, C.L. Evaluación de la calidad del agua del estero El Sauce, Laguna Verde: Impacto y Consecuencias. Bachelor's Thesis, Facultad de Ciencias Naturales y Exactas, Universidad de Playa Ancha, Valparaíso, Chile, 2014; 175p.
34. DGA. *Dirección General de Aguas, Gobierno de Chile. Evaluación de los Recursos Subterráneos de las Cuencas Costeras de la V Región*; Informe Técnico. Departamento de Recursos Hídricos. S.D.T. N°130; Dirección General de Aguas, Ministerio de Obras Públicas: Santiago, Chile, 2002; 43p.
35. Dirección Meteorológica de Chile (DMC). 2024. Available online: https://climatologia.meteochile.gob.cl/ (accessed on 26 July 2023).

36. Superintendencia de Servicios Sanitarios (SISS). Decreto Supremo N°90/2000 Establece Norma de Emisión para la Regulación de Contaminantes Asociados a las Descargas de Residuos Líquidos a Aguas Marinas y Continentales Superficiales. Ministerio Secretaría General de la Presidencia, Chile. 2000. Available online: https://bcn.cl/2esfo (accessed on 14 July 2024).
37. Reyes, A.; Ulises, F.; Carvajal, Y. *Guía Básica para la Caracterización Morfométrica de Cuencas Hidrográficas*; Universidad del Valle: Cali, Colombia, 2010; 139p, ISBN 9789587654011.
38. Villela, S.M.; Mattos, A. *Hidrologia Aplicada*; McGraw-Hill do Brasil: Sao Paulo, Brazil, 1975; 245p.
39. Strahler, A.N. Quantitative analysis of watershed geomorphology. *Eos Trans. Am. Geophys. Union* **1957**, *38*, 913–920. [CrossRef]
40. Alcántara, J.L. Caracterización Hidromorfométrica de la Microcuenca Puyllucana-Baños del Inca-Cajamarca, Mediante la Aplicación de ArcGis. Universidad Nacional de Cajamarca. 2008. Available online: https://alicia.concytec.gob.pe/vufind/Record/RUNC_800c627c183b752bafd2bd6892033979/Details (accessed on 15 August 2023).
41. Rizo, R.; Romero, L.; Zeledón, J. *Caracterización Biofísica y Socioeconómica de la Microcuenca La Jabonera, Perteneciente a la Subcuenca del Río Estelí*; Universidad Nacional Autónoma de Nicaragua: Managua, Nicaragua, 2011.
42. Gana, P.; Wall, R.; Gutiérrez, A.; Yañez, G. *Mapa Geológico Del Área Valparaíso–Curacaví, Región de Valparaíso y Región Metropolitana*; Esc. 1:100.000. Mapas Geológicos N° 1; Servicio Nacional de Geología y Minería: Santiago de Chile, Chile, 1994.
43. Gana, P.; Wall, R.; Gutiérrez, A. *Mapa Geológico del área de Valparaíso-Curacaví, Regiones de Valparaíso y Metropolitana [en línea]*; Mapas Geológicos N°001; SERNAGEOMIN: Santiago de Chile, Chile, 1996.
44. Encinas, A.; Le Roux, J.; Buatois, L.; Nielsen, S.; Finger, K.; Fourtanier, E.; Lavenu, A. Nuevo esquema estratigráfico para los depósitos marinos mio-pliocenos del área de Navidad (33°00′-34°30′ S), Chile central. *Rev. Geológica De Chile* **2006**, *33*, 221–246. [CrossRef]
45. Keller, B.Y.; Martin, A. Estudio de la Actividad de la Falla Laguna Verde para Zonificación de usos Urbanos y sus Restricciones. In Proceedings of the Congreso Geológico Chileno, Puerto Varas, Chile, 4 August 2000; pp. 77–78.
46. Instituto Privado de Investigación Sobre Cambio Climático (ICC). *Manual de Medición de Caudales*; ICC: Santa Lucía Cotzumalguapa, Guatemala, 2017; 18p, ISBN 978-9929-8241-4-0. Available online: https://icc.org.gt/wp-content/uploads/2023/03/064.pdf (accessed on 15 August 2024).
47. Richards, L. *Diagnosis and Improvement of Saline and Alkali Soils*; USDA Handbook 60; US Department of Agriculture: Washington, DC, USA, 1954.
48. Instituto Nacional de Normalización (INN). *Norma Chilena Oficial NCh 1333/78. Modificada en 1987. Requisitos de Calidad del agua para Diferentes Usos*; Gobierno de Chile: Santiago, Chile, 1987.
49. Fundación Centro Internacional de Hidrología Subterránea (FCIHS). *Hidrogeología*; Fundación Centro Internacional de Hidrología Subterránea: Barcelona, Spain, 2009.
50. Piper, A.M. A graphic procedure in the geochemical interpretation of water-analyses. *Trans. Am. Geophys. Union* **1944**, *25*, 914–928.
51. Custodio, E.; Llamas, M.R. *Hidrología Subterránea*; Ediciones Omega, S. A.: Barcelona, Spain, 1976; 2350p.
52. López-Geta, J.A.; Mena, J.M. *Aspectos Metodológicos en el Estudio de la Intrusión Salina. Documento Básico*; Geomecánica y Aguas S.A. para División de Aguas Subterráneas y Geotécnica del Instituto Geológico y Minero de España: Madrid, Spain, 1988; 230p.
53. Giménez, E.; Morell, I. Análisis hidrogeoquímico de los procesos de salinización en el acuífero costero de Oropesa (Castellón, España). *Environ. Geol.* **1997**, *29*, 118–131. [CrossRef]
54. DGA. *Dirección General de Aguas, Gobierno de Chile. Estimación de la Demanda Actual, Proyecciones Futuras y Caracterización de la calidad de los Recursos Hídricos en Chile*; Unión Temporal de Proveedores Hídrica Consultores SPA y Aquaterra Ingenieros Ltd.a. S.I.T. N°419; Ministerio de Obras Públicas, Gobierno de Chile: Santiago, Chile, 2017.
55. Freeze, R.A.; Cherry, J.A. *Groundwater*; Prentice-Hall: Hoboken, NJ, USA, 1979; 604p.
56. Craig, H. Isotopic variations in meteoric waters. *Science* **1961**, *133*, 1702–1703. [CrossRef]
57. Dansgaard, W. Stable isotopes in precipitation. *Tellus* **1964**, *16*, 436–468. [CrossRef]
58. Gat, J.R. Oxygen and hydrogen isotopes in the hydrologic cycle. *Annu. Rev. Earth Planet. Sci.* **1996**, *24*, 225–262. [CrossRef]
59. Bianchi Janetti, E.; Riva, M.; Guadagnini, A. Natural springs protection and probabilistic risk assessment under uncertain conditions. *Sci. Total Environ.* **2021**, *751*, 141430. [CrossRef] [PubMed]
60. Schiavo, M. Entropy, fractality, and thermodynamics of groundwater pathways. *J. Hydrol.* **2023**, *617*, 128930. [CrossRef]
61. Aziz, O.I.; Burn, D.H. Trends and variability in the hydrological regime of the Mackenzie River Basin. *J. Hydrol.* **2006**, *319*, 282–294. [CrossRef]
62. Scibek, J.; Gleeson, T.; McKenzie, J.M. The biases and trends in fault zone hydrogeology conceptual models: Global compilation and categorical data analysis. *Geofluids* **2016**, *16*, 782–798. [CrossRef]
63. EEA. European Environment Agency. Europe's Groundwater: A Key Resource under Pressure. Report. 2022. Available online: https://www.eea.europa.eu/publications/europes-groundwater (accessed on 5 July 2024).
64. Metcalf & Eddy, Inc. *Wastewater Engineering: Treatment, Disposal, and Reuse*; McGraw-Hill: New York, NY, USA, 1995.
65. Strack, O.D.L. A single-potential solution for regional interface problems in coastal aquifers. *Water Resour. Res.* **1976**, *12*, 1165–1174. [CrossRef]
66. Ataie–Ashtiani, B.; Volker, R.E.; Lockington, D.A. Tidal effects on sea water intrusion in unconfined aquifers. *J. Hydrol.* **1999**, *216*, 17–31. [CrossRef]
67. Levanon, E.; Yechieli, Y.; Haim, G.; Eyal, S. Tide-induced fluctuations of salinity and groundwater level in unconfined aquifers-field measurements and numerical model. *J. Hydrol.* **2016**, *551*, 665–675. [CrossRef]

68. Loáiciga, H.; Pingel, T.; Garcia, E. Sea Water Intrusion by Sea-Level Rise: Scenarios for the 21st Century. *Ground Water* **2012**, *50*, 37–47. [CrossRef]
69. Wang, J.; Tsay, T. Tidal Effects on Groundwater Motions. *Transp. Porous Media* **2001**, *43*, 159–178. [CrossRef]
70. Michael, H.; Mulligan, A.; Harvey, C. Seasonal oscillations in water exchange between aquifers and the coastal ocean. *Nature* **2005**, *436*, 1145–1148. [CrossRef] [PubMed]
71. Werner, A.D.; Simmons, C.T. Impact of sea-level rise on seawater intrusion in coastal aquifers. *Groundwater* **2009**, *47*, 197–204. [CrossRef]
72. Yechieli, Y.; Shalev, E.; Wollman, S.; Kiro, Y.; Kafri, U. Response of the Mediterranean and Dead Sea coastal aquifers to sea level variations. *Water Resour. Res.* **2010**, *46*, W12550. [CrossRef]
73. Ketabchi, H.; Mahmoodzadeh, D.; Ataie-Ashtiani, B.; Simmons, C. Sea-level rise impacts on seawater intrusion in coastal aquifers: Review and integration. *J. Hydrol.* **2016**, *535*, 235–255. [CrossRef]
74. Custodio, E. Coastal aquifers in Europe: An overview. *J. Hydrol.* **2010**, *18*, 269–280. [CrossRef]
75. Custodio, E.; Lázaro, M.; Martínez, R. Salinización de las aguas subterráneas en los acuíferos. *Rev. De La Soc. Española De Hidrogeol.* **2017**, *27*, 43–60.

land

MDPI

Article

Construction of Ecological Security Patterns Incorporating Multiple Types of Ecological Service Functions for Developed Coastal Regions: A Case Study in Jinjiang Watershed, China

Yuchen Zhu [1], Yaci Liu [1,2], Yong Xiao [3,*], Jiahuan Liu [4], Zihui Zhao [5], Yasong Li [1], Qichen Hao [1], Chunlei Liu [1] and Jianfeng Li [1,*]

1 Fujian Provincial Key Laboratory of Water Cycling and Eco-Geological Processes, Institute of Hydrogeology and Environmental Geology, CAGS: Chinese Academy of Geological Sciences, Xiamen 361000, China; zhuyuchen413@163.com (Y.Z.); liuyaci11@163.com (Y.L.); liyasong@mail.cgs.gov.cn (Y.L.); haoqichen@mail.cgs.gov.cn (Q.H.); liuchunlei@mail.cgs.gov.cn (C.L.)

2 Key Laboratory of Water and Soil Resources Conservation and Restoration in the Middle and Lower Reaches of Yellow River Basin, MNR: Ministry of Natural Resources, Zhengzhou 450016, China

3 Faculty of Geosciences and Engineering, Southwest Jiaotong University, Chengdu 611756, China

4 College of Resource Environment and Tourism, CNU, Beijing 100048, China; liujiahuan1105@163.com

5 North China Institute of Aerospace Engineering, Langfang 065000, China; 13770765697@163.com

* Correspondence: xiaoyong@swjtu.edu.cn (Y.X.); lichris99@foxmail.com (J.L.); Tel.: +86-180-1000-8002 (Y.X.); +86-156-3381-6081 (J.L.)

Abstract: An ecological security pattern is an essential natural solution for promoting the coordinated and sustainable development of urban growth and the ecological environment within a given region. This study aims to develop a comprehensive ecological security pattern (ESP) for the Jinjiang watershed by integrating multiple ecological service functions: ecological regulation, eco-product provisions, and human settlement services. Using a well-established methodology of "source identification—resistance surface construction—ecological corridor extraction", we identified 322 ecological source areas and delineated 321 ecological corridors spanning 1941.67 km, with key corridors connecting ecological regulation sources and ordinary corridors primarily linking eco-product provisions and human settlement sources. Our results show that the northern hilly regions are significant for ecological regulation, Anxi County is key for eco-product provisions, and human settlement services are scattered across urban areas and river valleys. This integration enhances regional ecological sustainability and provides a model for similar coastal regions globally. Our findings offer scientific guidance for balancing urban growth with ecological preservation, contributing to sustainable development worldwide.

Keywords: urban development; sustainable development; watershed management; ecosystem services; coastal regions; ecological security pattern

1. Introduction

Rapid urbanization has heightened the conflict among urban development, limited resources, and fragile ecosystems [1–4]. Global changes stemming from human activities have precipitated biodiversity loss, land degradation, and climate change, causing increasingly conspicuous structural contradictions in living, production, and ecological spaces [5–7]. This has accelerated the deterioration of already fragile ecosystems, weakened their functions, and heightened ecological vulnerability [8,9], thereby posing a grave threat to the ecological security of regions and compromising the sustainable development of cities [10,11].

Ecological security encompasses the integrity and health of the ecosystem, which are fundamental for maintaining human well-being [12]. Balancing urban development with ecological security has emerged as a paramount concern that must be addressed to foster healthy urban growth [13]. The People's Republic of China's 14th Five-Year Plan (2021–2025) for National Economic and Social Development outlines the need to enhance the coordination mechanism within the realm of ecological civilization. It underscores the necessity to construct an ecological civilization system and propel a comprehensive shift towards green economic and social development. In this context, it is required that urbanization should take the ecological environment's sustainable development into account [14].

As a link that facilitates the relationship between ecosystem services and the social and economic growth of people, the ecological security pattern (ESP), is considered as a pivotal element in ensuring regional ecological security and improving human well-being [15]. It is a crucial strategy for coordinating the sustainable development of the social and economic spheres with the ecological environment and for addressing issues with ecological security brought on by social development [16].

Globally, the construction of ESPs is essential, especially in developed coastal regions. For instance, coastal cities in Europe, North America, and Southeast Asia face similar challenges of urban expansion, resource limitations, and ecological degradation [17,18]. The establishment of ESP in these regions is essential for mitigating the adverse effects of urbanization on natural ecosystems, ensuring sustainable development, and improving the quality of life for urban populations [19]. By studying the ESP frameworks and methodologies applied in these regions, valuable insights can be gained for addressing the ecological challenges faced by rapidly developing coastal areas worldwide.

Early ESP research primarily focused on biodiversity conservation and ecosystem service value assessment [20]. As public awareness of ecological preservation grows and the connection between economic development and ecological protection strengthens, research on the pattern of ecological security is gradually transitioning into a systematic coupled analysis of natural ecology and economic and social development [21]. These research results have gradually formed the mainstream paradigm of creating an ecological security pattern, which includes "identification of ecological sources—resistance surface construction—ecological corridor extraction" [22].

According to the *National Ecological Function Zoning (Revised Edition)*, ecosystem services are categorized based on the natural attributes of the ecosystem and their dominant service functions into three main types: ecological regulation service, product provision service, and human settlement service (satisfaction of the residents' ecological needs) [23,24]. Most academics only focused on the ecological regulation function of ecosystems in recent years, relying on a "paradigm" to create an ecological security pattern [25,26]. They focused solely on ecological regulation functions while overlooking the characteristics of social and economic development, the population concentration in coastal cities, and the demand for ecosystem functions in human life and agricultural production. This often leads to the conclusion that ecology and social development are in opposition [27–31]. However, the ecological service functions in densely populated coastal developed areas are often multifaceted, and their ecological security patterns need to meet the unity and harmonious development of ecological needs, production needs, and living needs. Obviously, an analysis based solely on the single function of ecological regulation cannot meet the needs of harmonious development, and there is an urgent need to explore a method of creating ecological security patterns that account for multiple types of ecological service functions.

Considering the global relevance of this issue, the methodology and findings from ESP studies in regions such as the Mediterranean coast, the Gulf of Mexico, and the coastal areas of Japan provide crucial benchmarks [32–34]. These studies have demonstrated the effectiveness of integrated ecological approaches in mitigating the impacts of urbanization and climate change, enhancing biodiversity, and promoting sustainable land use practices. By incorporating lessons learned from these regions, the ESP framework can be further refined and adapted to diverse environmental and socio-economic contexts, ensuring broader applicability and effectiveness.

Globally, constructing ESPs in coastal cities like the Jinjiang watershed is crucial for balancing urban development with natural conservation. Jinjiang is the largest river in Quanzhou, a coastal city, and it serves as the primary source of drinking water in Quanzhou, as well as serving multiple functions such as irrigation, aquaculture, and tourism [35]. The Jinjiang watershed is located in the water source conservation area in southwestern Fujian Province [36]. Quanzhou, its affiliated city, is not only the most populous city in Fujian Province but is also renowned as a major production area for Chinese Wuyi rock tea [37]. Its ecosystem services encompass ecological regulation, product provisions, and human settlement services.

To address these challenges, this study employs a well-established methodology of "source identification—resistance surface construction—ecological corridor extraction" to develop a comprehensive ecological security pattern for the Jinjiang watershed. This approach includes identifying sources of ecological regulation, eco-product provisions, and human settlement services; constructing resistance surfaces to model ecological connectivity; and delineating ecological corridors to ensure landscape connectivity and ecological function.

The objective of this study is to develop a comprehensive ecological security pattern for the Jinjiang watershed, hypothesizing that incorporating multiple ecological service functions will significantly enhance regional ecological sustainability. We aim to address the complex interactions between ecological regulation, eco-product provisions, and human settlement services to create a holistic model that can be applied to similar coastal regions globally.

While this study focuses on the Jinjiang watershed in China, the methods and findings are applicable to other developed coastal regions globally, addressing similar ecological and urban development challenges. The insights gained from this study can provide valuable guidance for constructing ecological security patterns that balance urbanization and ecological preservation, promoting sustainable development worldwide. This research not only contributes to the theoretical framework of ecological security patterns but also offers practical implications for policymakers and urban planners aiming to harmonize development and conservation efforts.

2. Materials and Methods

2.1. Study Area

Jinjiang lies within the administrative boundaries of Quanzhou City, Fujian Province, and originates from the Daiyun Mountains in the central part of Fujian Province (Figure 1). The Jinjiang watershed encompasses 5629 km^2 of land. Dongxi and Xixi are the two major tributaries of Jinjiang. Dongxi originates from Jindou Town, Yongchun County, with a drainage area of 1917 km^2 and a length of 120 km. The Xixi originates from Gande Township, Anxi County, with a drainage area of 3101 km^2 and a length of 145 km. Its two tributaries flow into the sea from Ninety-nine Creek to Fengze District in Jinjiang City. The topography of the Jinjiang watershed gradually slopes down from the mountains in the northwest to the lower elevations in the southeast.

The basin's climate is classified as subtropical humid monsoon, which means that it has substantial monsoon features, an abundance of light and heat, and abundant precipitation, but with uneven seasonal and regional distribution. The majority of the landscape is mountainous and hilly, with steep slopes.

The vegetation in the Jinjiang watershed primarily consists of man-made and natural forests, including Masson pine (*Pinus massoniana*), bamboo, and various subtropical broad-leaved species. The region also features large tea plantations, particularly in the hilly areas of Anxi County. Despite significant human influence, forested areas remain crucial for the region's ecological health, providing essential services such as water conservation, soil retention, climate regulation, and biodiversity protection. According to the 2015 National Ecological Function Zoning, this region is designated as a key area for water conservation and biodiversity protection.

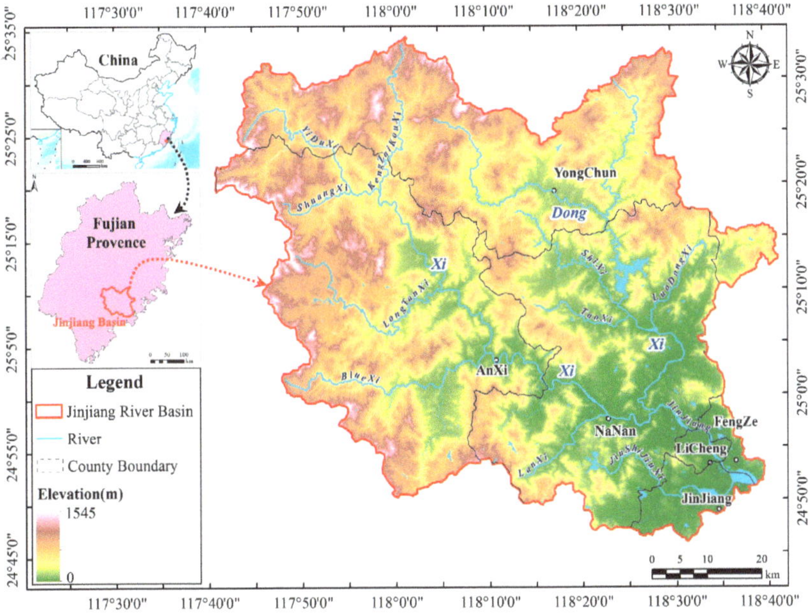

Figure 1. Location map of the study area.

Notably, the region is home to endangered species such as the Chinese crested tern (*Thalasseus bernsteini*) and the spoon-billed sandpiper (*Calidris pygmaea*), both of which are critically endangered and nationally protected.

The ecosystem types in the Jinjiang watershed cover urban ecosystems, forest ecosystems, and agricultural ecosystems, making it an ideal area for this study. The principal crops include rice, sweet potatoes, peanuts, sugarcane, and soybeans. Jinjiang is also the largest carrot-growing base in Fujian Province, with its unique climate and soil characteristics enabling off-season cultivation, making it a key area for winter and spring carrot production and export.

The selection of the Jinjiang watershed as the study area is due to its representative nature of the ecological challenges faced by developed coastal regions globally. The area is experiencing rapid urbanization, which intensifies the conflict between urban development and ecological preservation. Additionally, the watershed's significant role in providing various ecosystem services makes it an ideal case study for developing and testing comprehensive ecological security patterns. This region's characteristics—combining high population density, economic development, and ecological significance—mirror the challenges faced by many other coastal regions worldwide. Therefore, insights gained from this study can be broadly applicable, providing valuable guidance for similar regions globally striving to balance urbanization with ecological sustainability.

2.2. Data Collection

The data used in this study were collected from multiple sources to ensure comprehensive coverage of the Jinjiang watershed. The primary data sources and their characteristics are as follows:

(1) Satellite Imagery: Landsat TM images were obtained from the United States Geological Survey (USGS) with a spatial resolution of 30 m. The images cover the period from January 2019 to December 2020, capturing different seasons to account for seasonal variations in vegetation and land use.

(2) Meteorological Data: Monthly precipitation and evapotranspiration data were retrieved from the National Meteorological Science Data Center (https://data.cma.cn/, accessed on 10 September 2020) with a spatial resolution of 1 km. These data cover the same period as the satellite imagery, ensuring temporal consistency.

(3) Topographical Data: Digital elevation model (DEM) data from the Shuttle Radar Topography Mission (SRTM) with a resolution of 30 m. These data were used to generate topographic features such as slope and aspect.

(4) Land Use and Land Cover Data: These data were derived from Landsat TM imagery through supervised classification and field verification, focusing on identifying tea gardens, farmlands, and residential areas.

(5) Soil Data: Soil attributes data were sourced from the Soil Science Database (http://vdb3.soil.csdb.cn/, accessed on 23 September 2020). This includes soil types, texture, and other relevant characteristics.

All data were subjected to preprocessing steps, including geometric correction, atmospheric correction, and resampling, to ensure spatial and temporal consistency. The data were then integrated into a unified spatial reference system using ArcGIS software 10.8 for further analysis.

2.3. *Methods*

The methodology employed in this study to develop a comprehensive ecological security pattern for the Jinjiang watershed is shown in Figure 2. The approach integrates multiple ecosystem service functions—ecological regulation, eco-product provisions, and human settlement services—with ecological sensitivity assessment. This integrated approach addresses the conflict between urban development and ecological preservation in developed coastal cities. The key steps involve: (1) Integrated evaluation of ecosystem service functions and ecological sensitivity to identify source areas for ecological regulation, eco-product provisions, and human settlement; (2) Construction of a resistance surface to model ecological connectivity, incorporating land cover, slope, and human disturbances; (3) Extraction of ecological corridors using the minimum cumulative resistance (MCR) model to ensure landscape connectivity; (4) Development of the final ecological security pattern by integrating identified ecological sources, resistance surfaces, and ecological corridors.

2.3.1. Integrated Evaluation of Ecosystem Service Function

The integrated evaluation of ecosystem service functions is a key innovation of this study, addressing the complex interactions and conflicts between urban development and ecological preservation in developed coastal cities. The purpose of integrating multiple ecosystem service functions—ecological regulation, eco-product provisions, and human settlement services—is to develop a comprehensive understanding of how these services interact and impact the overall ecological security of the Jinjiang watershed.

By combining these services with ecological sensitivity assessment, this study aims to identify areas that are critical for maintaining ecological balance while supporting urban growth and agricultural productivity. This integrated approach ensures that the ecological security pattern is robust, sustainable, and reflective of the diverse needs of densely populated coastal regions. The identification of three types of ecological sources based on this integration is a novel contribution, providing a holistic framework that has not been previously applied in similar contexts.

Selecting indicators for each ecosystem service function is crucial for accurately capturing the respective contributions to ecological regulation, eco-product provisions, and human settlement services. The indicators were chosen based on their relevance to the specific environmental, economic, and social characteristics of the Jinjiang watershed, as well as their ability to provide measurable and meaningful data for the integrated evaluation (Table 1).

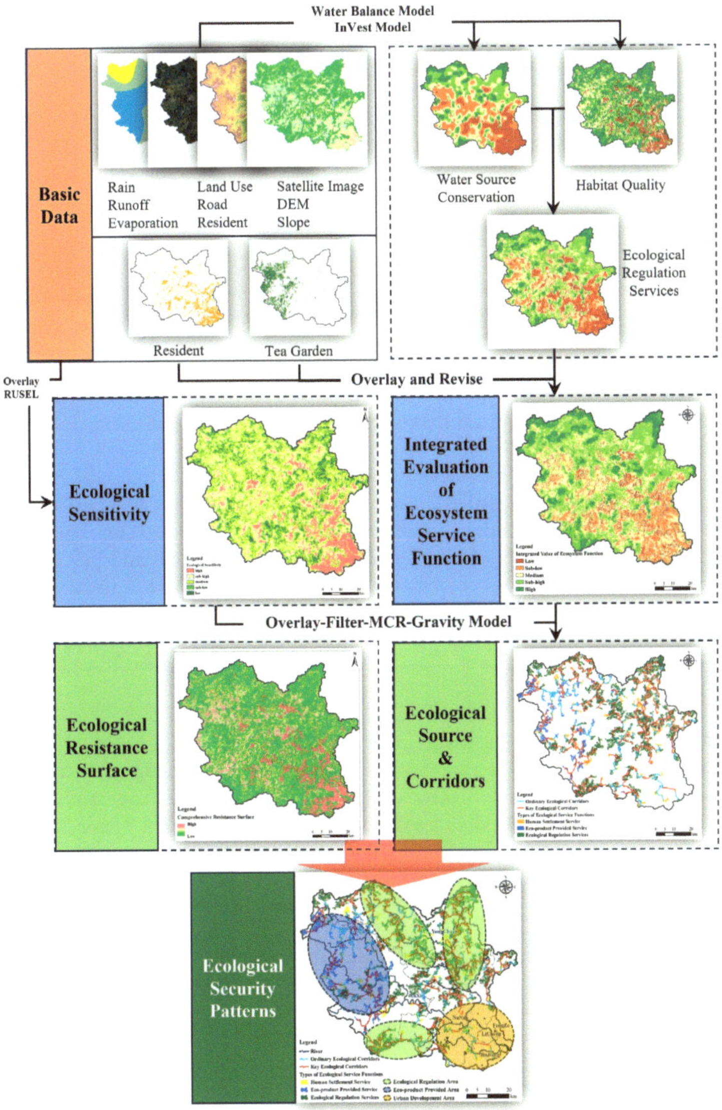

Figure 2. Workflow of construction of the ESP.

Table 1. Ecosystem service functions evaluated in the study.

Ecosystem Service Function	Definition and Components	Methods of Assessment
Ecological Regulation	The function of maintaining ecological balance and ensuring regional ecological security [38]	InVEST model for water conservation; RUSLE for soil retention; climate regulation assessment
Eco-product Provisions	Supply of agricultural and forestry products	High-resolution satellite imagery; field surveys to identify and map tea plantations and other crops

Table 1. *Cont.*

Ecosystem Service Function	Definition and Components	Methods of Assessment
Human Settlement Services	The reflection in the ecological leisure needs, specifically manifested in the need for residents to relax on weekdays and the demand for short-distance travel to and recreation on ecological land on weekends [39]	Mapping residential areas; creating accessibility buffers to delineate zones of human settlement services

(1) Ecological Regulation Services

The abundant water resources and over 60% forest coverage in the Jinjiang watershed provide a natural advantage for ecological regulation in the region. Habitat quality refers to the suitability of habitats within a given region for the distribution of organisms, the preservation of biodiversity, and the support of ecologically suitable areas for the sustainable development and existence of individuals and populations [40]. In view of this, this study employs water conservation and habitat quality as evaluation indicators to gauge the significance of ecological regulation.

This study employs the water balance equation to quantify the water conservation capacity of the study area [41]. The InVEST (Integrate Valuation of Ecosystem Service and Tradeoffs Tool) model's habitat quality module is one of the most commonly used methods for evaluating habitat quality [42]. Considering the natural resources and population distribution patterns of the Jinjiang watershed, habitats in this study are defined as bare land, water bodies, forest land, grassland, and cultivated land. Conversely, other land uses are categorized as non-habitat. The identified sources of threats to habitats include built-up areas, railways, and highways.

(2) Eco-product Provision Services

Tea is the most important agricultural product in the Jinjiang watershed [43], and together with farmland, they are the core indicators that reflect the function of providing ecological products in the region. In this study, high-resolution satellite imagery and field surveys were used to identify areas dedicated to agricultural production. The indicators captures the economic value and ecological contributions of agricultural lands, highlighting their role in providing eco-products.

(3) Human Settlement Services

Residential areas were mapped using high-resolution satellite imagery, and a 15 min accessibility buffer was created around these areas to delineate zones of human settlement services. This indicator ensures that the spatial extent of residents' recreational and ecological needs is considered in the ecological security pattern.

2.3.2. Ecological Sensitivity Assessment

The term ecological sensitivity describes how the ecosystem environment reacts to disruptions in natural and human activities within the region, indicating the intensity and likelihood of causing ecological environmental issues [44]. Through ecological sensitivity assessment, it is possible to screen out areas that are highly sensitive and not easily restored after a disturbance. Due to the complex terrain and uneven seasonal and regional distribution of precipitation in the Jinjiang watershed, geological disasters and soil erosion have caused serious social, economic, and ecological damage. In order to assess the ecological sensitivity in the basin, this study has chosen soil erosion and geological disaster susceptibility as its primary ecological environmental concerns.

The sensitivity of soil erosion is evaluated using the Revised Universal Soil Loss Equation (RUSLE). Due to the close relationship between geological disasters and landforms in the Jinjiang watershed, and their susceptibility to human-induced activities, elevation, slope, terrain undulation, and human activity intensity are selected as evaluation factors. The intensity of human activities can be partially characterized by the types of land cover.

In this study, a grading system is employed to assign values to different land use types for assessment, as depicted in Table 2. The sensitivity of soil erosion and geological hazards are weighted and overlapped to derive the evaluation outcomes of ecological sensitivity. The regions are then classified into five levels: highly sensitive, sensitive, relatively sensitive, moderately sensitive, and not sensitive.

Table 2. Division criteria of influencing factors and sensitivity of geological hazards.

Factor	Sensitivity Rate					Weight
	5	4	3	2	1	
Elevation	<300	300–800	800–1000	1000–1300	>1300	0.3
Slope	≤5	5–10	10–15	15–25	>25	0.35
Topographic Relief Amplitude	0–3	3–6	6–9	9–15	>15	0.25
Human Activity	Trees	Water and Wetland	Bare Ground and Tea Garden	Grass and Crops	Built Area	0.1

By integrating these indicators above all, the study provides a comprehensive and multi-dimensional evaluation of the Jinjiang watershed, ensuring that the resulting ecological security pattern effectively balances ecological regulation, eco-product provisions, and human settlement services.

2.3.3. Construction of Ecological Security Pattern

This study adopts the well-established standard paradigm for constructing ecological security patterns. The construction of the ecological security pattern for the Jinjiang watershed involves several key steps, with a particular focus on the innovative extraction of ecological sources. This process integrates multiple ecosystem service functions and ecological sensitivity to address the complex interactions between urban development and ecological preservation.

(1) Identification of an ecological source area

An ecological source area denotes a land use system characterized by a high ecological system service value and the capacity to consistently yield valuable ecological outputs in a sustainable manner [45]. It serves as a vital component in upholding the stability of the ecosystem. In addition, ecological source areas are habitats for indigenous species, exhibiting internal homogeneity and the capability to expand or attract outwardly, characterized by a high-quality ecological environment [46]. They primarily contribute to the supply of essential ecosystem services, the preservation of ecosystem stability, and the avoidance of ecosystem degradation, all of which are important characteristics for constructing the foundation of an ecological security pattern.

This study combines the importance of ecosystem service functions and ecological sensitivity with weights of 0.6 and 0.4, respectively, and uses the reclassification function of ArcGIS to classify the regions according to the natural break method and extract the top two levels of regions as alternative source areas. Some studies have shown that small and discontinuous ecological source patches may offer restricted ecosystem service functions [47]. Therefore, considering the different classifications of ecosystem service functions in the study region, ecological protection areas with patch areas larger than 1 km^2 are selected from the alternative source areas as ecological source areas.

The ecological source area derived from modified tea gardens and farmland indicates the product supply function, while the source area obtained through the modification of residential and buffer areas serves as an ecological source area for human habitation protection.

(2) Construction of ecological resistance surface

Ecological resistance is the hindrance that species and ecological functions encounter during the processes of spatial control and dispersion [48]. Land use and slope are the two major factors selected for the construction of resistance surfaces (Table 3). The primary resistance surface is obtained through weighted overlay. As urbanization progresses and economic and social activities intensify, human activities further enhance the degree of interference with ecological conditions. Under the same land cover type, there are differences in human interference. To further delineate the range of human activity impacts, night light data are employed to adjust the resistance surface. The calculation formula is as follows:

$$R_i = \frac{NL_i}{NL_a} \times R \tag{1}$$

Here, R_i is the modified resistance value of the grid; NL_i is the nightlight index of the grid; NL_a represents the average nightlight index of land cover type a corresponding to the grid; and R is the primary resistance coefficient of the grid.

Table 3. Factors of ecological resistance surface.

Factor	Weight	Type	Resistance Value
Land Cover	0.6	Crops	50
		Trees	10
		Tea Garden	30
		Water	20
		Wetland	25
		Bare Ground	100
		Built Area	120
Slope (Degree)	0.4	<8	10
		8–15	20
		15–25	30
		25–35	50
		>35	100

(3) Extraction of ecological corridors

Ecological corridors are crucial paths for matter exchange and energy flow within an ecosystem, with excellent material energy flow performance [49]. They can connect different areas to form an ecological network, maintaining ecosystem stability. They serve as vital mechanisms for upholding ecological functional services, ensuring effective circulation of ecological factors between source areas, and maintaining the stability of ecological processes.

To extract ecological corridors, we apply the minimum cumulative resistance (MCR) model. The model quantifies the cumulative resistance encountered from the ecological "source point" to the target landscape unit, reflecting species movement trends and the potential utilization of the ecological environment. It can better calculate the minimum cost route consumed between ecological source points, accurately reflecting the possibility and direction of energy flow and transfer within the ecosystem [50]. The formula is noted as follows:

$$MCR = f_{min} \sum_{j=n}^{i=m} D_{ij} \times R_i \tag{2}$$

MCR is the cumulative resistance value to reach a certain point in space, f reflects the positive correlation between cumulative resistance value and landscape ecological processes, min indicates the minimum cumulative resistance value, D_{ij} represents the spatial distance from spatial landscape i to spatial landscape j, and R_i represents the resistance coefficient of landscape i to a certain target ecological vitality.

To delve further into the important ecological corridors and enhance the ecological security pattern, ecological corridors with key connections and influences between spatial landscapes are defined as key corridors. These corridors connect landscapes at a local level are defined as potential corridors. The ecological systems connected by key corridors have relatively high landscape consistency, which is conducive to the material and information exchange. The intersections of potential corridors play a role in communicating and exchanging various ecological source points within the ecological structure, while also providing a resting function for biological migration.

The gravity model can construct interaction matrices among ecological source areas, identify key corridors and ordinary corridors, and build hierarchical ecological corridor networks.

$$G_{ab} = \frac{N_a N_b}{D_{ab}^2} = \frac{\left[\frac{1}{P_a} \times \ln(S_a)\right]\left[\frac{1}{P_b} \times \ln(S_b)\right]}{\left(\frac{L_{ab}}{L_{max}}\right)^2} = \frac{L_{max}^2 \ln(S_a) \ln(S_b)}{L_{ab}^2 P_a P_b} \tag{3}$$

G_{ab} represents the interacting force between source site a and source site b, N_a and N_b are the weight of two ecological source areas, D_{ab} is the standardized potential corridor resistance between patches a and b, P is the resistance value of the patch, S is the area of the patch, L is the accumulated resistance value of the corridor between patches, and L_{max} is the maximum accumulated resistance value of all corridors in the study region.

3. Results

3.1. Ecological Source Area in Jinjiang Watershed

In line with the study's objective to construct a comprehensive ecological security pattern, we first identified the ecological source areas within the Jinjiang watershed. This integration of multiple ecological service functions is a novel approach aimed at enhancing regional ecological sustainability.

3.1.1. Integrated Ecological Service Evaluation Results

Based on the evaluation of individual factors contributing to the ecological regulation service, areas with high-value water conservation function are predominantly situated in the hilly and mountainous regions characterized by high annual precipitation and dense vegetation coverage within the study area. This area also has high ecological regulation service characteristics, including the northwest and northeast parts of Yongchun County, and the western hilly areas of Anxi County (Figure 3a). High-value areas of habitat quality are widespread in forested mountainous regions less affected by human activities (Figure 3b), while low-value areas are primarily found in developed areas such as valleys and coastal plains (Figure 3c). Under the combined action of these two factors, regions characterized by minimal human impact and dense vegetation coverage in the study area exhibit robust ecological regulation functions (Figure 3d,e).

The comprehensive ecosystem service functions corrected by the range of residential areas, tea gardens, and farmland obtained through interpretation show the following: (a) The influencing factors of the high-value comprehensive ecosystem areas are still mainly ecological regulation, similar to the evaluation outcomes of ecological regulation services, distributed in forested hilly areas with high vegetation coverage. (b) Tea-covered areas have higher service value due to their product supply function. (c) Small parks and green spaces within the built-up areas of valleys and plains present as small high-value ecological service areas due to their service in providing human settlement services.

3.1.2. Results of Ecological Sensitivity Assessment

Ecological sensitivity, as depicted in Figure 4, is primarily distributed in the hilly tea garden regions in the eastern part of Anxi County, the concentrated residential areas in Yongchun County towns, and areas such as Licheng District. These areas are heavily influenced by human activities, have less vegetation, and are prone to soil erosion. It is worth mentioning that the excessive development of hilly areas due to the extensive cultivation of rock tea in Anxi County has exacerbated soil erosion [51].

The medium-value areas are dispersed geographically, primarily in areas such as Licheng District, the northern residential areas of the city, and the southern plains of Nan'an County. These areas have flat terrain and a lack of forest and grass cover, which exacerbates ecological sensitivity due to soil erosion. Dehua County in the southeast mountainous area, the hilly plateau in the north of Nan'an County, and Guanqiao Town in Anxi County exhibit relatively low ecological sensitivity. These areas feature intricate terrain and are subject to minimal human influence.

The combined area of highly sensitive and sensitive ecological zones in the Jinjiang River Basin covers 3091.77 km^2, representing 55.21% of the total study area. These zones are predominantly located in the eastern part of Anxi County and the residential areas of Yongchun County towns. The southern plain region, characterized by high ecological sensitivity, is particularly vulnerable to environmental damage due to economic development. Effective strategies to mitigate ecological sensitivity issues should be prioritized, focusing on balancing economic growth with ecological conservation.

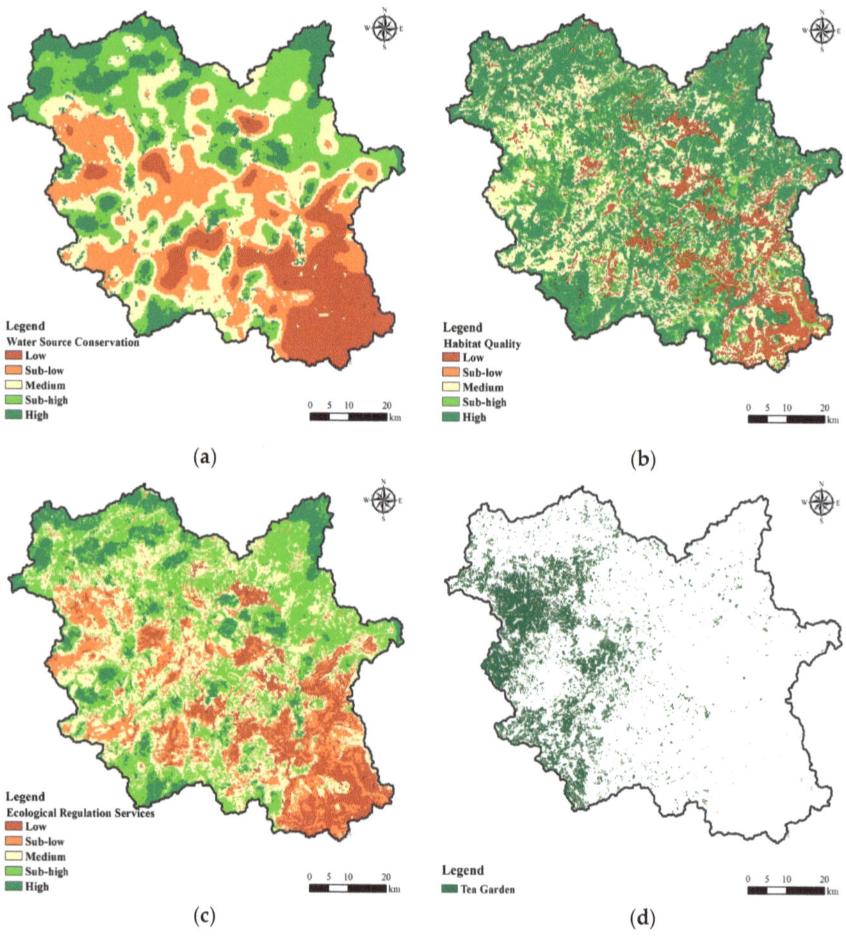

Figure 3. *Cont.*

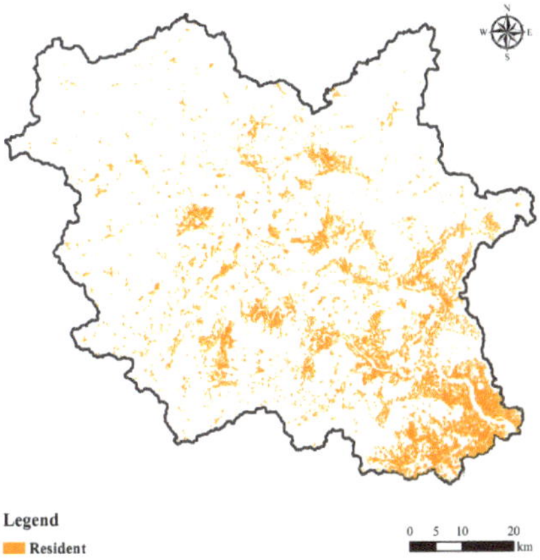

(e)

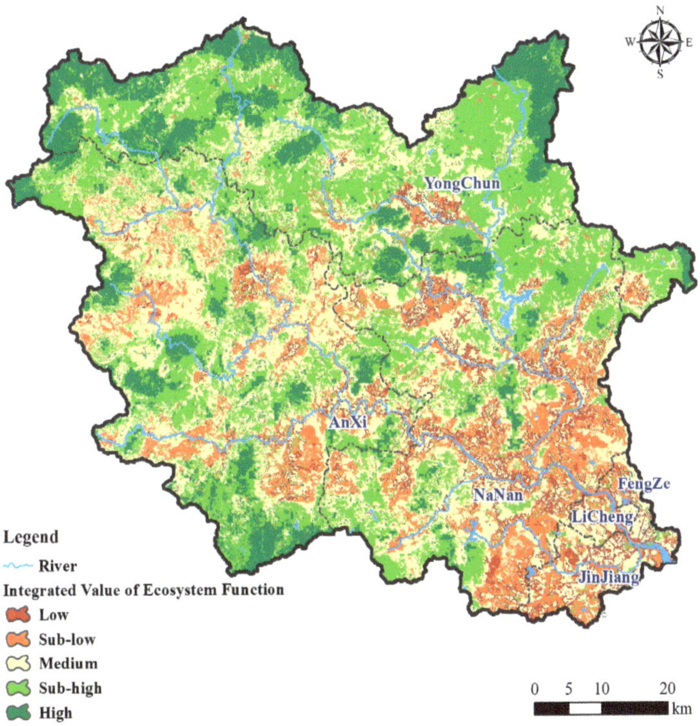

(f)

Figure 3. Assessment of the integrated value of ecosystem service function: (**a**) water source conservation; (**b**) habitat quality; (**c**) ecological regulation services; (**d**) distribution of tea gardens and crops; (**e**) distribution of residents; (**f**) integrated value of ecosystem service function.

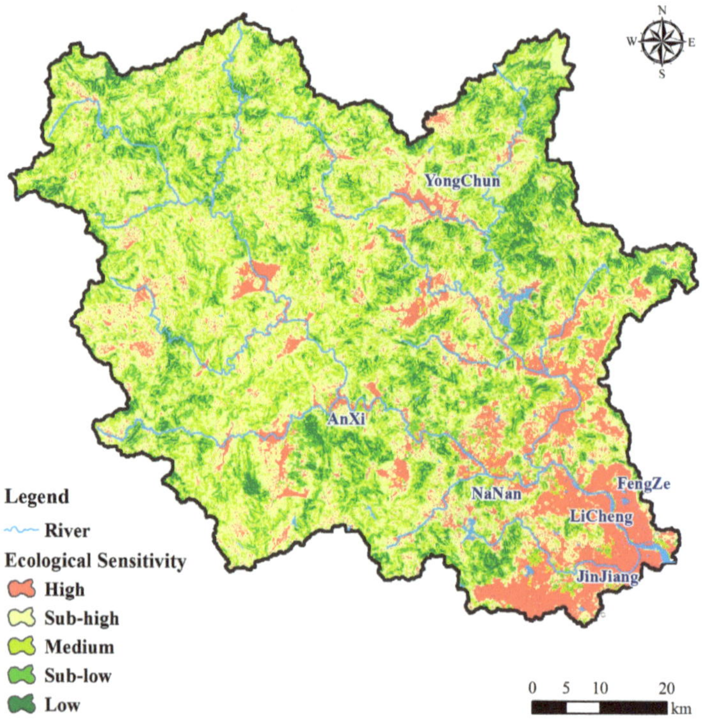

Figure 4. Assessment of ecological sensitivity.

3.1.3. Results of Ecological Source Area Identification

Through a thorough assessment of ecological service functions and sensitivity, utilizing weighted superimposition and screening methods, 322 ecological source areas are identified. After excluding scattered and fragmented patches, these areas cover a total area of 1287.18 km^2, representing 22.96% of the total study area. Among them, the ecological regulation service source areas with high habitat quality are mostly forests, primarily distributed in the hilly and mountainous regions in the north of Nan'an City, the forested area of Yongchuan County, and the hilly and mountainous areas in the south of Anxi County. The main source areas of eco-product providing services such as tea gardens are mainly distributed in Anxi County. These are the regions where economic development and environmental ecology are most closely correlated. The distribution of source areas for human settlement services is relatively dispersed. They are mainly located in the plain areas characterized by intensive human activities. Urban areas are situated in mountainous valleys, including urban green spaces in Quanzhou, Jinjiang city, and the county town of Yongchun. The distribution reflects a balance between ecological protection and the well-being needs of urban residents.

Considering the characteristics of the source areas, it can be seen from the area map of each county (Figure 5) that Yongchun County has the largest source area, followed by Anxi County, both of which are mainly ecological regulation service source areas. Among them, the area providing eco-products and services in Anxi County constitutes 40% of the total source area, the highest proportion among the five zones, and serves as the primary region for tea production. The ecological source areas for human settlement service are distributed in all five zones, with Jinjiang City having the largest distribution. On the other hand, the Quanzhou urban area allocates a relatively low proportion of its area for human settlement services. This indicates a scarcity of green space to meet the ecological requirements of urban residents during the process of urban development in the central urban area. With the exception of Jinjiang City, ecological regulation predominates in all other districts and counties, comprising the largest proportion of the source area, and making the highest contribution to it.

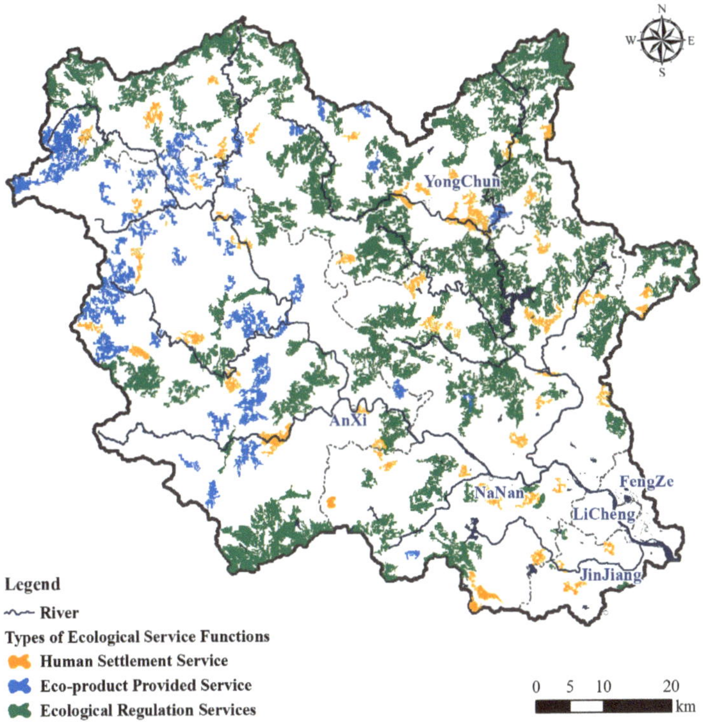

Figure 5. Classification and distribution of ecological source areas.

From the perspective of landscape distribution, forests emerge as the most crucial land cover type in ecological source areas, serving as the primary source areas for ecological regulation services, encompassing 76.36% of the ecological source area (Figure 6). This is followed by tea gardens and farmland, which serve as the main types of eco-product and service source areas, comprising 16.83% of the ecological source area.

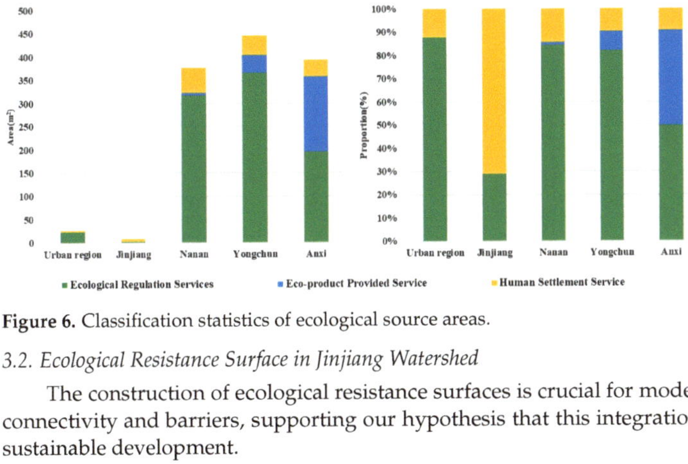

Figure 6. Classification statistics of ecological source areas.

3.2. Ecological Resistance Surface in Jinjiang Watershed

The construction of ecological resistance surfaces is crucial for modeling ecological connectivity and barriers, supporting our hypothesis that this integration will promote sustainable development.

According to the distribution of resistance surfaces adjusted for dark light data (Figure 7a–c), the resistance values are higher on both sides of the river valley in the mountainous areas and in the plain area to the southeast, presenting an overall pattern of zonal distribution in the mountainous areas and a concentrated distribution pattern in the plain areas (Figure 7d).

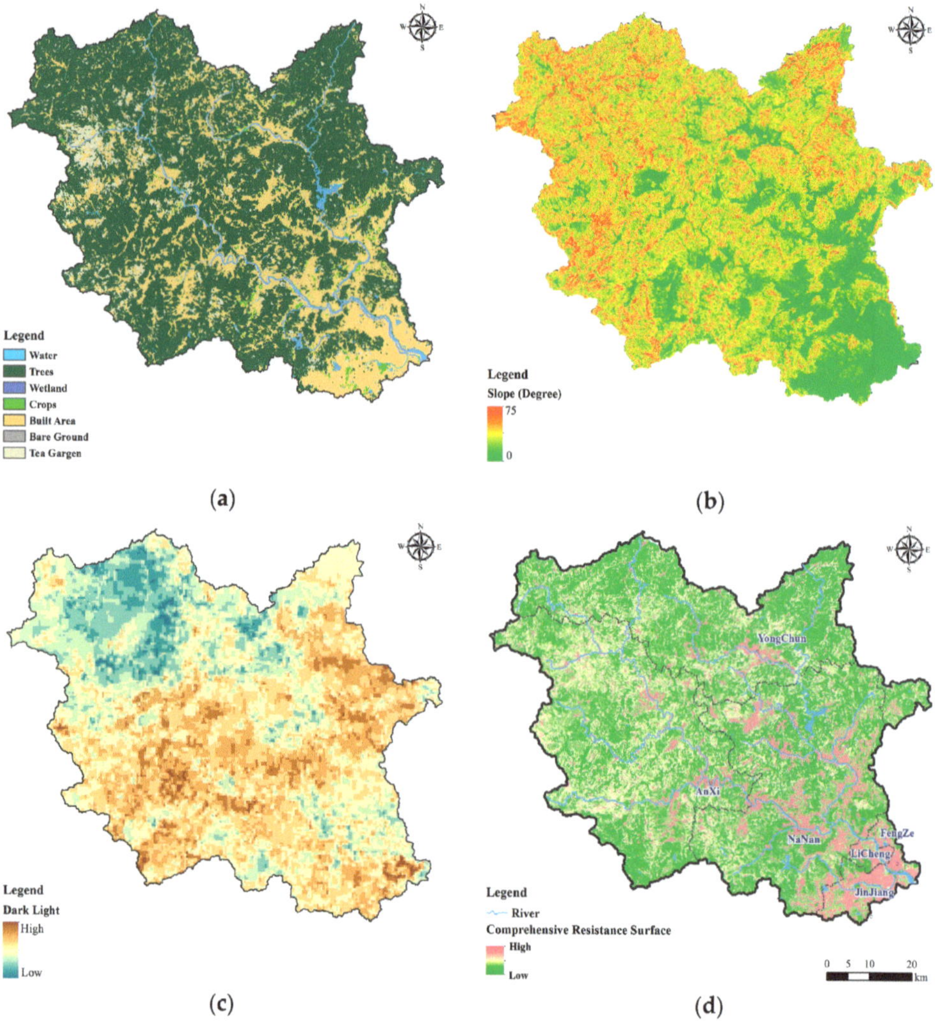

Figure 7. Construction of the comprehensive resistance surface: (**a**) land cover; (**b**) slope; (**c**) dark light; (**d**) comprehensive resistance surface.

Areas with higher resistance values are primarily located in urban regions such as Quanzhou city, Yongchun County, and Anxi County, characterized by high human activity frequency and predominant land use as built-up areas. In these areas, ecological flow is significantly restricted, exerting a resistance effect on species migration and material information diffusion in the surrounding regions. Surrounding the high-value area is the resistance median area, which is mainly affected by human activities, which negatively impact the transmission and exchange of ecological material information in this region. Areas with lower ecological resistance values are located in the mountainous and hilly regions in the northwest, mainly covered by forests and grasslands. This area has large undulating terrain, a good ecological environment with minimal impact from human activities, rich biological species, and a good ecological environment, which can produce a positive driving effect on the migration and diffusion of surrounding species.

The methodology for constructing resistance surfaces can be adapted to other developed coastal regions, providing a valuable framework for addressing similar ecological and urban challenges globally.

3.3. Ecological Corridors in Jinjiang Watershed

Following our objective to enhance regional ecological sustainability through the integration of multiple ecological service functions, we have identified and delineated ecological corridors that are crucial for maintaining ecological connectivity. This aligns with our hypothesis that such integration would support sustainable development.

According to the principle of minimum cost, 321 ecological corridors were identified, with a total length of 2420.98 km (Figure 8). Among them, 114 key corridors extend over a total length of 1941.67 km, primarily distributed in the Dongxi watershed, specifically in areas such as Yongchun County and the northern part of Nan'an City. These key corridors exhibit continuity and density, primarily serving to connect the source areas of ecological regulation functions. This type of corridor has high ecological type consistency and facilitates the communication of material energy within the region, making them more conducive to species migration.

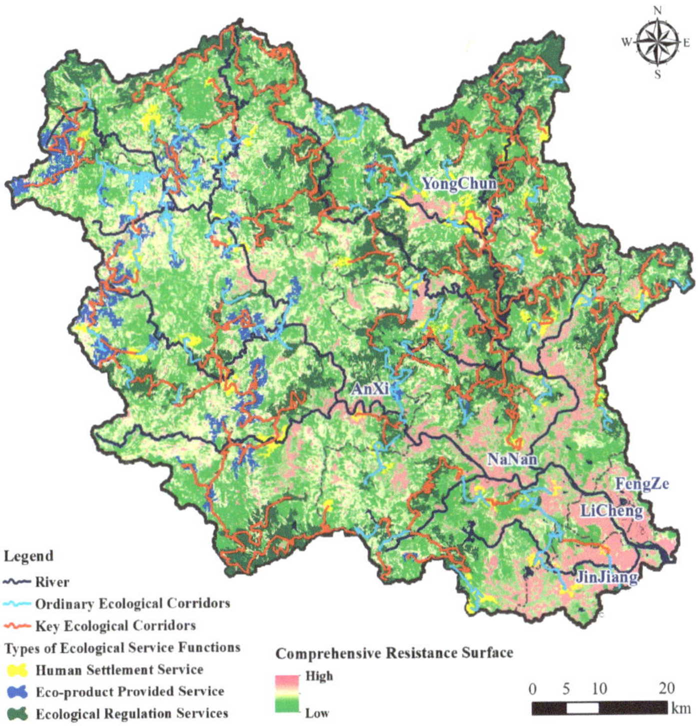

Figure 8. Distribution of ecological corridors.

There are 207 ordinary corridors, totaling 479.31 km in length. These corridors are primarily distributed in the Xixi watershed and the plain region, close to areas affected by human activities. They tend to be relatively shorter in length and more intermittent in nature, mainly serving to connect ecological source areas with product supply value and human living guarantee value, ensuring the connectivity of ecological source areas in small areas.

Ecological source areas that characterize the ecological regulation service function in the Jinjiang River Basin are primarily concentrated near the mainstream of Dongxi and Huyangxi (one of tributaries of Dongxi) and the upper reaches of Lanxi. Through important corridors connecting the source areas, the ecological source areas along the above-mentioned lines are interconnected, ensuring the overall integrity and consistency of the pattern of ecological regulation service. The ecological source areas that provide eco-production services are distributed in the mountainous areas of Anxi County in the Xixi watershed, mainly concentrated in the YiDuXi-ShuangXi-LongTanXi-BlueXi basin range, connected in a circular pattern through corridors, reflecting to a certain extent the clustering characteristics of ecological product provision. The scattered distribution of source areas of residents' ecological needs is primarily located in the county and urban areas, connected through potential corridors, and has relatively fragile ecological connectivity. The potential corridors are more for the purpose of ensuring the integrity and connectivity of the pattern. They are slightly more affected by resistance compared to the key corridors and are more threatened and influenced by human activities. To safeguard the overarching integrity and connectivity of the ecological security pattern in the Jinjiang watershed, more protection and restoration efforts are needed for potential corridors.

3.4. Ecological Security Pattern in Jinjiang Watershed

Drawing from the aforementioned research, a comprehensive ecological security pattern for the Jinjiang watershed was developed, incorporating three key service components (Figure 9). Based on the spatial distribution of ecological service functions, three types of ecological functional aggregation areas were delineated, comprising ecological regulation areas, eco-product provision areas, and urban development areas.

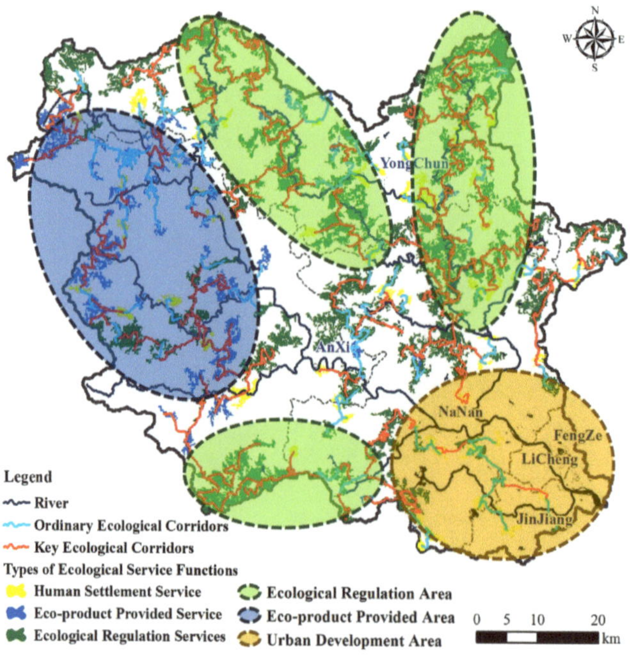

Figure 9. Ecological security pattern in the Jinjiang watershed.

(1) The ecological regulation area primarily encompasses regions where natural ecology maintenance and restoration efforts need to be intensified. It is predominantly situated in the upstream regions of the tributaries within the Jinjiang watershed, with low human impact and an overall good ecological environment and high forest coverage. The source water conservation area in the central and northern parts is mainly located within Yongchun County in the Shanmei Reservoir watershed. Yongchun County experiences a relatively high level of development. However, due to the Shanmei Reservoir's task of supplying water to Kinmen, the overall ecological environment is good. The local area mainly focuses on ecological development for tourism and fruit orchards.

(2) The eco-product provision area primarily serves the function of providing agricultural and forestry products and is mainly located within the Anxi County of the Xixi watershed. The local tea planting area is widely distributed, and there are many areas where forests have been converted into tea plantations, leading to serious soil erosion. Industrial and mining enterprises are clustered around Jiandou Town in Anxi County and significantly impact the water quality of rivers and groundwater in the region, making it the main source area of nitrogen and phosphorus pollutants in the Jinjiang watershed. To promote the harmonious and orderly co-development of production and ecology, it is imperative to enhance local environmental quality monitoring and regulate mining and mineral extraction enterprises.

(3) The urban development area is situated at the lower reaches of the Jinjiang watershed and based on the results of surveys and surface water sampling analysis, it is also one of the main non-point source pollutants in the Jinjiang watershed. The Jinjiang river system is the main source of drinking water in Quanzhou and is a typical river-type drinking water source. To ensure the quality of the water at the intake, targeted measures for basin water environment zoning classification protection and management are needed.

The comprehensive ecological security pattern developed for the Jinjiang watershed aligns with our objective to integrate multiple ecological service functions, thereby enhancing regional sustainability. This pattern not only addresses the specific needs of the Jinjiang watershed but also serves as a model for other developed coastal regions globally. By incorporating ecological regulation, eco-product provisions, and human settlement services, we provide a holistic framework that promotes sustainable urban development and ecological conservation. Practical applications of this pattern include enhanced environmental monitoring, regulation of industrial activities, and targeted conservation measures to maintain ecological connectivity and sustainability.

4. Discussion

4.1. Advantages of Ecological Source Identification

The scientific identification of ecological source areas is a fundamental prerequisite for constructing an ecological security pattern. The Jinjiang watershed is an area that integrates highly developed coastal cities with areas of prominent natural ecological contradictions. This study identifies ecological source areas that meet the diverse needs of various ecosystem services from different service function perspectives and broadens the rationality of ecological source areas in such conflicting regions. The ultimate goal of ecosystem services is to ensure healthy and stable development at the regional level. Therefore, at the basin scale, it is crucial to focus on human needs, while taking into account the influence of cultural and recreational activities on the quality of urban residents' lives, as well as the services of eco-product provided and microclimate regulation. These efforts not only ensure regional ecological security and uphold ecosystem stability but also fulfill the ecosystem service needs for enhancing the urban living environment, thereby greatly extending the framework of constructing an ecological security pattern.

The 14th Five-Year Plan emphasizes the robust growth of urban agglomerations, positioning them as the primary driving force behind the country's new urbanization and key contributors to national prosperity. These urban agglomerations have the historical duty of shifting the global economy center. Consequently, the development of cities will lead to the widespread emergence of regions characterized by similar ecological contradictions. In these areas, ecological security issues will not only concern protecting natural ecology but also strive for balanced development between humans and nature while addressing human necessities.

Therefore, in our exploration, we examine the influence of different types of ecological source areas on the establishment of an ecological security pattern. Our findings indicate that the areas designated for ecological regulation services, which are characterized by their continuous distribution, occupy the largest area among the three types of source areas and play a crucial role in mitigating the environmental pressures associated with urbanization. The concentrated distribution of areas dedicated to providing ecological products not only boosts the agricultural output and income for local farmers but also creates employment opportunities and bolsters rural economy growth, thereby facilitating the urbanization process. The dispersed distribution of human settlement service source areas provides valuable recreational and leisure spaces for urban residents and those residing along river valleys, contributing to the quality of life and well-being of the citizens. Furthermore, these captivating natural landscapes and recreational spaces play a significant role in attracting tourists, thus promoting the growth of local tourism and associated service sectors. Therefore, the identification and preservation of ecological source areas tailored to different functions can effectively reconcile urban construction, agricultural development, and natural ecology, thereby providing a solution to the contradictions between urban construction and ecological preservation that often arise during the urbanization process.

4.2. Development Suggestions Based on the Ecological Security Pattern

By combining the landscape distribution features of Jinjiang watershed, the ecological security pattern is optimized, forming an ecological corridor with the mainstream of the Dongxi serving as the axis and the Yunding Mountain-Tianzhu Mountain serving as the ecological security barrier. It consists of four ecological security development proposal patterns: the northern ecological regulation zone, the central and southern ecological regulation zone, the region used for growing tea in the west and center, and the rapidly developing urban area at the mouth of the river in the south (Figure 10). Within the national ecological planning area, the Jinjiang watershed is part of the ecological water conservation area in western Fujian. The northern and central-southern ecological regulation zones, characterized by their high forest coverage, are considered natural and high-quality water conservation areas. Therefore, these zones require priority ecological protection, which can be achieved through the establishment of forest parks, nature reserves, and other relevant measures.

The ecological security pattern reveals that core corridors primarily connect ecological regulation function source areas, offering a certain level of resilience compared to others and serving as the main conduit for ecological security in the Jinjiang watershed. Ordinary corridors, however, are positioned amidst minor watersheds of the Jinjiang River or spanning across various land cover types, making them more susceptible to disruptions arising from urban expansion. Therefore, ecological corridors should be designated as no-development zones, with coordinated planning at the provincial, municipal, and county levels, to ensure the connectivity of the corridor network. This entails advancing ecological construction projects like reclamation for afforestation and implementing forest closure programs, while establishing a comprehensive shelterbelt system integrating trees, shrubs, and grasslands into a multi-class, multi-layer natural protective layer.

Tea plantations are widely distributed in the central-western product provision areas, including regions where forests have been transformed into tea gardens. The presence of rich white quartz sand in the soil coupled with continuous weed removal throughout tea cultivation, results in the loss of fine soil through surface runoff following heavy rains. This process exposes the white sand and leading to a decline in soil quality. Moreover, the tea plants' selective absorption of aluminum compounds from the laterite leads to rapid acidification of the surface soil in the tea gardens, further exacerbating the degradation of the soil's surface fertility. Thus, while increasing the agricultural income of the region, the issue of soil erosion in hilly areas cannot be overlooked. It is essential to strike a balance between the economic benefits derived from agricultural products and the ecological restoration expenses incurred enhance soil erosion monitoring and foster the harmonious and orderly development of production ecology.

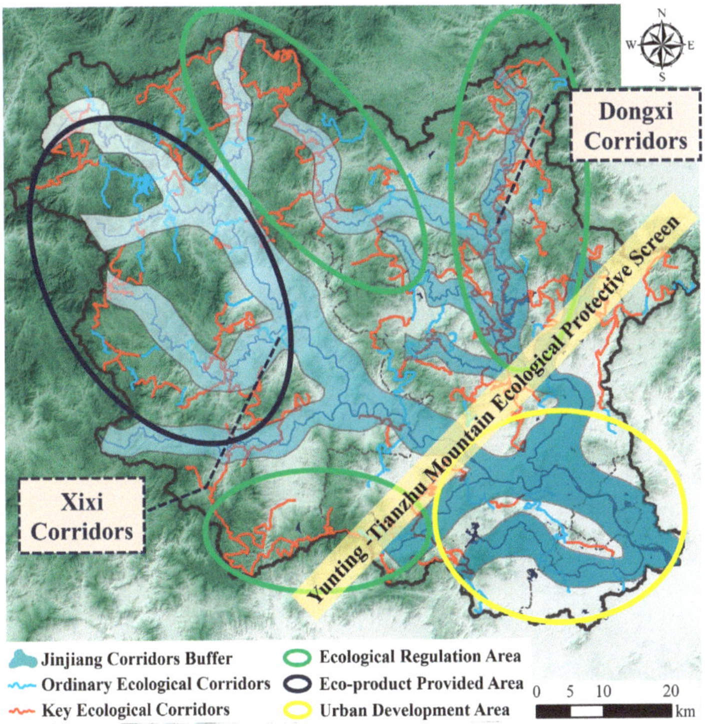

Figure 10. Advice based on ecological security patterns in Jinjiang watershed.

The southern urban development area, situated in the downstream plain area of the Jinjiang River basin, is the most densely populated area of Quanzhou City, with the highest demand for human settlement services. It is essential to protect existing ecological service green spaces while further constructing parks, green areas, and other public ecological service facilities to alleviate the ecological service demand pressure in the central urban development areas. Additionally, based on survey and surface water sampling analysis results, this area is also one of the main non-point sources of pollutants in the Jinjiang River basin. To prevent ecological degradation resulting from urbanization, targeted watershed water environment zoning protection and management are essential while ensuring the functionality of ecological services.

River ecological corridors are crucial components of the ecological security pattern, significantly impacting water conservation. Therefore, enhancing ecological protection of rivers and other water resources is particularly important. Specific plans should be adopted to further optimize water environment functional zoning, implement segmented river responsibility management, and strictly control industrial pollution to ensure water resource safety. Strengthening domestic sewage treatment, strictly regulating pollution from residential and industrial activities on water quality, and undertaking comprehensive restoration efforts for polluted rivers to rehabilitate ecologically damaged areas are examples of other recommendations.

4.3. Broader Implications and Practical Applications

The comprehensive ecological security pattern developed for the Jinjiang watershed aligns with our objective to integrate multiple ecological service functions, thereby enhancing regional sustainability. This pattern not only addresses the specific needs of the Jinjiang watershed but also serves as a model for other developed coastal regions globally. By incorporating ecological regulation, eco-product provisions, and human settlement services, we provide a holistic framework that promotes sustainable urban development and ecological conservation.

Practical applications of this pattern include enhanced environmental monitoring, regulation of industrial activities, and targeted conservation measures to maintain ecological connectivity and sustainability. For instance, local governments can use these findings to designate ecological corridors as no-development zones, ensuring the connectivity of the ecological network. This approach can also inform the planning and management of similar regions globally, contributing to the broader discourse on sustainable urban development and ecological preservation.

Moreover, the potential for scaling this methodology to other regions should be explored. Coastal regions worldwide face similar challenges of balancing urban development with ecological preservation. The approach used in this study can be adapted to different local contexts, providing valuable insights for global applications. Future research should focus on refining this model and testing its applicability in diverse environmental and socio-economic settings.

4.4. Critical Analysis of Results

While this study provides a robust framework for constructing ecological security patterns in developed coastal regions, there are some limitations and areas for improvement. The reliance on remote sensing data for land use classification introduces uncertainties due to the spatial and temporal resolution of the data. Incorporating ground truthing methods and higher resolution data can improve accuracy.

Additionally, the study's focus on the Jinjiang watershed means that certain localized factors may not be applicable to other regions without adjustments. The weighting of ecological service functions, for example, could be adapted to better reflect the specific ecological, economic, and social conditions of different regions.

Furthermore, the impact of climate change on ecological security patterns was not extensively addressed in this study. Future research should consider how changing climate conditions might affect ecological regulation, eco-product provisions, and human settlement services and how adaptive measures can be integrated into the ecological security patterns.

By addressing these limitations and refining the methodology, the ecological security pattern framework can be made more robust and widely applicable, offering a valuable tool for sustainable urban and regional planning.

5. Conclusions

This study addresses the critical intersections of ecology, production, and living spaces in rapidly developing coastal cities in China. Using the Jinjiang watershed as a case study, we applied the "source identification—resistance surface construction—corridor extraction" methodology to develop an ecological security pattern that integrates ecological regulation, eco-product provisions, and human settlement services.

The results identified 322 ecological source areas covering 1287.18 km^2 and 2420.98 km of ecological corridors. Ecological regulation areas are primarily in northern mountainous regions, eco-product provision areas are concentrated in Anxi County, and human settlement services are dispersed in densely populated plains and river valleys.

While the methodology provides a robust framework, limitations include potential inaccuracies from satellite imagery and the need for ground truthing. The weighting of ecological service functions should be tailored to local conditions.

This ecological security pattern offers a replicable model for other coastal regions globally. Practical applications include enhanced environmental monitoring, stricter regulation of industrial activities, and targeted conservation efforts. The following key recommendations are highlighted: (1) Enhanced Protection: Designate key corridors as no-development zones to preserve ecological integrity. (2) Soil Erosion and Acidification Mitigation: Optimize tea garden layouts and restore affected areas. (3) Urban Green Space Development: Construct green parks to meet ecological needs and boost tourism. (4) Comprehensive Monitoring: Implement continuous environmental monitoring to assess the effectiveness of the ecological security pattern and adapt policies as necessary.

The ecological security pattern developed for the Jinjiang watershed demonstrates the feasibility of integrating multiple ecological service functions to support sustainable urban development. These findings offer a replicable model for other coastal regions globally, promoting a balance between urban growth and ecological conservation. Practical applications include enhanced environmental monitoring, regulation of industrial activities, and targeted conservation measures. Future research should refine these methods and explore their applicability in diverse contexts.

Author Contributions: Conceptualization, Y.Z. and Y.X.; methodology, Y.Z.; software, J.L. (Jiahuan Liu); validation, Y.L. (Yaci Liu); formal analysis, Y.Z.; investigation, Y.Z. and J.L. (Jianfeng Li); resources, Y.Z.; data curation, Y.X.; writing—original draft preparation, Y.Z.; writing—review and editing, Y.X., Y.L. (Yasong Li), and Z.Z.; visualization, C.L.; supervision, Q.H. and C.L.; project administration, Y.L. (Yasong Li); funding acquisition, Y.X. and Y.L. (Yasong Li). All authors have read and agreed to the published version of the manuscript.

Funding: This research was supported by the Fujian Provincial Key Laboratory of Water Cycling and Eco-Geological Processes (SK202305KF01), the China Geological Survey (DD20190303 and DD20230459), the National Natural Science Foundation of China (42007183, 41702282), the Natural Science Foundation of Sichuan Province (2022NSFSC1084), MOE Key Laboratory of Groundwater Circulation and Environmental Evolution, China University of Geosciences (Beijing) (2023-004), the Student Research Training Program of Southwest Jiaotong University (241925), Open Fund of Key Laboratory (No.WSRCR-2023-01).

Data Availability Statement: The data that support the findings of this study are available upon reasonable request from the authors.

Conflicts of Interest: The authors declare no conflicts of interest. The funders had no role in the design of the study; in the collection, analyses, or interpretation of data; in the writing of the manuscript; or in the decision to publish the results.

References

1. Li, Q.; Zhou, Y.; Yi, S. An integrated approach to constructing ecological security patterns and identifying ecological restoration and protection areas: A case study of Jingmen, China. *Ecol. Indic.* **2022**, *137*, 108723. [CrossRef]
2. Liu, J.; Cheng, Y.; Zhang, F.; Wen, X.; Yang, L. Research hotspots and trends of groundwater and ecology studies: Based on a bibliometric approach. *J. Groundw. Sci. Eng.* **2023**, *11*, 20–36. [CrossRef]
3. Hu, J.; Huang, Y.; Du, J. The impact of urban development intensity on ecological carrying capacity: A case study of ecologically fragile areas. *Int. J. Environ. Res. Public Health* **2021**, *18*, 7094. [CrossRef] [PubMed]
4. Fang, Z.; Ding, T.; Chen, J.; Xue, S.; Zhou, Q.; Wang, Y.; Wang, Y.; Huang, Z.; Yang, S. Impacts of land use/land cover changes on ecosystem services in ecologically fragile regions. *Sci. Total Environ.* **2022**, *831*, 154967. [CrossRef] [PubMed]
5. Tong, S.; Beggs, P.J.; Davies, J.M.; Jiang, F.; Kinney, P.L.; Liu, S.; Yin, Y.; Ebi, K.L. Compound impacts of climate change, urbanization and biodiversity loss on allergic disease. *Int. J. Epidemiol.* **2023**, *52*, 655–663. [CrossRef] [PubMed]
6. Eswaran, H.; Lal, R.; Reich, P. Land degradation: An overview. In *Response to Land Degradation*; CRC Press: Boca Raton, FL, USA, 2019; pp. 20–35.
7. Lin, G.; Jiang, D.; Fu, J.; Zhao, Y. A review on the overall optimization of production–living–ecological space: Theoretical basis and conceptual framework. *Land* **2022**, *11*, 345. [CrossRef]

8. Shah, M.I.; Abbas, S.; Olohunlana, A.O.; Sinha, A. The impacts of land use change on biodiversity and ecosystem services: An empirical investigation from highly fragile countries. *Sustain. Dev.* **2023**, *31*, 1384–1400. [CrossRef]
9. Jiang, B.; Chen, W.; Dai, X.; Xu, M.; Liu, L.; Sakai, T.; Li, S. Change of the spatial and temporal pattern of ecological vulnerability: A case study on Cheng-Yu urban agglomeration, Southwest China. *Ecol. Indic.* **2023**, *149*, 110161. [CrossRef]
10. Li, X.; Xiao, L.; Tian, C.; Zhu, B.; Chevallier, J. Impacts of the ecological footprint on sustainable development: Evidence from China. *J. Clean. Prod.* **2022**, *352*, 131472. [CrossRef]
11. Kang, J.; Zhang, X.; Zhu, X.; Zhang, B. Ecological security pattern: A new idea for balancing regional development and ecological protection. A case study of the Jiaodong Peninsula, China. *Glob. Ecol. Conserv.* **2021**, *26*, e01472. [CrossRef]
12. Liu, C.; Li, W.; Xu, J.; Zhou, H.; Li, C.; Wang, W. Global trends and characteristics of ecological security research in the early 21st century: A literature review and bibliometric analysis. *Ecol. Indic.* **2022**, *137*, 108734. [CrossRef]
13. Zeng, P.; Wei, X.; Duan, Z. Coupling and coordination analysis in urban agglomerations of China: Urbanization and ecological security perspectives. *J. Clean. Prod.* **2022**, *365*, 132730. [CrossRef]
14. Cui, X.; Deng, W.; Yang, J.; Huang, W.; de Vries, W.T. Construction and optimization of ecological security patterns based on social equity perspective: A case study in Wuhan, China. *Ecol. Indic.* **2022**, *136*, 108714. [CrossRef]
15. Ye, H.; Song, Y.; Xue, D.; Xia, S.; Tang, Y.; Zhang, L. Construction of an ecological security pattern for the Guanzhong Plain urban agglomeration: Scale perspective and hierarchy demand. *Front. Environ. Sci.* **2023**, *11*, 1136630. [CrossRef]
16. Zhang, X.; Liu, G.; Zheng, Z. Construction of an ecological security pattern based on functional wetland theory: A case study in a landscape city. *Front. Environ. Sci.* **2022**, *10*, 955230. [CrossRef]
17. Rauch, H.; von der Thannen, M.; Raymond, P.; Mira, E.; Evette, A. Ecological challenges for the use of soil and water bioengineering techniques in river and coastal engineering projects. *Ecol. Eng.* **2022**, *176*, 106539. [CrossRef]
18. De La Cruz, C.; Manuel, P. The knowledge status of coastal and marine ecosystem services-challenges, limitations and lessons learned from the application of the ecosystem services approach in management. *Front. Mar. Sci.* **2021**, *8*, 684770. [CrossRef]
19. Hysa, A.; Löwe, R.; Geist, J. Ecosystem services potential is declining across European capital metropolitan areas. *Sci. Rep.* **2024**, *14*, 8903. [CrossRef] [PubMed]
20. Yutian, S.; Jin, F.; Liyan, X. Review of Ecological Infrastructure Planning Practices Based on the Ecological Security Pattern Theory in China (1997~2019). *Landsc. Archit. Front.* **2021**, *9*, 28–48.
21. Zhu, B.; Hashimoto, S.; Cushman, S.A. Navigating ecological security research over the last 30 years: A scoping review. *Sustain. Sci.* **2023**, *18*, 2485–2498. [CrossRef]
22. Fu, Y.; Shi, X.; He, J.; Yuan, Y.; Qu, L. Identification and optimization strategy of county ecological security pattern: A case study in the Loess Plateau, China. *Ecol. Indic.* **2020**, *112*, 106030. [CrossRef]
23. Ma, B.; Zhou, T.; Lei, S.; Wen, Y.; Htun, T.T. Effects of urban green spaces on residents' well-being. *Environ. Dev. Sustain.* **2019**, *21*, 2793–2809. [CrossRef]
24. Ke, Y.; Bai, Y.; Ali, M.; Ashraf, A.; Li, M.; Li, B. Exploring residents' perceptions of ecosystem services in nature reserves to guide protection and management. *Ecol. Indic.* **2024**, *158*, 111535. [CrossRef]
25. Yan, Y.; Ju, H.; Zhang, S.; Chen, G. The construction of ecological security patterns in coastal areas based on landscape ecological risk assessment—A case study of Jiaodong Peninsula, China. *Int. J. Environ. Res. Public Health* **2021**, *18*, 12249. [CrossRef] [PubMed]
26. Zhu, N.; Ai, J.; Zeng, Z.; Zhou, C. Exploring the Spatial Relationship between the Ecological Topological Network and Carbon Sequestration Capacity of Coastal Urban Ecosystems: A Case Study of Yancheng City, China. *Remote Sens.* **2023**, *15*, 4007. [CrossRef]
27. Yi, D.; Guo, X.; Han, Y.; Guo, J.; Ou, M.; Zhao, X. Coupling ecological security pattern establishment and construction land expansion simulation for urban growth boundary delineation: Framework and application. *Land* **2022**, *11*, 359. [CrossRef]
28. Wen, J.; Hou, K. Research on the progress of regional ecological security evaluation and optimization of its common limitations. *Ecol. Indic.* **2021**, *127*, 107797. [CrossRef]
29. Yuan, Y.; Bai, Z.; Zhang, J.; Xu, C. Increasing urban ecological resilience based on ecological security pattern: A case study in a resource-based city. *Ecol. Eng.* **2022**, *175*, 106486. [CrossRef]
30. Sadeghi, S.H.; Vafakhah, M.; Moosavi, V.; Pourfallah Asadabadi, S.; Sadeghi, P.S.; Khaledi Darvishan, A.; Bagheri Fahraji, R.; Mosavinia, S.H.; Majidnia, A.; Gharemahmudli, S. Assessing the health and ecological security of a human induced watershed in central Iran. *Ecosyst. Health Sustain.* **2022**, *8*, 2090447. [CrossRef]
31. Ma, L.; Bo, J.; Li, X.; Fang, F.; Cheng, W. Identifying key landscape pattern indices influencing the ecological security of inland river basin: The middle and lower reaches of Shule River Basin as an example. *Sci. Total Environ.* **2019**, *674*, 424–438. [CrossRef] [PubMed]
32. Sarker, A. Ecological perspectives on water, food, and health security linkages: The Minamata case in Japan. *Environ. Sci. Pollut. Res.* **2021**, *28*, 32177–32189. [CrossRef] [PubMed]
33. Kh, D.A. The Mediterranean Sea Basin as a Single Ecosystem: Problems and Prospects for International Cooperation. *Bull. Peoples' Friendsh. Univ. Russ. Ser. Int. Relat.* **2021**, *21*, 625–641.
34. Kiskaddon, E.; Bienn, H.; Hemmerling, S.A.; Dalyander, S.; Grismore, A.; Parfait, J.; Miner, M.D.; Cameron, C.; Hopkins, T.E.; Allen, Y. Supporting habitat restoration in the northern Gulf of Mexico through synthesis of data on multiple and interacting benefits and stressors. *J. Environ. Manag.* **2022**, *318*, 115589. [CrossRef] [PubMed]

35. Liu, Y.; Wu, L.; Shi, G.; Cao, S.; Li, Y. Characteristics and sources of microplastic pollution in the water and sediments of the Jinjiang River Basin, Fujian Province, China. *China Geol.* **2022**, *5*, 429–438. [CrossRef]
36. Li, S.; Cao, Y.; Liu, J.; Wang, S.; Zhou, W. Assessing spatiotemporal dynamics of land use and cover change and carbon storage in China's ecological conservation pilot zone: A case study in Fujian Province. *Remote Sens.* **2022**, *14*, 4111. [CrossRef]
37. Zhu, Y.; Yang, H.; Xiao, Y.; Hao, Q.; Li, Y.; Liu, J.; Wang, L.; Zhang, Y.; Hu, W.; Wang, J. Identification of Hydrochemical Characteristics, Spatial Evolution, and Driving Forces of River Water in Jinjiang Watershed, China. *Water* **2023**, *16*, 45. [CrossRef]
38. Dai, X.; Wang, L.; Tao, M.; Huang, C.; Sun, J.; Wang, S. Assessing the ecological balance between supply and demand of blue-green infrastructure. *J. Environ. Manag.* **2021**, *288*, 112454. [CrossRef] [PubMed]
39. Lubowiecki-Vikuk, A.; Đerčan, B.M.; de Sousa, B.M.B. Sustainable development and leisure services: Changes and trends. In *Handbook of Sustainable Development Leisure Services*; Springer: Cham, Switzerland, 2021; pp. 1–20.
40. Wu, L.; Sun, C.; Fan, F. Estimating the characteristic spatiotemporal variation in habitat quality using the invest model—A case study from Guangdong–Hong Kong–Macao Greater Bay Area. *Remote Sens.* **2021**, *13*, 1008. [CrossRef]
41. Li, M.; Liang, D.; Xia, J.; Song, J.; Cheng, D.; Wu, J.; Cao, Y.; Sun, H.; Li, Q. Evaluation of water conservation function of Danjiang River Basin in Qinling Mountains, China based on InVEST model. *J. Environ. Manag.* **2021**, *286*, 112212. [CrossRef]
42. Berta Aneseyee, A.; Noszczyk, T.; Soromessa, T.; Elias, E. The InVEST habitat quality model associated with land use/cover changes: A qualitative case study of the Winike Watershed in the Omo-Gibe Basin, Southwest Ethiopia. *Remote Sens.* **2020**, *12*, 1103. [CrossRef]
43. Zhang, W.; Yan, Y.; Yu, R.; Hu, G. The sources-specific health risk assessment combined with APCS/MLR model for heavy metals in tea garden soils from south Fujian Province, China. *Catena* **2021**, *203*, 105306. [CrossRef]
44. Kong, X.; Han, M.; Li, Y.; Kong, F.; Sun, J.; Zhu, W.; Wei, F. Spatial differentiation and formation mechanism of ecological sensitivity in large river basins: A case study of the Yellow River Basin, China. *Ecol. Indic.* **2024**, *158*, 111571. [CrossRef]
45. Cheng, H.; Zhu, L.; Meng, J. Fuzzy evaluation of the ecological security of land resources in mainland China based on the Pressure-State-Response framework. *Sci. Total Environ.* **2022**, *804*, 150053. [CrossRef] [PubMed]
46. Wei, B.; Kasimu, A.; Fang, C.; Reheman, R.; Zhang, X.; Han, F.; Zhao, Y.; Aizizi, Y. Establishing and optimizing the ecological security pattern of the urban agglomeration in arid regions of China. *J. Clean. Prod.* **2023**, *427*, 139301. [CrossRef]
47. Nie, W.; Xu, B.; Yang, F.; Shi, Y.; Liu, B.; Wu, R.; Lin, W.; Pei, H.; Bao, Z. Simulating future land use by coupling ecological security patterns and multiple scenarios. *Sci. Total Environ.* **2023**, *859*, 160262. [CrossRef] [PubMed]
48. Huang, L.; Wang, D.; He, C. Ecological security assessment and ecological pattern optimization for Lhasa city (Tibet) based on the minimum cumulative resistance model. *Environ. Sci. Pollut. Res.* **2022**, *29*, 83437–83451. [CrossRef] [PubMed]
49. Jia, Q.; Jiao, L.; Lian, X.; Wang, W. Linking supply-demand balance of ecosystem services to identify ecological security patterns in urban agglomerations. *Sustain. Cities Soc.* **2023**, *92*, 104497. [CrossRef]
50. Wei, W.; Liu, C.; Ma, L.; Zhang, X.; Xie, B. Ecological land suitability for arid region at river basin scale: Framework and application based on Minmum Cumulative Resistance (MCR) model. *Chin. Geogr. Sci.* **2022**, *32*, 312–323. [CrossRef]
51. Li, W.; Geng, J.; Bao, J.; Lin, W.; Wu, Z.; Fan, S. Analysis of spatial and temporal variations in ecosystem service functions and drivers in Anxi county based on the InVEST model. *Sustainability* **2023**, *15*, 10153. [CrossRef]

sustainability

MDPI

Article

Evolution of Land Use and Its Hydrological Effects in the Fenhe River Basin Under the Production–Living–Ecological Space Perspective

Junzhe Zhang [1], Azhar Ali Laghari [1,*], Qingxia Guo [1,*], Jiyao Liang [1], Akash Kumar [2], Zhenghao Liu [1], Yongheng Shen [1] and Yuehan Wei [1]

1 The College of Resources and Environment, Shanxi Agricultural University, Jinzhong 030801, China; journeyzjz@163.com (J.Z.); 15090073828@163.com (J.L.); qq240419050@163.com (Z.L.); 15803838621@163.com (Y.S.); wyh1206070992@163.com (Y.W.)

2 School of Civil Engineering, Guangzhou University, Guangzhou 510006, China; akash.kumar@gzhu.edu.cn

* Correspondence: azharlaghari18@sxau.edu.cn (A.A.L.); gqx696@sxau.edu.cn (Q.G.); Tel.: +86-158-2246-7087(A.A.L.); +86-138-3548-7451 (Q.G.)

Abstract: Analysing the patterns and impacts of land-use changes in the production–living–ecological space (PLES) of the Fenhe River Basin (FRB 39,721 km^2), China, is necessary to support sustainable development. Based on remote sensing images from 1990 to 2020, we aimed to analyse the PLES land-use changes. Industrial production and living spaces continuously encroached on the agricultural production and ecological spaces between 1990 and 2022 owing to industrialisation and urbanisation, and the ecological land area decreased by 699.21 km^2, while the industrial production land area increased by 521.32 km^2. We used the soil and water assessment tool (SWAT) model to quantitatively analyse the impact of PLES changes on runoff in the FRB. With the continuous expansion of production and living spaces, the extensive use of concrete in cities has led to ground hardening, making it difficult for precipitation to infiltrate, with surface runoff increasing by 0.3 mm annually. The reduction in ecological space has led to a reduction in forests and grasslands, weakening the water-holding capacity of the watershed and affecting groundwater storage. This study provides a scientific basis for watershed management and the integrated development of PLES.

Keywords: production–living–ecological space; land-use change; SWAT; runoff; water resource management

1. Introduction

Land-use and cover change (LUCC) refers to changes in surface cover caused by changes in human land use and management practices (e.g., urbanisation and deforestation) [1,2], which affect ecosystems and the water cycle at regional, local, and global scales [3,4]. Ecohydrological effects induced by human activities are a focal point in the field of regional and global change research. By observing and modelling changes in runoff, an important indicator of the terrestrial water cycle, regional land-use patterns can be assessed and adjusted to increase surface production and prevent or mitigate sinking and flooding, thereby supporting the sustainable development of watersheds [5].

Since its reform and opening up in 1978, China's urbanisation rate (proportion of the urban population to the total population in a country or region, used to measure the level and pace of urbanisation) has more than tripled [6], It increased from 17.9% in 1978 to 65.2% in 2022, with the resultant challenges of scarce land resources, increasing environmental pollution, and ecological degradation [7,8]. In 2012, the 18th Chinese Communist Party (CPC) National Congress proposed the construction of production–living–ecological space (PLES), which encompasses intensive and efficient production space, space for habitable living, and attractive ecological space, providing a direction for land optimisation [9].

However, defining PLES remains a challenge. Previous studies have determined that a dominant function typically emerges among the multiple functions of different land-use types [10]. Thus, the transformation of land use manifests as the transformation of the dominant function of land use. The change in dominant function reflects the socio-economic transformation and development, including production, living, or ecological space. Past studies have mainly focused on the ecological and environmental changes caused by land-use changes [11,12]. However, few studies have investigated the ecological and environmental problems caused by the transformation of the dominant functions of land use, especially the ecohydrological effects of changes to the structure of PLES at the watershed scale.

The Fenhe River is the second-largest tributary of the Yellow River and is located in Shanxi Province [13]. The Fenhe River Basin, known as the "granary of Shanxi", is relatively rich in water resources, has fertile land, and displays favourable light and heat conditions [14]. However, the land-use structure of the basin has changed considerably in recent decades due to mining, urbanisation, and agricultural production expansion [15,16]. These changes have led to problems such as increased surface runoff, over-exploitation of groundwater, and water pollution, which have seriously constrained the sustainable socioeconomic development of the basin [17–19]. Since 2019, the Chinese government has prioritised the ecological protection and high-quality sustainable development of the Yellow River Basin [20]. Therefore, exploring the ecohydrological effects of PLES land-use transformation in the Fenhe River Basin (FRB) will highlight the systematic, holistic, and synergistic nature of basin governance and promote the construction of an integrated ecological and environmental governance system upstream and downstream to ensure the safety of the Fenhe and Yellow rivers [21,22].

The semi-distributed soil and water assessment tool (SWAT) is a hydrological model that predicts river runoff, sediment, and pollutants by simulating hydrological processes in a watershed [23,24]. This model is mainly used to assess the impact of land-use changes and management measures on the hydrological cycle [25]. In China, many scholars have used the SWAT model to study the impact of land-use change on runoff. Chen (2002) [26] simulated different land-cover scenarios in the upper reaches of the Yangtze River using the SWAT model and found that improvement in land cover reduced the depth of runoff and increased evapotranspiration. The Sino-German Cooperation Project of the Chinese Academy of Sciences (2004) [27] applied the SWAT model to the middle and lower reaches of the Yangtze River Basin to explore the impact of land-use change on hydrological processes and flooding mechanisms, providing a scientific basis for sustainable land use and ecological management in the Yangtze River Basin.

In this study, we aimed to analyse the patterns of land-use change in the FRB with a PLES perspective and investigate the hydrological effects of PLES changes in the FRB from 1990 to 2022. To this end, we used a transfer matrix, land-use dynamics, and the SWAT model. Spatial and temporal changes in runoff were used as important visual indicators to quantitatively reveal the impacts of the changes in PLES on runoff in the FRB. The results of this study serve as an important reference for improving the quality of the ecological environment and the security of water resources in the basin.

2. Materials and Methods

2.1. Study Area

The Fenhe River, which is the largest waterway in Shanxi Province, flows from north to south through 6 cities and 29 counties, spanning a total length of 713 km and a watershed area of 39,721 km^2. The basin is located in the continental monsoon climate zone, with an average annual temperature of approximately 11 °C and annual precipitation of approximately 500 mm. The primary soil types in the FRB are alluvial, brown, and coarse-grained soils, and the main land-use types are woodlands, meadows, and croplands. The rich mineral resources and superior soil and water conditions have made the FRB an important industrial and agricultural production area in Shanxi Province [28,29]. The urbanisation level within the watershed is 66.6%, higher than the provincial average. The GDP of the watershed is CNY 796.227 billion, accounting for 45.4% of the provincial GDP. Uneven spatial and temporal distributions of precipitation and rapid industrialisation have led to serious water shortages in the FRB, and the per capita water resources (400 m^3 per capita) are lower than the average levels of the province (500 m^3 per capita) and countryside (2000 m^3 per capita) [30]. In the future, ecological protection and high-quality development will face severe challenges in the FRB (Figures 1 and 2).

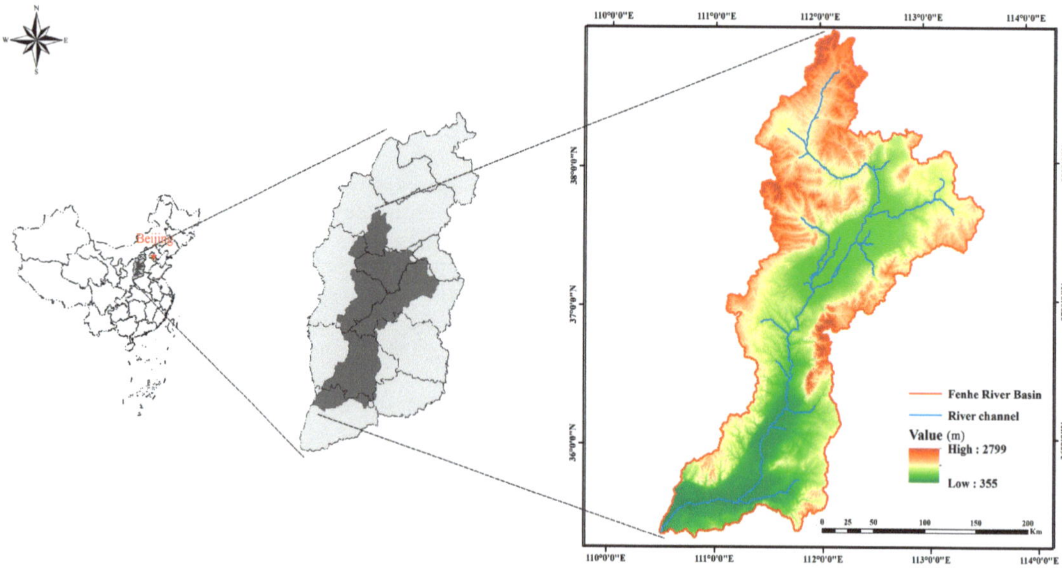

Figure 1. Location of the Fenhe River Basin, Shanxi Province, China.

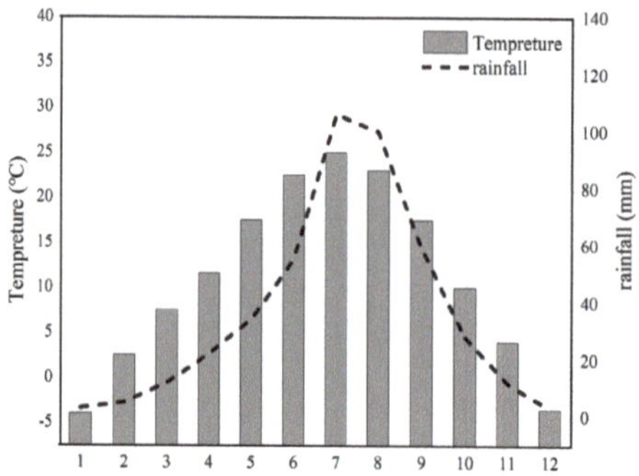

Figure 2. Monthly mean temperature and precipitation in the Fenhe River Basin. Data were obtained from eight meteorological stations in the basin (1990–2022), and values were calculated using the Tyson polygon method.

2.2. Data Sources

Data required to construct the SWAT model attribute database mainly consisted of elevation, land-use, soil, and meteorological data as well as measured runoff data for calibration verification. The digital elevation model (DEM) data were sourced from the geospatial data cloud platform, cropped, and radiometrically corrected. Land-use data (30 × 30 m resolution) were acquired from the Resource and Environment Science Data Centre of the Chinese Academy of Sciences and reclassified according to the PLES classification standard (described in detail in Section 2.3.3). Soil data were obtained from the World Soil Database, which was jointly created by the United Nations Food and Agriculture Organization and the Nanjing Soil Research Institute, and processed using soil, plant, atmosphere, and water (SPAW software 6.02.75). Meteorological data, including precipitation, temperature, wind speed, relative humidity, and solar radiation, were obtained from the National Meteorological Science Data Centre of China, and missing data were calculated and filled in using a weather generator. The measured runoff data used for model calibration were obtained from the Yellow River Conservancy Commission, with the Hejin station as the basin outlet (Table 1).

Table 1. Soil and water assessment tool (SWAT) model data sources.

Data	Source	Resolution
Digital evaluation model	Geospatial data cloud (http://www.gscloud.cn/, accessed on 16 September 2023)	30 m
Land-use/land-cover map Soil map	Resource and Environment Science and Data Centre (https://www.resdc.cn, accessed on 16 September 2023) Harmonized World Soil Database (V 1:1)	30 m × 30 m 30 arc-s
Observed stream flow	Yellow River Conservancy Commission	on a per-day basis
Meteorological data	National Meteorological Science Data Centre Daily Values of Surface Climatological Data for China (V 3.0)	on a per-day basis

2.3. Study Methodology

2.3.1. SWAT Model Principles

According to Figure 3, the SWAT model is a watershed-scale, semi-distributed hydrological model developed by the Agricultural Research Service of the United States Department of Agriculture [31]. It is based on the principle that hydrological processes, for instance, precipitation, runoff, evapotranspiration, and sediment transport, are simulated in each watershed by dividing the watershed into sub-basins or hydrological response units [32,33]. The SWAT model consists of modules for vegetation growth, hydrological processes, and land-management practices. Its parameters are directly related to surface characteristics; therefore, it can be used to simulate the water cycle under different land-use management measures, land-use structures, and climatic conditions, which has an important application value [34]. The SWAT model makes predictions based on the following balance equation:

$$SW_t = SW_{t-1} + \left(P - Q_{surf} - ET - W_{seep} - Q_{gw}\right) \quad (1)$$

where SW_t is the current soil water content, SW_{t-1} is the previous day's soil water content, P is precipitation, Q_{surf} is surface runoff, ET is evapotranspiration, W_{seep} is seepage, and Q_{gw} is groundwater.

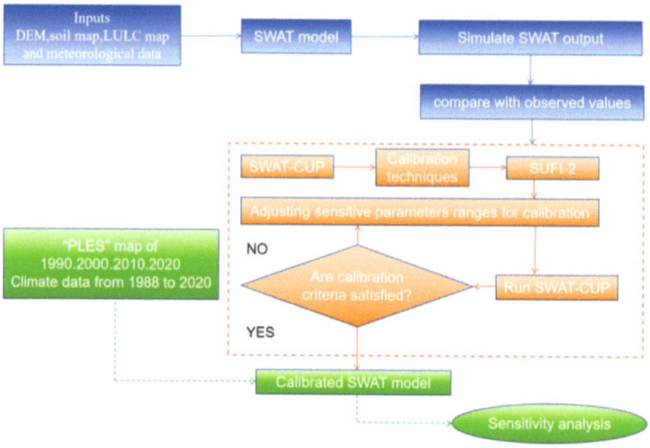

Figure 3. Soil and water assessment tool (SWAT) model flow chart.

2.3.2. SWAT Model Evaluation Criteria

In this study, we selected three commonly used metrics—the Nash efficiency coefficient (NSE), percentage bias (PBIAS), and coefficient of determination (R^2)—to assess the performance of the SWAT model in the study area [35]. The NSE value is used to measure the degree of match between the simulated data of the model and the observed data, with values ranging from $-\infty$ to 1; the closer the value approaches 1, the more closely the simulation results align with the observed data. The PBIAS value is used to determine whether the model tends to overestimate or underestimate the flow rate; positive and negative values of PBIAS indicate an overestimation and underestimation of the model, respectively. The R^2 value is used to measure the relationship between the modelled and measured values, with values ranging from 0 to 1; the closer the value is to 1, the better the linear fit between the simulated and observed values. These metrics were calculated using the following formulas:

$$PBIAS = 100\left(\frac{\sum_{i=1}^{N}(Q_{sim,i}-Q_{obs,i})}{\sum_{i=1}^{N}(Q_{obs,i})}\right) \tag{2}$$

$$NSE = 1-\frac{\sum_{i=1}^{N}(Q_{obs,i}-Q_{sim,i})^2}{\sum_{i=1}^{N}(Q_{obs,i}-Q_{obs})^2} \tag{3}$$

$$R^2 = \frac{\left[\sum_{i=1}^{N}(Q_{obs,i}-Q_{obs})(Q_{sim,i}-Q_{sim})\right]^2}{\sum_{i=1}^{N}(Q_{obs,i}-Q_{obs})^2\sum_{i=1}^{N}(Q_{sim,i}-Q_{sim})^2} \tag{4}$$

where $Q_{obs,i}$ and $Q_{sim,i}$ are the observed and simulated values, respectively; Q_{obs} and Qsim are the means of the observed and simulated values, respectively; n denotes the number of samples; and i denotes the ith sample.

2.3.3. PLES Classification System

Land is a multifunctional complex integrating production, living, and ecological functions [36]. Based on the definition of PLES, spatial functions of the national territory, land-use characteristics of the study area, practicability of the data, and PLES land-use classification method by Jianhong et al. (2021) [37], we established a new PLES classification system. This new system considers the fact that the primary function of PLES covers the secondary functions of different land-use types. For example, arable land has both production and ecological functions; however, because its primary function is crop production, it is classified as agricultural production land. Furthermore, industrial production land is classified as construction or special-use land, living space is classified as urban or rural residential land, and ecological space is classified as forest, grassland, water, or other ecological land. The Field Calculator tool in ArcGIS was used to classify the downloaded land-use data according to the PLES classification standard. A new field was created in the layer attribute table, to store the classification results (e.g., Land Use Type) (Table 2).

Table 2. Land-use classification system of the production–living–ecological space (PLES).

Primary Classification	Secondary Classification and Code	Secondary Classification of Land Use
Production land	Agricultural production land (1A) Industrial production land (1B)	Paddy field, dry land Other construction land, special use area
Living land	Urban living land (2A)	Urban land
	Rural living land (2B)	Rural residential area
Ecological land	Forest (3A) Grassland (3B) Water (3C) Other ecological land (3D)	Dredged woodland, shrub land woodland, and other woodland High-, medium-, and low-cover grassland Canals, reservoirs, ponds, and lakes Marshland, bare rock stony land, sandy land, beach land, saline–alkali land, and bare land

2.3.4. PLES Spatiotemporal Change Analysis Methods

Land-use dynamics characterise the rate of change in the quantity of land resources [38]. This metric consists of a single land-use momentum (K), which represents the transformation in the quantity of a particular land-use type, and a combined land-use momentum (LC), which indicates the transformation in the quantity of a combination of land-use types. Land-use dynamics were calculated using the following formulas:

$$K=\frac{U_b-U_a}{U_a}\times\frac{1}{T}\times 100\% \tag{5}$$

$$LC=\left[\frac{\sum_{i=1}^{n}\Delta LU_{i-j}}{2\sum_{i=1}^{n}LU_i}\right]\times\frac{1}{T}\times 100\% \tag{6}$$

where U_a and U_b denote the areas of the initial and final land-use types, respectively; T is the time period of the study; LU_{i-j} denotes the absolute value of the area converted from land-use type i to land-use type j; and LU_i is the area of land-use type i at the beginning of the study.

A land-use transfer matrix was used to characterise the structure of regional land-use change and the direction of change across categories [39]. The formula used is as follows:

$$S_{ij}=\begin{vmatrix} S_{11} & \cdots & S_{1n} \\ \vdots & \cdots & \vdots \\ S_{n1} & \cdots & S_{nn} \end{vmatrix} \tag{7}$$

where S is the area, n is the number of land-use types, and i and j denote the land-use types at the beginning and end of the study, respectively.

3. Results

3.1. SWAT Model Suitability Evaluation

Following the input of the DEM data, the SWAT model automatically extracts river networks and generates relevant watershed information. Based on the attributes of soil, land use, and slope (all with a threshold setting of 10%), the FRB was divided into 45 sub-watersheds and 585 hydrological response units. Subsequently, the 24 parameters of the catchment were ranked for sensitivity, and the optimum values were determined in the SWAT-CUP using the SUFI-2 algorithm (see Table 3). The parameters in the hydrological model, VCH_K2.rte, R__SOL_AWC(..).sol, and V__GW_REVAP.gw, which have a large impact on runoff, control processes such as water flow, evaporation, infiltration, and recharge, indicating that groundwater infiltration and flow had a great impact on hydrological processes in the watershed.

Table 3. Calibration parameters of the SWAT model in the Fenhe River Basin, ranked in descending order of sensitivity, and the optimal values of the parameters.

Rank	Parameter	Description	t-Stat	*p*	Fitted Value
1	V_CH_K2.rte	Effective hydraulic conductivity in main channel alluvium	−27.84	0.00	13.57
2	R__SOL_AWC(..).sol	Available water capacity of the soil layer	4.37	0.00	0.058
3	V__GW_REVAP.gw	Groundwater revamp coefficient	−1.96	0.05	−0.21
4	R__CN2.mgt	SCS runoff curve number for moisture condition II	1.76	0.08	0.077
5	V__ESCO.hru	Soil evaporation compensation factor	−1.72	0.09	0.070
6	V__REVAPMN.gw	Threshold depth of water in the shallow aquifer for revap to occur (mm)	−1.55	0.12	226.48
7	V__GW_DELAY.gw	Groundwater delay (days)	−1.36	0.18	362.86
8	R__HRU_SLP.hru	Average slope steepness	−1.32	0.19	−0.46
9	V__SPEXP.bsn	Exponential coefficient of sediment transport	1.27	0.20	1.44
10	R__SLSUBBSN.hru	Average slope length	1.10	0.27	−0.33
11	R__SURLAG.bsn	Lag coefficient of surface runoff	−0.82	0.41	0.33
12	V__SPCON.bsn	Linear coefficient of sediment transport	0.73	0.47	0.01
13	V__SMFMN.bsn	Minimum melt rate for snow during year	−0.69	0.49	6.18
14	V__SFTMP.bsn	Snowfall temperature	−0.64	0.52	0.64
15	V__ALPHA_BF.gw	Baseflow alpha-factor	−0.60	0.55	0.96
16	V__TIMP.bsn	Temperature drop rate	−0.59	0.56	0.08
17	V__GWQMN.gw	Threshold water level in shallow aquifer for base flow (mm)	−0.50	0.62	0.19
18	V__USLE_P.mgt	USLE soil and water conservation measures factor	0.48	0.63	0.36
19	R__SOL_K.sol	Saturated hydraulic conductivity	−0.23	0.82	−0.33
20	V__CH_N2.rte	Manning's "n" value for the main channel	−0.20	0.84	0.01
21	V__EPCO.hru	Plant uptake compensation factor	−0.15	0.88	0.24
22	R__SOL_BD(..).sol	Moist bulk density	−0.03	0.97	1.88
23	R__OV_N.hru	Manning's n value for overland flow	0.02	0.98	0.08
24	R_CANMX.hru	Maximum canopy storage	0.00	1.00	−0.51

Figure 4 shows a time-series comparison of the observed and simulated values based on the SWAT model for the calibration and validation periods. The results indicate that the SWAT model accurately reproduced the dynamics of flow in the basin, although some deviations were observed in the simulation of peak flows (e.g., some of the peaks were not fully captured throughout the calibration period). The deviation in simulating peak flows is primarily due to the model's temporal resolution, as SWAT is calibrated using daily data, which may not fully capture the rapid changes associated with extreme, short-duration rainfall events. The NSE values exceeded 0.7, and R^2 was between 0.7 and 0.8 during the calibration and validation periods, indicating that the model is effective for the analysis and prediction of hydrological processes in the FRB and that the results are credible.

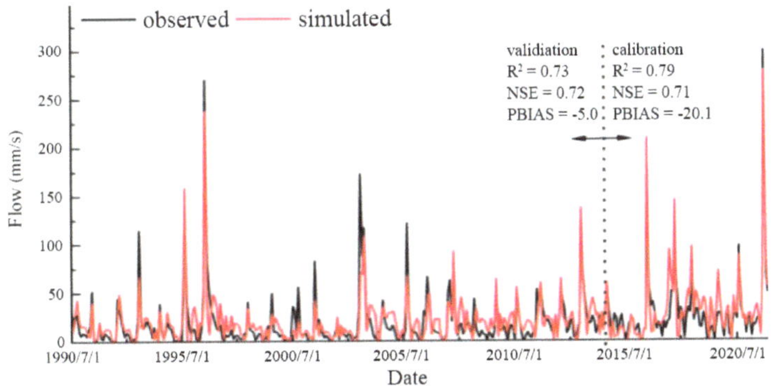

Figure 4. Observed and SWAT-simulated monthly stream flow for the calibration (January 2014–December 2022) and validation (January 1990–December 2014) periods in the Fenhe River Basin, Shanxi Province, China.

3.2. Analysis of PLES Temporal and Spatial Changes

3.2.1. Spatial and Temporal Distribution of and Changes in PLES

Figures 5 and 6 show the distribution of PLES in the FRB in 1990, 2000, 2010, and 2020, as visualised using ArcGIS 10.5 (Figure 6 is specific to secondary land classes). The primary land-use types in the basin were ecological and production spaces, accounting for 52.8% and 40.9% of the basin area, respectively. The ecological space was distributed in an enclosed figure eight shape, with a ring in the north and a semi-ring in the south. From 1990 to 2020, the ecological space gradually decreased, with a total decrease of 632.52 km^2. In particular, the forest area first decreased and then increased, which was related to the national policy of returning farmland to forest that was implemented in 2000. However, the grassland area gradually decreased from 10,135 km^2 in 1990 to 9408 km^2 in 2020, and this condition mainly occurred at the edge of agricultural production land. Meanwhile, production land includes agricultural and industrial production lands, and was mainly concentrated in Taiyuan Basin in the north and Linfen Basin in the south, both of which are dense urban areas. The agricultural production land area was larger, and its proportion gradually decreased from 41.86% to 39.38% from 1990 to 2020. In contrast, industrial production land increased from 91 km^2 in 1990 to 610 km^2 in 2020, representing an increase of approximately seven times. Living space consists of urban and rural residential lands. As shown in Figure 5, urban living space was block-shaped and concentrated around the provincial capital of Taiyuan City and the counties, whereas rural living space was more dispersed and distributed in a point-like pattern, with both areas showing an upward trend.

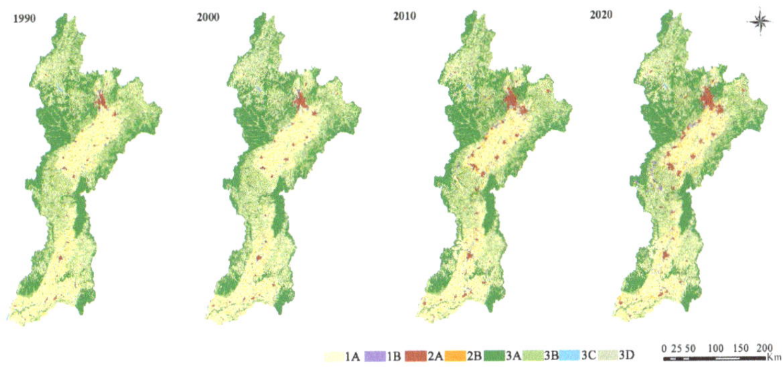

Figure 5. Development of production–living–ecological space (PLES) over 1990–2020 in Fenhe River Basin.

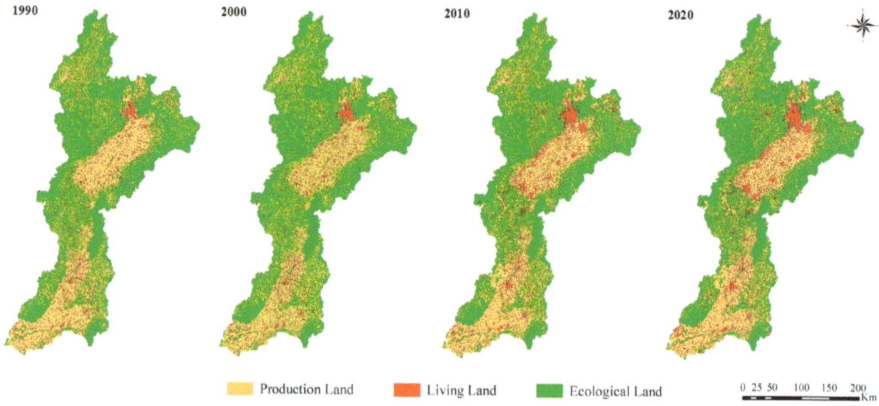

Figure 6. Secondary class distribution of PLES in Fenhe River Basin.

3.2.2. Changes in PLES Land-Use Dynamics

As shown in Table 4, agricultural production land in the FRB experienced negative growth between 1990 and 2020, and the rate of decline slowed after 2010. Industrial production land showed the greatest change, reaching its highest value between 2000 and 2010. Both urban and rural residential lands increased; however, urban residential land increased faster than rural residential land owing to urbanisation. Changes in ecological space were irregular and broadly aligned with the implementation of national or local ecological policies. The combined land-use dynamics of the PLES secondary classification increased from 1990 to 2020, with an average value of 0.46%, and changes between different land-use categories were generally frequent and dramatic. From 1990 to 2000, the transformation between PLES land-use types was relatively slow (0.07%), and the human–land relationship remained fairly consistent. From 2000 to 2010, change in land use was significant (1.73%), the human–land relationship was unstable, and the spatial pattern of PLES was unstable. From 2010 to 2020, the change in land use declined to 0.69%, the changes in PLES land-use types were consistent, human–land relationship became more stable, and development became more balanced.

Table 4. Single PLES land-use dynamic in Fenhe River Basin.

Type	1990–2000	2000–2010	2010–2020	1990–2020
1A	−0.0322%	−0.4332%	−0.1342%	−0.1973%
1B	1.7920%	34.6945%	2.7017%	18.9809%
2A	3.0315%	8.0249%	1.3801%	5.5770%
2B	0.2276%	3.7893%	0.6201%	1.6592%
3A	−0.0146%	0.2043%	−1.6912%	−0.5113%
3B	−0.0770%	−0.5880%	1.8923%	0.3690%
3C	0.0614%	−1.6180%	0.1185%	−0.4889%
3D	1.9673%	−5.0830%	−0.1571%	−1.4027%

3.2.3. PLES Land-Use Transfer Matrix

As shown in Figure 7, the biggest change in PLES was the decrease in agricultural production space. Approximately 14,272 km^2 of agricultural land was transferred to other land-use types (almost half was transferred to living space), accounting for 36% of the total transfer area. From 1990 to 2000, increases in the agricultural production space mainly resulted from the transfer of ecological land, with a much smaller area of agricultural production land converted to forests and grasslands than vice versa, indicating that deforestation and land clearing were serious issues during this period. At the beginning of the 21st century, large areas of agricultural land unsuitable for farming were gradually restored to forests and grasslands, and ecological environmental protection was increasingly emphasised, resulting in a greater balance between inflow and outflow ecological land areas. The largest source of industrial production land was agricultural land, and from 2000 to 2010, agricultural land area converted into industrial land was 276 km^2.

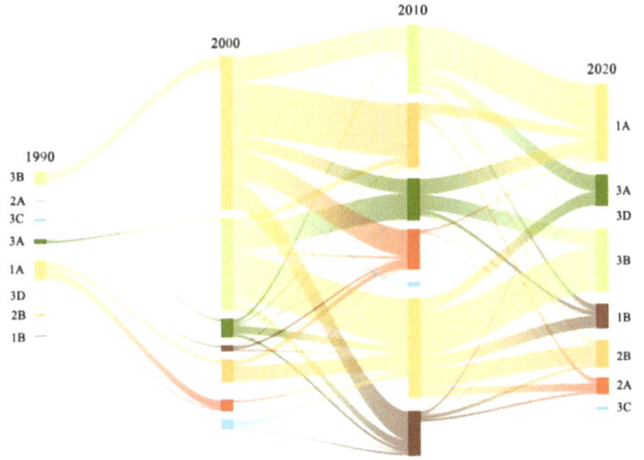

Figure 7. Trajectories of spatial transfer changes of PLES in the Fenhe River Basin. Different coloured trajectory lines show the direction of transfer between land classes, and the thickness of the trajectory lines represents the amount of transformation.

Figure 8 shows spatial changes in the transfer of PLES categories in the FRB. The greatest changes resulted from the transfers of production land to living land and ecological land to production land. The transfer of production land to living land was mainly concentrated in Taiyuan Basin in the north and gradually expanded to Linfen City and its neighbouring cities and counties in the south. This change was related to the national "Rise of Central China Strategy" and the strong support of Shanxi Province. Between 2010 and 2020, there was a significant shift from ecological land to production land; however, a large amount of production land was simultaneously converted to ecological land under national ecological civilisation construction measures.

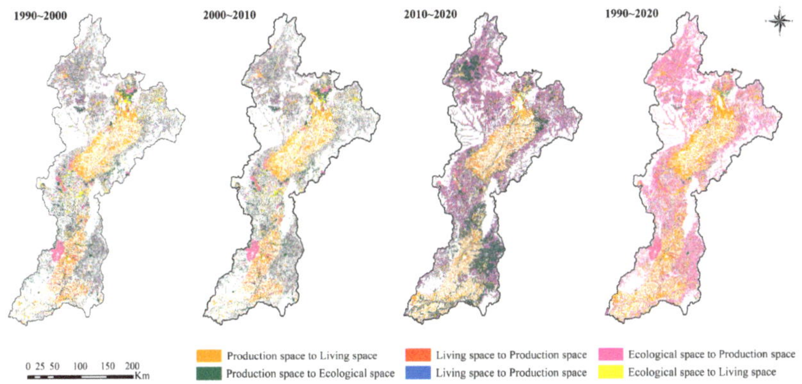

Figure 8. Spatial transfer of PLES land-use types in Fenhe River Basin.

3.3. Impact of PLES Changes on Runoff

3.3.1. Temporal Variation in Runoff

Runoff processes under different PLES scenarios for 1990, 2000, 2010, and 2020 were simulated using the calibrated SWAT model, and the results are shown in Figure 9. The annual runoff in the watershed decreased slightly and then increased substantially, with an overall upward trend, especially during 2011–2015, when the increase in runoff was most significant. Table 5 lists the average, maximum, and minimum runoff, as well as the rate of change in runoff during 1990–2020. Runoff increased by 7.98 mm from 1990 to 2020, with the highest runoff occurring in 1996 and the lowest in 2000.

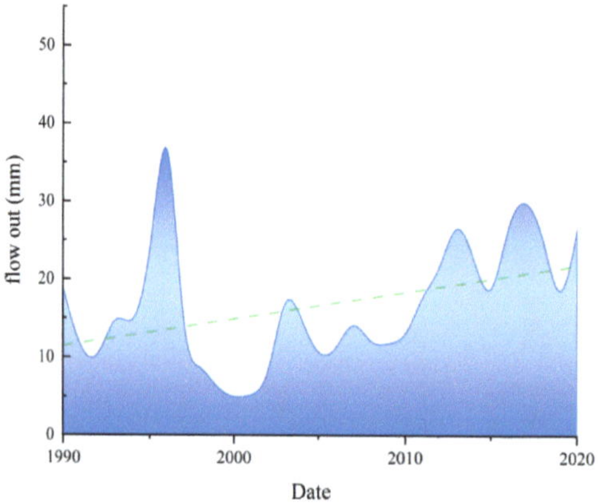

Figure 9. Simulated runoff changes in the Fenhe River Basin between 1990 and 2020. (The green dushed line represents the overall trend of precipitation changes).

Table 5. Runoff analysis for different time periods.

Annual Interval	Average Precipitation	Mean Runoff	Maximum Runoff	Year	Minimum Runoff	Year	Rate of Change
1990–1995	26.67	14.83	21.18	1995	9.00	1992	-
1996–2000	44.35	16.25	37.36	1996	4.75	2000	0.096%
2001–2005	24.29	10.16	20.12	2003	5.12	2001	−0.36%
2006–2010	20.39	11.91	15.72	2007	10.46	2006	0.17%
2011–2015	33.20	19.79	29.00	2013	15.64	2015	0.66%
2016–2020	39.19	23.81	28.17	2016	14.69	2019	0.2%

Based on Table 5, through further analysis, the Pearson correlation coefficient (p) between precipitation and runoff is 0.912, indicating a strong positive correlation between the two. This means that an increase in precipitation generally leads to an increase in runoff, and this relationship is statistically significant (with a p-value of 0.022, which is less than 0.05). This suggests that precipitation is an important driving factor behind changes in runoff. However, despite the strong correlation between precipitation and runoff, we should also consider other factors, such as land-use changes and surface cover, which may have significant impacts on runoff during certain periods. For example, short-term extreme precipitation events may lead to dramatic fluctuations in runoff, even if precipitation increases, as runoff may also be influenced by these additional factors.

To further explore the impacts of PLES changes on runoff, we classified the monthly average runoff by season during 1990–2020. As shown in Figure 10, runoff increased significantly in summer (June–August) and autumn (September–November), with peaks in July and August. This indicates that during the flood season (summer), runoff increased significantly, whereas during winter (December–February) and spring (March–May), runoff was lower. These results reflect the characteristics of the monsoon climate, which brings heavy rainfall in summer, leading to seasonal runoff patterns. Notably, the runoff volume in 2020 increased over several months compared to that in other years and was higher during the flood season. Since meteorological and soil data were controlled for consistency in the modelling, it is reasonable to assume that the increase in runoff was due to PLES changes. Rapid urbanisation, the expansion of industrial and domestic land use, the reduction in vegetation cover, and the decrease in land permeability likely resulted in more precipitation forming runoff rather than infiltrating into the ground.

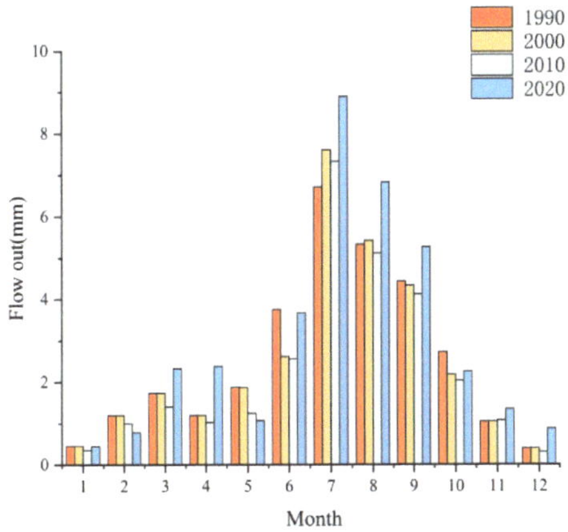

Figure 10. Average monthly runoff under different PLES scenarios.

3.3.2. Spatial Variation in Runoff

To further visualise the response of runoff to PLES changes, we performed a spatial visualisation of the annual average surface runoff and groundwater in the FRB using ArcGIS. As shown in Figure 11, during 1990–2020, the highest values of surface runoff occurred near sub-basin 16 in the north and sub-basins 43, 44, and 45 in the south, which are located in the vicinity of the provincial capital, Taiyuan, and several large cities in the south, consistent with the living and production spaces. Furthermore, the maximum value of surface runoff increased from 1990 to 2020, and the area of the high value range gradually expanded. For example, in sub-basin 16, the modelled surface runoff increased from 2.52 mm in 1990 to 9.91 mm in 2020, which was directly related to the encroachment of production and living spaces into the ecological space due to rapid economic and social development and the expansion of the city. Over 1990–2020, the proportion of ecological land has decreased from 42.8% to 37.2%, while urban living land has increased 2.13 times. The growth rate of rural residential land was approximately 134.28 km^2 per year.

The trend in groundwater change was consistent with the allocation of ecological areas, especially in the northwestern and south-central parts of the country where groundwater storage is higher. Groundwater recharge in agricultural production land was approximately 10.36 mm, which was lower than that in ecological land (16.28 mm), suggesting that ecological spaces, such as forests and grasslands, play a key role in storing and replenishing water resources. In contrast, the contribution of agricultural production land to groundwater decreased due to high-intensity development.

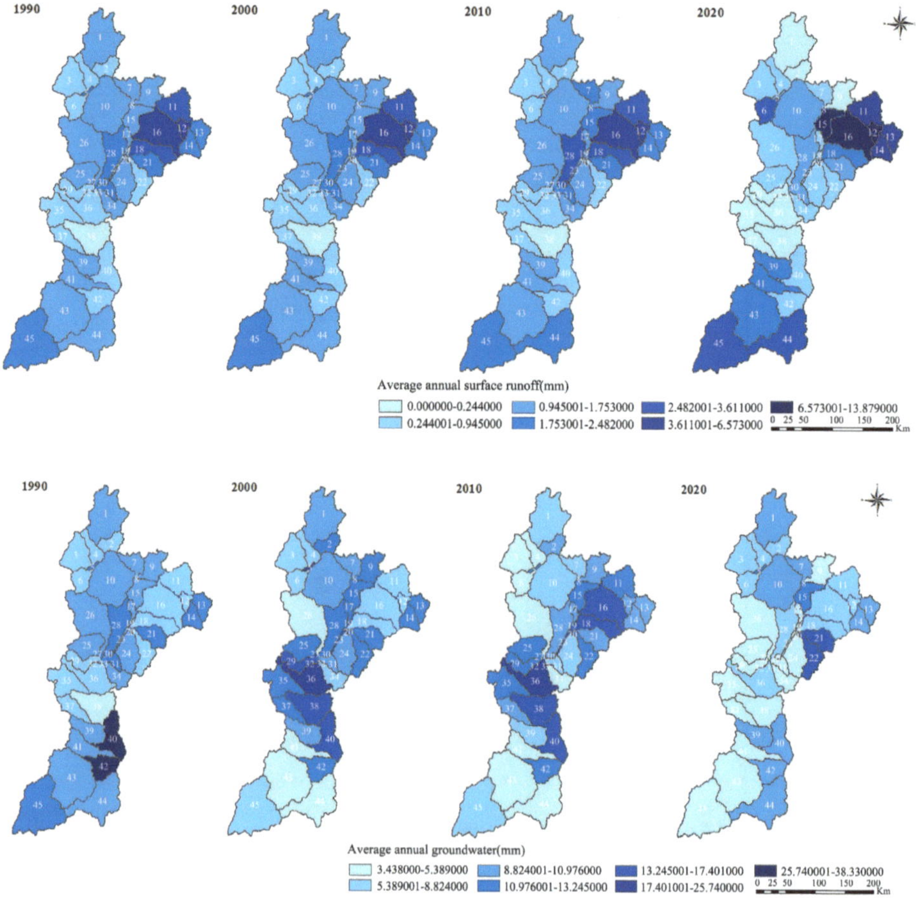

Figure 11. Average annual surface runoff (SURQ) and groundwater (GWQ) under different PLES scenarios.

3.3.3. Quantitative Analysis of the Impact of PLES Changes on Runoff

The trend in groundwater change was consistent with the distribution of ecological space, As illustrated in Figure 12, agricultural production land remains the largest land-use type, with a self-sustained area of about 198.29 km^2. Significant portions were transferred to industrial production and urban living lands, reflecting industrialisation and urbanisation. Industrial land self-sustained area increased to 33.3 km^2, encroaching on some agricultural land, while urban residential land expanded by 19.2 km^2, mainly from agricultural land and grassland. Forest and grassland largely retained their areas but underwent some transformation to urban and industrial land. These changes align with economic development and urban expansion trends.

Figure 13 demonstrates the significant impact of PLES changes on runoff. From 1990 to 2020, as agricultural and ecological land was extensively converted to industrial and urban land, surface runoff in sub-basin 42 increased from approximately 1.5 mm to 4.0 mm, while groundwater recharge decreased from approximately 7.0 mm to 3.5 mm. Similarly, in sub-basin 43, surface runoff increased from approximately 1.0 mm to 3.0 mm, and groundwater recharge declined from approximately 6.0 mm to 3.0 mm. In sub-basin 44, surface runoff increased from approximately 0.5 mm to 2.5 mm, while groundwater recharge decreased from approximately 4.0 mm to 2.0 mm. This trend indicates that the expansion of industrial and urban land, which increases impervious surfaces, significantly reduces rainfall infiltration, leading to higher surface runoff and lower groundwater recharge. Sub-basin 42 experienced the most pronounced changes. The above results provide reliable data for our quantitative study of the relationship between PLES changes and runoff.

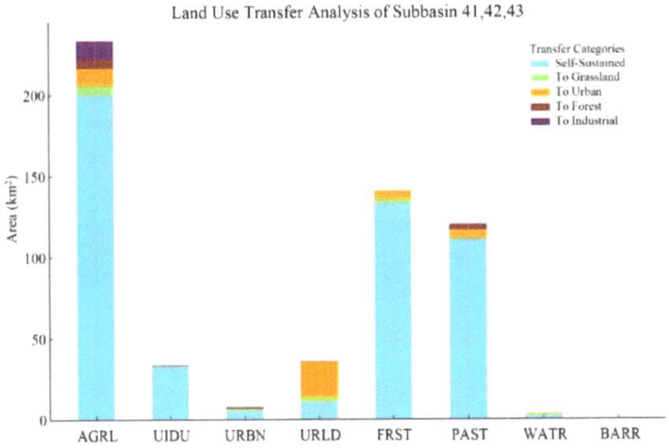

Figure 12. Stacked chart of PLES spatial transfer in sub-basins 42, 43, and 44 from 1990 to 2020.

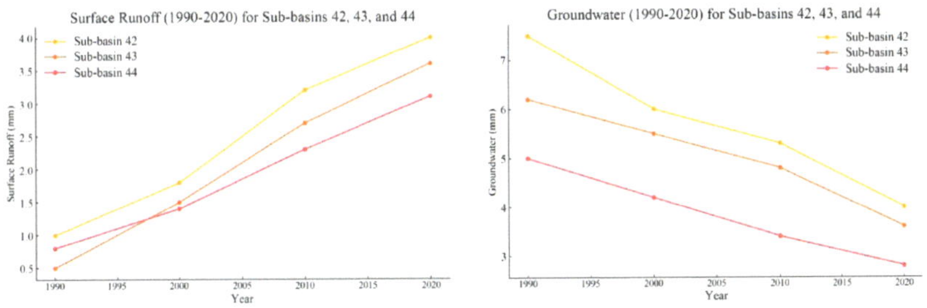

Figure 13. Temporal variations of surface runoff and groundwater in sub-basins 42, 43, and 44.

3.4. *Impact of PLES Changes on Pollutants*

In the Fenhe River Basin, as the PLES changes, especially with the expansion of production and living spaces and the reduction in ecological space, the changes in water quality pollutants are evident. This study used the SWAT model to simulate the trends in pollutants such as nitrogen (N), phosphorus (P), suspended solids (SS), and heavy metals under land-use changes. The results are outlined in Table 6.

Table 6. Changes in pollutant concentrations in the Fenhe River Basin (1990–2020).

Time Period	Nitrogen (mg/L)	Phosphorus (mg/L)	Suspended Solids (mg/L)	Heavy Metal (mg/L)
1990	2.51	0.82	15.2	0.01
2000	2.72	0.95	16.0	0.01
2010	3.08	1.08	18.0	0.02
2020	3.57	1.21	20.0	0.03

Changes in nitrogen and phosphorus pollutants are due to the expansion of production and living spaces caused by rapid industrialisation and urbanisation, which has led directly to the reduction in agricultural production spaces. This in turn has reduced the use of fertilisers and pesticides, thereby decreasing the leaching of nitrogen and phosphorus. However, with the increase in industrial production land, particularly the expansion of living space brought about by urbanisation, nitrogen and phosphorus concentrations increased by 40% and 50%, respectively, between 1990 and 2020. This change is closely related to the increased emissions from urban and industrial development, especially during the period from 2000 to 2010, when the pace of production space expansion accelerated, leading to intensified of nitrogen and phosphorus pollution.

The suspended solids concentration also showed an increasing trend during PLES change related to the expansion of urbanisation and industrialisation, especially construction activities and increased soil erosion. The reduction in agricultural land and the increase of industrial land could have led to soil exposure and erosion, which in turn increased SS loss. For instance, in the Fenhe River Basin, SS concentration increased by approximately 5% from 2010 to 2020. This change was primarily influenced by construction activities and land hardening during the urbanisation process.

Changes in heavy metal pollutants increased gradually with the expansion of industrial production, particularly during the 2000–2010 period. Heavy metal and wastewater emissions led to a steady rise in heavy metal concentrations in water bodies. According to model predictions, the concentration of heavy metals in 2020 increased by approximately 30% compared to 1990, especially in the industrial concentration areas along the Fenhe River.

In summary, the impact of PLES changes on pollutants is significant. The expansion of production and living spaces directly leads to an increase in the concentrations of nitrogen and phosphorus pollutants, while the reduction in agricultural and ecological spaces weakens the self-purification capacity of water bodies, increasing the loss of suspended solids and worsening water quality. Additionally, heavy metal pollution during the industrialisation and urbanisation processes shows a year-on-year increasing trend.

4. Discussion

4.1. *Impact of PLES Changes on Runoff in the Fenhe River Basin*

This study investigated the impact of PLES transformation on runoff in the FRB from 1990 to 2022. The results show that significant changes in runoff have occurred with changes in land-use structure, particularly the expansion of production space and the reduction in ecological space.

Specifically, the increase in production space is closely linked to urbanisation. During the study period, the production space in the FRB increased by approximately 25%, which directly resulted in a 12% increase in the annual average runoff of the basin. Urban construction and agricultural expansion were the primary drivers of this change. For example, the surface runoff depth in urbanised areas was about 20% higher than in ecological areas, while the increase in runoff depth in agricultural lands was approximately 15%. This indicates that the expansion of production space is typically accompanied by an increase in impermeable surface area, which prevents rainwater from infiltrating into the groundwater layer, thereby exacerbating surface runoff [40].

Simultaneously, the reduction in ecological space has also had a significant impact on runoff. According to the results of this study, ecological space in the FRB decreased by about 18% over 1990–2020. The reduction in ecological space led to a decline in vegetation cover, weakening the soil, water conservation, and water retention functions, which further intensified surface runoff. In areas where ecological space was reduced, the increase in runoff depth typically ranged between 10 and 18%, and in heavily developed regions, the increase reached up to 25%. This phenomenon is consistent with findings in both domestic and international studies [41], which suggest that the degradation of ecological space directly impacts the hydrological regulation capacity of a watershed.

In conclusion, the impact of PLES transformation on runoff is significant and complex. The expansion of production and living spaces often leads to the reduction in ecological space, disrupting the original hydrological balance and resulting in increased runoff [42]. According to the SWAT model simulations in this study, the annual runoff in the FRB increased by about 12% from 1990 to 2020, a change closely related to the transformation of PLES spaces. This finding provides a new perspective for understanding the relationship between land-use change and hydrological processes and offers a quantitative basis for watershed water resource management and ecological protection [43].

4.2. Combined Effect of Precipitation Patterns and PLES Changes on Runoff

The impacts of PLES changes and precipitation on runoff are both significant and closely interconnected. PLES changes influence runoff by altering land-use patterns [44]. For example, the expansion of agricultural and industrial land increases surface hardening, reducing water infiltration and thus increasing runoff. The expansion of urban residential areas typically comes with more impervious surfaces and drainage systems, further exacerbating runoff. The intensity and duration of precipitation have a more direct impact on runoff; heavy or prolonged rainfall can rapidly increase runoff within the watershed, especially in urbanised areas. Changes in precipitation volume also affect the balance between groundwater recharge and surface runoff, further influencing the hydrological process. The interaction between PLES changes and precipitation patterns determines the watershed's hydrological response. As rainfall intensity increases, impervious urban or industrial areas may exacerbate runoff, while the protection of ecological land can mitigate this issue. Overall, PLES changes act as a secondary redistribution of precipitation, influencing runoff, and this study aims to better analyse this "redistribution" to achieve sustainable watershed management.

4.3. Limitations and Future Research Prospects

This study has thoroughly demonstrated the impact of PLES changes on runoff, but there are certain limitations. In particular, the study adopts a five-year analysis cycle, which does not fully account for the sudden impact of the extreme rainfall event in 2020 on runoff. Since extreme precipitation can cause significant fluctuations in runoff, future research should consider higher time-resolution data to capture the immediate effects of short-term extreme rainfall events on runoff. In addition, although the SWAT model showed satisfactory simulation results in both the calibration and validation phases, it tended to underestimate the peak flows (this deviation was within acceptable limits), which suggests that the model has a limited ability to accurately simulate extreme hydrological events (e.g., floods or droughts). Thus, the accuracy of the model requires further optimisation. Moreover, the results showed that PLES changes in the FRB had the greatest effect on surface runoff and less impact on other components of the water balance. This aligns with the findings of Kaellsson et al. (2016) [45], who showed that land-use change has a limited effect on mean annual runoff, especially when forest change is below 15–20%, which has almost no effect on annual runoff. However, as we used a realistic scenario of PLES change rather than an extreme land-use change scenario in our simulation, the land-use pattern did not change substantially in the short term, leading to limited changes in runoff. Thus, future studies should further explore the impacts of extreme land-use changes on the water cycle to determine the role of different land-use patterns in the hydrological balance. Identifying the most efficient land-use configurations will support the maintenance of the hydrological balance and ecosystem services.

5. Conclusions

Based on the research perspective of PLES, we systematically analysed the spatial and temporal changes in production, living, and ecological spaces in the FRB from 1990 to 2020. Additionally, we conducted quantitative research on the impacts of PLES changes on runoff. The main conclusions are as follows:

(1) From 1990 to 2020, PLES underwent marked changes in the FRB. Urbanisation led to the continuous expansion of cities, with industrial production and living lands encroaching on agricultural production and ecological lands. However, the conflict between humans and nature over land use was alleviated through national efforts to promote ecological protection and construction. Structural changes in PLES resulted

from the joint action of socioeconomic and natural factors, and the interaction between socioeconomic factors had a strong explanatory power for PLES land-use changes.

(2) The extensive use of reinforced concrete has led to the hardening of the ground and a reduction in water infiltration. Additionally, the area covered by vegetation has decreased, negatively impacting water conservation. Consequently, surface runoff has increased annually. Specifically, from 1990 to 2020, surface runoff in the FRB increased at a rate of 1.06%, garnering the attention of managers. In addition, our findings revealed that the water-holding capacity of ecological land exceeds that of agricultural production land; therefore, the protection of ecological space is crucial for maintaining the hydrological balance of the basin. In future watershed management, the protection and restoration of ecological land should be prioritised to reduce the risk of extreme events, such as floods and droughts.

(3) The impact of PLES changes on pollutants was found to be significant. The expansion of production and living spaces directly leads to an increase in nitrogen and phosphorus pollutant concentrations, while the reduction in agricultural and ecological spaces weakens water bodies' self-purification capacity, consequently increasing SS loss and water quality deterioration. Additionally, heavy metal pollution during the industrialisation and urbanisation processes showed a year-on-year increasing trend. In the future, watershed water quality management should place greater emphasis on the protection and restoration of ecological spaces, as the restoration of ecological spaces is crucial for water quality purification. This will help mitigate the negative impact of pollutants and promote the sustainable development of the watershed.

(4) The uniqueness of this study resides in our application of the PLES perspective to hydrology. This approach yielded comprehensive analyses, interdisciplinary integration, and practical guidance for policy formulation. The PLES concept supports prioritising ecological protection while maintaining a dynamic balance between production and living spaces, thereby supporting sustainable development. This provides a comprehensive framework for understanding the complex relationship between land-use changes and hydrological processes. Our results serve as a valuable reference for watershed management and water resource protection.

Author Contributions: Software, J.Z., J.L., A.K., Z.L., Y.S. and Y.W.; Validation, J.Z.; Investigation, J.Z.; Writing—original draft, J.Z.; Writing—review & editing, A.A.L. and Q.G.; Project administration, Q.G.; Funding acquisition, Q.G. All authors have read and agreed to the published version of the manuscript.

Funding: This research was funded by [Research Project Supported by Shanxi Scholarship Council of China], grant number [2022-111].And The APC was funded by [Qingxia Guo].

Institutional Review Board Statement: Not applicable.

Informed Consent Statement: Not applicable.

Data Availability Statement: Data are contained within the article.

Conflicts of Interest: The authors declare no conflict of interest.

References

1. Abdulkareem, J.H.; Pradhan, B.; Sulaiman, W.N.A.; Jamil, N.R. Prediction of spatial soil loss impacted by long-term land use/land-cover change in a tropical watershed. *Geosci. Front.* **2019**, *10*, 389–403. [CrossRef]
2. Risal, A.; Parajuli, P.B.; Dash, P.; Ouyang, Y.; Linhoss, A. Sensitivity of Hydrology and Water Quality to Variation in Land Use and Land Cover Data. *Agric. Water Manag.* **2020**, *241*, 106366. [CrossRef]
3. Gebremicael, T.G.; Mohamed, Y.A.; Van der Zaag, P. Attributing the Hydrological Impact of Different Land Use Types and Their Long-Term Dynamics through Combining Parsimonious Hydrological Modeling, Alteration Analysis and PLSR Analysis. *Sci. Total Environ.* **2019**, *660*, 1155–1167. [CrossRef] [PubMed]
4. Patil, N.S.; Nataraja, M. Effect of land use land cover changes on runoff using hydrological model: A case study in Hiranyakeshi watershed. *Model. Earth Syst. Environ.* **2020**, *6*, 2345–2357. [CrossRef]

5. Wang, Y.; Li, X.; Zhang, F.; Wang, W.; Xiao, R. Effects of Rapid Urbanization on Ecological Functional Vulnerability of the Land System in Wuhan, China: A Flow and Stock Perspective. *J. Clean. Prod.* **2020**, *248*, 119284. [CrossRef]
6. Jiang, X.; Zhai, S.; Liu, H.; Chen, J.; Zhu, Y.; Wang, Z. Multi-Scenario Simulation of Production-Living-Ecological Space and Ecological Effects Based on Shared Socioeconomic Pathways in Zhengzhou, China. *Ecol. Indic.* **2022**, *137*, 108750. [CrossRef]
7. Chim, K.; Tunnicliffe, J.; Shamseldin, A.Y.; Bun, H. Assessment of land use and climate change effects on hydrology in the upper Siem Reap River and Angkor Temple Complex, Cambodia. *Environ. Dev.* **2021**, *39*, 100615. [CrossRef]
8. Xia, N.; Hai, W.; Tang, M.; Song, J.; Quan, W.; Zhang, B.; Ma, Y. Spatiotemporal Evolution Law and Driving Mechanism of Production–Living–Ecological Space from 2000 to 2020 in Xinjiang, China. *Ecol. Indic.* **2023**, *154*, 110807. [CrossRef]
9. Yang, Y.; Bao, W.; Liu, Y. Coupling Coordination Analysis of Rural Production-Living-Ecological Space in the Beijing-Tianjin-Hebei Region. *Ecol. Indic.* **2020**, *117*, 106512. [CrossRef]
10. Bewket, W.; Sterk, G. Dynamics in land cover and its effect on stream flow in the Chemoga watershed, Blue Nile basin, Ethiopia. *Hydrol. Process. Int. J.* **2005**, *19*, 445–458. [CrossRef]
11. Liao, G.; He, P.; Gao, X.; Lin, Z.; Huang, C.; Zhou, W.; Deng, O.; Xu, C.; Deng, L. Land Use Optimization of Rural Production–Living–Ecological Space at Different Scales Based on the BP–ANN and CLUE–S Models. *Ecol. Indic.* **2022**, *137*, 108710. [CrossRef]
12. Liu, H.; Wang, Z.; Zhang, L.; Tang, F.; Wang, G.; Li, M. Construction of an Ecological Security Network in the Fenhe River Basin and Its Temporal and Spatial Evolution Characteristics. *J. Clean. Prod.* **2023**, *417*, 137961. [CrossRef]
13. Yang, Z.; Zhan, J.; Wang, C.; Twumasi-Ankrah, M.J. Coupling Coordination Analysis and Spatiotemporal Heterogeneity between Sustainable Development and Ecosystem Services in Shanxi Province, China. *Sci. Total Environ.* **2022**, *836*, 155625. [CrossRef] [PubMed]
14. Wang, Z.; Fu, X. Scheme Simulation and Predictive Analysis of Water Environment Carrying Capacity in Shanxi Province Based on System Dynamics and DPSIR Model. *Ecol. Indic.* **2023**, *154*, 110862. [CrossRef]
15. Nourani, V.; Gholinia, A.; Abbaszadeh, P.; Ahmadi, R.; Ke, C.-Q. Unravelling the impact of climate change and anthropogenic activities on streamflow: The benefit of newly developed evapotranspiration data. *Hydrol. Sci. J.* **2024**, *69*, 2026–2043. [CrossRef]
16. Yuan, D.; Du, M.; Yan, C.; Wang, J.; Wang, C.; Zhu, Y.; Wang, H.; Kou, Y. Coupling Coordination Degree Analysis and Spatiotemporal Heterogeneity between Water Ecosystem Service Value and Water System in Yellow River Basin Cities. *Ecol. Inform.* **2024**, *79*, 102440. [CrossRef]
17. Zhang, D.; Zhao, Z.-Q.; Peng, Y.; Fan, B.; Zhang, L.; Li, J.; Chen, A. Sulfur Cycling in the Yellow River and the Sulfate Flux to the Ocean. *Chem. Geol.* **2020**, *534*, 119451. [CrossRef]
18. Hassen, E.E.; Assen, M. Land use/cover dynamics and its drivers in Gelda catchment, Lake Tana watershed, Ethiopia. *Environ. Syst. Res.* **2018**, *6*, 4–13. [CrossRef]
19. Su, Y.; Ma, X.; Feng, Q.; Liu, W.; Zhu, M.; Niu, J.; Liu, G.; Shi, L. Patterns and Controls of Ecosystem Service Values under Different Land-Use Change Scenarios in a Mining-Dominated Basin of Northern China. *Ecol. Indic.* **2023**, *151*, 110321. [CrossRef]
20. Wang, H.; Xu, Y.; Li, C. Coupling Coordination and Spatio-Temporal Pattern Evolution between Ecological Protection and High-Quality Development in the Yellow River Basin. *Heliyon* **2023**, *9*, e21089. [CrossRef]
21. Chai, N.; Yi, X.; Xiao, J.; Liu, T.; Liu, Y.; Deng, L.; Jin, Z. Spatiotemporal Variations, Sources, Water Quality and Health Risk Assessment of Trace Elements in the Fen River. *Sci. Total Environ.* **2021**, *757*, 143882. [CrossRef] [PubMed]
22. Arnold, J.G.; Fohrer, N. SWAT2000: Current Capabilities and Research Opportunities in Applied Watershed Modelling. *Hydrol. Process.* **2005**, *19*, 563–572. [CrossRef]
23. Shi, W.; Huang, M. Predictions of Soil and Nutrient Losses Using a Modified SWAT Model in a Large Hilly-Gully Watershed of the Chinese Loess Plateau. *Int. Soil Water Conserv. Res.* **2021**, *9*, 291–304. [CrossRef]
24. Zhang, H.; Li, T.-T.; Han, J.-J. Quantifying the Relationship between Land Use Features and Intra-Surface Urban Heat Island Effect: Study on Downtown Shanghai. *Appl. Geogr.* **2020**, *125*, 102305. [CrossRef]
25. Moges, E.; Demissie, Y.; Larsen, L.; Yassin, F. Review: Sources of Hydrological Model Uncertainties and Advances in Their Analysis. *Water* **2020**, *13*, 28. [CrossRef]
26. Chen, J. The Impact of Land Cover Change on Watershed Hydrology: A Case Study of the Suo Mo River Basin in the Upper Yangtze River. Ph.D. Dissertation, Institute of Geographic Sciences and Natural Resources Research, Chinese Academy of Sciences, Changchun, China, 2002.
27. Fu, J.-X. Land use function change and its driving force of the "production-living-ecological" space in Fenhe River Basin from 1980 to 2020. *Chin. J. Appl. Ecol.* **2023**, *34*, 1901–1911.
28. Gong, C.; Lyu, F.; Wang, Y. Spatiotemporal Change and Drivers of Ecosystem Quality in the Loess Plateau Based on RSEI: A Case Study of Shanxi, China. *Ecol. Indic.* **2023**, *155*, 111060. [CrossRef]
29. Shen, W.; Li, Y.; Qin, Y.; Cheng, J. Influencing Mechanism of Climate and Human Activities on Ecosystem Health in the Middle Reaches of the Yellow River of China. *Ecol. Indic.* **2023**, *150*, 110191. [CrossRef]
30. Zhang, H.; Wang, B.; Liu, D.L.; Zhang, M.; Leslie, L.M.; Yu, Q. Using an Improved SWAT Model to Simulate Hydrological Responses to Land Use Change: A Case Study of a Catchment in Tropical Australia. *J. Hydrol.* **2020**, *585*, 124822. [CrossRef]
31. Arnold, J.G.; Moriasi, D.N.; Gassman, P.W.; Abbaspour, K.C.; White, M.J.; Srinivasan, R.; Santhi, C.; Harmel, R.D.; Van Griensven, A.; Van Liew, M.W.; et al. SWAT: Model Use, Calibration, and Validation. *Trans. ASABE* **2012**, *55*, 1491–1508. [CrossRef]
32. Li, Z.; Liu, W.-Z.; Zhang, X.-C.; Zheng, F.-L. Impacts of land use change and climate variability on hydrology in an agricultural catchment on the Loess Plateau of China. *J. Hydrol.* **2009**, *377*, 35–42. [CrossRef]

33. Shi, P.; Chen, C.; Srinivasan, R.; Zhang, X.; Cai, T.; Fang, X.; Qu, S.; Chen, X.; Li, Q. Evaluating the SWAT Model for Hydrological Modeling in the Xixian Watershed and a Comparison with the XAJ Model. *Water Resour. Manag.* **2011**, *25*, 2595–2612. [CrossRef]
34. Saha, P.P.; Zeleke, K.; Hafeez, M. Streamflow Modeling in a Fluctuant Climate Using SWAT: Yass River Catchment in South Eastern Australia. *Env. Earth Sci.* **2014**, *71*, 5241–5254. [CrossRef]
35. Wei, S.; Lu, R.; Zhang, L.; Lin, X.; Lu, Y. Transition of production—Living-ecological space and carbon storage effect in Guangxi border area under carbon neutrality goal. *J. Bull. Soil Water Conserv.* **2023**, *43*, 399–408. [CrossRef]
36. Zhang, Y.; Zheng, M.; Qin, B. Optimization of Spatial Layout Based on ESV-FLUS Model from the Perspective of "Production-Living-Ecological": A Case Study of Wuhan City. *Ecol. Model.* **2023**, *481*, 110356. [CrossRef]
37. Dong, J.H.; Zhang, Z.B.; Da, X.J.; Zhang, W.B.; Feng, X.L. Eco-environmental effects of land use transformation and its driving forces from the perspective of "production-living-ecological" spaces: A case study of Gansu Province. *Acta Ecol. Sin.* **2021**, *41*, 5919–5928. [CrossRef]
38. Wang, A.; Liao, X.; Tong, Z.; Du, W.; Zhang, J.; Liu, X.; Liu, M. Spatial-Temporal Dynamic Evaluation of the Ecosystem Service Value from the Perspective of "Production-Living-Ecological" Spaces: A Case Study in Dongliao River Basin, China. *J. Clean. Prod.* **2022**, *333*, 130218. [CrossRef]
39. Duan, Y.; Wang, H.; Huang, A.; Xu, Y.; Lu, L.; Ji, Z. Identification and Spatial-Temporal Evolution of Rural "Production-Living-Ecological" Space from the Perspective of Villagers' Behavior—A Case Study of Ertai Town, Zhangjiakou City. *Land Use Policy* **2021**, *106*, 105457. [CrossRef]
40. Wang, D.; Fu, J.; Xie, X.; Ding, F.; Jiang, D. Spatiotemporal Evolution of Urban-Agricultural-Ecological Space in China and Its Driving Mechanism. *J. Clean. Prod.* **2022**, *371*, 133684. [CrossRef]
41. Chen, Y.; Xu, F. The Optimization of Ecological Service Function and Planning Control of Territorial Space Planning for Ecological Protection and Restoration. *Sustain. Comput. Inform. Syst.* **2022**, *35*, 100748. [CrossRef]
42. Xu, H.; Zhang, F.; Li, W.; Shi, J.; Johnson, B.A.; Tan, M.L. Spatial-Temporal Pattern of Change in Production-Living-Ecological Space of Nanchong City from 2000 to 2020 and Underlying Factors. *Env. Monit. Assess.* **2024**, *196*, 94. [CrossRef] [PubMed]
43. Fu, J.; Bu, Z.; Jiang, D.; Lin, G.; Li, X. Sustainable Land Use Diagnosis Based on the Perspective of Production–Living–Ecological Spaces in China. *Land Use Policy* **2022**, *122*, 106386. [CrossRef]
44. Lihong, L.; Wenyuan, D.; Juan, C.; Wanli, H.; Fangqi, J.; Qiufeng, H. Spatial conflict between ecological- production- living spaces on Pingtan Island during rapid urbanization. *Resour. Sci.* **2017**, *39*, 1823–1833. [CrossRef]
45. Karlsson, I.B.; Sonnenborg, T.O.; Refsgaard, J.C.; Trolle, D.; Børgesen, C.D.; Olesen, J.E.; Jeppesen, E.; Jensen, K.H. Combined effects of climate models, hydrological model structures and land use scenarios on hydrological impacts of climate change. *J. Hydrol.* **2016**, *535*, 301–317. [CrossRef]

water

MDPI

Article

Evaluation of Soil-Water Characteristic Curves for Different Textural Soils Using Fractal Analysis

Chunliu Yang [1,2,3], Jianhua Wu [1,2,3,*], Peiyue Li [1,2,3,*], Yuanhang Wang [1,2,3] and Ningning Yang [1,2,3]

1 School of Water and Environment, Chang'an University, No. 126 Yanta Road, Xi'an 710054, China
2 Key Laboratory of Subsurface Hydrology and Ecological Effects in Arid Region of the Ministry of Education, Chang'an University, No. 126 Yanta Road, Xi'an 710054, China
3 Key Laboratory of Eco-hydrology and Water Security in Arid and Semi-arid Regions of the Ministry of Water Resources, Chang'an University, No. 126 Yanta Road, Xi'an 710054, China
* Correspondence: wjh2005xy@126.com or wujianhua@chd.edu.cn (J.W.); lipy2@163.com or peiyueli@chd.edu.cn (P.L.)

Abstract: The soil-water characteristic curve (SWCC) is an essential tool to determine hydraulic and mechanical properties of unsaturated soils. As an inherent influencing factor, soil texture controls the characteristics of SWCCs. Fractal theory can quantitatively describe the physical characteristics of soil. This study used particle size distribution data and water content data contained in the UNSODA2.0 database to explore the fractal characteristics of 12 soil types with different textures under different matrix suctions. The SWCC fractal model was adopted to characterize the hydraulic properties of soil with various soil textures. The findings revealed that the mass fractal dimensions of particles from these 12 different soil types significantly differed and were closely related to the clay content. Fractal dimension increased with increasing clay content. The fractal dimension established a good relationship between soil structure and hydraulic properties. Fractal analysis can be used to determine the connection between physical properties and soil hydraulic parameters. The estimated results of the SWCC fractal model indicated that it had a good performance regarding the description of SWCCs for the 12 soil textures. The soil structure could be described through fractal dimensions, which can effectively indicate soil hydraulic characteristics. The estimated fractal dimension of this model could be obtained by particle size distribution. Furthermore, using the SWCC fractal model, we found that the SWCC of coarse textured soil changed sharply in the low suction stage and its residual water content was small, and the SWCC of fine textured soil changed gently with a large residual water content. The water retention capacity followed the order clay > silty clay > sandy clay > clay loam > silty clay loam > sandy clay loam > loam > silt loam > sandy loam > silt > loamy sand > sand.

Keywords: soil water; hydraulic property; soil texture; soil particle; unsaturated soil; soil moisture; fractal theory

1. Introduction

Unsaturated soils have been extensively studied in environmental science and geotechnical engineering for revealing their hydraulic and mechanical characteristics [1–6]. Some of the major properties of unsaturated soil could be represented by the soil-water characteristic curve (SWCC), which describes the complex relationship between matrix suction and water content in the soil [7–10]. Scientists have carried out numerous studies on the SWCC to better understand water infiltration, soil erosion, shear strength, and slope stability of soils [11]. For example, Zeng et al. [12] measured the SWCC of disintegrated carbonaceous mudstone (DCM) for highway embankments in southwest China to determine stability problems. Wang et al. [13] established the SWCC model of mine wastes to assess the incidents of tailing dams under conditions of rainfall or flood. Moazeni-Noghondar et al. [14] predicted the soil water conditions after rainfall in eastern Iran by transforming the SWCC from the soil water-suction curve to the suction-time curve to study soil erosion. These studies indicated that the strength and hydraulic properties of soil can be described quantitatively by the SWCC, which plays a crucial role in engineering construction and environmental improvement.

The SWCC of unsaturated soils is greatly influenced by different conditions, such as drying and wetting cycles [15] soil texture [16], soil dry density [17,18], initial water content, particle sizes [19], and pore structure [20,21]. Iyer et al. [22] measured the two finely grained soils' SWCCs (kaolin soil and bentonite soil) under different initial water contents by changing the ratio of water weight to dry soil weight, and found that its SWCCs showed significant differences only when soil suction was below 500 kPa, whereas the SWCCs of soils overlap when soil matrix suction was above 1000 kPa. Li et al. [23], analyzing the SWCCs of 120 soil samples of seven soil types, concluded that the available water content (AW) of finely textured soil was higher, whereas the values for sandy soils were lower. Mahmoodlu et al. [24] employed the van Genuchten model to estimate the mixed sand's SWCC curves with different particle sizes and found that the shape parameters of the SWCC (α) decreased with decreasing porosity, which indicated that sand with a larger porosity had a smaller air-entry value. Shen et al. [25] explored SWCCs of calcareous silty sand, and they showed some differences in the boundary effect stage (0–5 kPa) and the residual stage (10–750 kPa), whereas the SWCCs were similar in the transition stage. Fine particle content affects SWCCs by mainly controlling the number of soil pores. When fine particle content increases, the pore size formed in the soil decreases, but the porosity may increase [26]. Zhu et al. [27] researched the impacts of soil particle size on SWCCs for the Gaomiaozi (GZM) bentonite pellet-contained material, a Na-bentonite found from Inner Mongolia, China, under different suction ranges. Based on their results, the SWCC was less affected by soil particle distribution under high suction, but was more sensitive to particle size under low suction. These findings showed that the SWCCs of soils with different textures are significantly different, which will further affect the soil water retention capacity. Soil texture was universally divided into 12 types based on the weight ratio of clay fraction (<2 μm), silt fraction (2–50 μm), and sand fraction (>50 μm) [28], and any soil sample can be classified into one of these types. However, there are few studies on the differences of the SWCCs for all soil textures, and identifying the differences of SWCC performance in all soil textures can help in understanding the variations in the water retention capacity of soils and the problems of soil pollution and soil stability.

Considering the importance of SWCCs in geotechnical engineering, the SWCC has been estimated using experimental methods and mathematical models [29–32]. For example, Matlan et al. [33] evaluated the performance of four empirical models (Van Genuchten model, Kosugi model, Brooks and Corey model, and Modified Gardner model) to forecast the SWCCs of sand, clay, and silt. Soltani et al. [34] developed a new method to determine two parameters describing SWCC changes using soil of different textures, including air-entry value and residual matrix suction. Tao et al. [35] optimized two prediction methods of the SWCC for sandy loam with different dry densities considering soil water evaporation, reducing the workload of measuring SWCCs. However, these above empirical models contain parameters whose physical meanings are unclear. In 1967, Mandelbrot [36] proposed the fractal theory to specifically describe complex and irregular objects. The application of this theory solves the problem of the unclear physical meanings of parameters in the empirical models. In this theory, soil is considered a complex natural fractal medium with self-similarity, and its particles and pores both have fractal characteristics [37–40]. Fractal dimension, a parameter characterizing the fractal property of an object, can describe soil structure quantitatively, such as pore and particle of soil [41–46]. Furthermore, fractal dimension can also establish a link between soil structure and soil-water characteristic curves (SWCC) [47,48], thus providing a convenient method to settle the problem of time-consuming and laborious SWCC measurement. For example, Tyler and Wheatcraft [49] deduced a fractal model of SWCC to represent the volumetric distribution of soil pores through the Sierpinski carpet. Bird et al. [50] established the relationship of the SWCC with the fractal characteristics of soil particle mass distribution. Huang et al. [51] analyzed the fractal dimensions of the soil structure with different textures based on fractal theory and established an equation for the connection between the fractal dimensions of soil particle mass and the clay content. Using the SWCC fractal model, Tao et al. [52] conducted a study on the impacts of clay content and initial porosity on hydraulic properties of soil and provided a theoretical explanation for the effects of the two elements on the SWCCs. Zhou et al. [53] provided a straightforward physical model using fractal theory to evaluate the effect of initial porosity on the soil-water characteristic curve and the hydraulic conductivity of unsaturated soil. Zhang et al. [54] applied this theory to loess and reported that the fractal dimension of soil particles closely depended on soil water distribution, which was an effective tool to explain soil hydraulic characteristics in the loess region. The application of fractal theory in soil property studies simplifies the quantitative expression of soil complex structures (particles and pores) and provides a reliable parameter for accurately predicting soil water holding capacity and for revealing soil-water transport law by establishing the relationship between soil's physical properties and hydraulic properties. In general, fractal theory is an effective analytical method in soil property studies and should be worthy of extensive application.

The water retention capacity and hydraulic parameters of soil are different for soil with different textures. Traditionally, the hydraulic properties can be expressed directly by the SWCC, the determination of which, however, is time-consuming. The SWCC can also indirectly reflect the property of soil structure distribution (pore and particle). Considering this, the fractal theory, which is a simple tool to combine soil structure with hydraulic properties, was used in this research to study the SWCCs for various types of soil, hoping to finding an easier way to quantify the hydraulic properties of soil. In this sense, the purposes of this article are (1) to analyze fractal properties of soil particles to study the structural differences of soil with various textures, and (2) to establish an SWCC fractal model to study the difference of hydraulic characteristics for soil with various textures. The research could provide a basis for accurate prediction of SWCCs and contribute to solving some engineering problems and the management of the ecological environment.

2. Materials and Methods

2.1. Fractal Theory

Fractal theory can quantitatively characterize soil structural characteristics, such as soil particles and pore structures, by the fractal dimension (D), according to Mandelbrot [36]. There are two methods to figure out the fractal dimension of soil particles, including the particle mass fractal dimension and the particle volume fractal dimension. Generally, soil particle mass can be obtained easily and is therefore generally used for calculation. Fractal objects have two main features. One is its magnitude $M(L)$, in which the measured scale L obeys the following scaling relationship:

$$M(L) \sim L^{D_f} \tag{1}$$

In the above equation, D_f is the fractal parameter, $M(L)$ can represent the mass, volume, or length of a fractal object, and L is the scale.

The second feature is that the cumulative number N can be expressed by its size distribution, using the following relation:

$$N(L \geq \lambda) = \left(\frac{\lambda_{\max}}{\lambda}\right)^{D_f} \tag{2}$$

For soil particles, λ and $\lambda_{\max}$ are the grain size and the maximum particle size, respectively. Following Turcotte [55], a fractal model of the soil particle number was proposed to characteristic soil structure, which can be shown as follows:

$$N(> R) = CR^{-D_1} \tag{3}$$

In Equation (3), N (> R) indicates the number of particles with grain size greater than R, C is a constant dependent upon the soil properties, and D_1 is the particles' fractal dimension of soil particles' number. If the sum number of soil particles of different particle sizes and each particle size are obtained, the fractal dimension of the soil particle size can be solved by using Equation (3). However, it is rather difficult to obtain the number of each soil particle, because it is often subject to large error in the actual operation process.

Generally, the particle size distribution and its cumulative percentage curve can be used to represent a soil's particle size distribution. Tyler and Wheatcraft [56] believed that the distribution of soil particle mass could account for fractal dimension in the model, which is easy to obtain, to derive a fractal model of particle mass distribution.

Mandelbrot [36] gave the fractal formula for particle surface area, in which the particle size is larger than a certain scale R:

$$A(r > R) = C_a\left[1 - \left(\frac{R}{\lambda_a}\right)^{2-D_1}\right] \tag{4}$$

In this formula, A ($r > R$) is the particle surface area with a particle size greater than R, R is particle size, C_a and λ_a are both constants related to soil particle shape and size, and D_1 is the same as in Equation (3). The formula indicates that area A tends to be constant as the particle size decreases. By introducing it into a three-dimensional domain, we can obtain the expression of soil particle size and particle volume as follows:

$$V(r > R) = C_V\left[1 - \left(\frac{R}{\lambda_V}\right)^{3-D_1}\right] \tag{5}$$

where, C_V and λ_V describe the particle shape and size. If we assume soil density ρ being a constant, the particle mass with particle size greater than R is expressed as follows:

$$M(r > R) = \rho C_V\left[1 - \left(\frac{R}{\lambda_V}\right)^{3-D_1}\right] \tag{6}$$

Therefore, the total soil particle mass is:

$$M_T = M(r > 0) = \rho C_V\left[1 - \left(\frac{0}{\lambda_V}\right)^{3-D_1}\right] = \rho C_V \tag{7}$$

The combination of Equations (5) and (6) can yield:

$$\frac{M(r > R)}{M_T} = 1 - \left(\frac{R}{\lambda_V}\right)^{3-D_1} \tag{8}$$

Here, λ_V can be calculated through the upper limit of the maximum particle scale R_L, and when R is equal to R_L, we could get M ($r > R)/M_T = 0$, and λ_V equals R_L. Consequently, Equation (8) could also be expressed by:

$$\frac{M(r > R)}{M_T} = 1 - \left(\frac{R}{R_L}\right)^{3-D_1} \tag{9}$$

$$\frac{M(r < R)}{M_T} = \left(\frac{R}{R_L}\right)^{3-D_1} \tag{10}$$

Equation (10) is the fractal model based on particle mass distribution established by Tyler and Wheatcraft [56], and the D_1 in Equation (10) is the mass particle fractal dimension. Through the linear fitting calculation of Equation (10), taking M ($r > R)/M_T$ and R/R_L

as the vertical and horizontal coordinates, respectively, the mass fractal dimensions with different soil textures can be obtained.

2.2. Fractal Model of the SWCC

Many fractal models for understanding SWCCs were established. A widely used SWCC fractal model was derived by Bird [50] using the soil particle mass distribution and combined the pore-solid fractal (PSF) model with the Young–Laplace equation. The Pore-Solid Fractal (PSF) model was established by Perrier et al. [57], which was a generalized fractal model describing soil particle mass distribution. This model describes soil complexity and divides the soil into three parts, i.e., pore set (P), solid set (S), and fractal set (F). The proportions of these three parts satisfy the relation of $p + s + f = 1$. Assuming that there is a fractal object with a characteristic length of L, the fractal set (F) consists of N_f sub-regions of size $r_{f(1)}$. Correspondingly, the pore set (P) and the solid set (S) are also made up of N_p sub-regions of size $r_{p(1)}$ and Ns sub-regions of size $r_{s(1)}$, respectively. Based on the assumption, Bird et al. [50] presented a fractal model to predict SWCC as follows:

$$\theta = \Phi - \frac{p}{p+s}\left\{1 - \left(\frac{h}{h_{\min}}\right)^{D_2-d}\right\} \quad (11)$$

where, Φ is soil total porosity, D_2 is fractal dimension, h is matrix suction, and $h_{\min}$ is air-entry value of soil. θ and θ_s represent the volumetric water content and saturated volumetric water content, respectively.

$$\Phi = \frac{p}{p+s} \quad (12)$$

Consequently, Equation (11) suits the three dimensions:

$$\theta = \theta_s\left(\frac{h}{h_{\min}}\right)^{D_2-3} \quad (13)$$

Equation (13) is a generalized fractal model of SWCCs according to the PSF model. D_2 is the fractal dimension of mass or volume, and the other parameters have the same meanings as Equation (11). Compared to other complex SWCC models, such as the Van Genuchten model and the Fredlund & Xing model [58,59], this fractal model is simple in its expression form. The model is fully consistent with the empirical Brooks-Corey model [30] and the Campbell model [60]. The advantage of this model is that it gives clear physical meaning to the fitting parameters and closely links the soil's basic physical properties with its hydraulic properties.

Three main parameters are important in Equation (13), namely saturated volumetric water content (θ_s), fractal dimension (D_2), and air-entry value ($h_{\min}$). The UNSODA 2.0 database provided the values of saturated volumetric water content (θ_s) for different types of soil, and the fractal dimension (D_2) could be theoretically computed using Equation (10). The air-entry value ($h_{\min}$) is also researched by many researchers, and refers to the difference value between the water pressure and air pressure in soil when the maximum pore starts to drain water [61]. It is a threshold on the SWCC and can display soil pore structure features and water infiltration characteristics in soil [62]. The determination of soil air-entry value is of important meaning to explore the permeability of unsaturated soil. Soils with different textures have different air-entry values; for example, the air-entry value of clay is larger than sand [63]. Rawls et al. [64] provided a series of typical air-entry values for soils with different textures (Table 1).

Table 1. Empirical values of the air-entry values for all soil textures.

Soil Texture	Clay	Silty Clay	Sandy Clay	Clay Loam	Silty Clay Loam	Sandy Clay Loam	Loam	Silt Loam	Sandy Loam	Silt	Loamy Sand	Sand
Value	37.30	34.19	29.17	25.89	32.56	28.08	11.15	20.76	14.66	20.00	8.69	7.26

2.3. Data Sources

The U.S. Salinity Laboratory has developed a database of UNSODA 2.0 on unsaturated soil from many regions of the world (http://www.ussl.ars.usda.gov/, accessed on 10 November 2022). The database has 790 soil samples and provides details about the basic properties and hydraulic properties of each soil sample, including saturated hydraulic conductivity (K_s), particle size distribution data, and volumetric water content (θ) under different pressure (h) values [65]. Soil particle size distribution (PSD) was measured via dry sieving, and the relationship between volumetric water content (θ) and pressure head (h) was measured using pressure plate extractors. As per the American Soil Texture Classification Standard, the soil types in the database can be divided into sand (0.05–2 mm), silt (0.002–0.05 mm), and clay (< 0.002 mm) [66]. Therefore, these 790 soil samples from the UNSODA 2.0 database can be classified into 12 types of soil textures, namely sandy clay, silty clay, silty clay loam, sandy clay loam, sand, loamy sand, sandy loam, silt, silt loam, clay loam, loam, and clay.

In this research, 357 soil samples covering 12 soil textures from the UNSODA 2.0 database were randomly selected to study the fractal characteristics of soils of different textures. Moreover, the SWCC fractal model was established by using the water retention data (θ–h) of these 357 samples contained in the database. To do this, the mass fractal dimension (D_1) of each soil texture was initially obtained by using soil particle distribution data and the logarithmic fitting of Equation (10). Subsequently, the changes in the SWCC with different soil textures were analyzed using the above same data (θ–h) and the SWCC fractal model.

For the SWCCs for different soil textures, we used statistical parameter (R^2) to quantify the performance of this SWCC fractal model, and the expression of goodness of fitting (R^2) could be shown as follows:

$$R^2 = \left[\frac{\sum_{i=1}^{n} (\theta_i - \overline{\theta}) \left(\theta_i' - \overline{\theta}'\right)}{\sqrt{\sum_{i=1}^{n} (\theta_i - \overline{\theta})^2} \sqrt{\sum_{i=1}^{n} \left(\theta_i' - \overline{\theta}'\right)^2}} \right]^2 \tag{14}$$

where, n is the sample number, θ_i is measured soil volumetric water content, θ_i' is estimated soil water content, $\overline{\theta}$ is average of measured soil water content, and $\overline{\theta}'$ is the average of estimated soil water content.

3. Results and Discussion

3.1. Fractal Dimension of Different Textural Soil

The results of the fractal dimensions for soils with different textures, obtained using the particle mass distribution data of 357 soil samples, were shown in Figure 1 and Table 2. There was a good linear correlation between log (M ($r < R$)/M_T) and log(R/R_L), and we can know that the value goodness of fitting (R^2) varied from 0.6503 for silt to 0.9120 for sandy clay loam (Table 2). Figure 1 shows that the slopes of the fitting lines (k) for soil with different soil textures were different. The mass fractal dimensions (D_1) of soils with different textures were calculated by D_1 = 3-k, and the mass fractal dimension (D_1) decreased as the slope of this fitting line increased.

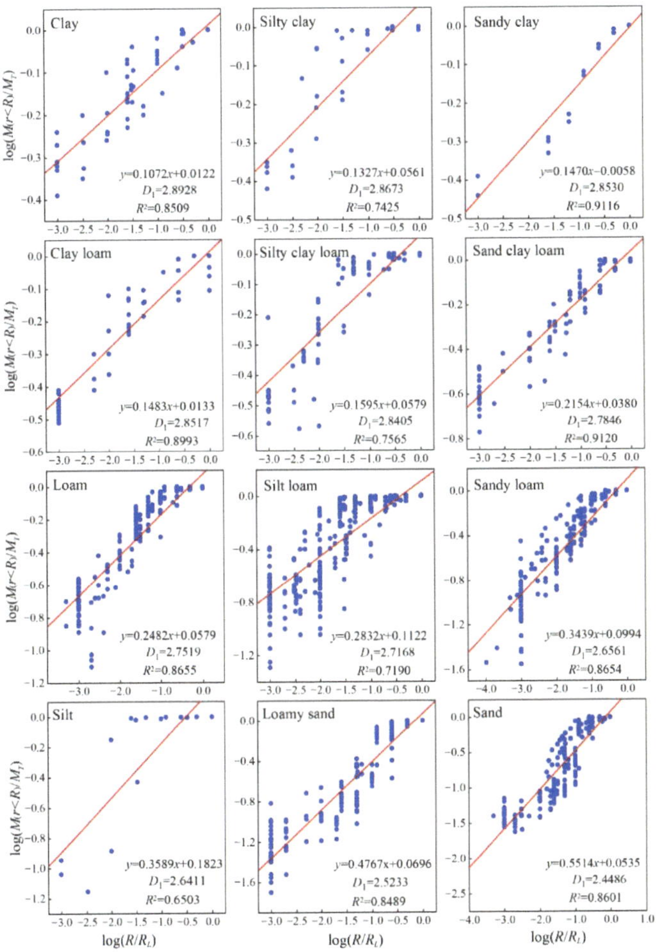

Figure 1. Fitting relationships between log $(M(r < R)/M_T)$ and $\log(R/R_L)$ for all soil textures in UNSODA 2.0.

Table 2. Fractal dimension (D_1) values of soils with different textures.

Soil Textures	Number	Slope	Fractal Dimension	Clay Content	R^2
Clay	15	0.1072	2.8928	51.01	0.8509
Silty clay	12	0.1327	2.8673	40.33	0.7425
Sandy clay	3	0.1470	2.8530	39.13	0.9116
Clay loam	14	0.1483	2.8517	37.32	0.8993
Silty clay loam	20	0.1595	2.8405	36.65	0.7565
Sandy clay loam	24	0.2154	2.7846	24.60	0.9120
Loam	50	0.2482	2.7519	21.49	0.8655
Silt loam	78	0.2832	2.7168	18.59	0.7190
Sandy loam	50	0.3439	2.6561	13.16	0.8654
Silt	3	0.3589	2.6411	9.10	0.6503
Loamy sand	34	0.4767	2.5233	6.72	0.8489
Sand	53	0.5976	2.4024	3.02	0.7875

As shown in Figure 1, clay has the steepest fitting line with the largest D_1, whereas the sand has the smallest D_1. Overall, the values of D_1 for the soils with different textures ranged between 2.4024 and 2.8928 (Table 2) and followed the order (from largest to the smallest) clay, silty clay, sandy clay, clay loam, silty clay loam, sandy clay loam, loam, silt loam, sandy loam, silt, loamy sand, and sand. Huang and Zhang [41] obtained some similar results by investigating the fractal characteristics with 12 textural soils.

Correlation analysis was applied to study the relationship between D_1 and three particle size fractions (clay content, silt content, and sand content) of 12 different soil textures. As shown in Table 3, the mass fractal dimension (D_1) had different degrees of correlation with three particle contents. Specifically, the value of D_1 was correlated with clay content and silt content positively ($p < 0.01$), while it was correlated with sand content negatively ($p < 0.01$). Particularly, fractal dimension was strongly dependent on clay content. The finding of this research is similar to the results of Millán et al. [67] and Zhao et al. [68]. Millán et al. [67] even explored a linear relationship between the fractal dimensions of soils with different textures and the clay content.

Table 3. Correlation between fractal dimension of soils and soil mechanical composition.

Index	D_1	Clay Content	Silt Content	Sand Content
D_1	1.000			
Clay content	0.943 **	1.000		
Silt content	0.449 **	0.474 **	1.000	
Sand content	−0.678 **	−0.726 **	−0.917 **	1.000

** indicates significant correlation at $p < 0.01$; $n = 357$.

A soil's fractal dimension reflects its capacity to fill pore spaces with soil particles. Smaller pore size is usually observed in soil associated with a higher ratio of finer particles (i.e., clay content), and the soil structure with a higher ratio of finer particles would be more compacted, thus showing a larger fractal dimension [69]. Conversely, a looser soil structure leads to a smaller fractal dimension. Therefore, it can be concluded that the fractal dimension is a representative index for soil particle composition. Some scholars have used the fractal dimension to describe the spatial variability of soil textures and to monitor the degree of soil erosion and desertification [70,71].

3.2. Fractal Modeling of the SWCC

The SWCC fractal model presented in Equation (13) was proposed by Bird [50] to estimate soil water retention data with different textures. Table 4 showed the results of estimated SWCCs with different soil textures. The average goodness of fitting (R^2) ranged from 0.8584 to 0.9419, showing that the SWCC fractal model exhibited a good prediction performance. The estimated fractal dimension (D_2) of soils is identical with that of the mass fractal dimension (D_1) calculated. However, we found that the mass fractal dimension (D_1) is not equivalent to the estimated fractal dimension (D_2), and the D_1 value calculated by using PSD data was slightly lower than the fitted D_2 value (Table 4).

The relative error range of these two fractal dimensions was 1.59%–7.34%, and the mean relative error of 4.09% was less than 5%. Similarly, Huang and Zhan [72], investigating the relationship between mass fractal dimension (D_1) and estimated fractal dimension (D_2), suggested that the D_2 of the SWCC fractal model can be replaced by D_1. In the subsequent research on the differences in hydraulic properties in soils with different textures, the SWCC fractal model expressed in Equation (13) was adopted, and the values of D_1 calculated in Section 3.1 were used as parameters in this model. Considering the applicability of the fractal model in this study, the air-entry values (h_{min}) in Equation (13) were selected from the empirical values recommended by Rawls et al. [64], as shown in Table 1.

Table 4. Average estimated fractal dimension (D_2) and mass fractal dimension (D_1) values for 12 soil textures.

Soil Textures	Number	D_1	D_2	Relative Error (%)	R^2
Clay	15	2.8928	2.9497	1.92	0.8911
Silty clay	12	2.8673	2.9152	1.64	0.9365
Sandy clay	3	2.8530	2.9113	2.00	0.8539
Clay loam	14	2.8517	2.9105	2.02	0.9419
Silty clay loam	20	2.8405	2.8865	1.59	0.9222
Sandy clay loam	24	2.7846	2.8854	3.49	0.9083
Loam	50	2.7519	2.8836	4.56	0.9208
Silt loam	78	2.7168	2.8798	5.66	0.9008
Sandy loam	50	2.6561	2.8650	7.29	0.8786
Silt	3	2.6411	2.8504	7.34	0.8584
Loamy sand	34	2.5233	2.6751	5.03	0.9304
Sand	53	2.4024	2.5712	6.56	0.8625

Overall, we selected 12 representative soil samples to evaluate the differences in SWCCs (Table 5). These 12 representative samples' basic physical properties were similar, with particle densities between 2.54 and 2.78 and porosity between 0.34 and 0.56. None of these physical properties of sandstone were provided in the database. We introduced the three parameters (θ_s, D_1, and $h_{\min}$) of each sample into Equation (13) to obtain the expression of the SWCC and generated the corresponding graph using Origin 8.0. Figure 2 shows the SWCCs for these 12 representative soil samples.

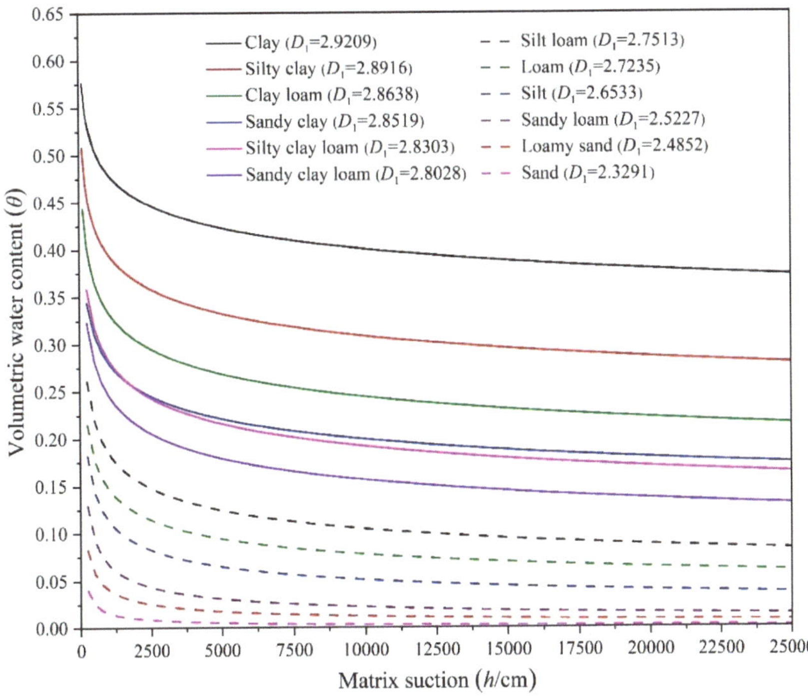

Figure 2. Soil-water characteristic curves (SWCCs) of 12 soil textures predicted using the SWCC fractal model.

Table 5. Basic physical parameters of representative soil samples.

Soil Textures	Code	Particle Density	Porosity	D_1	Clay Content
Clay	2362	2.65	0.56	2.9249	63.0%
Silty clay	3030	2.66	0.50	2.8916	42.0%
Sandy clay	1135	—	—	2.8568	41.0%
Clay loam	2701	2.61	0.34	2.8325	31.2%
Silty clay loam	3110	2.54	0.47	2.8195	31.0%
Sandy clay loam	2630	2.72	0.53	2.8028	25.8%
Loam	2591	2.65	0.43	2.7809	21.8%
Silt loam	2671	2.78	0.49	2.7513	17.3%
Sandy loam	2560	2.61	0.48	2.6567	9.4%
Silt	4670	2.65	0.46	2.4931	9.0%
Loamy sand	2763	2.67	0.43	2.4852	2.8%
Sand	4660	2.56	0.46	2.3291	2.0%

As seen in Figure 2, the volumetric water content decreased gradually as matrix suction increased. In the low matrix suction section ($h < 2500$ cm), the SWCC changed sharply. It indicates that a smaller increase of matrix suction will lead to a rapid decline in volumetric water content. However, the curve changed gently in the high suction stage ($h > 2500$ cm), and the volumetric water content changed slightly with the increase of matrix suction. The change trend of coarse textured soil (sand, loamy sand, sandy loam) was the strongest in all 12 types of soils, followed by medium textured soil (silt, loam, silt loam, sandy clay loam, silty clay loam) and fine textured soil (sandy clay, clay loam, silty clay, clay). Generally, soil water can be retained by both soil adsorption force and capillary force. However, the capillary force usually attracts water molecules in the low suction stage, while water molecules were retained by soil particles mainly by the adsorption force in the high suction stage [73]. Therefore, soil with different textures showed different trends in SWCCs.

There was an apparent inflection point in SWCCs of all soil samples, which is in connection with residual water content [74]. The corresponding volumetric water content and matrix suction at the inflection points of soils with different textures significantly differed. The order of the volumetric water content at the inflection point was consistent with the order of the fractal dimension of different textures. In addition, under the same water content, soil with a larger fractal dimension had a larger matrix suction. Similarly, the soil volumetric water content was higher for a larger fractal dimension of soil under the same suction condition. This suggests that the fractal dimension may control the shape of the SWCCS with fractal characteristics [75]. The SWCC reflects the water retention capacity of soil. In this study, clay had the strongest water retention capacity while sand had the weakest, and the order of other textures was consistent with the fractal dimension order. Rusell [76] put forward that the water retention capacity of soil was mostly achieved through the adsorption of soil particles and that smaller particles have a larger surface area and a better capacity to attract water molecules. Consequently, soil with higher clay content has better water retention ability. Correspondingly, the clay contents in silt, sandy loam, loamy sand, and sand are less than 10%, and the SWCC, therefore, reflects the poor soil water retention capacity. The water retention capacities for the 12 soil textures were consistent with soil's fractal characteristics. It could draw a conclusion that the fractal dimension is sensitive to soil texture and hydraulic properties [77]. The shape and parameter changes of SWCCs with different textures could be determined by the value of fractal dimension. Specifically, soil with larger fractal dimension indicated a higher soil clay content, which corresponded to higher residual water content. This is actually caused by the different compositions and structures of soil particles. The specific surface area of soil particles gradually increased with the thinning of soil particles, but the number of macropores in soil decreased accordingly. Thus, water molecules were retained in small pores and were not easy to seepage. Therefore, fine textured soil usually has higher water retention capacity, which can be indicated using fractal dimension. Furthermore, the SWCC changed gently in the high matrix suction stage. In short, the SWCC changes are obviously due to the difference of soil textures, resulting in different water retention capacities. This finding can be particularly helpful for understanding the changes in hydraulic properties in loess undergoing tremendous human disturbance, such as land creation projects currently underway in China [78], facilitating the sustainable management of water resources and soil erosions in loess areas [79]. In addition, SWCC is also a key tool to understand the infiltration of rainfall and irrigation water, which helps in establishing various hydraulic models for understanding the laws and mechanisms of soil water seepage [80].

4. Conclusions

In the present study, the fractal characteristics of soil particles and the SWCCs of soils with different soil textures were analyzed with the help of the fractal theory. Many soil samples from UNSODA2.0 database containing particle size distribution data and SWCC data were adopted to check the performance of the fractal model. The following conclusions can be achieved:

- The fractal characteristic of soil with different textures was significantly different, and fractal dimension was strongly dependent on the clay content in the soil. The average fractal dimensions of the 12 different textures ranged from 2.4024 to 2.8928. Clay had the largest fractal dimension, whereas sand had the lowest one. The particle size composition of soil will significantly change the fractal dimension of soil. The fractal dimension of finely textured soil was larger than that of medium and coarse textured soil. Correlation analysis also suggests that the fractal dimension of soil particles is intensively related to the contents of clay, silt, and sand ($p < 0.01$). A higher ratio of clay content in the soil can produce a greater mass fractal dimension of soil particles. Fractal theory can quantitatively describe the features of soil particle composition.
- The SWCC of unsaturated soil was strongly dependent on soil texture. The relationship of soil structure with hydraulic properties can be established using fractal analysis. The fractal model representing SWCC has good fitting results for soils of different textures, and the estimated fractal dimension (D_2) in this fractal model can be obtained by particle size distribution. The fitting results of the SWCC fractal model showed that soil with different textures had different changes in SWCC. The soil water retention capacity gradually increased with increasing fractal dimension. In the low suction stage, the changes in the SWCC of coarse textured soil were steeper than those of fine textured soil. Fine textured soil had a larger residual moisture content, while coarse textured soil had a smaller one.

This study will provide a theoretical and practical basis for scholars to explore soil-related issues such as soil water infiltration, pollution transport, and soil stability, and will be helpful for water and soil resource management and protection. It will also deepen the application of fractal theory in soil science, benefiting the development of soil science.

Author Contributions: Research conceptualization, C.Y., J.W., P.L.; data curation, C.Y., Y.W. and N.Y.; methodology, C.Y.; writing—original draft, C.Y.; writing—review and editing, J.W., P.L., Y.W., N.Y., supervision, J.W., P.L. All authors have read and agreed to the published version of the manuscript.

Funding: The financial support from the National Natural Science Foundation of China (42272302, 42072286, 41761144059 and 42090053), the Qinchuangyuan "Scientist + Engineer" Team Development Program of the Shaanxi Provincial Department of Science and Technology (2022KXJ-005), the Fok Ying Tong Education Foundation (161098), and the National Ten Thousand Talent Program (W03070125).

Data Availability Statement: All the raw data are available in the database of UNSODA 2.0 developed by the U.S. Salinity Laboratory, which is available at http://www.ussl.ars.usda.gov/, accessed on 10 November 2022.

Acknowledgments: We are grateful for the useful comments and suggestions rendered by the editors and reviewers, which are essential for us to further improve the quality of the manuscript. Finally, we thank the U.S. Salinity Laboratory for sharing the database of UNSODA 2.0 on unsaturated soils.

The first author is also grateful for the educational and scientific resources provided by Chang'an University, which enabled her to conduct this research.

Conflicts of Interest: The authors declare no conflict of interest.

References

1. Xu, J.; Yang, X. Three-dimensional stability analysis of slope in unsaturated soils considering strength nonlinearity under water drawdown. *Eng. Geol.* **2018**, *237*, 102–115. [CrossRef]
2. Rahardjo, H.; Kim, Y.; Satyanaga, A. Role of unsaturated soil mechanics in geotechnical engineering. *Int. J. Geo-Eng.* **2019**, *10*, 8. [CrossRef]
3. Tian, K.; Yang, A.; Nie, K.Y.; Zhang, H.; Xu, J.; Wang, X. Experimental study of steady seepage in unsaturated loess soil. *Acta Geotech.* **2020**, *15*, 2681–2689. [CrossRef]
4. Wang, C.; Li, S.Y.; He, X.; Chen, Q.; Zhang, H.; Liu, X.Y. Improved prediction of water retention characteristic based on soil gradation and clay fraction. *Geoderma* **2021**, *404*, 115293. [CrossRef]
5. Karakan, E. Comparative Analysis of Atterberg Limits, Liquidity Index, Flow Index and Undrained Shear Strength Behavior in Binary Clay Mixtures. *Appl. Sci.* **2022**, *12*, 8616. [CrossRef]
6. Fiori, A.; Pomarico, I.; Zarlenga, A.; Catani, V.; Leone, G. Extending the Overlay and Index: A Simple Method for Assessing Aquifer Vulnerability in a Combined Vadose Zone—Groundwater Flow System. *Water* **2023**, *15*, 364. [CrossRef]
7. Leong, E.C.; Rahardjo, H. Review of Soil-Water Characteristic Curve Equations. *Geotech. Geol. Eng.* **1997**, *123*, 1106–1117. [CrossRef]
8. Rudiyanto; Sakai, M.; van Genuchten, M.T.; Alazba, A.A.; Setiawan, B.I.; Minasny, B. A complete soil hydraulic model accounting for capillary and adsorptive water retention capillary and film conductivity and hysteresis. *Water Resour. Res.* **2015**, *51*, 8757–8772. [CrossRef]
9. Wu, T.; Wu, G.; Wang, D.; Shi, Z. Soil-hydrological properties response to grazing exclusion in a steppe grassland of the Loess Plateau. *Environ. Earth Sci.* **2013**, *71*, 745–752. [CrossRef]
10. Li, Y.; Liu, Y.; Nie, W.; Ma, X. Inverse Modeling of Soil Hydraulic Parameters Based on a Hybrid of Vector-Evaluated Genetic Algorithm and Particle Swarm Optimization. *Water* **2018**, *10*, 84. [CrossRef]
11. Chen, P.; Wei, C. Numerical procedure for simulating the two-phase flow in unsaturated soils with hydraulic hysteresis. *Int. J. Geomech.* **2016**, *16*, 04015030. [CrossRef]
12. Zeng, L.; Li, F.; Liu, J.; Gao, Q.; Bian, H. Effect of initial gravimetric water content and cyclic wetting-drying on soil-water characteristic curves of disintegrated carbonaceous mudstone. *Trans. Saf. Environ.* **2019**, *1*, 230–240. [CrossRef]
13. Wang, H.; Chen, Q.; Zhang, L.; Li, X. A framework to determine soil-water retention relation for mine wastes and its applications in emergency risk assessment. *Hydrol. Res.* **2021**, *52*, 389–413. [CrossRef]
14. Moazeni-Noghondar, S.; Golkarian, A.; Azari, M.; Lajayer, B.A. Study on soil water retention and infiltration rate: A case study in eastern Iran. *Environ. Earth Sci.* **2021**, *80*, 474. [CrossRef]
15. Wen, T.; Shao, L.; Guo, X.; Zhao, Y. Experimental investigations of the soil water retention curve under multiple drying-wetting cycles. *Acta Geotech.* **2020**, *15*, 3321–3326. [CrossRef]
16. Liu, Q.; Yasufuku, N.; Omine, K.; Hazarika, H. Automatic soil water retention test system with volume change measurement for sandy and silty soils. *Soils Found.* **2012**, *52*, 369–380. [CrossRef]
17. Fattah, M.Y.; Salim, N.M.; Irshayyid, E.J. Determination of the soil-water characteristic curve of unsaturated bentonite-sand mixtures. *Environ. Earth Sci.* **2017**, *76*, 201. [CrossRef]
18. Dang, M.; Chai, J.; Xu, Z.; Qin, Y.; Cao, J.; Liu, F. Soil water characteristic curve test and saturated-unsaturated seepage analysis in Jiangcungou municipal solid waste landfill, China. *Eng. Geol.* **2019**, *264*, 105374. [CrossRef]
19. Zhang, F.; Zhao, C.; Lourenco, S.D.N.; Dong, S.; Jiang, Y. Factors affecting the soil-water retention curve of Chinese loess. *Bull. Eng. Geol. Environ.* **2021**, *80*, 717–729. [CrossRef]
20. Chen, P.; Liu, J.; Wei, C.; Xue, W.; Tian, H. Approach to Rapidly Determining the Water Retention Curves for Fine-Grained Soils in Capillary Regime Based on the NMR Technique. *J. Eng. Mech.* **2017**, *143*, 04017032. [CrossRef]
21. Lu, P.; He, Y.; Zhang, Z.; Ye, W. Predicting chemical influence on soil water retention curves with models established based on pore structure evolution of compacted clay. *Comput. Geotech.* **2021**, *138*, 104360. [CrossRef]
22. Iyer, K.; Jayanth, S.; Gurnani, S.; Singh, D.N. Influence of Initial Water Content and Specimen Thickness on the SWCC of Fine-Grained Soils. *Int. J. Geomech.* **2013**, *13*, 894–899. [CrossRef]
23. Li, D.; Gao, G.; Shao, M.; Fu, B. Predicting available water of soil from particle-size distribution and bulk density in an oasis-desert transect in northwestern China. *J. Hydrol.* **2016**, *538*, 539–550. [CrossRef]
24. Mahmoodlu, M.G.; Raoof, A.; Sweijen, T.; van Genuchten, M.T. Effects of Sand Compaction and Mixing on Pore Structure and the Unsaturated Soil Hydraulic Properties. *Vadose Zone J.* **2016**, *15*, 1–11. [CrossRef]
25. Shen, J.; Hu, M.; Wang, X.; Zhang, C.; Xu, D. SWCC of Calcareous Silty Sand Under Different Fines Contents and dry Densities. *Front. Environ. Sci.* **2021**, *9*, 682907. [CrossRef]
26. Jiang, X.; Wu, L.; Wei, Y. Influence of Fine Content on the Soil-Water Characteristic Curve of Unsaturated Soils. *Geotech. Geol. Eng.* **2020**, *38*, 1371–1378. [CrossRef]

27. Zhu, J.; Su, Z.; Zhang, H. Soil-water characteristic curves and hydraulic conductivity of Gaomiaozi bentonite pellet-contained materials. *Environ. Earth Sci.* **2022**, *81*, 92. [CrossRef]
28. Bockheim, J.G. Genesis of soils with an abrupt textural contrast in the United States. *Catena* **2016**, *137*, 422–431. [CrossRef]
29. Gardner, W.R. Some steady state solutions of the unsaturated moisture flow equation with application to evaporation from a water table. *Soil Sci.* **1958**, *85*, 228–232. [CrossRef]
30. Brooks, R.H.; Corey, A.T. Hydraulic properties of porous medium and Their Relation to Drainage Design. *Trans. ASAE* **1964**, *7*, 26–28. [CrossRef]
31. Castellini, M.; Prima, S.D.; Iovino, M. An assessment of the BEST procedure to estimate the soil water retention curve. *Geoderma* **2018**, *320*, 82–94. [CrossRef]
32. Gao, Y.; Li, Z.; Sun, D.; Yu, H. A simple method for predicting the hydraulic properties of unsaturated soils with different void ratios. *Soil Till. Res.* **2021**, *209*, 104913. [CrossRef]
33. Matlan, S.J.; Taha, M.R.; Mukhlisin, M. Assessment of Model Consistency for Determination of Soil-Water Characteristic Curves. *Arab. J. Sci. Eng.* **2016**, *41*, 1233–1240. [CrossRef]
34. Soltani, A.; Azimi, M.; Deng, A.; Taheri, A. A simplified method for determination of the soil–water characteristic curve variables. *Int. J. Geotech. Eng.* **2019**, *13*, 316–325. [CrossRef]
35. Tao, G.; Lei, D.; Liu, L.; Li, Y.; Zhu, X. Prediction of Soil Water Characteristic Curve Based on Soil Water Evaporation. *Adv. Civ. Eng.* **2021**, *2021*, 6686442. [CrossRef]
36. Mandelbrot, B.B. How long is the coast of Britain? Statistical self-similarity and fractional dimension. *Science* **1967**, *156*, 636–638. [CrossRef]
37. Yu, B. Analysis of Flow in Fractal Porous Media. *Appl. Mech. Rev.* **2008**, *61*, 050801. [CrossRef]
38. Wang, L.; Yang, R.; Zhang, S.; Zhang, G. Fractal features and infiltration characteristics of the soil from different land uses in a small watershed in a rocky, mountainous area in Shandong Province. *J. For. Res.* **2019**, *31*, 1017–1024. [CrossRef]
39. Zhu, J. Impact of fractal characteristics on evaporation and infiltration in unsaturated heterogeneous soils. *Hydrol. Sci. J.* **2020**, *65*, 1872–1878. [CrossRef]
40. Echogdali, F.Z.; Boutaleb, S.; Abioui, M.; Aadraoui, M.; Bendarma, A.; Kpan, R.B.; Ikirri, M.; El Mekkaoui, M.; Essoussi, S.; El Ayady, H.; et al. Spatial Mapping of Groundwater Potentiality Applying Geometric Average and Fractal Models: A Sustainable Approach. *Water* **2023**, *15*, 336. [CrossRef]
41. Huang, G.; Zhang, R. Evaluation of soil water retention curve with the pore-solid fractal model. *Geoderma* **2005**, *127*, 52–61. [CrossRef]
42. Peng, G.; Xiang, N.; Lv, S.; Zhang, G. Fractal characterization of soil particle-size distribution under different land-use patterns in the Yellow River Delta Wetland in China. *J. Soil Sediment.* **2014**, *14*, 1116–1122. [CrossRef]
43. Zolfaghari, A.A.; Taghizadeh-Mehrjardi, R.; Asadzadeh, F.; Hajabbasi, M.A. Soil structure changes due to different land-use practices in the central Zagruos region, Iran. *Arch. Agron. Soil Sci.* **2016**, *62*, 163–175. [CrossRef]
44. Qi, F.; Zhang, R.; Liu, X.; Niu, Y.; Zhang, H.; Li, H.; Li, J.; Wang, B.; Zhang, G. Soil particle size distribution characteristics of different land-use types in the Funiu mountainous region. *Soil Till. Res.* **2018**, *184*, 45–51. [CrossRef]
45. Qiu, S.; Yang, M.; Xu, P.; Rao, B. A new fractal model for porous media based on low-field nuclear magnetic resonance. *J. Hydrol.* **2020**, *586*, 124890. [CrossRef]
46. Dai, L.; Wang, G.; He, Y. Assessing soil fractal and sorting characteristics based on geostatistics and modeling approaches in a typical basin of North China plain. *Earth Sci. Inform.* **2021**, *14*, 819–829. [CrossRef]
47. Russell, A.R.; Buzzi, O. A fractal basis for soil-water characteristics curves with hydraulic hysteresis. *Géotechnique* **2012**, *62*, 269–274. [CrossRef]
48. Niu, X.; Gao, P.; Wang, B.; Liu, Y. Fractal Characteristics of Soil Retention Curve and Particle Size Distribution with Different Vegetation Types in Mountain Areas of Northern China. *Int. J. Environ. Res. Public Health* **2015**, *12*, 15379–15389. [CrossRef]
49. Tyler, S.W.; Wheatcraft, S.W. Fractal processes in soil water retention. *Water Resour. Res.* **1990**, *26*, 1047–1056. [CrossRef]
50. Bird, N.R.A.; Perrier, E.; Rieu, M. The water retention function for a model of soil structure with pore and solid fractal distributions. *Eur. J. Soil Sci.* **2000**, *51*, 51–63. [CrossRef]
51. Huang, G.; Zhang, R.; Huang, Q. Modeling Soil Water Retention Curve with a Fractal Method. *Pedosphere* **2006**, *16*, 137–146. [CrossRef]
52. Tao, G.; Chen, Y.; Kong, L.; Xiao, H.; Chen, Q.; Xia, Y. A Simple Fractal-Based Model for Soil-Water Characteristic Curves Incorporating Effects of Initial Void Ratios. *Energies* **2018**, *11*, 1419. [CrossRef]
53. Zhou, A.; Fan, Y.; Cheng, W.; Zhang, J. A Fractal Model to Interpret Porosity-Dependent Hydraulic Properties for Unsaturated Soils. *Adv. Civ. Eng.* **2019**, *2019*, 3965803. [CrossRef]
54. Zhang, X.; Zhao, W.; Wang, L.; Liu, Y.; Liu, Y.; Feng, Q. Relationship between soil water content and soil particle size on typical slopes of the Loess Plateau during a drought year. *Sci. Total Environ.* **2019**, *648*, 943–954. [CrossRef]
55. Turcotte, D.L. Fractals in geology and geophysics. *Pure Appl. Geophys.* **1989**, *131*, 171–196. [CrossRef]
56. Tyler, S.W.; Wheatcraft, S.W. Fractal scaling of soil particle-size distributions: Analysis and limitations. *Soil Sci. Soc. Am. J.* **1992**, *56*, 362–369. [CrossRef]
57. Perrier, E.; Bird, N.R.A.; Rieu, M. Generalizing the fractal model of soil structure: The pore-solid fractal approach. *Geoderma* **1999**, *88*, 137–164. [CrossRef]

58. Van Genuchten, M.T. A closed-form equation for predicting the hydraulic conductivity of unsaturated soils. *Soil Sci. Soc. Am. J.* **1980**, *44*, 892–898. [CrossRef]
59. Fredlund, D.G.; Xing, A. Equations for the soil-water characteristic curve. *Can. Geotech. J.* **1994**, *31*, 521–532. [CrossRef]
60. Campbell, G.S. A simple method for determining unsaturated hydraulic conductivity from moisture retention data. *Soil Sci.* **1974**, *177*, 311–314. [CrossRef]
61. Pasha, A.Y.; Khoshghalb, A.; Khalili, N. Pitfalls in Interpretation of Gravimetric Water Content-Based Soil-Water Characteristic Curve for Deformable Porous Media. *Int. J. Geomech.* **2016**, *16*, D4015004. [CrossRef]
62. Soltani, A.; Azimi, M.; Boroomandnia, A.; O'Kelly, B.C. An objective framework for determination of the air-entry value from the soil-water characteristic curve. *Results Eng.* **2021**, *12*, 100298. [CrossRef]
63. Tao, G.; Wu, Z.; Li, W.; Li, Y.; Dong, H. Simplified Relation Model of Soil Saturation Permeability Coefficient and Air-Entry Value and Its Application. *Fractal Fract.* **2021**, *5*, 180. [CrossRef]
64. Rawls, W.J.; Brakensiek, D.L.; Saxtonn, K.E. Estimation of Soil Water Properties. *Trans. Am. Soc. Agri. Eng.* **1982**, *25*, 1316–1320. [CrossRef]
65. Nemes, A.; Schaap, M.G.; Leij, F.J. Description of the unsaturated soil hydraulic database UNSODA version 2.0. *J. Hydrol.* **2001**, *251*, 151–162. [CrossRef]
66. Soil Survey Staff. *Keys to Soil Taxonomy*, 12th ed.; USDA: Washington, DC, USA; National Resources Conservation Service: Washington, DC, USA; National Soil Survey Center: Lincoln, NE, USA, 2014.
67. Millán, M.; González-Posada, M.; Aguilar, M.; Domínguez, J.; Céspedes, L. On the fractal scaling of soil data. Particle-size distributions. *Geoderma* **2003**, *117*, 117–128. [CrossRef]
68. Zhao, Y.; Feng, Q.; Yang, H. Soil salinity distribution and its relationship with soil particle size in the lower reaches of Heihe River, Northwestern China. *Environ. Earth Sci.* **2016**, *75*, 810. [CrossRef]
69. Ding, D.; Zhao, Y.; Feng, H.; Si, B.; Hill, R.L. A user-friendly modified pore-solid fractal model. *Sci. Rep.* **2016**, *6*, 39029. [CrossRef]
70. Xia, D.; Deng, Y.; Wang, S.; Ding, S.; Cai, C. Fractal features of soil particle-size distribution of different weathering profiles of the collapsing gullies in the hilly granitic region, south China. *Nat. Hazards* **2015**, *79*, 455–478. [CrossRef]
71. Mohammadi, M.; Shabanpour, M.; Mohammadi, M.H.; Davatgar, N. Characterizing Spatial Variability of Soil Textural Fractions and Fractal Parameters Derived from Particle Size Distributions. *Pedosphere* **2019**, *29*, 224–234. [CrossRef]
72. Huang, G.; Zhan, W. Modeling the soil water retention curve with fractal theory. *Adv. Water Sci.* **2002**, *13*, 55–60. (In Chinese)
73. Peng, Y.; Yin, Z.; Zhou, C.; Ding, X. Micromechanical analysis of capillary suction effect on bearing capacity of unsaturated fine granular foundation soil using coupled CFD-DEM method. *Comput. Geotech.* **2022**, *153*, 105092. [CrossRef]
74. Li, X.; Hu, C.; Li, F.; Gao, H. Determining soil water characteristic curve of lime treated loess using multiscale structure fractal characteristic. *Sci. Rep.* **2020**, *10*, 21569. [CrossRef]
75. Tao, G.; Chen, Y.; Xiao, H.; Chen, Q.; Wan, J. Determining Soil-Water Characteristic Curves from Mercury Intrusion Porosimeter Test Data Using Fractal Theory. *Energies* **2019**, *12*, 752. [CrossRef]
76. Rusell, A.R. How water retention in fractal soils depends on particle and pore sizes, shapes, volumes and surface areas. *Geotechnique* **2014**, *64*, 379–390. [CrossRef]
77. Fuentes, C.; Chavez, C.; Brambila, F. Relating Hydraulic Conductivity Curve to Soil-Water Retention Curve Using a Fractal Model. *Mathematics* **2020**, *8*, 2201. [CrossRef]
78. Li, P.; Qian, H.; Wu, J. Accelerate research on land creation. *Nature* **2014**, *510*, 29–31. [CrossRef]
79. Li, P.; Qian, H. Water in loess. In *Encyclopedia of Sustainability Science and Technology*; Meyers, R.A., Ed.; Springer: New York, NY, USA, 2018; pp. 1–17. [CrossRef]
80. Su, F.; Wu, J.; Wang, D.; Zhao, H.; Wang, Y.; He, X. Moisture movement, soil salt migration, and nitrogen transformation under different irrigation conditions: Field experimental research. *Chemosphere* **2022**, *300*, 134569. [CrossRef]

water

MDPI

Article

Investigating the Hydrological Relationship between the North Taihang Tunnel and Tianshengqiao Nine Falls

Xinfeng Wang [1,2], Lei Gong [3], Yuanqing Liu [1,3], Yan Wang [2], Qingang Wang [4], Mian Song [3], Pan Xiao [5] and Zheming Shi [1,*]

1 School of Water Resources and Environment, China University of Geosciences (Beijing), Beijing 100083, China; wangxinfeng@mail.cgs.gov.cn (X.W.); liuyuanqing@mail.cgs.gov.cn (Y.L.)
2 Langfang Natural Resources Comprehensive Survey Center, China Geological Survey, Langfang 065000, China; wangyan@mail.cgs.gov.cn
3 Center for Hydrogeology and Environmental Geology, China Geological Survey, Baoding 071051, China; gonglei@mail.cgs.gov.cn (L.G.); songmian@mail.cgs.gov.cn (M.S.)
4 China Railway Design Corporation, Tianjin 300308, China; wangqingang@crdc.com
5 Wuhan Geological Survey Center, China Geological Survey, Wuhan 430205, China; xiaopanfree2008@163.com
* Correspondence: szm@cugb.edu.com

Abstract: The impact of a tunnel construction on the groundwater system depends on various parameters and cannot be easily predicted. Along these lines, a deep understanding of the hydrological relationship between tunnels and surface water is considered of vital importance for ensuring safety during railway construction. Upon completion, the North Taihang Tunnel will be one of four extra-long railway tunnels running through the natural ecotope and level-3 protection areas of the Tianshengqiao National Geological Park in Fuping County, Hebei Province. It will be 1 km away from the Tianshengqiao Nine Falls, which is known as a breathtaking landscape feature in Northern China. Local government, societies, and railway design units have raised concerns about whether the construction and operation of the North Taihang Tunnel will affect the Tianshengqiao Nine Falls. To effectively address this issue, in this work, hydrogeological mapping and hydraulic potential-energy calculations were performed in conjunction with hydrochemical and geological structure analyses. The groundwater system units in the study area were divided and the water source of the nine-level waterfall was determined retrospectively. In addition, the recharge of groundwater to the nine-level waterfall was calculated, the hydrogeological properties of the linear structure were analyzed, and the dominant channels of underwater discharge in surface water were compared and studied. The extracted results indicated that: (1) The Tianshengqiao Nine Falls represent a seasonal fall landscape, which is mainly supplied by surface water formed by precipitation and a low proportion of groundwater supply. (2) The water bodies of the North Taihang Tunnel project and Tianshengqiao Nine Falls belong to two independent groundwater systems. (3) No linear structure that connects these two groundwater systems has yet been discovered. It is widely accepted that a minor possibility of hydraulic connection might be present between the North Taihang Tunnel and Tianshengqiao Nine Falls. This work analyzed the water quantity of Nine Falls, determined the hydraulic relationship between the tunnel project and the waterhead of the Nine Falls, and addressed all stakeholder concerns. The conclusions could provide technological support for the scheduled construction projects.

Keywords: North Taihang Tunnel; engineering hydrogeological problem; headwater recognition; hydrological relationship; groundwater system; comprehensive hydrogeological survey

1. Introduction

1.1. The Origin of the Research Question

The planned railway is a crucial part of the Beijing–Kunming Corridor, a major high-speed railway of "eight horizontal and eight longitudinal" networks in the mid-term and long-term railway plan. It could significantly improve the high-speed railway network. The North Taihang Tunnel is one of four extra-long tunnels [1]. It is 12.5 km long and runs through the denudation Zhongshan Region of Taihang Mountains, which has complicated terrain and steep mountains, as well as ravines and gullies to cross. Both the inlet and outlet have clinoform valley slopes. As shown in Figure 1, the North Taihang Tunnel runs 5.2 km through the Tianshengqiao National Geological Park, where the upslope is 29‰ [1]. Moreover, the section next to the head groundwater units of Tianshengqiao Nine Falls is 1.6 km long. The shortest distance of the tunnel to Tianshengqiao and Tianshengqiao Nine Falls is 977 m and the buried depth in this section is higher than 200 m [2].

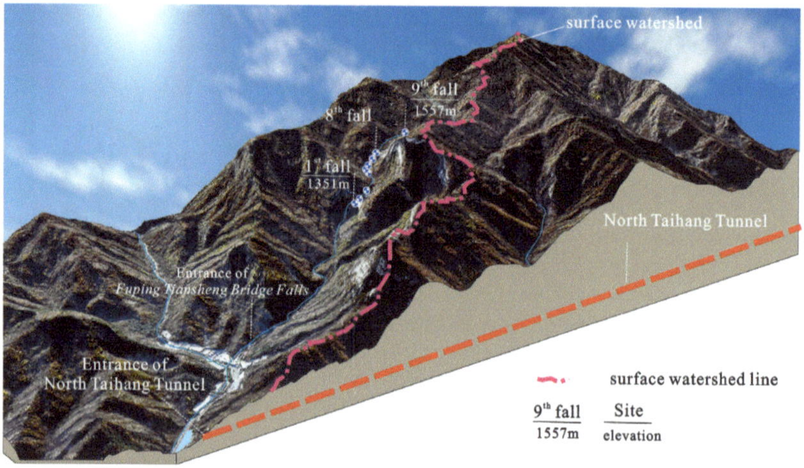

Figure 1. Schematic diagram of the location relationship between the proposed North Taihang Tunnel and the Tianshengqiao Nine Falls.

The Tianshengqiao Nine Falls in Fuping County is a geological landscape of a type that was discovered for the first time in China [3]. To date, it is also the largest known landscape feature created by an inherent stone bridge of gneiss. Water continuously falls beneath the Tianshengqiao, forming the Yaotai Fall with a fall head of 112.5 m. Due to the valley terrain control, another eight falls (Yinhe Fall, Qinglv Fall, Mawei Fall, Sandie Fall, Wangkui Fall, Qiongjiang Fall, Tianmen Fall, and Shuanglong Fall) are formed straight upward within 800 m [4]. With the Yaotai Fall, these form the Tianshengqiao Nine Falls. The planned railway tunnel is less than 1 km away from the Yaotai Fall [1]. Whether the tunnel construction and operation will affect the falls landscape is a key concern of local government, people from all walks of life, and railway design units. Currently, the above-mentioned issues have not been addressed. Therefore, the goal of this work was to provide a systematic analysis of the underlying origins of this effect.

1.2. *Selection of Research Methods*

A tunnel can be described as a new groundwater catchment area or drainage channel. A thorough understanding of the hydrological relationship between the tunnel and surface water is considered a hydrogeological necessity in railway construction. In addition to endangering the safety of workers, an inflow of water in the tunnel during the construction could cause exhaustion of shallow groundwater and surface water. This effect could lead to the acceleration of the water circulation and hydrogeochemical processes, and destroy the water-system balance [5,6]. Certain established research methods such as sodium fluorescein, have been extensively used to examine the source of inflow water in tunnels [7–9], enabling an understanding of the groundwater system distribution and determining the groundwater flow paths. Additionally, the variation characteristics of river flow have been measured and the infiltration groundwater supply has been analyzed by performing multi-phase flow measurements [10]. The controllable-source audio magnetotelluric sounding method has also been used to investigate the hydrological characteristics of tunnel-excavation regions and determine the possible water inflow areas [11]. According to the literature, the water bursting points in tunnels can be accurately spotted and the inflow water volume can be predicted. Although the application of the above methods provides data based on a single professional perspective, they are not compared or validated taking into account other technological methods. On top of that, the majority of the reported works are based on comprehensive methods. For example, Yan Changhe et al. [12] disclosed the hydrological relationship between the South Taihang Tunnel and the groups of springs that developed in the region. More specifically, the authors performed a comprehensive analysis of the geomorphological features, tectonic connection, lithological characteristics, and spring levels and supply, as well as runoff and drainage conditions. It was concluded that tunnel construction cut off the groundwater seepage channel, resulting in a significant inflow of water in certain sections. Luo Mingming et al. [13] analyzed the existence of various water inflow sources and processed them by combining hydrogeologic survey, hydrology, and hydrogeochemical approaches. The authors found that the concentrated inflow of water in tunnels mainly resulted from karstic water. Fracture water of clastic rocks was the major water head during the stable drainage phase, which increased the accuracy of predicting the inflow water head. In another interesting work, Zhou Xuejun [14] determined the karst zoning in the Dafangshan Karst Tunnel by combining engineering geology, hydrogeological mapping, physical prospecting, and drilling. The water inflow based on season change and horizontal circulation zones was also calculated. Based on the groundwater flow system theory, Zeng Bin et al. [15] explored hydraulic connection karst groundwater systems between the Shuanglong Cave and the tunnel site and estimated the possibility of water inflow by combining hydrogeochemical test results. The authors found that the tunnel construction only affected the groundwater in the finite range of the outlet section. Nonetheless, no threats to karst water in the Shuanglong Cave were demonstrated. Wei Huapeng et al. [16] built a three-dimensional unsteady-flow numerical model of groundwater by using exploration drilling data. In particular, the variation amplitude of the diversion fissure zone was predicted and the degree to which groundwater affected the water inflow under different working conditions was analyzed. Based on the exploration of drilling data, Liu Mou et al. [17] developed a relatively vivid three-dimensional unsteady flow numerical model by adding geophysical prospecting, pumping tests, and long-term observation of groundwater. This model was also used to predict mine inflow and groundwater drawdown under coal-seam mining conditions, while calculating groundwater evaporation, overflow, and aquifer storage under the attack of new water inflow. From the above-mentioned analysis, it can be inferred that a combination of hydrodynamic and hydrochemical fields has been employed to study water inflow properties. Based on the overall characteristics of the frequent transformation between shallow surface water and groundwater in the study area and the limited connection between deep and shallow water bodies, the water flow of Tianshengqiao Nine Falls was determined in this work. Moreover, the characteristics of the groundwater system were revealed and the hydrological relationship between the North Taihang Tunnel and Tianshengqiao Nine Falls was investigated. The concerns of all stakeholders are also addressed and a guarantee of the proper construction of vital projects is provided. The latter was achieved by combining groundwater system division, head recognition of the Tianshengqiao Nine Falls, and calculation of the surface water–groundwater transformation, as well as by performing cross-system water diversion channel and hydrodynamic analyses.

2. Geological Background of the Study Area

2.1. Natural Geography

The study area has been in a state of continuous stable rising since the Mesozoic era, accompanied by intermittent pauses, and local uneven settlement. Thus, the dominant role of high, middle, and low mountains, as well as complicated and changing landforms was taken into consideration. The terrain is generally high in the south and low in the north, high in the west and low in the east. Elevations in most regions are higher than 1200 m. The highest elevation is at the Baicaotuo (2144.5 m), which is the main peak of Tianshengqiao Ditch and the lowest is at the mouth (800 m) of Tianshengqiao Ditch, showing a relative difference of 1344.5 m. In the study area, the landform consists of peak valleys alternating with cliffs, walls, and platforms. The major landform combination consists of overlapping cliff platforms, deep-cutting gorges, U-shaped valleys, and enclosed valleys. Major landforms in the study area include erosional, glaciated, tectonic erosion, and gravity landforms [18].

The study area belongs to the semi-arid region of the central Hebei Province and has a distinctive continental climate. It has a cold, dry, and long winter with little rain and snow. Furthermore, it has a hot and wet summer, which is the main precipitation season and the main period to appreciate the Tianshengqiao Nine Falls. The annual average precipitation is 594.0 mm and the rainy season takes mainly place from June to September. The annual evaporation capacity is 2024.6 mm, reaching a maximum in June and a minimum in January [18].

2.2. Geological Conditions

The study area is located at the foreland on the east of the central orogenic belt in North China [19] and is mainly composed of sediments at passive continental margins, accompanied by tectonic activation or reconstruction of a basement complex [20]. It is less than 2 km from the Longquanguan shear zone, which is the largest shear zone in North China. The study area has undergone several tectonic changes. Structures such as folds and fractures formed by the Yanshan Movement form the major tectonic skeleton of the study area [20]. The Precambrian folds and fractures in the study area mainly originated from the Fuping tectonic movement [21]. A series of folds and fault structures centered at ductile shear zones have been also formed. The overall tectonic line in the study is related to the NWW–NW trend [22].

2.3. Hydrogeological Conditions

Groundwater type is mainly weathering fracture water of metamorphic rocks. Considering the local hydrogeological survey results and the water abundance of the aquifer (Figure 2), the study area could be divided into a medium-water regions and a poor-water regions [23]. The medium-water regions are mainly controlled by the degree of weathered crust growth and the thickness of the quaternary alluvial diluvium. Some are also controlled by faults and their distribution is scattered. The groundwater is generally at a depth of 1.46–5 m and the water inflow per well is about 100 m^3/d. The water quality is good and $HCO_3{\cdot}SO_4$-Ca is the dominant water type. The total dissolved solids is 260 mg/L. The poor-water region is mainly controlled by landforms, and there are bare bed rocks and steep terrain. The water inflow per well is lower than 10 m^3/d.

Groundwater basic units contain water volume, quality and energy input, and migration and output, as well as a combination of these. These are independent units of common groundwater circulation in temporal and spatial distributions [24]. According to the Guideline for Groundwater System Division and Geology · Minerals · Environment of Hebei Province, the study area is located in the level-2 groundwater system of Hai River, level-3 groundwater system of Daqing River, level-4 groundwater system of the Dasha River–Ci River fracture karst, and level-5 fracture groundwater system of Beiliu River.

Tianshengqiao Nine Falls belongs to the Tianshengqiao fracture groundwater system, which is the subzone of level-5 fracture groundwater system of the Beiliu River. In contrast, the North Taihang Tunnel is located in the subzone of the Sandaoling fracture groundwater system. A distinctive border exists between these two groundwater systems (Figure 3). The boundary separating the Camel Yingke Peak and the Yangshuta Ridge forms the edge of the surface watershed. The formation lithology of the study area is dominated by

Fuping gneiss and groundwater mainly consists of shallow weathering crust fracture water. Therefore, the groundwater division is consistent with the surface water division.

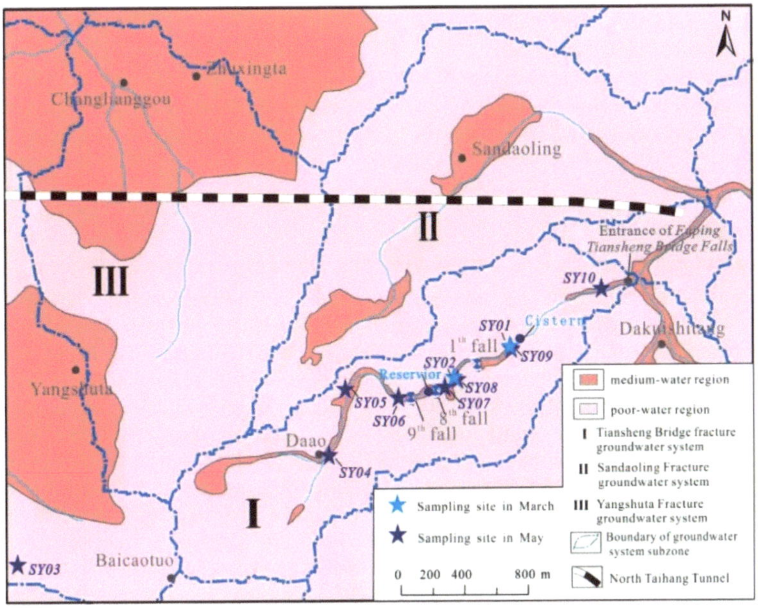

Figure 2. Hydrogeology and groundwater system division map.

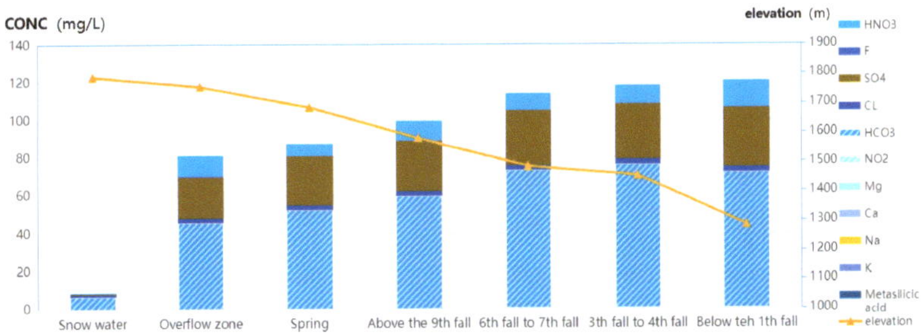

Figure 3. Components of precipitation, surface water, and groundwater in Tianshengqiao Gully.

The Tianshengqiao and Sandaoling fracture groundwater system subzones both belong to the low–middle mountains, which are sliced shallowly by the etching structure, as well as by their spreading ditch and valley zones. They have the same groundwater system characteristics. Precipitation and ice-snow melted water form the groundwater supply sources. The groundwater occurrence and migration depth are controlled by weathering of gneiss in the crust. It presents layered aquifer characteristics and has tectonic control locally, showing veined enrichment or extreme belt shortage. According to the existing statistical results of hydrogeological survey data in similar regions of Fuping, the development depth of the weathering crust of metamorphic rocks is closely related to the landform. The weathering degree is highest in the valleys. Large weathering depths are 9.00–25.80 m and moderate weathering depths are 15.00–48.00 m [14]. Hence, the maximum depth of weathering crust fracture water is less than 50 m. The groundwater flow direction in the system is controlled by the terrain; it sinks in valleys along mountain slopes, and drains as discharge flows or springs. According to the survey data in similar regions of Fuping, the common spring flow is 0.02–0.12 L/s [4].

There are also some differences between the Tianshengqiao and Sandaoling fracture groundwater system subzones. The former has narrow and long valleys, which are mainly developed at steep cliffs, forming the group of falls. The latter has narrow and low valleys. There is a thick stacking of alluvial sediments, as well as frequent transformation between the groundwater and surface water.

Based on the above-mentioned analysis, Tianshengqiao Nine Falls and the North Taihang Tunnel can be divided into two independent groundwater systems. Moreover, the groundwater systems mainly consist of weathering crust fracture water with a buried depth of less than 50 m. A weak hydraulic connection also exists between the two groundwater systems.

3. Data Acquisition and Results

3.1. Waterheads and Volume Information of Tianshengqiao Nine Falls

No records or descriptions of waterheads or volumes of Tianshengqiao Nine Falls have yet been found in any of local hydrogeological data; hydrometeorological data; special data like the Special Research Report for Tianshengqiao National Geological Park in Fuping, Hebei; the Overall Tour Plan in Fuping Tianshengqiao Nine Falls; or the Fuping County Annals from the last 30 years. Therefore, the total flow rate of the nine falls and the flow rates of each fall were measured in May 2022 by using the section survey method with a Flowatch portable current meter. The total flow (1105 m^3/d) of the nine falls during ice melting and snow during the middle ten days of May was obtained for the first time. Based on a survey of paths and source tracking, a comprehensive analysis of landforms, high-definition images of vegetation coverage, formation lithology, weathering crust thickness, surface watershed, temperature, and elevation differences, it can be concluded that the groundwater overflow belt has been formed below the alpine meadow cliff along Baicaotuo, Liaodaobei, and Yangshuta and is the groundwater source of Tianshengqiao Nine Falls.

3.2. Hydrochemical Information of Tianshengqiao Nine Falls

To analyze the headwater of Tianshengqiao Nine Falls, as well as its composition and water circulation depth, snow water, spring water, and water samples from the nine falls were collected from upstream to downstream of fracture groundwater in Tianshengqiao in May 2022. In addition, a water composition test and deuterium–oxygen isotope test were carried out. The water samples were tested at the Water Environment Laboratory, Ministry of Natural Resources. The properties of cations and anions were examined by an ionization emission spectrometer ICP-OES (ICP-OES: AVIO 500, Perkinelmer Co., Ltd., Waltham, MA, USA) and ion chromatograph (ICS-2100) (model: ICS-2100, Metrohm 883, Metrohm, Herisau, Inc., Switzerland), with a test accuracy of 0.001 mg/L. The analytical errors of cations and anions were both lower than 5%. The contents and features of major hydrochemical components from precipitation at the major peak to the first fall along the Tianshengqiao Ditch are presented in Figure 3 and Table 1. The deuterium−oxygen isotopes were tested with a liquid-water stable-isotope analyzer (Picarro L2130-i, Picarro, Inc. Santa Clara, CA, USA.), and expressed as δD and $\delta^{18}O$ based on VSMOW standards. The typical accuracies of the test were 0.3‰ and 0.08‰, respectively.

Table 1. Main hydrochemical indexes of precipitation, surface water, and groundwater in the Tianshengqiao Gully.

Parameters	Snow Water SY03	Surface Water					Groundwater SY05
		SY06	SY07	SY08	SY09	Average	
pH	6.46	7.51	7.57	7.49	7.63	7.55	7.22
TR		88.05	97.28	98.36	100.12	95.95	71.34
TDS (mg/L)	12.3	118.7	127.7	130.3	137.3	128.50	100.70

Table 1. *Cont.*

Parameters	Snow Water SY03	Surface Water					Groundwater SY05
		SY06	SY07	SY08	SY09	Average	
HCO_3 (mg/L)	6.70	59.68	73.08	76.12	71.86	70.19	49.02
CI (mg/L)	0.95	2.49	2.64	2.67	2.92	2.68	2.39
SO_4^2 (mg/L)	0.50	26.58	29.05	29.30	31.26	29.05	23.59
F (mg/L)	0.01	0.12	0.13	0.13	0.14	0.13	0.32
NO_3 (mg/L)	0.23	10.68	8.90	9.82	14.45	10.96	8.84
NO_2 (mg/L)		-	0.106	0.079	0.026	0.211	0.005
Turbidity (mg/L)	3.65	0.15	0.25	0.49	0.21	0.28	0.29
Metasilicic acid (mg/L)	1.63	11.01	10.21	8.62	10.77	10.15	10.25
K^+ (mg/L)		1.46	1.69	1.70	2.21	1.77	1.06
Na^+ (mg/L)	0.70	2.60	2.64	2.49	2.74	2.62	2.76
Ca^{2+} (mg/L)		27.66	30.95	31.13	31.25	30.25	21.43
Mg^{2+} (mg/L)		4.53	4.78	4.93	5.28	4.88	4.27
^{18}O (‰)	−5.71	−10.21	−9.87	−9.82	−9.1	−9.75	−10.33
D (‰)	−25.84	−68.48	−66.80	−66.41	−61.85	−65.88	−69.74

3.3. Fracture Measurement

By combining the statistical window method, network yield survey, network area survey, etc., we recorded, in detail, the serial number of each fracture, coordinates of the endpoints, relationships between the fracture and the survey network (inclusion, cutting, intersecting), yield, trace length, undulation, roughness, openness, filling, groundwater information, and fracture characteristics.

Using the equirectangular projection method, the measured data of the fractures were organized to generate a polar-density map, and then the grouping of the fractures was completed according to the different densities of the production projections.

4. Discussion

4.1. Proportion Analysis of the Surface-Water Source and Groundwater Source of Tianshengqiao Nine Falls

The water sources of Tianshengqiao Nine Falls differ in rainy and dry seasons. During the rainy season, surface water collected from rainwater is the major source, supplemented by groundwater, which is supplied continuously along the path above Da'ao of the major vein. During the dry season, the nine falls experience ice falls and they begin to melt after March. The flowing stream is mainly supplied by ice and snow water.

Precipitation is the major water source for Tianshengqiao Nine Falls, which is attributed to the high mountains, abundant water, rich precipitation, and limited thickness of the weathered crust of the hard metamorphic rocks. It lacks deep space and cannot accept precipitation or surface water infiltration for a prolonged time. As a result, most precipitation is transformed into surface water and sinks into the falls.

Considering the test data of the collected snow water, spring water, and Tianshengqiao Nine Falls water samples, the mixing ratio of different water bodies was calculated according to the binary mixing theory and differences in hydrochemical components of supply and runoff, and different drainage positions. The specific formula was

$$\delta_m = \lambda \times \delta_A + (1 - \lambda) \times \delta_B \quad (1)$$

where δ_m is the mixed sample, δ_A and δ_B refer to two end elements, and λ stands for the mixing ratio.

Specifically, the snow water sample and spring water sample of Liaodaobei were chosen as the two mixing end elements to represent the isotope composition of precipitation and groundwater in the study area, respectively. The surface water sample was used as the mixing sample. Based on the above formula, the supply proportion of groundwater to water of the nine falls in the study area was calculated. The calculated results are presented in Table 2.

Table 2. Calculation results of the mixing ratio of groundwater and Tianshengqiao Nine Falls.

Location	Proportion of Groundwater Mixed with the Nine Falls (%)
Above the 9th fall	2.67
6th fall to 7th fall	10.04
3rd fall to 4th fall	11.14
Below the 1st fall	26.57
Stream outlet	30.05

Based on the above-mentioned analysis, surface water was found to be the major water source of the nine falls and the tunnel construction had a small possibility of affecting the falls formed by the catchment of surface water.

4.2. Water Source Analysis of Tianshengqiao Nine Falls Based on Water Volume and Hydrochemical Information

By performing hydrogeologic surveys and analyzing the remote-sensing high-definition images, it can be concluded that the groundwater overflow belt formed below the alpine meadow cliff along Baicaotuo, Liaodaobei, and Yangshuta is the groundwater source of Tianshengqiao Nine Falls. This is supported by a comprehensive analysis of landform, vegetation coverage, formation lithology, weathering crust thickness, surface watershed, temperature, and elevation differences.

Hundreds of thousands of mu of original ecological park are present in the alpine meadow. As a result, a water conservation area with 6000 mu of artificial larch forest and 70,000 mu of white birch forest was formed, creating favorable conditions for a continuous supply of groundwater. The groundwater accumulates continuously along gullies. This result indicated that the falls are supplied by ice and snow-melted water moving downward along gullies. The stream flow was tested on 13 May 2022 after the melting of snow. The results showed that the water flow from the head to the mouth of the gully first increased and then decreased (Figure 4). The stream flow continuously increased from Da'ao to the nine falls and the flow peak was formed at the 8th fall, while the stream flow slightly decreased from the 7th fall to Jiangjun Stone. The water flow at the exit of the landscape decreased to 1/3 of that at the 7th fall due to the impact of the impounding reservoir below Jiangjun Stone. The water flow between the 6th and the 7th falls reached 16.1 L/s. The water flow between the 3rd and the 4th falls was 13.5 L/s and the water flow at Jiangjun Stone, below the 1st fall, was 12.792 L/s. Nevertheless, the 8th fall reservoir and impounding reservoir had a significant impact on the stream flows.

High elevation is present in the catchment area of Tianshengqiao Nine Falls. There was ice and snow coverage in shady areas with an elevation of 1600 m during the first ten days of June. Ice and snow melted water is the primary water source of the nine falls during the dry seasons. According to the test results of stream flows at Jiangjun Stone and stream flows between the 3rd and the 4th falls, the deuterium–oxygen isotope test data of the water samples in the two phases differed significantly. Deuterium and oxygen both showed enrichment characteristics in May compared with March. The data in March reflected that the ^{18}O and D values of stream water at Jiangjun Stone were −10.19‰ and −69.92‰, respectively, which were lower than those between the 3rd and the 4th falls (−10.12‰ and −68.59‰). However, the numerical values were almost identical. This outcome showed that the increase in water supply from the 4th fall to the 1st fall was mainly attributable to ice and snow melting. According to data in May, the ^{18}O and D values of stream water at Jiangjun Stone were −9.10‰ and −61.85‰, respectively, which was higher than between the 3rd and the 4th falls (−9.82‰ and −66.41‰). This proved that with the thawing of the active layer, the groundwater had participated in transformation after the melting of ice and snow in May. A comparison of the deuterium–oxygen isotopic characteristics of snow water, groundwater, and stream water is displayed in Figure 5. The deuterium and oxygen of top-down stream water from Da'ao to the landscape exit were continuously enriched and were above the global precipitation line. In addition, they showed some differences in the local average precipitation. However, the fitting trend line intersected with the local average precipitation. This effect demonstrates that stream water and spring water are both supplied by ice and snow-melted water. They also exhibited obvious elevation-based zoning characteristics. In addition, the downstream stream water is supplied not only by snow in May but also by snow in winter. The supply period is longer and the supply scope is wider while having continuous accumulation.

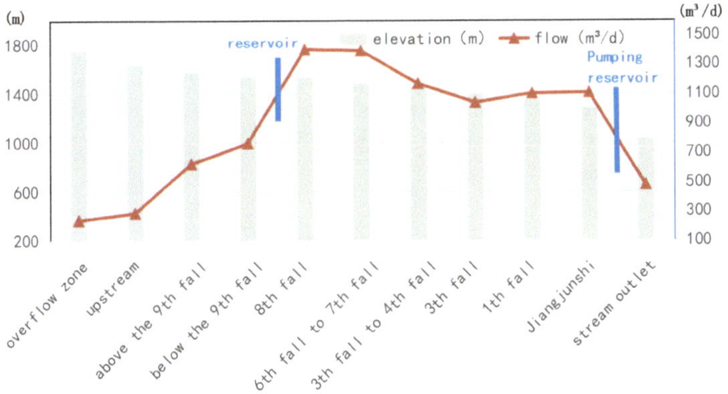

Figure 4. Schematic diagram of the water quantity change in Tianshengqiao Creek (measured on 13 May 2022).

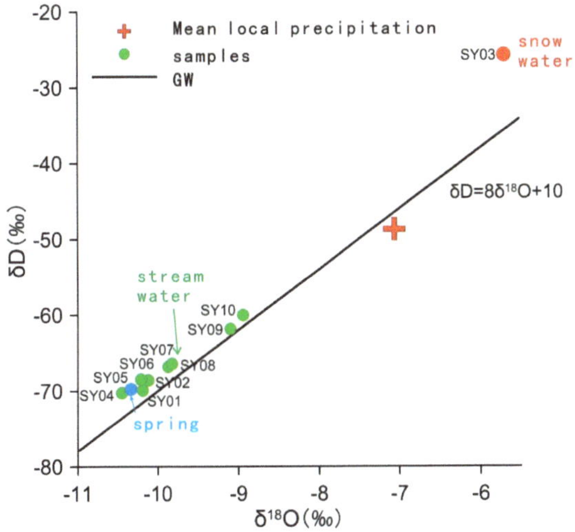

Figure 5. Deuterium and oxygen isotope characteristics of snow water, groundwater, and stream water.

Furthermore, the collected hydrochemical data reflected small differences in the hydrochemical composition between the Tianshengqiao spring and stream water from the head. However, the hydrochemical components of stream water were also continuously enriched from upstream to downstream (Figure 6). $\gamma Cl/\gamma HCO_3$ and $\gamma Mg/\gamma Ca$ are the most commonly used methods to identify the source of water. According to the hydrochemical data of stream water at Jiangjun Stone, TDS was extremely close between the 3rd and 4th falls in May. The values of $\gamma Cl/\gamma HCO_3$ were both 0.05, whereas the values of $\gamma Mg/\gamma Ca$ were 5.55 and 5.62, respectively, which was also extremely close. This outcome also verified that the quality and source of these two water samples was the same. During this period, the stream water mainly originated from ice and snow-melted water, indicating that the shallow groundwater in this interval was extremely deficient or was still in the freezing period. A common characteristic of the waterhead for all falls during dry seasons could be found—according to the hydrochemical data of stream water in May, the values of $\gamma Cl/\gamma HCO_3$ also remained at 0.04, while $\gamma Mg/\gamma Ca$ decreased quickly to less than 0.2. Based on this analysis, it can be inferred that the water sources of each waterfall are homologous. Nevertheless, due to the thawing of ice water in the gravel–cobble layer and the weathering crust in the gully, the stream water is continuously transformed with groundwater along the way and the water−rock interaction is continuously enhanced. The content of calcium ions significantly increased compared with that of magnesium ions as the former could be dissolved faster. This phenomenon rapidly decreased the $\gamma Mg/\gamma Ca$ values.

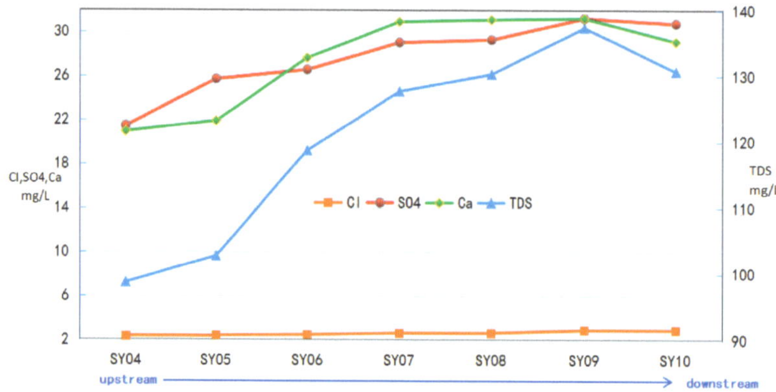

Figure 6. The variation characteristics of the main hydrochemical components in creek water.

Additionally, nitrite was detected at multiple water-sampling sites. This effect further proved that that surface water or shallow groundwater has a closer relationship with human activities as the major water source. Based on the above-mentioned analysis, it can be argued that the water of Tianshengqiao Nine Falls mainly originates from surface water formed by precipitation and the tunnel's construction will hardly affect the nine falls formed by the catchment of surface water.

4.3. Connectivity Analysis between the Tunnel and Waterhead of Tianshengqiao Nine Falls Based on the Linear Structure

Strong tectonic water control is present in the bedrock mountain areas, and linear structures often become the hydraulic channels that connect the different groundwater systems. To guarantee hydraulic relationships between the tunnel and the nine falls, which are located in different systems, there are two prerequisites: (1) there must be a water-head difference and (2) continuous hydraulic connection channels must be present.

Even though the water head difference is present, the energy consumption for downward drainage is significantly less than that for the infiltration supply channel. The shortest distance from the tunnel to the 1st fall is 977 m. The elevation of the tunnel is 1005 m and the elevation of the 1st fall is 1351 m. After the tunnel's construction, a water head difference of 346 m could be formed between the 1st fall and the tunnel. However, there is a 112.5 m-long cliff downward along the 1st fall and the hydraulic slope is nearly 90°. The groundwater flow coincides with the theory of minimum energy consumption rate. Therefore, the discharge force of surface water from the 1st fall along the cliff is far stronger than that in the 977 m-long rock block region with a groundwater gradient of 346 m. The shortest distance from the upstream of the 1st fall (e.g., the 7th fall) to the tunnel is 1185 m. The elevation of the tunnel is 1014 m and the elevation of the 7th fall is 1543 m. After the tunnel's construction, a water head difference of 531 m could be formed between the 1st and 7th fall. The tunnel follows the hypothesis that there are hydraulic connection channels between them. In addition, a surface slope of 304.5 m is present at the bottom of the 7th and the 1st falls. According to the minimum energy consumption of fluid (Figure 7), the hydraulic slope from the 7th fall to the 1st fall is 77°. Without considering the infiltration resistance, the hydraulic slope from the 7th fall to the nearest tunnel point is 25° and the gradient is 0.98, which is much higher than 0.45. Hence, the water may first choose the surface channel. Therefore, the water-head difference still exists. However, flowing water consumes more energy than the downward discharge along the slope. Hence, the water-head difference is not anticipated to affect the water flow of Tianshengqiao Nine Falls.

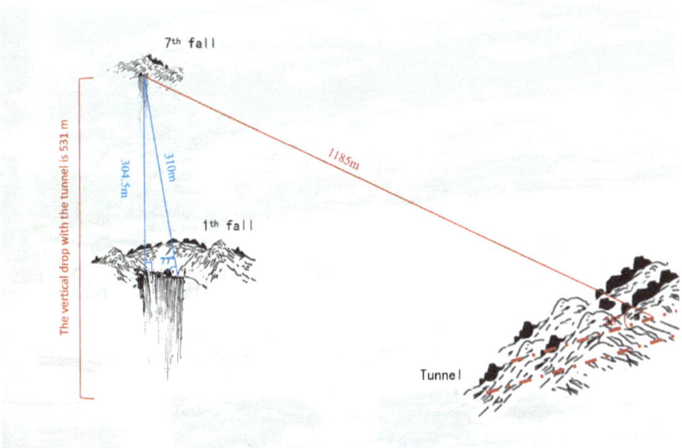

Figure 7. Analysis of the hydraulic slopes of the 7th fall and 1ths fall to the tunnel.

The rocks in the region of the North Taihang Tunnel mainly consist of weakly weathered biotite monzonite gneiss, biotite plagioclase gneiss, leptite, granulite, and plagioclase amphibolite. The statistical-window method was used to measure 188 fractures in 170 m^2, and the fracture development characteristics in the study area were obtained. According to the fracture measurement results, the dominant fracture group is 200~230°∠75~90° (Figure 8). Among these, 69.7% of the total fractures are closed and 54% of the fractures are shorter than 1 m.

Considering the local tectonic background analysis, the dominant fracture group is associated with a secondary, short, and small tectonic group of the regional tectonics, which is mainly developed in shallow weathering crust. It has no cross-system connection ability for groundwater flows.

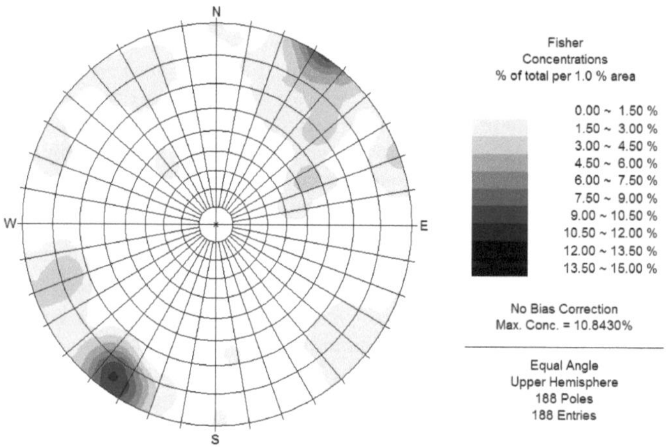

Figure 8. Measured fracture polar density map.

Taking into account the local geological background and landform analysis, no connection is present between the surface ravines and valleys of the nine falls and the tunnel since they have different surface watersheds with no hydraulic connection. According to lithological analysis of the formation, the weathering degree of metamorphic crust in ravines and valleys is the highest with a weathering depth of 48.00 m. The buried depths at the points that have the shortest distance from the North Taihang Tunnel to Tianshengqiao and Tianshengqiao Nine Falls are all larger than 200 m. None of these points have hydraulic connections with groundwater in the Sandaoling weathering crust and the water of Tianshengqiao Nine Falls is in a different watershed. According to the analysis of geological structures, no linear structure that can connect the nine falls and the tunnel was found in the 1:50,000 geologic map or interpretation of the current remote sensing imagery. Based on the above analysis, the water of the nine falls would have first used surface channels to form the falls. Moreover, the lack of hydraulic connection between the Tianshengqiao Nine Falls and the tunnel means that there will not be the conditions to generate hydraulic connections between them.

5. Conclusions

In this work, the water head of Tianshengqiao Nine Falls was investigated by performing hydrogeologic surveys coupled with hydrogeochemical and tectonic analyses. The hydraulic relationship between the Tianshengqiao Nine Falls and the North Taihang Tunnel was thoroughly discussed. Some major conclusions are as follows:

(1) The main water of Tianshengqiao Nine Falls is surface water. Shallow groundwater is the supplementary water source at downward discharge along ravines in rainy seasons. For the first time, the total flow of Tianshengqiao Nine Falls was measured and, during the middle 10 days of May 2022, it was 1400 m^3/d.

(2) Based on the division of the surface watershed and groundwater system, weathering crust thickness, and the buried depth of the tunnel, and from connectivity analysis of linear structures, it could be concluded that the planned North Taihang Tunnel and Tianshengqiao Nine Falls could be divided into two independent groundwater systems. A small possibility for the generation of hydraulic connections remains.

(3) By analyzing the groundwater system division and taking into account the hydrochemical and isotopic approaches, it can be argued that the tectonic-stress field and potential-energy field are effective ways to recognize hydraulic relationships between different water sources. These should also be included in future associated studies.

Author Contributions: Conceptualization, X.W. and Z.S.; methodology, X.W. and L.G.; software, X.W.; validation, Y.L., Y.W. and X.W.; formal analysis, X.W. and Z.S.; investigation, Q.W.; resources, X.W.; data curation, M.S. and P.X.; writing—original draft preparation, X.W. and Z.S.; writing—review and editing, X.W. and Z.S.; supervision, X.W.; project administration, X.W.; funding acquisition, X.W. All authors have read and agreed to the published version of the manuscript.

Funding: This research was funded by the project of the China Geological Survey entitled "Investigation of lakes in the northern part of the plain lake region in eastern China" (DD20230505).

Data Availability Statement: Data used in the manuscript can be obtained by request to the corresponding author.

Conflicts of Interest: Author Qin-Gang Wang was employed by the company China Railway Design Corporation. The remaining authors declare that the research was conducted in the absence of any commercial or financial relationships that could be construed as a potential conflict of interest.

References

1. China Railway Design Corporation. *Environmental Impact Report on the New Xiongan New Area to Xinzhou High-Speed Railway (Draft for Comment)*; China Railway Design Corporation: Tianjin, China, 2022.
2. Hebei Geological and Environmental Monitoring Institute. *Report on the Impact Assessment of the New Xiongan New Area to Xinzhou High-Speed Railway Passing through the Geological Remains of Hebei Fuping Tianshengqiao National Geological Park*; Hebei Geological and Environmental Monitoring Institute: Shijiazhuang, China, 2021.
3. Wang, W. Fuping Tianshengqiao National Geopark. *Teach. Ref. Middle Sch. Geogr.* **2005**, *10*, 14.
4. Hydrogeological and Environmental Geological Survey Center, China Geological Survey. *Research Report on the Impact of the Construction of the North Taihang Mountain Tunnel of the Xiongxin High-Speed Railway on the Water Source of the Tianshengqiao Nine Waterfalls*; Hydrogeological and Environmental Geological Survey Center, China Geological Survey: Baoding, China, 2022.
5. Wu, J. Water Inflow Prediction and Water-Reducing Effect of Grouting for a Deep Buried Diversion Tunnel Crossing Faults. Ph.D. Thesis, China University of Geosciences, Wuhan, China, 2017.
6. Zhang, Y.; Guo, C.; Li, X.; Bi, J.; Ma, J.; Liu, F. Key problems on hydro-engineering-environmental geology along the Sichuan-Tibet Railway corridor: Current status and development direction. *Hydrogeol. Eng. Geol.* **2021**, *48*, 1–12.
7. Chang, W.; Tan, J.; Huang, K.; Cheng, X.; Huang, Z.; Wan, J. Application of groundwater multielement tracing test in water damage prediction of karst Tunnel: Taking the orchid tunnel of Zhangjiajie-Jishou-Huaihua high-speed railway as an example. *Carsologica Sin.* **2020**, *39*, 400–408.
8. Chen, Y.; Zhu, S.; Xiao, S. Discussion on controlling factors of hydrogeochemistry and hydraulic connections of groundwater in different mining districts. *Nat. Hazards* **2019**, *99*, 689–704. [CrossRef]
9. Peng, H.; Yuan, Y.; Li, T.; Dong, G.; Liu, Y.; Ba, R. Identification and quantitative analysis of groundwater discharged from New Guanjiao Tunnel in Tianjun, Qinghai. *Bull. Geol. Sci. Technol.* **2022**, *41*, 60–70.
10. Ma, J.; Li, X.; Li, C.; Zhang, C.; Fu, C.; Bai, Z.; Wang, Z. Transformation characteristics of the large-flow river and groundwater in the fault zone in the glacier-covered area of Bomi in Tibet. *Hydrogeol. Eng. Geol.* **2021**, *48*, 23–33.
11. Luo, Y.; Cheng, G.; Yi, L.; Deng, H.; Tan, N.; Liu, Z. Prediction and prevention of sudden water gushing in the Hongtuzhang tunnel. *Hydrogeol. Eng. Geol.* **2020**, *47*, 64–72.
12. Yan, C.; Sun, G.; Wang, Y.; Yuan, X. Analysis of Regional Spring Basin Characteristics and Hydrological Conditions of Southern Taihang Tunnel. *J. Railw. Eng. Soc.* **2017**, *34*, 81–85, 123.
13. Luo, M.; Zhou, H.; Guo, X.; Chen, Q.; Qi, L.; Kuang, Y. Processes and sources identification of intermittent karst water inrush in Xiakou Tunnel. *Bull. Geol. Sci. Technol.* **2021**, *40*, 246–254.
14. Zhou, X. Comprehensive analysis of karst water zone in the Dafang mountain tunnel. *J. Railw. Eng. Soc.* **2019**, *36*, 67–73.
15. Zeng, B.; Chen, Z.; Shao, C.; Peng, D. Analysis of source and path of water inrush in karst tunnel based on the theory of groundwater flow system. *Bull. Geol. Sci. Technol.* **2022**, *41*, 99–108.
16. Wei, H.; Luo, Q.; Kang, W.; Zhang, Z. Prediction of coal mine water inflow by different mining methods and environment impact analyses. *Hydrogeol. Eng. Geol.* **2023**, *50*, 21–31.
17. Liu, M.; Wang, J.; Wu, G.; Zhou, J.; Luo, Q.; Kang, W. Prediction of mine water inflow and analyses of its influence on desert vegetation. *Hydrogeol. Eng. Geol.* **2023**, *50*, 65–75.
18. Hebei Geological Environment Monitoring Institute. *The New Xiong a'n New Area-Xinzhou High-Speed Railway Passes through the Tianshengqiao National Geopark in Fuping, Hebei Province. The Impact Assessment Report on Geological Relics*; Hebei Geological Environment Monitoring Institute: Shijiazhuang, China, 2021.
19. Kusky, T.M.; Li, J.H. Paleoproterozoic tectonic evolution of the north China craton. *J. Asian Earth Sci.* **2003**, *22*, 383–397. [CrossRef]
20. Li, J.; Niu, X.; Chen, Z.; Kusky, T.; Polat, A. Discovery of deep nappe structure in Taihang Mountains and its geological significance. *Prog. Nat. Sci.* **2004**, *14*, 39–48.
21. Ren, L.; Geng, Y.; Du, L.; Wang, Y.; Liu, P.; Guo, J. Anatexis and migmatization of the Fuping Complex, North China Craton. *Acta Petrol. Sin.* **2011**, *27*, 1056–1066.

22. Zhang, S.; Jin, L.; Xiao, Q. Tectonic style and deformation history of the Fuping Paleozoic dome complex fold group. *Chin. Reg. Geol.* **1983**, *6*, 97–110.
23. Meng, S. *1:50000 Hydrogeological Survey Report of Taihang Mountain Area*; Hydrogeological Environmental Geological Survey Center of China Geological Survey: Baoding, China, 2016.
24. Zhang, R.; Liang, X.; Jin, M.; Wan, L.; Yu, Q. *Fundamentals of Hydrogeology, 7th ed*; Geological Publishing House: Beijing, China, 2018.

MDPI AG
Grosspeteranlage 5
4052 Basel
Switzerland
Tel.: +41 61 683 77 34

MDPI Books Editorial Office
E-mail: books@mdpi.com
www.mdpi.com/books

www.ingramcontent.com/pod-product-compliance
Lightning Source LLC
Chambersburg PA
CBHW041424210926
51850CB00013B/3349

* 9 7 8 3 7 2 5 8 4 5 1 9 4 *